ISCHIA GROUP THEORY 2010

ISCHIA GROUP THEORY 2010

Proceedings of the Conference

Ischia, Naples, Italy 14 – 17 April 2010

editors

Mariagrazia Bianchi

University of Milano, Italy

Patrizia Longobardi • Mercede Maj

University of Salerno, Italy

Carlo Maria Scoppola

University of L'Aquila, Italy

World Scientific

NEW JERSEY • LONDON • SINGAPORE • BEIJING • SHANGHAI • HONG KONG • TAIPEI • CHENNAI

Published by

World Scientific Publishing Co. Pte. Ltd.

5 Toh Tuck Link, Singapore 596224

USA office: 27 Warren Street, Suite 401-402, Hackensack, NJ 07601

UK office: 57 Shelton Street, Covent Garden, London WC2H 9HE

British Library Cataloguing-in-Publication Data
A catalogue record for this book is available from the British Library.

Photo on the cover by Claudio Pipolo (Ischia)

ISCHIA GROUP THEORY 2010
Proceedings of the Conference

ISBN-13 978-981-4350-38-9
ISBN-10 981-4350-38-9

Printed in Singapore by World Scientific Printers.

Contents

Preface

It is a great pleasure for us to present the Proceedings of the Conference "Ischia Group Theory 2010", the fourth in a series that is becoming an important periodic appointment for the Group Theory international community. The Conference took place at the NH Ischia Thermal SPA Resort, Ischia (Naples, Italy) from April 14th to April 17th, 2010.

The scientific committee was formed by Mariagrazia BIANCHI (Università di Milano), Patrizia LONGOBARDI (Università di Salerno), Mercede MAJ (Università di Salerno), Carlo Maria SCOPPOLA (Università dell'Aquila), while the local organizing committee consisted of Costantino DELIZIA, Annamaria LUCIBELLO, Chiara NICOTERA, Carmela SICA, Antonio TORTORA, Maria TOTA (all from Università di Salerno) and Diana IMPERATORE (Università di Napoli Federico II).

This volume includes scientific contributions by speakers and participants of the conference.

The span of the research topics represented in these Proceedings includes finite p-groups, character and representation theory, finiteness conditions, profinite groups and pro-p-groups, subgroup structure, linear groups, conjugacy classes, automorphisms, varieties of groups, graphs connected with groups, combinatorial group theory, radical rings.

The schedule of the talks and the list of all participants are included in these Proceedings. A whole day was dedicated to the memory of Karl Gruenberg, who in 2006 was an active participant of the Ischia series, and that passed away in October 2007. Another day was dedicated to the memory of Silvia Lucido, who left us much too early in March, 2008.

A poster session on various research topics augmented the scientific program.

A very smooth and pleasant stay for all guests is now part of the tradition of the Ischia Group Theory Conferences; for this, once more, we are pleased to thank our colleagues from the Università di Salerno. We thank the speakers, who greatly contributed to the success of the conference; the participants, who created a perfect atmosphere for doing mathematics; the authors of all papers, for their interesting contributions; the referees of the papers, for their effective and accurate effort in providing the editors with reliable information; Costantino Delizia and Antonio Tortora, for the remarkable work they have done in taking care of the editing of these Proceedings, and the Publisher, for their professional and timely production.

We also thank the staff of the NH Ischia Thermal SPA Resort for their patience and support.

Last but not least, we thank our Sponsors, listed in the next page.

The Editors

Sponsors

INdAM – GNSAGA
Istituto Nazionale di Alta Matematica "F. Severi" – Gruppo Nazionale per le Strutture Algebriche, Geometriche e le loro Applicazioni

PRIN – Programmi di Ricerca di Interesse Nazionale
Progetto dal titolo: "Gruppi, Algebre di Lie, Crittografia"
Progetto dal titolo: "Teoria dei Gruppi e Applicazioni"

Dipartimento di Fisica "E. R. Caianiello" dell'Università di Salerno

Dipartimento di Ingegneria dell'Informazione e Matematica Applicata dell'Università di Salerno

Dipartimento di Matematica "F. Enriques" dell'Università di Milano

Dipartimento di Matematica e Applicazioni "R. Caccioppoli" dell'Università di Napoli "Federico II"

Dipartimento di Matematica e Informatica dell'Università di Salerno

Dipartimento di Matematica Pura ed Applicata dell'Università dell'Aquila

Facoltà di Scienze Matematiche, Fisiche e Naturali dell'Università di Salerno

Università degli Studi dell'Aquila

Università degli Studi di Milano

Università degli Studi di Salerno

Conference Program

WEDNESDAY, APRIL 14

09.20 Welcome Greetings

<u>Chairman: Roger Bryant</u>

09.30 G. GLAUBERMAN
Finite p-groups with a maximal elementary abelian subgroup of rank 2

10.15 E. KHUKHRO
Group-theoretic applications of Lie rings with finite cyclic grading

10.55 Coffee break

11.20 D.J.S. ROBINSON
Sylow permutability in locally finite groups

12.05 G. NAVARRO
Advances on Brauer's height zero conjecture

12.50 E. SPINELLI
Group identities for symmetric units of group algebras

13.10 Picture

13.10 Lunch break

<u>Chairman: Andrea Del Centina</u>

15.15 P. SHUMYATSKY
Commutators in residually finite groups

16.00 M. LEWIS
Lifts and generalized vertices for Brauer characters of solvable groups

16.45 M. TOTA
On symplectic nil-algebras

17.05 Coffee break

17.30 F. DE GIOVANNI
Groups with finiteness conditions on conjugates and commutators

18.15 A. MAROTI
Average dimension of fixed point spaces

18.50 V. MONTI
Pro-p groups with waists

21.30 Recital of classical Neapolitan songs

THURSDAY, APRIL 15

09.30 Social trip to the gardens "La Mortella"

13.30 Lunch break

dedicated to the memory of Silvia Lucido

Chairman: Andrea Caranti

15.15 D. DIKRANJAN
The Markov-Zariski topology of an infinite group

16.00 M. HERZOG
On infinite Camina groups

16.45 Wujie SHI
Arithmetical properties of finite groups

17.15 Coffee break

17.40 D. CHILLAG
Semi-rational groups

18.25 N. O'SULLIVAN
Localizations of finitely generated soluble groups

18.50 D. BUBBOLONI
Normal coverings of the symmetric and alternating group

FRIDAY, APRIL 16

dedicated to the memory of Karl Gruenberg

Chairman: Wolfgang Herfort

09.00 A. MANN
Uniform growth of free products and related constructions

9.45 M. EVANS
Some stability operations in Group Theory

10.30 M.C. TAMBURINI
Uniform $(2, k)$-generation of matrix groups of small rank

11.00 Coffee break

11.25 O. TALELLI
On complete resolutions

12.10 A. PREVITALI
Galois invariance, trace and subfield subcodes

12.35 G. MALLE
Uniform triples

13.30 Lunch break

Chairman: Niamh O'Sullivan

15.15 J.S. WILSON
Diameters of soluble groups, and applications

16.00 A. TURULL
The Brauer-Clifford group of G-rings

16.45 G.A. FERNANDEZ-ALCOBER
Uniform conciseness of outer commutator words

17.05 Coffee break

17.30 A. JUHASZ
Twisted conjugacy in certain Artin groups and Coxeter groups

17.55 B. AMBERG
Periodic groups saturated with dihedral groups

18.20 M. DE FALCO
Groups with many abelian subgroups

18.45 E. BANNAI
A survey on spherical designs, with emphasis on the aspects related to finite groups

20.30 Social dinner

SATURDAY, APRIL 17

Chairman: Federico Menegazzo

09.00 A. ZALESSKI
On the decomposition numbers for projective characters of Chevalley groups

9.45 N.YU. MAKARENKO
Groups and Lie rings with Frobenius groups of automorphisms

10.30 A. LUCCHINI
Profinite groups with polynomially bounded Möbius numbers

11.00 Coffee break

11.25 Y. SYSAK
On the adjoint group of radical rings and related questions

12.10 P. MORAVEC
Unramified Brauer groups of finite and infinite groups

12.35 M. PELLEGRINI
Hurwitz generation of the universal covering of $Alt(n)$

13.30 Lunch break

Chairman: Andrea Lucchini

15.15 G. TRAUSTASON
Right Engel subgroups

16.00 V. COLOMBO
Subgroups with nontrivial Möbius number in the alternating and symmetric groups

16.25 M. PATASSINI
On the irreducibility of the Dirichlet polynomial of a simple group of Lie type

16.50 P.P. PALFY
Exterior self-quotient modules

17.30 Coffee break

18.45 Concert: Musiche del XVIII e XIX secolo con strumenti d'epoca

Registered Participants

Cristina Acciarri, *Università dell'Aquila, Italy*
Bernhard Amberg, *Universität Mainz, Germany*
Antonio Auletta, *Università di Napoli "Federico II", Italy*
Marina Avitabile, *Università di Milano-Bicocca, Italy*
Eiichi Bannai, *Kyushu University, Japan*
Etsuko Bannai, *Kyushu University, Japan*
Marco Barlotti, *Università di Firenze, Italy*
Luigia Berardi, *Università dell'Aquila, Italy*
Mariagrazia Bianchi, *Università di Milano, Italy*
Victor Bovdi, *University of Debrecen, Hungary*
Rolf Brandl, *Universität Würzburg, Germany*
Francesco Brenti, *Università di Roma "Tor Vergata", Italy*
Roger Bryant, *University of Manchester, U.K.*
Daniela Bubboloni, *Università di Firenze, Italy*
Andrea Caranti, *Università di Trento, Italy*
Carlo Casolo, *Università di Firenze, Italy*
Tullio Ceccherini-Silberstein, *Università del Sannio, Italy*
Lucio Centrone, *Università di Bari, Italy*
David Chillag, *Technion-Israel Institute of Tecnology, Haifa, Israel*
Valentina Colombo, *Università di Padova, Italy*
Gabriella Corsi, *Università di Firenze, Italy*
Mauro Costantini, *Università di Padova, Italy*
Eleonora Crestani, *Università di Padova, Italy*
Alma D'Aniello, *Università di Napoli "Federico II", Italy*
Francesca Dalla Volta, *Università di Milano-Bicocca, Italy*
Stephen Dalton, *Stony Brook University, U.S.A.*
Ulderico Dardano, *Università di Napoli "Federico II", Italy*
Maria De Falco, *Università di Napoli "Federico II", Italy*
Francesco de Giovanni, *Università di Napoli "Federico II", Italy*
Andrea Del Centina, *Università di Ferrara, Italy*
Costantino Delizia, *Università di Salerno, Italy*
Eloisa Detomi, *Università di Padova, Italy*
Dikran Dikranjan, *Università di Udine, Italy*
Vincent Emery, *Università di Milano-Bicocca, Italy*
Gulin Ercan, *Middle East Technical University, Turkey*
Kivanc Ersoy, *Mimar Sinan University, Turkey*

Martin Evans, *University of Alabama, U.S.A.*
Giovanni Falcone, *Università di Palermo, Italy*
Gustavo A. Fernandez-Alcober, *Universidad del Pais Vasco, Spain*
Agota Figula, *University of Debrecen, Hungary*
Carla Fiori, *Università di Modena e Reggio Emilia, Italy*
Martino Garonzi, *Università di Padova, Italy*
Norberto Gavioli, *Università dell'Aquila, Italy*
Anna Gillio, *Università di Milano, Italy*
Anna Luisa Gilotti, *Università di Bologna, Italy*
George Glauberman, *University of Chicago, U.S.A.*
Karise Gonçalves Oliveira, *Instituto Federal de Educação, Brazil*
Wolfgang Herfort, *Technische Universität Wien, Austria*
Marcel Herzog, *Tel Aviv University, Israel*
Paz Jiménez Seral, *Universidad de Zaragoza, Spain*
Diana Imperatore, *Università di Napoli "Federico II", Italy*
Arye Juhasz, *Technion-Israel Institute of Tecnology, Haifa, Israel*
Evgenii Khukhro, *Institute of Mathematics-Novosibirsk, Russia*
Antonella Leone, *Università di Napoli "Federico II", Italy*
Mark Lewis, *Kent State University, U.S.A.*
Patrizia Longobardi, *Università di Salerno, Italy*
Maria Loukaki, *University of Crete, Greece*
Andrea Lucchini, *Università di Padova, Italy*
Annamaria Lucibello, *Università di Salerno, Italy*
Mercede Maj, *Università di Salerno, Italy*
Natalia Yu. Makarenko, *Institute of Mathematics-Novosibirsk, Russia*
Université de Haute Alsace, France
Dmitry Malinin, *University of Vlora, Albania*
Gunter Malle, *Universität Kaiserslautern, Germany*
Avinoam Mann, *Hebrew University of Jerusalem, Israel*
Selena Marinelli, *Università di Pisa, Italy*
Attila Maroti, *Hungarian Academy of Sciences, Hungary*
Juan Medina, *Universidad Politecnica de Cartagena, Spain*
Federico Menegazzo, *Università di Padova, Italy*
Maurizio Meriano, *Università di Salerno, Italy*
Valerio Monti, *Università di Roma "La Sapienza", Italy*
Primoz Moravec, *University of Ljubljana, Slovenia*
Alexander Moreto, *Universidad de Valencia, Spain*
Marta Morigi, *Università di Bologna, Italy*
Carmela Musella, *Università di Napoli "Federico II", Italy*

Gabriel Navarro, *Universidad de Valencia, Spain*
Chiara Nicotera, *Università di Salerno, Italy*
Niamh O'Sullivan, *Dublin City University, Ireland*
Emanuele Pacifici, *Università di Milano, Italy*
Peter P. Palfy, *Hungarian Academy of Sciences, Hungary*
Massimiliano Patassini, *Università di Padova, Italy*
Marco Pellegrini, *Università di Milano-Bicocca, Italy*
Andrea Previtali, *Università Insubria-Como, Italy*
Flavia F. Ramos Zapata, *University of Brasilia, Brazil*
Silvana Rinauro, *Università della Basilicata, Italy*
Derek J.S. Robinson, *University of Illinois, U.S.A.*
Emanuela Romano, *Università di Salerno, Italy*
Rosaria Rota, *Università di Roma Tre, Italy*
Nicola Sambonet, *Università di Milano, Italy*
Lucia Sanus, *Universitat de València, Spain*
Carlo Scoppola, *Università dell'Aquila, Italy*
Luigi Serena, *Università di Firenze, Italy*
Wujie Shi, *Suzhou University, P.R. China*
Pavel Shumyatsky, *University of Brasilia, Brazil*
Carmela Sica, *Università di Salerno, Italy*
Salvatore Siciliano, *Università del Salento, Italy*
Ernesto Spinelli, *Università del Salento, Italy*
Yaroslav Sysak, *Ukrainian National Academy of Sciences, Ukraine*
Olympia Talelli, *University of Athens Panepistimioupolis, Greece*
M. Chiara Tamburini, *Università Cattolica-Brescia, Italy*
Tommaso Terragni, *Università di Milano-Bicocca, Italy*
Antonio Tortora, *Università di Salerno, Italy*
Maria Tota, *Università di Salerno, Italy*
Gunnar Traustason, *University of Bath, U.K.*
Alexandre Turull, *University of Florida, U.S.A.*
Angela Valenti, *Università di Palermo, Italy*
Giovanni Vincenzi, *Università di Salerno, Italy*
Thomas Weigel, *Università di Milano-Bicocca, Italy*
John S. Wilson, *University College-Oxford, U.K.*
Alexandre Zalesski, *University of East Anglia, U.K.*
Theo Zapata, *University of Brasilia, Brazil*
Università di Milano-Bicocca, Italy

POSITIVE LAWS ON GENERATORS IN POWERFUL PRO-p GROUPS

C. ACCIARRI

Dipartimento di Matematica Pura ed Applicata
Università degli Studi dell'Aquila
I-67010 Coppito, L'Aquila (Italy)
Email: acciarricristina@yahoo.it

G. A. FERNÁNDEZ-ALCOBER

Matematika Saila
Euskal Herriko Unibertsitatea
48080 Bilbao (Spain)
Email: gustavo.fernandez@ehu.es

If G is a finitely generated powerful pro-p group satisfying a certain law $v \equiv 1$, and if G can be generated by a normal subset T of finite width which satisfies a positive law, we prove that G is nilpotent. Furthermore, the nilpotency class of G can be bounded in terms of the prime p, the number of generators of G, the law $v \equiv 1$, the width of T, and the degree of the positive law. The main interest of this result is the application to verbal subgroups: if G is a p-adic analytic pro-p group in which all values of a word w satisfy positive law, and if the verbal subgroup $w(G)$ is powerful, then $w(G)$ is nilpotent.

Keywords: positive laws; powerful pro p groups; verbal subgroups.

1. Introduction

If α and β are two group words, we say that a group G satisfies the *law* $\alpha \equiv \beta$ if every substitution of elements of G by the variables gives the same value for α and for β. If the words α and β are positive, i.e. if they do not involve any inverses of the variables, then we say that $\alpha \equiv \beta$ is a *positive law*. We can similarly speak about a law holding on a subset T of G, if we only substitute elements of T by the variables. Groups satisfying a positive law have received special attention in the past decade. The main result is due to Burns and Medvedev, who proved in Ref. 2 that a locally graded group G satisfies a positive law if and only if G is nilpotent-by-(locally finite of finite exponent). This applies in particular to residually finite groups.

A similar kind of problem has been considered by Shumyatsky and the second author in Ref. 5. If G is a finitely generated group and T is a set of generators satisfying a positive law, they ask whether the whole of G will also satisfy a (possibly different) positive law, provided that T is sufficiently large in some sense. In this direction, they obtain a positive answer if T is a normal subset of G which is closed under taking commutators of its elements (*commutator-closed* for short), under the assumption that G satisfies an arbitrary law and is residually-p for some prime p. More precisely, the result is proved for all primes outside a finite set $P(n)$ depending only on the *degree* n of the law (that is, the maximum of the lengths of α and β).

The result in the previous paragraph can be applied to verbal subgroups $w(G)$ in a group G, where T is considered to be the set G_w of all values of the word w in G. Note that G_w is always a normal subset. Among other results, Shumyatsky and the second author provc that, if G is a p-adic analytic pro-p group and $p \notin P(n)$ then, for every word w such that G_w is commutator-closed, a positive law on G_w implies a positive law on the whole of $w(G)$. Now two questions naturally arise: (i) can we get rid of the restriction $p \notin P(n)$?; (ii) can we get rid of the condition that G_w should be commutator-closed? If we can give a positive answer to both these questions, then the result will hold in p-adic analytic pro-p groups for all primes and for all words.

If G is a p-adic analytic pro-p group, Jaikin-Zapirain has proved (see Theorem 1.3 of Ref. 7) that the set G_w has finite width for every word w, and then, by Proposition 4.1.2 of Ref. 9, the verbal subgroup $w(G)$ is closed in G. (See Section 2 for the definition of width.) Thus $w(G)$ is again a p-adic analytic pro-p group and, according to Interlude A of Ref. 3, it contains a powerful subgroup of finite index. One of the main results of this paper is the solution of the problem raised in the last paragraph in the case when $w(G)$ itself is powerful.

Theorem 1.1. *Let G be a p-adic analytic pro-p group, and let w be any word. If all values of w in G satisfy a positive law and the verbal subgroup $w(G)$ is powerful, then $w(G)$ is nilpotent.*

Observe that the conclusion in the previous theorem that $w(G)$ is nilpotent is actually stronger than $w(G)$ satisfying a positive law.

Following the approach of Ref. 5, we obtain Theorem 1.1 from a more general result not involving directly word values. In this case, we work with G a finitely generated powerful pro-p group for an arbitrary prime p, and the set of generators T has to be normal and of finite width, but not necessarily commutator-closed. Recall that, as mentioned above, if G is a p-adic analytic pro-p group, then G_w has finite width for every word w.

Theorem 1.2. *Let G be a powerful d-generator pro-p group which satisfies a certain law $v \equiv 1$. Suppose that G can be generated by a normal subset T of width m that satisfies a positive law of degree n. Then G is nilpotent of bounded class.*

Here, and in the remainder of the paper, when we say that a certain invariant of a group is bounded, we mean that it is bounded above by a function of the parameters appearing in the statement of the corresponding result. Thus, in Theorem 1.2, the nilpotency class of G is bounded in terms of the prime p, the number d of generators of G, the law $v \equiv 1$, the width m of T, and the degree n of the positive law. If we want to make explicit the set S of parameters in terms of which a certain quantity is bounded, then we will use the expression 'S-bounded'.

We want to remark that, contrary to what happens in Theorem 1.2, we cannot guarantee that the nilpotency class of $w(G)$ is bounded in Theorem 1.1. The reason is that we are using the above-mentioned result of Jaikin-Zapirain, which provides the finite width of G_w, but not bounded width for that set.

2. The action on abelian normal sections

Our first step is to translate the positive law on the normal generating set T into a condition about the action of the elements of T on the abelian normal sections of G. More precisely, we have the following consequence of Lemma 2.1 in Ref. 5. (Let $f(X)$ be the product of the polynomials $f_1(X)$ and $f_{-1}(X)$ in the statement of that lemma.)

Lemma 2.1. *Let T be a normal subset of a group G, and assume that T satisfies a positive law of degree n. Then there exists a monic polynomial $f(X) \in \mathbb{Z}[X]$ of degree $2n$, depending only on the given positive law, which satisfies the following property: if A is an abelian normal section of G, then $f(t)$, viewed as an endomorphism of A, is trivial for every $t \in T \cup T^{-1}$.*

If T is a subset of a group G, we say that T has *finite width* if there exists a positive integer m such that every element of the subgroup $\langle T \rangle$ can be expressed as a product of no more than m elements of $T \cup T^{-1}$. The smallest possible value of m is then called the *width* of T.

In our next theorem, we show how some properties of the generating set T of G are hereditary for the natural generating set of $\gamma_k(G)$ which can be constructed from T. For simplicity, if $A = K/L$ is a normal section of a group G, we say that two elements $g, h \in G$ commute modulo A if gL and hL commute modulo A (or, equivalently, if g and h commute modulo K).

Theorem 2.1. *Let G be a d-generator finite p-group, and let T be a normal generating set of G. Then*

$$T_k = \{[t_1, \ldots, t_k] \mid t_i \in T\}$$

is a normal generating set of $\gamma_k(G)$, and furthermore:

(i) If T has finite width m, then the width of T_k is at most md^{k-1}.
(ii) If T satisfies a positive law of degree n, then there exists a monic polynomial $h(X) \in \mathbb{Z}[X]$ of n-bounded degree such that $h(t_k)$ annihilates $\gamma_{k+1}(G)/\gamma_{k+1}(G)'$ for every $t_k \in T_k \cup T_k^{-1}$.

Proof. Of course, T_k is a normal subset of G, and the proof that T_k generates $\gamma_k(G)$ is routine.

(i) We argue by induction on k. The result is obvious for $k = 1$, so we assume next that $k \geq 2$. By the Burnside Basis Theorem, we can choose $t_1, \ldots, t_d \in T$ such that $G = \langle t_1, \ldots, t_d \rangle$. If y is an arbitrary element of $\gamma_k(G)$, we can write

$$y = [g_1, t_1] \ldots [g_d, t_d], \quad \text{for some } g_i \in \gamma_{k-1}(G), \tag{1}$$

by using Proposition 1.2.7 of Ref. 9. Now, if g is an arbitrary element of $\gamma_{k-1}(G)$, then by the induction hypothesis, we have $g = u_1 \ldots u_s$ for some $u_i \in T_{k-1} \cup T_{k-1}^{-1}$, where $s \leq md^{k-2}$. Then, for every $t \in T$, we have

$$[g, t] = [u_1, t]^{u_2 \ldots u_s} \ldots [u_{s-1}, t]^{u_s} [u_s, t].$$

If $u_i \in T_{k-1}$, then $[u_i, t] \in T_k$; on the other hand, if $u_i \in T_{k-1}^{-1}$ then

$$[u_i, t] = \left([u_i^{-1}, t]^{u_i}\right)^{-1}$$

is an element of T_k^{-1}. Thus $[g, t]$ is a product of at most s elements of $T_k \cup T_k^{-1}$, and it follows from (1) that y is a product of no more than ds elements of $T_k \cup T_k^{-1}$. This completes the proof of (i).

(ii) Set $A = \gamma_{k+1}(G)/\gamma_{k+1}(G)'$. By Lemma 2.1, there exists a monic polynomial $f(X) \in \mathbb{Z}[X]$ of degree $2n$ such that $f(t)$ annihilates A for every $t \in T \cup T^{-1}$.

Let I be the ideal of $\mathbb{Z}[X_1, X_2]$ generated by $f(X_1)$ and $f(X_2)$. Since f is monic, the quotient ring $R = \mathbb{Z}[X_1, X_2]/I$ is a finitely generated $\mathbb{Z}$-module, generated by the images of the monomials $X_1^i X_2^j$ with $0 \leq i, j \leq 2n-1$. By

Theorem 5.3 in Chapter VIII of Ref. 6, R is integral over $\mathbb{Z}$. In particular, there exists a monic polynomial $h(X) \in \mathbb{Z}[X]$ such that $h(X_1X_2) \in I$. Also, by examining the proof of that result in Ref. 6, it is clear that the degree of $h(X)$ is at most $(2n)^2$.

Now, let $[u,t]$ be an arbitrary element of T_k, where $u \in T_{k-1}$ and $t \in T$. Since $(t^u)^{-1}$ and t commute modulo A, these elements define commuting endomorphisms of A, and hence we can define a ring homomorphism

$$\begin{aligned} \varphi : \mathbb{Z}[X_1, X_2] &\longrightarrow \mathrm{End}\,(A) \\ X_1 &\longmapsto (t^u)^{-1} \\ X_2 &\longmapsto t. \end{aligned}$$

Since $f((t^u)^{-1})$ and $f(t)$ are both the null endomorphism of A, it follows that $f(X_1)$ and $f(X_2)$ are contained in the kernel of φ, and so the same holds for the ideal I. Hence $h(X_1X_2) \in \ker\varphi$, which means that $h([u,t])$ is the null endomorphism of A.

We can similarly prove that $h([t,u]) = 0$ in $\mathrm{End}\,(A)$, by defining $\psi : \mathbb{Z}[X_1, X_2] \longrightarrow \mathrm{End}\,(A)$ via the assignments $X_1 \mapsto t^{-1}$ and $X_2 \mapsto t^u$. Thus $h(t_k)$ annihilates A for every $t_k \in T_k \cup T_k^{-1}$. □

Finally, for a certain k, we are able to get an Engel action of all k-th powers of the elements of G on some abelian normal sections of G.

Theorem 2.2. *Let G be a finite p-group generated by a normal subset T which has width m. Suppose that A is an abelian normal section of G such that the elements of T commute pairwise modulo A, and that for some monic polynomial $f(X) \in \mathbb{Z}[X]$, $f(t)$ annihilates A for all $t \in T \cup T^{-1}$. Then:*

(i) There exists an $\{m, f\}$-bounded integer r such that $[A,_r g] \leq A^p$ for every $g \in G$.
(ii) There exist $\{m, f\}$-bounded integers n and k such that $[A,_n g^k] = 1$ for every $g \in G$.

Proof. The first part of the proof is similar to the proof of (ii) in the last theorem. Let us write n for the degree of $f(X)$. Consider the quotient ring $R = \mathbb{Z}[X_1, \ldots, X_m]/I$, where I is the ideal generated by the polynomials $f(X_1), \ldots, f(X_m)$. Then R is integral over $\mathbb{Z}$, and there exists a monic polynomial $h(X) \in \mathbb{Z}[X]$ of degree at most n^m such that $h(X_1 \ldots X_m) \in I$. Now let g be an arbitrary element of G. Since T generates G and has width m, we can write $g = t_1 \ldots t_m$ for some $t_i \in T \cup T^{-1}$. The map $X_1 \mapsto t_1, \ldots, X_m \mapsto t_m$ extends to a ring homomorphism

$\varphi : \mathbb{Z}[X_1, \ldots, X_m] \longrightarrow \mathrm{End}\,(A)$, since the elements of T commute pairwise modulo A. Since $f(t_1) = \cdots = f(t_m) = 0$, it follows that $I \subseteq \ker \varphi$. Consequently, $h(g) = h(t_1 \ldots t_m) = \varphi(h(X_1 \ldots X_m)) = 0$. Thus we have found a monic polynomial $h(X) \in \mathbb{Z}[X]$ such that $h(g)$ annihilates A for all $g \in G$. Note that the polynomial $h(X)$ only depends on $f(X)$ and m, but not on the particular element g or on the section A.

(i) Since G is a finite p-group, we have $[A,_c G] = 1$ for some c. Let $(X-1)^r$ be the greatest common divisor of $(X-1)^c$ and $h(X)$, when these polynomials are considered in $\mathbb{F}_p[X]$. Since $r \le \deg h$, it follows that r is $\{m, f\}$-bounded. By Bézout's identity, we can write

$$(X-1)^r = p(X)(X-1)^c + q(X)h(X),$$

for some $p(X), q(X) \in \mathbb{F}_p[X]$. If we consider an element $g \in G$, and substitute g for X in the previous expression, then, as endomorphisms of the $\mathbb{F}_p$-vector space A/A^p, we get $(g-1)^r = 0$. This means that $[A,_r g] \le A^p$, as desired.

(ii) Let J be the ideal of $\mathbb{Z}[X]$ generated by all polynomials $h(X^i)$ with $i \ge 1$. Then, if $j(X) \in J$, it follows that $j(g) = 0$ for every $g \in G$. By Lemma 3.3 of Ref. 10, there exist positive integers q, k and ℓ such that

$$qX^\ell (X^k - 1)^\ell \in J,$$

where q, k, ℓ depend only on $h(X)$, so only on $f(X)$ and m. Then

$$A^{qg^\ell (g^k-1)^\ell} = 1, \quad \text{for every } g \in G.$$

If p^s is the largest power of p which divides q, then $A^q = A^{p^s}$, since A is a finite p-group. Also, we have $A^g = A$. Hence

$$A^{p^s (g^k-1)^\ell} = 1$$

or, what is the same,

$$[A^{p^s}, g^k, \overset{\ell}{\ldots}, g^k] = 1 \tag{2}$$

for every $g \in G$.

Now, it follows from part (i) that

$$[A^{p^i},_r g] \le A^{p^{i+1}}, \quad \text{for every } i \ge 0, \text{ and for every } g \in G.$$

This, together with (2), shows that

$$[A,_n g^k] = 1, \quad \text{for all } g \in G,$$

where $n = sr + \ell$. □

3. Proof of the Main Theorems

We will begin by proving Theorem 1.2. In order to show that the powerful pro-p group G is nilpotent, we will rely on the following two lemmas. The first one is a classical result of Philip Hall (see, for example, Theorem 3.26 of Ref. 8), and the other one says that for a finitely generated powerful pro-p group 'nilpotent-by-finite' is the same as 'nilpotent'.

Lemma 3.1. *Let G be a group, and let N be a normal subgroup of G. If N is nilpotent of class k and G/N' is nilpotent of class c, then G is nilpotent of $\{k,c\}$-bounded class.*

Lemma 3.2. *Let G be a finitely generated powerful pro-p group. If G has a normal subgroup N of finite index which is nilpotent of class c, then G itself is nilpotent of $\{c,e\}$-bounded class, where e is the exponent of G/N.*

Proof. We prove the result for $p > 2$. For $p = 2$, the same proof applies with some little changes.

It follows from the hypotheses that G^e is nilpotent of class at most c. By Proposition 3.2 and Corollary 3.5 in Ref. 4, we get

$$[G^{e^{c+1}}, G, \overset{c}{\ldots}, G] = [G, \overset{c+1}{\ldots}, G]^{e^{c+1}} = [G^e, \overset{c+1}{\ldots}, G^e] = 1. \tag{3}$$

On the other hand, since G is powerful, we have $\gamma_{i+1}(G) \leq G^{p^i}$ for all $i \geq 1$. As a consequence, for some $\{c,e\}$-bounded integer k we have $\gamma_{k+1}(G) \leq G^{e^{c+1}}$. This, together with (3), shows that G is nilpotent of class at most $k + c$, and we are done. □

Note that we could have written the previous lemma under the apparently weaker assumption that the exponent of G/N is finite, rather than N being of finite index in G. However, if G is a finitely generated powerful pro-p group, these two conditions are equivalent: if $\exp G/N = p^k$, then G^{p^k} is contained in N, and then by Theorem 3.6 of Ref. 3, we have $|G : N| \leq |G : G^{p^k}| \leq p^{kd}$, where d is the minimum number of generators of G as a topological group. (In fact, the assumption that G should be powerful is not necessary for this equivalence, since $|G : G^{p^k}|$ is finite for every finitely generated pro-p group. But this is a much deeper result, which needs Zelmanov's positive solution of the Restricted Burnside Problem.)

We also need the following result of Black (see Corollary 2 in Ref. 1).

Theorem 3.1. *Let G be a finite group of rank r satisfying a law $v \equiv 1$. Then, there exists an $\{r,v\}$-bounded number k such that $\gamma_k((G^{k!})') = 1$. In particular, if G is soluble, then the derived length of G is $\{r,v\}$-bounded.*

Note that the positive solution to the Restricted Burnside Problem is needed for the conclusion in the soluble case: thus we know that the quotient $G/G^{k!}$ has bounded order, and so also bounded derived length.

We can now proceed to the proof of Theorem 1.2.

Proof of Theorem 1.2. Suppose that the result is known for G a finite p-group, so that all finite p-groups satisfying the conditions of the theorem have nilpotency class at most c, for some bounded number c. Now if G is a pro-p group as in the statement of the theorem, and N is an arbitrary open normal subgroup of G, it follows that $\gamma_{c+1}(G) \leq N$. Thus necessarily $\gamma_{c+1}(G) = 1$ and the result is valid also for pro-p groups.

Hence we may assume that G is a d-generator finite powerful p-group. By Theorem 11.18 of Ref. 8, it follows that G has rank d, i.e. that every subgroup of G can be generated by d elements. Since G satisfies the law $v \equiv 1$, by Theorem 3.1 we have $G^{(s)} = 1$ for some bounded number s. We argue by induction on s.

If $s \leq 2$, i.e. if G is metabelian, then the elements of T commute pairwise modulo G'. Choose generators $g_1, \ldots, g_d$ of G. By Lemma 2.1 and Theorem 2.2, since T satisfies a positive law, we know that there exist bounded numbers n and k such that $[G',_n g_i^k] = 1$ for all $i = 1, \ldots, d$. As a consequence, the subgroups $\langle g_i^k, G' \rangle$ have bounded nilpotency class. Thus $G^k G' = \langle g_1^k, \ldots, g_d^k, G' \rangle$ is the product of d normal subgroups of bounded class, and so has bounded class itself. Since $|G : G^k G'| \leq k^d$, it follows from Lemma 3.2 that G has bounded nilpotency class. This concludes the proof in the metabelian case.

Assume now that $s \geq 3$. We claim that the nilpotency class of $G/\gamma_{k+1}(G)'$ is bounded for all $k \geq 1$ (here, we must also take k into account for the bound). The result is true for $k = 1$, according to the last paragraph. Now we argue by induction on k. By Theorem 2.1, T_k is a normal set of generators of $\gamma_k(G)$ of bounded width. Also, the elements of T_k commute pairwise modulo $\gamma_{k+1}(G)$. On the other hand, by (ii) of Theorem 2.1, there exists a monic polynomial $h(X) \in \mathbb{Z}[X]$ such that $h(t_k)$ annihilates the abelian normal section $A = \gamma_{k+1}(G)/\gamma_{k+1}(G)'$ for every $t_k \in T_k$. Thus we may argue as in the metabelian case above with the group $Q = \gamma_k(G)/\gamma_{k+1}(G)'$ and deduce that Q has bounded nilpotency class. Since $G/\gamma_k(G)'$ has also bounded class by the induction hypothesis, the claim follows from Lemma 3.1.

Now that the claim is proved, the result easily follows. Indeed, since $G/G^{(s-1)}$ has bounded class by induction, there exists a bounded integer ℓ

such that $\gamma_{\ell+1}(G) \leq G^{(s-1)}$. Hence $\gamma_{\ell+1}(G)' = 1$, and G has bounded class by the previous claim. □

Now Theorem 1.1 follows readily from Theorem 1.2.

Proof of Theorem 1.1. As already mentioned, the set G_w of all values of w in G is a normal subset of G, and in particular of $w(G)$. Also, by Theorem 1.3 of Ref. 7, G_w has finite width, say m.

Let $\alpha \equiv \beta$ be the positive law satisfied by the set G_w, and suppose that the number of variables used in the law $\alpha \equiv \beta$ and in the word w is k and ℓ, respectively. Now, consider kl arbitrary elements $g_1, \ldots, g_{kl}$ of G. Since the k elements $w(g_1, \ldots, g_\ell), \ldots, w(g_{(k-1)\ell+1}, \ldots, g_{k\ell})$ satisfy the law $\alpha \equiv \beta$, it follows that

$$\alpha(w(g_1, \ldots, g_\ell), \ldots, w(g_{(k-1)\ell+1}, \ldots, g_{k\ell})) = \beta(w(g_1, \ldots, g_\ell), \ldots, w(g_{(k-1)\ell+1}, \ldots, g_{k\ell})).$$

This means that the group G satisfies a law $v \equiv 1$, where v is a word which depends only on w and on the positive law $\alpha \equiv \beta$. In particular, the law $v \equiv 1$ is also satisfied by $w(G)$.

Now, we can apply directly Theorem 1.2 to the group $w(G)$ and the generating set G_w, in order to conclude that $w(G)$ is nilpotent. □

Acknowledgments

The authors are supported by the Spanish Government, grant MTM2008-06680-C02-02, partly with FEDER funds, and by the Department of Education, Universities and Research of the Basque Government, grants IT-252-07 and IT 460-10. The second author is also supported by a grant of the University of L'Aquila.

References

1. S. Black, 'A finitary Tits alternative', *Arch. Math.* 72 (1999), 86–91.
2. R.G. Burns and Yu. Medvedev, 'Group laws implying virtual nilpotence', *J. Austral. Math. Soc.* 74 (2003), 295–312.
3. J.D. Dixon, M.P.F. du Sautoy, A. Mann, and D. Segal, *Analytic Pro-p Groups*, second edition, Cambridge Studies in Advanced Mathematics 61 (Cambridge University Press, 1999).
4. G.A. Fernández-Alcober, J. González-Sánchez, and A. Jaikin-Zapirain, 'Omega subgroups of pro-p groups', *Israel J. Math.* 166 (2008), 393–412.
5. G.A. Fernández-Alcober and P. Shumyatsky, 'Positive laws on word values in residually-p groups', preprint.

6. T.W. Hungerford, *Algebra* (Springer-Verlag, 1974).
7. A. Jaikin-Zapirain, 'On the verbal width of finitely generated pro-p groups', *Revista Matemática Iberoamericana* 24 (2008), 617–630.
8. E.I. Khukhro, *p-Automorphisms of Finite p-Groups*, LMS Lect. Notes, Vol. 246 (Cambridge University Press, 1998).
9. D. Segal, *Words: Notes on Verbal Width in Groups*, LMS Lect. Notes, Vol. 361 (Cambridge University Press, 2009).
10. J.F. Semple and A. Shalev, 'Combinatorial conditions in residually finite groups I', *J. Algebra* 157 (1993), no. 1, 43–50.

PERIODIC GROUPS SATURATED BY DIHEDRAL SUBGROUPS

B. AMBERG

Fachbereich Mathematik, der Universität Mainz,
D-55099 Mainz, Germany
E-mail: amberg@mathematik.uni-mainz.de

L. KAZARIN*

Department of Mathematics, Yaroslavl State University,
150000 Yaroslavl, Russia
E-mail: kazarin@uniyar.ac.ru

It is shown that an infinite periodic group is locally dihedral if every finite subgroup S of G is contained in a subgroup of G which is isomorphic to a finite dihedral subgroup. If a periodic group $G = AB$ is a product of two infinite locally dihedral subgroups A and B, then G is soluble.

1. Introduction

A group G is *saturated by finite dihedral subgroups* if every finite subgroup S of G is contained in a finite dihedral group. Periodic groups G with this property have been studied by A.K. Shlyopkin and A.G. Rubashkin in Ref. 6. They show that under certain conditions G must be a locally dihedral group.

It is not difficult to see that every locally finite group G which is saturated by finite dihedral subgroups is locally dihedral (see Ref. 6, Lemma 11). In Ref. 6 it is proved that this also holds for certain classes of periodic groups, for instance when the elements of G have bounded period or when each pair of conjugate elements of prime order generate a finite group. In other cases it is shown that G has a factorization as $G = ABC = ACB = BCA = CBA$ where A is (locally) finite dihedral and

*The second author is grateful to the Department of Mathematics of the University of Mainz for its warm hospitality during the time when this research was done. He also likes to thank the Deutsche Forschungsgemeinschaft for financial support.

B, C are (locally) cyclic subgroups.

The question arises whether every periodic group which is saturated by finite dihedral groups must be a locally dihedral group. In the following we answer this question positively.

Theorem 1.1. *Let G be an infinite periodic group saturated by finite dihedral subgroups. Then G is a locally dihedral group.*

Recall that an infinite periodic locally dihedral group X is the union of an infinite ascending chain of finite dihedral subgroups. Such a group has the form $X = X_0\langle i\rangle$, where i is an involution and X_0 is a locally cyclic normal subgroup of X such that $x^i = x^{-1}$ for every $x \in X_0$. Every (locally) cyclic subgroup of X whose order is greater than 4, belongs to X_0 and each non-(locally) cyclic subgroup of X contains its centralizer in X.

The main idea of the proof is to reduce the general case to the special situation, when G has a factorization of the form $G = AB$ with locally dihedral subgroups A and B. A similar situation occured in Ref. 1, where A and B are assumed to be Chernikov groups. We will use the technique developed there. Note that the solubility of a product of two finite dihedral groups was proved by V. Monakhov in Ref. 4. Here we we extend this result as follows.

Theorem 1.2. *Let $G = AB$ be a periodic group, which is a product of two locally dihedral groups A and B. Then G is soluble.*

In fact, the methods in Ref. 1 allow to prove the preceeding result even for non-periodic groups, but we present here only the simplified version for periodic groups.

2. Reduction

In this section G is a periodic group saturated by finite dihedral groups. If S is a Sylow-2-subgroup of G then by Ref. 6, Lemma 14, S is a locally dihedral group. Let τ denote a central involution in S.

If S has only two elements, then G is a locally dihedral group by Ref. 6, Lemma 16. Therefore we may assume that there exists an involution $\mu \neq \tau$ in S. Put $A = C_G(\tau)$ and $B = C_G(\mu)$. Also let $V =< \mu, \tau >$ and observe that also $\nu = \mu\tau$ is an involution in S.

The following simple properties of the group G were already observed in Ref. 6.

Lemma 2.1.

(1) The centralizer $C_G(\gamma)$ of any involution $\gamma \in G$ is a finite or locally finite dihedral group,
(2) The Sylow-2-subgroups of G are conjugate,
(3) $N_G(S) = S$ and if two elements in S are conjugate in G, then they are conjugate in S,
(4) The intersection of every two different Sylow subgroups of G is a cyclic group.
(5) If G has a finite or locally finite normal subgroup, then G is a locally finite dihedral group.

Proof. Statement (1) is Ref. 6, Lemma 15. Statement (2) is Ref. 6, Lemma 18, Statement (3) follows from Ref. 6, Lemmas 23 and 26. Statement (4) is Ref. 6, Lemma 25. Statement (5) is Ref. 6, Lemma 19. □

Lemma 2.2. *The subgroup $V =< \tau, \mu >$ of order 4 is a Sylow 2-subgroup of $B = C_G(\mu)$. Thus $B = O(B) \rtimes V$.*

Proof. Let T be a Sylow-2-subgroup of G containing $C_S(\mu) = V$ (in every locally dihedral 2-group the centralizer of a non-central involution is of order 4). Since $T \cap S$ is not cyclic, it follows from Lemma 2.1 (4) that $T = S$. Therefore $V = C_S(\mu) \in Syl_2(B)$. The lemma is proved. □

Lemma 2.3. *If S is non-abelian, then all involutions in $V- < \tau >$ are conjugate in S.*

Proof. Indeed, the normalizer $N_S(V)$ is a non-abelian dihedral group of order 8. The result follows. □

Lemma 2.4. $G = AB$.

Proof. Let $g \in G$ and $V =< \mu, \tau >$. Since μ^g and τ are not conjugate by Lemma 2.1, the subgroup $D =< \mu^g, \tau >$ is dihedral of order divisible by 4. Hence there exists an involution $\delta \in Z(D)$. Clearly $\delta \in C_G(\tau) = A$ and $\mu^g \in C_G(\delta)$.

Suppose first that S is non-abelian. Then all involutions in $V_1 =< \delta, \tau > - < \tau >$ are conjugate by Lemma 2.3. By Lemma 2.2, the Sylow 2-subgroups of $C_G(\delta)$ are elementary-abelian and $V_1 \leq C_S(\delta)$. Since $\mu^g \in C_G(\delta)$ this implies that μ^g is contained in a subgroup which is conjugate to V_1. Therefore μ^g is conjugate to δ. Since $\delta \in S$ is conjugate to μ, it follows from Lemma 2.1 (3) that $\delta = \mu^a$ for some $a \in A$. Hence $[\mu^g, \mu^a] = 1$. This implies that $\mu^{ga^{-1}}$ centralizes V.

Since $V \in Syl_2(B)$ and $B = O(B)V$ is a dihedral group, it follows that $\mu^{ga^{-1}} \in V$. Since the elements μ^a and $\mu^{ga^{-1}} \in A$ are conjugate, there exists an element $a_1 \in A$ such that $\mu^{ga^{-1}a_1} = \mu$. Hence $ga^{-1}a_1 \in B$ and $g \in BA$. Therefore $G = BA = AB$ in this case.

Assume now that S is abelian. Then $S = V$ is elementary abelian of order 4. As above, let $\delta \in Z(< \mu^g, \tau >)$. Since δ is conjugate to μ or $\mu\tau$ in $A = C_G(\tau)$, we may choose an element $a \in A$ such that $< \delta^a, \tau >= V$.

Since $[\mu^{ga}, \delta^a] = 1$, this implies that μ^{ga} centralizes V and $\mu^{ga} \in V$. Recall that $V = \{1, \tau, \mu\tau, \mu\}$. Since μ and $\mu\tau$ are not conjugate, this implies that $\mu^{ga} = \mu$, so that $ga \in C_G(\mu) = B$ and $g \in BA$. Since this holds for arbitrary $g \in G$, it follows again that $G = BA = AB$. The lemma is proved. □

3. Proof of Theorems 1.1 and 1.2

Consider a periodic group $G = AB$ which is a product of two locally dihedral subgroups A and B. Clearly, $A = A_0\langle c\rangle$, $B = B_0\langle d\rangle$ for some involutions $c \in A \setminus A_0$ and an involution $d \in B \setminus B_0$.

The following lemma will be used repeatedly in the next section.

Lemma 3.1. *Let N be a normal subgroup of a group G. Suppose that G has two abelian subgroups A and B such that $N \subseteq AB$. Then N is metabelian.*

Proof. Obviously, $NA = A(NA \cap B)$. By a well-known theorem of N. Itô the group NA is metabelian (see for instance Ref. 3, VI, Satz 4.4, p. 674). □

Lemma 3.2. *Let $G = AB$ be a finite group, which is a product of a subgroups A and B, where A is dihedral and B is either cyclic or a dihedral group. Then $G^{(k)} = 1$ for some positive integer k not depending on A and B.*

Proof. Suppose first that $G = AB$ is a 2-group.

Case 1. A is dihedral, $B = B_0$ is cyclic.

Let K be a normal subgroup of G contained in the set A_0B_0. By Lemma 3.1 K is metabelian. We use the standard bar-notation for $\bar{G} = G/K$ and its subgroups. Since $\bar{G}$ has no normal subgroups belonging to $\bar{A}_0\bar{B}_0$, it follows that there exists a subgroup $\bar{T}$ in $\bar{G}$ of order 2 not contained in $\bar{A}_0\bar{B}_0$. Then $\bar{T}$ is generated by an element $t = a\bar{c}b$ such that $a \in \bar{A}_0$, $b \in \bar{B}$ and $\bar{c} = Kc$. It follows that $a\bar{c} \in \bar{A} \setminus \bar{A}_0$. Thus we may assume that $c_1 = a\bar{c}$ is contained in $\bar{A} \cap \bar{B}$. Hence $\bar{G}/\bar{T}$ is a product of the subgroups $\bar{A}_0\bar{T}/\bar{T}$ and $\bar{B}_0\bar{T}/\bar{T}$. Therefore the derived length of G does not exceed 5.

Case 2. A and B are dihedral subgroups of G.

Now A_0 and B_0 are the cyclic subgroups of A and B with index 2 respectively.

If $H = A^2B^2 = B^2A^2$ is a subgroup of G, then the index of H in G is at most 16, and $H \leq \Phi(G)$, the Frattini subgroup of G. If $H = \Phi(G)$, then $G' \leq H$ and the derived length of G is at most 3. Clearly, $H \leq G'$ and by similar reasons we may assume that $|G : G'| \leq 8$. If $|G : G'| = 4$, then by a theorem of O.Taussky (see Ref. 3, III, Satz 11.9, p. 339) G is a 2-group of maximal class, so that G is metabelian. Hence $|G : G'| = 8$. Since G/G' is generated by elements of order 2, the factor-group G/G' is elementary abelian, so that $G' = \Phi(G)$.

Let K be the largest normal subgroup of G belonging to the set A_0B_0. Assume that $K = 1$.

It follows that there is a normal subgroup T of order 2 in G such that T is generated by the element $g \in G$, which is of the form $g = acb$, adb, or $acdb$ with $a \in A_0$, $b \in B_0$. Since $ac \in A \setminus A_0$, $db \in B \setminus B_0$, we may replace ac by c and db by d, so that $g = cb$, ad, or cd.

If $g = cb$, then in the group $\bar{G} = G/T$ the subgroups $\bar{A} = AT/T$ and $\bar{B} = BT/T$ have nontrivial intersection containing $bT = cT$. Hence the normal closure in $\bar{G}$ of the element cT is contained in $\bar{A}$. However, the normal closure in $\bar{A}$ of cT is a subgroup $\bar{L}$ of index 2 in $\bar{A}$. The full preimage of $\bar{L}$ in G is of the form $A_0^2\langle c, g\rangle$ or $A\langle g\rangle$ with $g \in Z(G)$. In each case the subgroup A_0^4 is normal in G. Hence $A_0^4 = 1$. This implies that L is an abelian normal subgroup of G. The factor-group G/L is either dihedral, or has a dihedral subgroup of index 2. In each case G/L has derived length at most 2. Therefore if $K = 1$ we have $G^{(3)} = 1$. Moreover, this implies that $G^{(5)} = 1$ in the general case $K \neq 1$.

By symmetry, the case $g = ad$ leads to the conclusion $G^{(5)} = 1$. Assume now that $g = cd$, and let $T = \langle cd\rangle$.

Then in the factor-group $\bar{G} = G/T$ the subgroups $\bar{A}_0^2 = A_0^2T/T$ and $\bar{B}_0^2 = B_0^2T/T$ permute and generate a normal subgroup $\bar{M}$ in $\bar{G}$. The full preimage M of $\bar{M}$ in G is a product of two abelian subgroups and has derived length at most 2. Clearly, $g \in Z(G) \cap G'$, so that $M \leq G'$. Since G/M is of order 8, this implies by the above considerations that $G' = M$. Hence the derived length of G in the case $K = 1$ is at most 3. Hence, also when $K \neq 1$ the derived length of a 2-group $G = AB$ is at most 5.

Thus in all cases the derived length of a 2-group, which is a product of two dihedral subgroups, is bounded by 5.

Consider now the case, when G is not a 2-group. It is well-known that in every finite group $G = AB$, which is the product of two subgroups A and B, for every prime p there exists a Sylow p-subgroup of G which is the product of a Sylow p-subgroup of A and a Sylow p-subgroup of B (see for instance Ref. 3, VI. Satz 4.7, p. 676). In our case the Sylow p-subgroups of G for an odd prime p are products of two cyclic subgroups. This implies that the Sylow p-subgroups of G are metacylic by a result of B. Huppert (see Ref. 3, III, Satz 11.5, p. 338). Since G is soluble by Ref. 4, even the Hall $2'$-subgroup of G are metacyclic. Therefore a positive integer k as stated in the lemma exists. $\square$

From now on we consider a counterexample $G = AB$ for Theorem 1.2. In particular, G is nonsoluble.

Lemma 3.3. *We may assume that G has no nontrivial soluble normal subgroups.*

Proof. Suppose that $N \neq 1$ is a soluble normal subgroup of G. Then $R = NA = A(NA \cap B)$ by Dedekind's modular law. Also R is a soluble group and so locally finite. If L and S are finite normal subgroups of A and B respectively, then $H = \langle L, S\rangle$ is finite and $K = N_R(\langle L, S\rangle) = (K \cap A)(K \cap B)$ by Ref. 1, Lemma 2.2. Therefore $K/C_K(H)$ is a finite group which is the product of two subgroups of dihedral groups. By Lemma 3.2 we have that $H/C_K(H)^{(k)} = 1$. This implies that $H^{(k+1)} = 1$. Since $|R : R_0| \leq 4$, it follows that $R^{(k+2)} = 1$. In particular, N is of derived length at most $k + 2$. Therefore the product T of all soluble normal subgroups of G is a soluble group of derived length at most $k + 2$.

The factor group $G/T = (AT/T)(BT/T)$ is the product of two locally dihedral subgroups, and G/T has no nontrivial soluble normal subgroups. The lemma is proved. $\square$

Lemma 3.4. $A \cap B = 1$.

Proof. Suppose that $A \cap B_0 = D \neq 1$. Clearly D is normal in B and since $A_G = \cap_{g \in G} A^g = \cap_{b \in B} A^b$ contains D, the group A_G is a soluble normal subgroup of G. This contradicts Lemma 3.3. Similary, also $A_0 \cap B = 1$. Assume that $E = A \cap B \neq 1$. Since $E \cap A_0 = 1$ and $|A : A_0| = 2$, we have that $EA_0 = A$. By similar reasons it follows that $EB_0 = B$ and $E = \langle t\rangle$

for an involution $t \in E$. Note that in this case $G = A_0B_0E$. In particular, if $g \in G$, then $g = a_0b_0$ or a_0b_0t with $a_0 \in A_0, b_0 \in B_0$.

We prove that the involution t has a finite centralizer in G. Indeed, if $x = a_0b_0 \in C_G(t)$ with $a_0 \in A_0, b_0 \in B_0$ then $a_0^t = a_0^{-1}$. Similarly $b_0^t = b_0^{-1}$. Therefore $a_0b_0 = a_0^{-1}b_0^{-1}$. Since $A_0 \cap B_0 = 1$, this implies $a_0^2 = 1, b_0^2 = 1$. Since B_0 and A_0 have not more than one involution, it follows that $|C_G(t)| \leq 8$.

By a well-known result of V.P. Shunkov (see Ref. 7) every periodic group G containing an involution with finite centralizer is soluble-by-finite and so locally finite. Since also a finite group, which is a product of two dihedral groups is soluble by Ref. 4, the group G is soluble. □

As in Ref. 1, for an element x of a group Y we denote the extended centralizer of x by $C_Y^*(x) = \{u \subset Y \mid x^u = x^{\pm 1}\}$.

Lemma 3.5. *For every $a \in A_0^{\#}$ we have that $C_G^*(a) \cap B_0 = 1$ and $C_G^*(b) \cap A_0 = 1$ for every $b \in B_0^{\#}$.*

Proof. Assume that $C_G^*(a) \cap B_0 \neq 1$. Since $A \leq C_G^*(a)$ it follows that $K = C_G^*(a) = AB_1$ for some $B_1 \leq B$ and $L = B_0 \cap B_1 \neq 1$. This implies that $M = K_G = \cap_{g \in G} K^g = \cap_{b \in B} K^b$ contains L and is a normal subgroup of G. If $M \cap B_1$ contains an element in $B \setminus B_0$, then $B^2 \leq M$ and $G/M = (AM/M)(BM/M)$ is a product of a dihedral or abelian group AM/M and a group of order at most 4. Therefore G/M and so G/M' are soluble. Hence also $C_G(M')M'/M'$ is soluble and $Z(M') = M \cap C_G(M')$ is normal in G. Since a centralizes K', it follows that $C_G(M') \neq 1$. Hence G has a nontrivial soluble normal subgroup. This is a contradiction by Lemma 3.3.

Therefore, $M \cap B \leq B_0$. If $a \in M$, then $|M : C_M(a)| \leq 2$ and $a \in Z(M')$. By Lemma 3.3 this is a contradiction. If $M = (A \cap M)L$ and $A \cap M \leq A_0$, then G is metabelian by Itô's theorem, again contradicting Lemma 3.3. If $A \cap M$ contains an element in $A \setminus A_0$, then $A^2 \leq M$ and G/M is soluble. This implies that either $C_G(M')$ is soluble, or $M' \cap C_G(M') = Z(M')$ is non-trivial. Both cases are impossible.

Let G/M be non-soluble. Then there is no element of M in $A \setminus A_0$ in M and in $B \setminus B_0$. Since M is not factorized, there exists an element $g \in M$ of one of the forms $g = a_1cb$, $g = a_1db$ or $g = a_1cdb$ with $a_1 \in A_0$, $b \in B_0$. Recall that the elements $c \in A \setminus A_0$ and $d \in B \setminus B_0$ were defined in the beginning of this section. Since A and B are dihedral, we may replace c by a_1c and d by db. Thus we may assume that $g = cb$, $g = a_1d$, or $g = cd$.

If $g = a_1d$, then $a_1M = dM$ and in the factor-group $\bar{G} = G/M$, the subgroups $\bar{A} = AM/M$ and $\bar{B} = BM/M$ have a non-trivial intersection $\langle dM \rangle \leq \bar{A}_0 \cap \bar{B}$. It follows that the normal closure of dM in $\bar{G}$ contains in $\bar{B}$, hence it coincides with $\bar{B}$. In this case G/M is soluble. By the symmetry between A and B, we the same conclusion follows, when $g = cb$.

Suppose that $g = cd$. Then $cM = dM$ and $\bar{G} = G/M$ is the product of two dihedral subgroups $\bar{A} = AM/M$ and $\bar{B} = BM/M$ such that $\bar{A} \cap \bar{B} = \langle cM \rangle = \langle dM \rangle$ and an involution cM has finite centralizer in $\bar{G}$. By Ref. 7 the group $\bar{G}$ is soluble-by-finite and therefore is soluble. This contradiction proves the lemma. □

Lemma 3.6. *Either $A_0 = O(A)$, or $B_0 = O(B)$, where $O(A)$ and $O(B)$ are the maximal normal subgroups of A and B without involutions.*

Proof. Assume that Lemma 3.6 is false. Then there exists an involution $\tau \in Z(A)$ and an involution $\mu \in Z(B)$. By Lemma 3.4 we have that $A \cap B = 1$, so that τ and μ are not conjugate in G. Let $D = \langle \tau, \mu \rangle$ be the subgroup generated by τ and μ. Clearly, D is a finite dihedral group and there exists an involution $\nu \in Z(D)$, different from τ and μ. By Ref. 1, Lemma 2.2 (iii) we have $H = N_G(D) = A_1 B_1$, where $A_1 \leq A$, $B_1 \leq B$. By Lemma 3.3 it follows that $H \neq G$. Since A_1 normalizes D, this implies that $A_1 D = A_1 B_2$ for some $B_2 \leq B_1$. If A_1 has a subgroup T of odd order at least 3, the group T centralizes some Sylow 2-subgroup $\langle \tau, \nu \rangle$ of D. By Lemma 3.4 this is a contradiction. Hence A_1 is a 2-group.

By similar reasons B_1 is a 2-group. Therefore D is also a 2-group. By Ref. 1, Lemma 2.12, it follows that $H = A_1 B_1$ is a dihedral group of order 8. Moreover, in this case the subgroups A_1 and B_1 are elementary abelian and $H = D$.

Without loss of generality we may assume that $A_1 = \langle \tau \rangle \times \langle c \rangle$ and $B_1 = \langle \mu \rangle$. Clearly, $\nu \in A_1$. Since $\nu \in A \setminus A_0$ we can put $\nu = c$. Using Lemma 3.5, we conclude that $C_G(\mu) = B\langle c \rangle$. Now, applying Lemma 3.2 in Ref. 1, we obtain a contradiction. The lemma is proved. □

To conclude the proof of Theorem 1.2 we proceed as in sections 5 and 6 of Ref. 1. In particular we use Lemma 5.1 in Ref. 1, and Lemmas 3.5 and 3.6 from above. Observe that this part of the proof does not depend on the main assumption in Ref. 1 that A and B are Chernikov groups. Theorem 1.2 is proved.

Now also Theorem 1.1 follows from Theorem 1.2 using Lemma 2.4.

Remark. It was proved in Ref. 2 that there exists a 2-group $G = AB$ with derived length 3, which is a product of two dihedral subgroups. In the proof of Lemma 3.2 it is shown that the derived length of a 2-groups, which is a product of two dihedral subgroups is at most 5. It seems be unknown at present whether this bound is best possible.

References

1. B. Amberg, L. Kazarin, On the product of two Chernikov subgroups, Israel J. Math. 175 (2010), 363 - 389.
2. B. Amberg, Ya. Sysak, Local nearrings with dihedral multiplicative group, J. Algebra 273 (2004), 700 - 717.
3. B. Huppert, Endliche Gruppen I, Springer, Berlin (1967).
4. V.S. Monakhov, The product of two groups, one of which contains a cyclic subgroup of index ≤ 2, Math. Notes, 16:2(1974), 757 - 762.
5. A.K. Shlyopkin , On certain periodic groups saturated by simple groups, Mat. Trudy 1 (1998), 129-138 (In Russian).
6. A.K. Shlyopkin, A.G. Rubashkin, A class of periodic groups, Algebra i Logika 44 (2005), 114-125.
7. V.P. Shunkov, On periodic groups with an almost regular involution, Algebra i Logika 9 (1970), 484-496.

A NOTE ON FINITE GROUPS IN WHICH THE CONJUGACY CLASS SIZES FORM AN ARITHMETIC PROGRESSION*

M. BIANCHI, A. GILLIO

Dipartimento di Matematica "F. Enriques",
Università di Milano,
Via Saldini, 50
I-20133 Milano, Italy
e-mail: mariagrazia.bianchi@unimi.it, anna.gillio@unimi.it

P. P. PÁLFY

Alfréd Rényi Institute of Mathematics
H-1364 Budapest, P.O. Box 127
Hungary
e-mail: ppp@renyi.hu

Many structural properties of a finite group G are encoded in the set of its conjugacy class sizes. We determine here those groups in which non-central conjugacy class sizes form the arithmetic progression 2, 4, 6, 8.

In Ref. 5 Huppert proves that if the set of character degrees of a finite group G is $\{1, 2, 3, ..., k\}$ then $k = 1, 2, 3, 4$, or 6 and a description of the groups for each possible k is given. In Ref. 2 an analogue for conjugacy class sizes is presented. More generally, groups in which the two largest sizes of conjugacy classes are consecutive numbers were studied by Mann (see Ref. 8).

Our distant goal is to study those finite groups in which the non-central conjugacy class sizes form an arithmetic progression. Clearly, if the group G has at most two conjugacy class sizes greater than 1, then it satisfies our asumption. These groups were studied by Ito (see Refs. 6,7) as groups of type either $[n_1, 1]$ or $[n_2, n_1, 1]$. In the first case n_1 is a power of a prime p and $G = P \times A$ where P is a p-group having the same property and A is

*Research supported by the Hungarian National Research Fund (OTKA), grant no. NK72523 and PRIN program "Teoria dei gruppi e applicazioni".

abelian; in the second case G is soluble.

Examining the list of groups of order less than 100 we observed that two groups of order 96 satisfy our assumption. In particular, the non-central conjugacy class sizes in these groups are 2, 4, 6, 8. Motivated by these examples we classify here those finite groups whose non-central conjugacy class sizes are exactly these numbers.

Theorem 1.1. *Let G be a finite group in which the conjugacy class sizes are 1, 2, 4, 6 and 8. Then*

$$G = ((Q \rtimes C) * B) \times A,$$

where

Q is the 8-element quaternion group,
C is a cyclic group of order 3^k $(k \geq 1)$,
B is a 2-group with $|B'| = 2$,
A is an abelian group,
$Q \rtimes C$ is a semidirect product with C acting on Q nontrivially, and
$(Q \rtimes C) * B$ is a central product with $(Q \rtimes C) \cap B = Z(Q) = B'$.

Conversely, for every group G of the given structure the class sizes are exactly 1, 2, 4, 6 and 8.

Remark 1.1. We have $|G| = 4 \cdot 3^k \cdot |B| \cdot |A|$, hence the smallest group satisfying the conditions in the Theorem has order 96 with $|C| = 3$, $|B| = 8$, $|A| = 1$. There are two isomorphism types of such groups of order 96, corresponding to $B \simeq Q$ and $B \simeq D_4$, the dihedral group of order 8.

First we formulate three lemmas that will be used in the proof of Theorem 1.1.

Lemma 1.1. *Let H be a nonabelian finite group, $M < H$ a proper subgroup such that $|H : C_H(x)| \leq 2$ for all $x \in H \backslash M$. Then $|H'| = 2$.*

Proof. Let $Z = \bigcap_{x \in H \backslash M} C_H(x)$. Since $H \backslash M$ generates H, we see that $Z = Z(H)$. Moreover, H/Z is an elementary abelian 2-group. Hence H is nilpotent of class 2, and the commutator is a homomorphism in both variables. Let $X = H \backslash (M \cup Z(H))$. If $x \in X$, then $|[x, H]| = 2$. We are going to show that $[x, H]$ does not depend on the choice of $x \in X$.

Indeed, if $x, y \in X$ do not commute, then $1 \neq [x, y] \in [x, H] \cap [y, H]$ implies $[x, H] = [y, H]$.

Suppose now that $x, y \in X$ commute. We claim that $C_H(x) \cup C_H(y)$ does not cover X. Otherwise, we have $H = C_H(x) \cup C_H(y) \cup M$. If the two centralizers coincide, then H would be the union of two proper subgroups, which is impossible. If the two centralizers are different, then their intersection is a normal subgroup N of index 4 and $H \setminus (C_H(x) \cup C_H(y))$ is a coset of N. Hence M must contain this coset, and therefore it contains N as well. However, since x and y commute, they both belong to $N < M$, contrary to $x, y \in X = H \setminus (M \cup Z(H))$. So there is a $w \in X$ not commuting with any of x and y. Then $[x, H] = [w, H] = [y, H]$.

Let $D = [x, H]$ ($x \in X$ arbitrary). We know $|D| = 2$. Now $H \backslash M$ is mapped into the center of H/D, hence H/D is abelian and $D = H'$. □

Lemma 1.2. *Let G be a finite group with a quotient $G/N \simeq S_3$. Suppose that the only class size of G divisible by 3 is 6. Then the largest class size of G is 6.*

Proof. Let $N < H < G$ be a subgroup with $|H/N| = 2$. If $x \in H \setminus N$, then $C_G(x) \leq H$, so $|G : C_G(x)|$ is divisible by 3, hence by our assumption $|G : C_G(x)| = 6$ thus $|H : C_H(x)| = 2$. By Lemma 1.1 we have $|H'| = 2$. The subgroup H has three conjugates in G, but there is no conjugacy class of size 3 of elements in G, hence we obtain $H' \lhd G$. Let $\overline{G} = G/H'$. Then H/H' and its conjugates are abelian subgroups of $\overline{G}$, and they together generate $\overline{G}$. Hence their intersection, N/H' is the center of $\overline{G}$. Now every element of $\overline{G}$ belongs to an abelian subgroup of index 2 or 3, hence the class sizes of $\overline{G}$ are $1, 2$, and 3. Therefore any class size of G is at most $3 \cdot |H'| = 6$. □

Lemma 1.3. *Let $G = G_1 * G_2$ be a central product $(G_1 \times G_2)/Z$, where $Z = \langle (z_1, z_2) \rangle$ with $z_i \in Z(G_i)$ of prime order p ($i = 1, 2$). Let $g_i \in G_i$ ($i = 1, 2$). Then the size of the conjugacy class of $(g_1, g_2)Z$ in G is either $|g_1^{G_1}| \cdot |g_2^{G_2}|$ or $|g_1^{G_1}| \cdot |g_2^{G_2}|/p$, where the latter case occurs exactly when both $g_1 z_1 \in g_1^{G_1}$ and $g_2 z_2 \in g_2^{G_2}$.*

Proof. The class size of $(g_1, g_2) \in G_1 \times G_2$ is $|g_1^{G_1}| \cdot |g_2^{G_2}|$. In $(G_1 \times G_2)/Z$ it is mapped to a class of the same size, unless there exist (h_1, h_2), $(h_1 z_1^k, h_2 z_2^k) \in (g_1, g_2)^{G_1 \times G_2}$ with some $0 < k < p$. In the latter case we have $t_i \in G_i$ such that $h_i^{t_i} = h_i z_i^k$ ($i = 1, 2$). Then for $0 \leq m < p$, $s_i \in G_i$ we get $h_i^{t_i^m s_i} = h_i^{s_i} z_i^{km}$, so with every element $(x_1, x_2) \in (g_1, g_2)^{G_1 \times G_2}$ all $(x_1 z_1^n, x_2 z_2^n)$, $n = 0, 1, ..., p-1$ belong to the conjugacy class of (g_1, g_2) in $G_1 \times G_2$. Hence the class size of the coset $(g_1, g_2)Z$ in $(G_1 \times G_2)/Z$ is $|g_1^{G_1}| \cdot |g_2^{G_2}|/p$ in this case. □

Proof of Theorem 1.1.

Since both the condition and the conclusion of the theorem does not change if the group has an abelian direct factor, we will assume that G does not have any nontrivial abelian direct factor.

Let $p > 3$ be a prime. Since p does not divide any class size, it follows that the Sylow p-subgroup is central (see Ref. 3, Proposition 4), and hence it is an abelian direct factor. So G is a $\{2,3\}$-group, and consequently it is solvable.

Clearly, G is not nilpotent, otherwise the existence of a conjugacy class of size 6 would imply that there is a class of size 3 as well.

Since class sizes of quotient groups and of normal subgroups are divisors of the class sizes of the group, we know that the class sizes of all quotients and normal subgroups of G belong to the set $\{1, 2, 3, 4, 6, 8\}$.

Consider $\overline{G} = G/\Phi(G)$. Then by the theorem of Gaschütz (see Ref. 4, Satz III.4.4, III.4.5]) $\overline{G} = F \rtimes K$ with $F = F(G)/\Phi(G)$ being the direct product of minimal normal subgroups of $\overline{G}$ and the complement K acting faithfully on F. Let $F = F_2 \times F_3$, where F_p ($p = 2, 3$) denotes the Sylow p-subgroup of F.

We first investigate the case when K acts nontrivially on F_3. Let $L/C_K(F_3)$ be a minimal normal subgroup of $K/C_K(F_3)$. Since F_3 can be considered as a vector space over the 3-element field and K acts completely reducibly, it follows that $L/C_K(F_3)$ is an elementary abelian 2-group. Let P be a Sylow 2-subgroup of K, then $P \nleq C_K(F_3)$. If $x \in P \setminus C_P(F_3)$, then $|\overline{G} : C_{\overline{G}}(x)|$ is divisible by $|F_3 : C_{F_3}(x)|$, hence $|F_3 : C_{F_3}(x)| = 3$. Since the order of x is coprime to $|F_3|$, we have $F_3 = C_{F_3}(x) \times U$ with a subgroup U of order 3 inverted by x. Hence the determinant of the linear transformation induced by x on F_3 is -1. This holds for every element in $P \setminus C_P(F_3)$, hence $|P : C_P(F_3)| = 2$. We obtain that $L = PC_K(F_3)$ and $L/C_K(F_3)$ has order 2, hence it is central in $K/C_K(F_3)$. Then $C_{F_3}(x)$ is invariant under K. It follows that $\overline{G}/(F_2 \times C_{F_3}(x))C_K(U) \simeq S_3$, contrary to Lemma 1.2. So we have shown that K acts trivially on F_3. Since G is not nilpotent, K must act nontrivially on F_2.

Let $M \leq F_2$ be a minimal normal subgroup on which K acts nontrivially. Since $|M|$ is a power of 2, M must contain a nontrivial conjugacy class of odd size. By our assumption this class has size 3, so let this conjugacy class be $\{a, b, c\}$. Now the product abc is fixed by K, so, by the minimality of M, we have $abc = 1$. The elements $a, b, c = ab$ generate M, hence $|M| = 4$. Now $K/C_K(M)$ is a subgroup of $Aut(M) = GL(2, 2) = S_3$. By Lemma 1.2, it cannot be S_3, so we have $|K/C_K(M)| = 3$.

Now $|K/C_K(M)| = 3$ holds for every noncentral minimal normal subgroup in F, and K acts faithfully and completely reducibly on F, hence K is an elementary abelian 3-group. If $1 \neq x \in K$, then

$$8 \geq |\overline{G} : C_{\overline{G}}(x)| = |\overline{G} : KC_F(x)| = |F : C_F(x)| > 1$$

is a power of 2, and $|F \setminus C_F(x)| = (|F : C_F(x)| - 1)|C_F(x)|$ is divisible by 3. It follows that $|F : C_F(x)| = 4$ for every $1 \neq x \in K$. If K does not act trivially on $C_F(x)$, then let M_1 be a noncentral minimal normal subgroup contained in $C_F(x)$, and let M_2 be a K-invariant complement of $C_F(x)$ in F. Then $C_K(M_1)$ and $C_K(M_2)$ are distinct proper subgroups of K, so there exists an $x' \in K \setminus (C_K(M_1) \cup C_K(M_2))$. Then $|F : C_F(x')| \geq |M_1 M_2| = 16$, contrary to our previous observation. Hence K acts trivially on $C_F(x)$, so $|K| = 3$ and $C_F(K) = Z(\overline{G})$ has index 4 in F.

Let us summarize what we have obtained so far. Let S be the Sylow 2-subgroup of G, and let $B < S$ be the largest subgroup with $B\Phi(G)/\Phi(G) \leq Z(G/\Phi(G))$. Then S is normal in G, $|S : B| = 4$ and B is also normal in G. Let R be a Sylow 3-subgroup of G. Then $G = S \rtimes R$, $|R : C_R(S)| = 3$ and R acts nontrivially on the 4-element group S/B.

Take $H = BR$, $M = BC_R(S)$. Let $x \in H \setminus M$, so $x = bu$ with $b \in B$, $u \in R \setminus C_R(S)$. If $y = sr$ ($s \in S$, $r \in R$) centralizes x in G, then srB and uB commute in G/B, so $[s, u] \in B$, hence $s \in B$. Thus for every $x \in H \setminus M$ we have $C_G(x) \leq H$, hence $|H : C_H(x)| \leq 2$, as $|G : C_G(x)| \leq 8$. So by Lemma 1.1 we have $|H'| \leq 2$. Hence $H = B \times R$ with R abelian and $|B'| \leq 2$. In particular, it follows that $C_S(R) = B$.

Now we have $S = [S, R]C_S(R)$ (see Ref. 1, 24.4). Using the Three-Subgroup-Lemma $[[R, B], S] = 1$ and $[[B, S], R] \leq [B, R] = 1$ implies $[[S, R], B] = 1$, so S is the central product of $[S, R]$ and $B = C_S(R)$. Let $Z = [S, R] \cap B \leq Z(S)$. Since $|[S, R] : Z| = 4$, we get $|[S, R]'| \leq 2$. If $x \in [S, R] \setminus Z$, then $C_G(x) \leq S \times C_R(S)$, hence we must have $|G : C_G(x)| = 6$, and $|S : C_S(x)| = 2$, so $[S, R]$ is nonabelian, $|[S, R]'| = 2$. Let $A = [S, R]/[S, R]'$. Then A is an abelian 2-group on which the 3-group R acts, so $A = [A, R] \times C_A(R)$ (see Ref. 1, 24.6). Since $[[S, R], R] = [S, R]$ (see Ref. 1, 24.5), we have $C_A(R) = 1$, so $[S, R]' = Z$. Now $[S, R]$ is a nonabelian group of order 8 and has a nontrivial automorphism of order 3, hence $[S, R] \simeq Q$, the quaternion group. We have already established that $G = QR * B$ is a central product with $QR \cap B = Z$ of order 2. The class sizes of QR are $1, 4$, and 6. If B were abelian, then the same would hold for G as well, which is not the case. Hence B is nonabelian with $|B'| = 2$, so the class sizes of B are 1 and 2. If $B' \neq Z$ then the central product has a class of size 12 (see Lemma 1.3), so we must have $B' = Z = Z(Q)$ as it is claimed in the statement of the theorem.

Finally, we are going to show that R is cyclic. If an element r of maximal order in R belongs to $C_R(S)$, then, using a standard proof of the fundamental theorem of finite abelian groups, we obtain a direct decomposition $R = \langle r\rangle \times R_1$, that yields a direct decomposition $G = SR_1 \times \langle r\rangle$, what we have excluded at the beginning of the proof. So every element of maximal order in R acts nontrivially on S.

Let $R = \langle r_1\rangle \times \cdots \times \langle r_k\rangle$ be a decomposition of R into the direct product of cyclic groups, with r_1 of maximal order. If some r_j, $2 \leq j \leq k$, also acts nontrivially on S, then either $r_1 r_j$ or $r_1 r_j^{-1}$ is an element of maximal order acting trivially on S, and that has already been excluded. Hence $r_2, \dots, r_k \in C_R(S)$, and so we obtain a direct decomposition $G = S\langle r_1\rangle \times (\langle r_2\rangle \times \cdots \times \langle r_k\rangle)$. Since no abelian direct factor of G is allowed, we have that $k = 1$ and R is cyclic, indeed.

The converse easily follows from Lemma 1.3. □

Acknowledgments

We thank the referee for his useful comments.

References

1. M. Aschbacher, *Finite group theory*, Cambridge studies in advanced mathematics 10, Cambridge University Press, 1986.
2. M. Bianchi, A. Gillio, D. Chillag, M. Herzog, C. Scoppola, *Application of a graph related to conjugacy classes in finite groups*, Arch. Math. 58, 126–132, 1992.
3. D. Chillag, M. Herzog, *On the length of the conjugacy classes of finite groups*, J. Algebra 131, 110–125, 1990.
4. B. Huppert, *Endliche Gruppen I*, Grundlehren der mathematischen Wissenschaften 134, Springer-Verlag, Berlin, 1967.
5. B. Huppert, *Character theory of finite groups*, De Gruyter, Berlin, 1998.
6. N. Ito, *On finite groups with given conjugate types I*, Nagoya Math. J. 6, 17–28, 1953.
7. N. Ito, *On finite groups with given conjugate types II*, Osaka J. Math. 7, 231–251, 1970.
8. A. Mann, *A note on conjugacy class sizes*, Algebra Colloq. 4, 169–174, 1997.

A SURVEY OF RECENT PROGRESS ON NON-ABELIAN TENSOR SQUARES OF GROUPS

R. D. BLYTH

Department of Mathematics and Computer Science, Saint Louis University, St. Louis, MO 63103, U.S.A.
Email: blythrd@slu.edu

F. FUMAGALLI

Dipartimento di Matematica "Ulisse Dini", Università di Firenze, Viale Morgagni, 67/a, 50134 Firenze, Italy
Email: fumagal@math.unifi.it

M. MORIGI

Dipartimento di Matematica, Università di Bologna, Piazza di Porta San Donato 5, 40127 Bologna, Italy
Email: mmorigi@dm.unibo.it

We discuss some of the progress that has occurred since the survey article "Non-abelian tensor products of groups: the commutator connection" by L.C. Kappe appeared in the Proceedings of Groups St Andrews 1997 in Bath, with a particular emphasis on non-abelian tensor squares of groups. We survey methods for computing non-abelian tensor squares, results that support such calculations, and recent general structure results.

Keywords: non-abelian tensor square; non-abelian exterior square.

1. Introduction

Thirteen years ago at "Groups St. Andrews 1997 in Bath" in her talk and paper "Nonabelian tensor products of groups: the commutator connection," L.-C. Kappe[31] reported on developments in non-abelian tensor products since the appearance of the 1987 paper "Some computations of non-abelian tensor products of groups" by Brown, Johnson and Robertson[14]. Her talk was an explicit invitation to others to join in investigations of non-abelian tensor products, and was focussed in large part on the status of a series of questions posed in Ref. 14. The first author was intrigued by and eventually accepted Kappe's invitation, with the additional guidance of Robert F. Morse. During the past three years he convinced the second and third authors to contribute to the enterprise. Our emphasis in this article is on the more specialized non-abelian tensor squares of groups and related constructions, although we sometimes state definitions and results in the more general non-abelian tensor product context. We survey some of the progress made on this topic over the past thirteen years. Our survey is by no means complete (as a search of the literature will quickly indicate), but we intend to give an overview sufficient to show the continued development of the topic over the past thirteen or so years.

As L.-C. Kappe[31] mentions, the notation most used in this area is contrary to that most often used in much of group theory in that actions are written on the left. Thus for elements g, g', h, h' of a group G, we write ${}^h g$ for the conjugate hgh^{-1} of g by h, and we write $[h, g]$ for the commutator $hgh^{-1}g^{-1}$. Familiar (right-action) commutator formulas have their appropriate left-action counterparts.

Definition 1.1. Two groups G and H are said to act compatibly on each other if

$$^{({}^g h)}g' = {}^g({}^h({}^{g^{-1}}g')) \quad \text{and} \quad ^{({}^h g)}h' = {}^h({}^g({}^{h^{-1}}h'))$$

for all $g, g' \in G$ and for all $h, h' \in H$.

Definition 1.2. Let G and H be two groups that act compatibly on each other and on themselves by conjugation. Then the non-abelian tensor product $G \otimes H$ of G and H is the group generated by the symbols $g \otimes h$ for $g \in G$ and $h \in H$ subject to the relations

$$gg' \otimes h = ({}^g g' \otimes {}^g h)(g \otimes h) \quad \text{and} \quad g \otimes hh' = (g \otimes h)({}^h g \otimes {}^h h')$$

for all $g, g' \in G$ and all $h, h' \in H$. If $G = H$ and all actions are conjugation, then $G \otimes G$ is called the non-abelian tensor square of G.

As Kappe points out, there is a striking resemblance between the defining relations for the non-abelian tensor product and the (left versions of the) standard commutator expansion formulas, a resemblance that has been examined and often exploited.

For completeness, we briefly mention that non-abelian tensor products originally arose in connection with algebraic K-theory and topology. The foundations of the topic can be found in the work of Whitehead[43] and

Miller[33]. Dennis[17] essentially discusses the non-abelian tensor square without using the name. Brown and Loday[15,16] are the first to define the non-abelian tensor product of two groups in connection with their work on generalized van Kampen theorems. In addition to the 1987 paper[14] by Brown, Johnson and Robertson, other work appeared by Aboughazi[1], Ellis[21] and Johnson[30] in the same year.

Several objects and maps related to the non-abelian tensor square of a group G have been used to study $G \otimes G$ since the beginning, and are of independent interest. The definition of the non-abelian tensor square ensures the existence of an epimorphism $\kappa : G \otimes G \longrightarrow G'$, defined on the generators by $\kappa(g \otimes h) = [g, h]$ for all $g, h \in G$. Let $J(G)$ be the kernel of the map κ. Also, let $\nabla(G)$ be the normal subgroup of $G \otimes G$ generated by the elements $g \otimes g$, for all $g \in G$.

Definition 1.3. The group $(G \otimes G)/\nabla(G)$ is called the non-abelian exterior square of G, denoted $G \wedge G$.

The map κ factorizes modulo $\nabla(G)$, inducing an epimorphism $\kappa' : G \wedge G \longrightarrow G'$. The kernel of the map κ' is isomorphic to the Schur multiplicator $M(G)$ of G ([15,16]). Let $\Gamma(G/G')$ be Whitehead's quadratic functor[43]. Results in Ref. 15,16 yield a useful commutative diagram with exact rows and central extensions as columns:

$$\begin{array}{ccccccccc}
 & & & & 1 & & 1 & & \\
 & & & & \downarrow & & \downarrow & & \\
 & & \Gamma(G/G') & \longrightarrow & J(G) & \longrightarrow & M(G) & \longrightarrow & 1 \\
 & & \downarrow & & \downarrow & & \downarrow & & \\
1 & \longrightarrow & \nabla(G) & \longrightarrow & G \otimes G & \longrightarrow & G \wedge G & \longrightarrow & 1 \\
 & & \downarrow & & {\scriptstyle\kappa}\downarrow & & {\scriptstyle\kappa'}\downarrow & & \\
 & & 1 & \longrightarrow & G' & \xrightarrow{\text{id}} & G' & & \\
 & & & & \downarrow & & \downarrow & & \\
 & & & & 1 & & 1 & &
\end{array} \tag{1}$$

For example, Brown, Johnson and Robertson[14] easily show the following two results.

Proposition 1.1 (see also Ref. 21). *If G is a finite group, then so is $G \otimes G$. If G is a finite p-group, then so is $G \otimes G$.*

Proposition 1.2. *If G is a free group, then $G \otimes G \cong G' \times \Gamma(G^{\mathrm{ab}})$.*

We return later to the question of when the middle row of the commutative diagram (1) splits.

2. Computing non-abelian tensor squares

In Ref. 14 Brown, Johnson and Robertson computed the non-abelian tensor squares of all groups of order up to 30 using the defining generators and relations given by Definition 1.2. By "computing" a non-abelian tensor square we mean that the non-abelian tensor square is identified in a generally recognized structural form or presentation. Their method quickly becomes untenable for most larger finite groups, since the definition provides $|G|^2$ generators and $2|G|^3$ relators for $G \otimes G$. Crossed pairings, used in Ref. 14 to demonstrate general properties of non-abelian tensor products, may also be used to compute specific non-abelian tensor squares.

Definition 2.1. Let G, H and L be groups. A function $\phi : G \times H \longrightarrow L$ is called a crossed pairing if for all $g, g' \in G$ and all $h, h' \in H$,

$$\phi(gg', h) = \phi({}^g g', {}^g h)\phi(g, h),$$

$$\phi(g, hh') = \phi(g, h)\phi({}^h g, {}^h h').$$

A crossed pairing determines a unique homomorphism of groups $\phi^* : G \otimes H \longrightarrow L$ so that $\phi^*(g \otimes h) = \phi(g, h)$ for all $g \in G$ and $h \in H$. To compute a non-abelian tensor square using a crossed pairing one conjectures a group L as the result, as well as a map $\phi : G \times G \longrightarrow L$. One must then show that ϕ is a crossed pairing and that the induced map ϕ^* is an isomorphism. This method was used successfully, for example, by Bacon, Kappe and Morse[5] and by Blyth, Morse and Redden[12] to determine the structures of the tensor squares of the free 2-Engel groups of finite rank.

Theorem 2.1 ([12]). *The non-abelian tensor square of the free 2-Engel group of rank $n > 2$ is a direct product of a free abelian group of rank $\frac{1}{3}n(n^2 + 2)$ and an $n(n - 1)$-generated nilpotent group of class 2 whose derived subgroup has exponent 2.*

The non-abelian tensor square of the free 2-Engel group of rank 2 is free abelian of rank 6[5]. The results of Refs. 5,12 depend on detailed computations, some computer-assisted, that are sufficiently intricate to dissuade the use of the crossed pairing method for computing the non-abelian tensor squares of other classes of groups, for example, of finite-rank free nilpotent groups of class 3. Nonetheless, Magidin and Morse[34] recently used crossed pairings to provide lower bounds on the orders of generators of the non-abelian tensor squares of 2-generator p-groups of class 2.

A third approach, which admits the study of non-abelian tensor squares via commutator calculations, involves the group $\nu(G)$ first studied independently by Rocco[39] and Ellis and Leonard[25].

Definition 2.2. Let G and let G^{φ} be a group isomorphic to G via the isomorphism $\varphi : G \longrightarrow G^{\varphi}$. Define $\nu(G)$ to be the group

$$\left\langle G, G^{\varphi} \middle| \mathcal{R}, \mathcal{R}^{\varphi}, {}^{g_3}[g_1, g_2^{\varphi}] = [{}^{g_3}g_1, ({}^{g_3}g_2)^{\varphi}] = {}^{g_3^{\varphi}}[g_1, g_2^{\varphi}],\ \forall g_1, g_2, g_3 \in G \right\rangle,$$

where $\mathcal{R}$ is the set of defining relations of G.

Rocco shows that $\nu(G)$ contains a commutator subgroup, namely $[G, G^{\varphi}]$, isomorphic to the non-abelian tensor square of G.

Proposition 2.1 ([39]). *Let G be a group. Then the map $\tau : G \otimes G \longrightarrow [G, G^{\varphi}]$, defined on the generators of $G \otimes G$ by $\tau : g_1 \otimes g_2 \mapsto [g_1, g_2^{\varphi}]$, is an isomorphism.*

At first glance the presentation given for $\nu(G)$ in Definition 2.2 appears to offer no improvement in complexity over using Definition 1.2 for computing non-abelian tensor squares. In Ref. 35 McDermott extends results of Ref. 25 to obtain for a finite group G a simplified presentation for $\nu(G)$ in terms of a generating sequence $\mathcal{L}_G$ relative to a subnormal series for G. From such a presentation one can compute a concrete representation of $\nu(G)$ and thence $[G, G^{\varphi}]$. The following properties of $\nu(G)$ are also useful.

Proposition 2.2 (see Ref. 39). *Let G be a group. Then the following hold.*

(i) If G is finite, then $\nu(G)$ is finite.
(ii) If G is a p-group then $\nu(G)$ is a p-group
(iii) If G is nilpotent of class c then $\nu(G)$ is nilpotent of class at most $c+1$.
(iv) If G is soluble of derived length d then $\nu(G)$ is soluble of derived length at most $d+1$.

For example, if G is nilpotent or soluble, then a nilpotent or soluble quotient algorithm can be used to find a concrete presentation of $\nu(G)$. This approach was implemented by Ellis and Leonard[25] using CAYLEY, in Ellis[24] using Magma[13], and by McDermott[35] and Rocco[40] using GAP[28].

Blyth and Morse show that McDermott's approach can be extended to infinite groups. They specialize their general analysis to polycyclic groups.

Theorem 2.2 ([11]). *Let G be a polycyclic group with finite presentation $\langle \mathcal{G} \mid \mathcal{R} \rangle$ and polycyclic generating set $\mathfrak{G}$. Then:*

(i) The groups $G \otimes G$ and $\nu(G)$ are polycyclic.
(ii) The group $\nu(G)$ has a presentation that depends only on $\mathcal{G}, \mathcal{R}$ and $\mathfrak{G}$.
(iii) The group $G \otimes G$ is generated by the set $\{\mathfrak{g}^{\pm 1} \otimes \mathfrak{h}^{\pm 1} \mid \mathfrak{g}, \mathfrak{h}$ in $\mathfrak{G}\}$.

This result provides the foundation for computing $G \otimes G$ for a polycyclic group G by computing a consistent polycyclic presentation of $\nu(G)$ using a polycyclic quotient algorithm and then computing the subgroup $[G, G^{\varphi}]$ of $\nu(G)$ using standard algorithms for polycyclic groups, for example using the GAP package `Polycyclic`[18], even if G is infinite. In practice, if G is nilpotent then the nilpotent quotient algorithm `nq`[38] can be used to good effect. Eick and Nickel[19] have developed another approach to computing the non-abelian tensor and non-abelian exterior squares that exploits the facts that both non-abelian squares are central extensions. Their method has been implemented in the `Polycyclic` package, and can effectively compute these squares for many infinite polycyclic groups, although for nilpotent groups their method tends to take significantly more time than the approach that uses `nq` (see Ref. 11).

Blyth and Morse[11] were originally motivated by an attempt to compute the non-abelian tensor squares of the free nilpotent groups of finite rank without attempting to use a crossed pairing approach.

Theorem 2.3. *Let $\mathcal{N}_{n,c}$ denote the free nilpotent group of class c and rank $n > 1$, and denote the free abelian group of rank n by F_n^{ab}.*

(i) [3] For $c = 2$, $\mathcal{N}_{n,2} \otimes \mathcal{N}_{n,2} \cong F^{\mathrm{ab}}_{f(n)}$, where

$$f(n) = \frac{n(n^2 + 2n - 1)}{3}.$$

(ii) [11] For $c = 3$, $\mathcal{N}_{n,3} \otimes \mathcal{N}_{n,3}$ is the direct product of W_n and F_n^{ab}, where W_n is nilpotent of class 2, minimally generated by $n(n-1)$ elements, and

$$h(n) = \frac{n(3n^3 + 14n^2 - 3n + 10)}{24}.$$

The problem of computing the non-abelian tensor squares for $\mathrm{GL}(n, q)$ and other linear groups was first posed by Brown, Johnson and Robertson[14], and appears also as Question 7 in Kappe's paper[31]. It was solved by

Hannebauer[29] for the groups SL$(2, q)$, PSL$(2, q)$, GL$(2, q)$, and PGL$(2, q)$ for all $q \geq 5$ and $q \neq 9$; later Erfanian, Rezaei and Jafari[26] extended Hannebauer's result to all linear groups SL(n, q), PSL(n, q), GL(n, q), and PGL(n, q) except for some small values of n and q.

We now mention two other relevant lines of investigation. One is the computation of the non-abelian tensor squares of metacyclic groups; this is mentioned as Question 6 in Kappe's paper[31] and was studied by Johnson in tha case of finite split metacyclic groups[30] and by Beuerle and Kappe[6] in the case of infinite metacyclic groups. The other is the computation of the non-abelian tensor squares of 2-generator p-groups of class 2. This last case requires the classification of 2-generator p-groups of class 2; it was first dealt with by Bacon, Kappe, Sarmin and Visscher in a series of papers[4,32,41], and presented in the final form by A. Ahmad, A. Magidin and R.F. Morse in a recent work[2], where they correct some mistakes appearing in the previous papers.

3. Some properties "inherited" by tensor products

We already mentioned some group theoretical properties which are closed with respect to forming the non-abelian tensor square, such as finiteness, nilpotency, solvability (see Propositions 1.1 and 2.2) or the fact of being polycyclic (see Proposition 2.2).

More generally, one main theme of research has been to address the question of determining which group theoretical properties are closed with respect to forming the non-abelian tensor products of groups.

Actually, it was Ellis[21] in 1987 who first showed that the tensor product of two finite groups is finite, Nakaoka[37] and Visscher[42] have studied solvability and nilpotency, but despite their contributions Questions 3 and 4 of Kappe's paper[31] are still open:

Question 3.1. If G is solvable of derived length $l(G)$, then $G \otimes G$ is solvable and $l(G \otimes G) = l(G)$ or $l(G) - 1$ (see Proposition 2.2). Is there any intrinsic characterization of solvable groups of either type?

Question 3.2. If G is nilpotent of class $cl(G)$, then $G \otimes G$ is nilpotent and $cl(G \otimes G) = cl(G')$ or $cl(G') + 1$ (see Ref. 5). Can either of these types be characterized internally?

In 2008 Moravec[36] investigated the non-abelian tensor product of locally finite groups of finite exponent, obtaining the following main results.

Theorem 3.1 ([36]). *Let the groups M and N be locally finite, acting compatibly on each other. Then the group $M \otimes N$ is locally finite. If, furthermore, M and N have finite exponents that are π-numbers, then the exponent of $M \otimes N$ is also a π-number that can be bounded by a function depending only on the exponents of M and N.*

Theorem 3.2 ([36]). *Let G be a finite nilpotent group of class at most 3. Then the exponent of $G \otimes G$ divides the exponent of G.*

We also mention here that Erfanian, Russo and Sarmin[27] studied the tensor square of a polycyclic group G with trivial centre, giving bounds on the Hirsch length of $G \otimes G$ in terms of the Hirsch length of G.

4. More structural results

The commutative diagram (1) suggests that computing the non-abelian tensor square of a group can be aided by knowing the structures of all or some of the other functors $\Gamma(G^{\mathrm{ab}}), J(G), M(G), \nabla(G), G \wedge G$ and G'; in fact we already saw this to be the case in Proposition 1.2. It turns out that the middle row of the commutative diagram (1) splits under fairly general conditions (see Theorem 4.1 (v) below). Points (*i*) and (*ii*) of the following Theorem are improvements of Proposition 3.3 of Rocco[40] and show that the structure of $\nabla(G)$ depends on G^{ab}.

Theorem 4.1 ([8]). *Let G be a group such that G^{ab} is finitely generated. Assume that G^{ab} is the direct product of the cyclic groups $\langle x_i G' \rangle$, for $i = 1, \ldots, s$ and set $E(G)$ to be $\langle [x_i, x_j^{\varphi}] | i < j \rangle [G', G^{\varphi}]$. Then the following hold.*

(i) *$\nabla(G)$ is generated by the elements of the set* $\{[x_i, x_i^{\varphi}], [x_i, x_j^{\varphi}][x_j, x_i^{\varphi}] | 1 \leq i < j \leq s\}$.

(ii) $[G, G^{\varphi}] = \nabla(G) E(G)$.

(iii) *The restriction $f|_{\nabla(G)} : \nabla(G) \longrightarrow \nabla(G^{ab})$ of the projection $f : G \longrightarrow G^{ab}$ onto G^{ab} has kernel $N = E(G) \cap \nabla(G)$. Moreover, N is a central elementary abelian 2-subgroup of $[G, G^{\varphi}]$ of rank at most the 2-rank of G^{ab}.*

(iv) $[G, G^{\varphi}]/N \simeq \nabla(G^{ab}) \times (G \wedge G)$.

(v) *Suppose either that G^{ab} has no elements of order two or that G' has a complement in G. Then $\nabla(G) \simeq \nabla(G^{ab})$ and $G \otimes G \simeq \nabla(G) \times (G \wedge G)$.*

The kernel N of the map $f|_{\nabla(G)}$ thus provides an obstruction to the splitting of the middle row of the commutative diagram (1) unless $N = 1$. However, if $N = 1$, then the structure of $G \otimes G$ can be found by determining the structures of G^{ab} and of $G \wedge G$. The proofs of Theorem 4.1 and similar such results use the commutator calculus within $\nu(G)$. The following lemma reports often-used facts.

Lemma 4.1 (Lemma 2.1 in Refs. 9,39, Lemma 3.1 in Ref. 40). *Let G be any group. The following relations hold in $\nu(G)$.*

(i) ${}^{[g_3,g_4^{\varphi}]}[g_1,g_2^{\varphi}] = {}^{[g_3,g_4]}[g_1,g_2^{\varphi}] = {}^{[g_3^{\varphi},g_4]}[g_1,g_2^{\varphi}]$ *for all* $g_1,g_2,g_3,g_4 \in G$.
(ii) $[g_1^{\varphi},g_2,g_3] = [g_1,g_2^{\varphi},g_3] = [g_1,g_2,g_3^{\varphi}] = [g_1^{\varphi},g_2^{\varphi},g_3] = [g_1^{\varphi},g_2,g_3^{\varphi}] = [g_1,g_2^{\varphi},g_3^{\varphi}]$ *for all* $g_1,g_2,g_3 \in G$.
(iii) $[g_1,[g_2,g_3]^{\varphi}] = [g_2,g_3,g_1^{\varphi}]^{-1}$ *for all* $g_1,g_2,g_3 \in G$.
(iv) $[g,g^{\varphi}]$ *is central in* $\nu(G)$ *for all* $g \in G$.
(v) $[g,g^{\varphi}] = 1$ *for all* $g \in G'$.
(vi) If $g_1 \in G'$ *or* $g_2 \in G'$, *then* $[g_1,g_2^{\varphi}]^{-1} = [g_2,g_1^{\varphi}]$.

We mention in passing that since by Proposition 2.8 of Ref. 40 the Schur multiplicator $M(G)$ is isomorphic to the quotient $J(G)/\nabla(G)$, Theorem 4.1 has as a consequence the following generalization of Proposition 3.5 of Ref. 9:

Corollary 4.1. *Let G be a group such that G^{ab} is a finitely generated group with no elements of order 2. Then $J(G) \cong \Gamma(G^{\mathrm{ab}}) \times M(G)$.*

Corollary 4.2 ([9]). *Let $G = \mathcal{N}_{n,c}$ be the free nilpotent group of class c and rank $n > 1$. Then $J(G) \cong \Gamma(G^{\mathrm{ab}}) \times M(G)$ is free abelian of rank $\binom{n+1}{2} + M(n,c+1)$, where $M(n,c)$ denotes the number of basic commutators in n symbols of weight c.*

For more applications of the structural results of Theorem 4.1 see also the already cited papers by Morse and Magidin[34] and by Erfanian, Russo and Sarmin[27].

In Ref. 14 Brown, Johnson and Robertson show that if the Schur multiplicator $M(G)$ of G is finitely generated, then $G \wedge G$ is isomorphic to the derived subgroup of any covering group $\hat{G}$ of G.

Definition 4.1. A covering group $\hat{G}$ of a group G is a central extension

$$1 \longrightarrow M(G) \overset{\iota}{\longrightarrow} \hat{G} \longrightarrow G \longrightarrow 1,$$

where the image of ι is a subgroup of $\hat{G}'$.

Thus, for example, if G is the free nilpotent group $\mathcal{N}_{n,c}$, then $\hat{G} \cong \mathcal{N}_{n,c+1}$, so that $G \wedge G \cong \mathcal{N}'_{n,c+1}$. We thus recover the main result of Ref. 9.

Theorem 4.2 ([9]). *Let $G = \mathcal{N}_{n,c}$ be the free nilpotent group of class c and rank $n > 1$. Then*

$$G \otimes G \cong \nabla(G) \times (G \wedge G) \cong F^{\text{ab}}_{\binom{n+1}{2}} \times \mathcal{N}'_{n,c+1}.$$

The structure of the derived subgroup $\mathcal{N}'_{n,c+1}$ is further examined in Ref. 10, resulting in a more refined description of $\mathcal{N}_{n,c} \otimes \mathcal{N}_{n,c}$.

The proof of the main result of Miller's 1952 paper[33] can be generalized to show:

Theorem 4.3 ([20]). *Let G be a group and let F be a free group such that $G \cong F/R$ for some normal subgroup R of F. Then $G \wedge G \cong F'/[F, R]$.*

This follows directly from Proposition 1 and Proposition 2 of Ellis's paper[20] (see also Ref. 8). The authors thank the referee for pointing this out.

In addition to Theorem 4.2, Theorems 4.1 and 4.3 have as immediate consequences Theorem 1.2, and the corresponding result for free solvable groups.

Corollary 4.3 ([8]). *Let F be the free group of finite rank $n > 1$, let d be a natural number, and let $G = F/F^{(d)}$ be the free solvable group $\mathcal{S}_{n,d}$ of derived length d and rank $n > 1$. Then*

$$G \otimes G \simeq \mathbb{Z}^{n(n+1)/2} \times F'/[F, F^{(d)}]$$

is an extension of a nilpotent group of class ≤ 3 by a free solvable group of derived length $d - 2$ and infinite rank. In particular, if $d = 2$, then $G \otimes G$ is a nilpotent group.

5. The subgroups $Z^{\otimes}(G)$ and $Z^{\wedge}(G)$

There are two central subgroups of G that naturally arise in connection with the non-abelian exterior square and the non-abelian tensor square of a group G, and they are defined as follows.

Definition 5.1 ([22]). *Let G be a group.*

(i) The non-abelian tensor centre $Z^{\otimes}(G)$ of G is

$$Z^{\otimes}(G) = \{g \in G | g \otimes x = 1, \forall x \in G\}.$$

(ii) The non-abelian exterior centre $Z^{\wedge}(G)$ of G is

$$Z^{\wedge}(G) = \{g \in G | g \otimes x \in \nabla(G), \forall x \in G\}.$$

Ellis showed in Ref. 22 that $Z^{\otimes}(G)$ is a characteristic central subgroup of G and is the largest normal subgroup L of G such that $G \otimes G \simeq G/L \otimes G/L$. Similarly, the non-abelian exterior centre $Z^{\wedge}(G)$ is a characteristic central subgroup of G and is the largest normal subgroup H of G such that $G \wedge G \simeq G/H \wedge G/H$.

Theorem 19 of Ref. 11 provides a way to compute $Z^{\otimes}(G)$ and $Z^{\wedge}(G)$ by looking at the natural isomorphic copy of G in $\nu(G)$ (see Definition 2.2) or in a suitable quotient $\tau(G)$ of $\nu(G)$. Under general conditions the non-abelian tensor centre and non-abelian exterior centre are closely related.

Corollary 5.1 ([8]). *Let G be any group such that G^{ab} is finitely generated. With the notation of Theorem 4.1, if $N = 1$ then $Z^{\otimes}(G) = Z^{\wedge}(G) \cap G'$. In particular, the conclusion holds if G is a finite group of odd order.*

A group G is said to be capable if it is isomorphic to $H/Z(H)$ for some group H, that is, if G is isomorphic to the group of inner automorphisms of some group H. The epicentre $Z^*(G)$ of a group G is the uniquely determined central subgroup of G that is minimal subject to being the image in G of some central extension of G. Therefore, a group G is capable if and only if it has trivial epicentre[7].

In Ref. 23 Ellis showed that $Z^{\wedge}(G)$ coincides with epicentre $Z^*(G)$ of G and therefore a group G is capable if and only if $Z^{\wedge}(G) = 1$.

The other extreme case which might occur is when $Z^*(G) = Z(G)$. In that case the group G is called unicentral. Recently Magidin and Morse[34] computed the tensor centre and the exterior centre of all 2-generator p-groups of class 2 and thus they determine which of those groups are capable and which are unicentral.

References

1. R. Aboughazi, Produit tensoriel du group d'Heisenberg, *Bull. Soc. Math. France* **115** (1987), 95–106.
2. A. Ahmad, A. Magidin, R.F. Morse, A new classification of 2-generator p-groups of class 2. Preprint.
3. M.R. Bacon, On the nonabelian tensor square of a nilpotent group of class 2, *Glasgow Math. J.* **36** (1994), 291–296.
4. M.R. Bacon, L.-C. Kappe, The nonabelian tensor square of a 2-generator p-group of class 2. *Arch. Math.* **61** (1993), 508–516.
5. M.R. Bacon, L.-C. Kappe, R.F. Morse, On the nonabelian tensor square of a 2−Engel group, *Arch. Math.* **69** (1997), 353–364.
6. J.R. Beuerle, L.-C. Kappe, Infinite metacyclic groups and their non-abelian tensor squares, *Proc. Edinburgh Math. Soc.* **43** (2000), 651–662.

7. F.R. Beyl, U. Felgner, P. Schmid, On groups occurring as center factor groups, *J. Algebra* **61** (1979), 161–177.
8. R.D. Blyth, F. Fumagalli, M. Morigi, Some structural results on the non-abelian square of groups, *J. Group Theory* **13** (2010), 83–94.
9. R.D. Blyth, P. Moravec, R.F. Morse, On the nonabelian tensor squares of free nilpotent groups of finite rank, in *Computational Group Theory and the Theory of Groups* (L.-C. Kappe et al., eds.), Contemporary Mathematics **470** (American Mathematical Society, 2008), 27–44.
10. R.D. Blyth, P. Moravec, R.F. Morse, On the derived subgroups of the free nilpotent groups of finite rank, in *Aspects of Infinite Groups* (B. Fine et al., eds.), Algebra and Discrete Math. **1** (World Scientific, 2008), 45–53.
11. R.D. Blyth, R.F. Morse, Computing the nonabelian tensor squares of polycyclic groups, *J. Algebra* **321** (2009), 2139–2148.
12. R.D. Blyth, R.F. Morse, J.L. Redden, On computing the non-abelian tensor squares of the free 2−Engel groups, *Proc. Edinb. Math Soc.* **47** (2004), 305–323.
13. W. Bosma, J. Cannon, C. Playoust, The Magma algebra system. I. The user language, in Computational Algebra and Number Theory, London, 1993, *J. Symbolic Comput.* **24** (3–4) (1997), 235–265.
14. R. Brown, D.L. Johnson, E.F. Robertson, Some computations of nonabelian tensor products of groups, *J. Algebra* **111** (1987), no. 1, 177–202.
15. R. Brown, J.L. Loday, Excision homotopique en basse dimension, *C.R. Acad. Sci. Pars Sér. I Math* **298** (1984), no. 15, 353–356.
16. R. Brown, J.L. Loday, Van Kampen theorems for diagram of spaces., *Topology* **26** (1987), no. 3, 311–335. With an appendix by M. Zisman.
17. R.K. Dennis, In search of "new" homology functors having a close relationship to K-theory, Preprint, Cornell University, Ithaca, NY (1976).
18. B. Eick, W. Nickel, Polycyclic — Computing with polycyclic groups (2002). A GAP package, see Ref. 28.
19. B. Eick, W. Nickel, Computing the Schur multiplicator and the nonabelian tensor square of a polycyclic group, *J. Algebra* **320** (2008), no. 2, 927–944.
20. G.J. Ellis, Non-abelian exterior products of groups and exact sequences in the homology of groups, *Glasgow Math. J.* **29** (1987), 13–19.
21. G.J. Ellis, The nonabelian tensor product of finite groups is finite, *J. Algebra* **111** (1987), 203–205.
22. G.J. Ellis, Tensor products and q-crossed modules, *J. London Math. Soc.* **51** (1995), no. 2, 243–258.
23. G.J. Ellis, On the capability of groups, *Proc. Edinburgh Math. Soc.* **41** (1998), no. 3, 487–495.
24. G.J. Ellis, On the computation of certain homotopical functors, *London Math. Soc. J. Comput. Math.* **1** (1998), 25–41 (electronic).
25. G.J. Ellis, F. Leonard, Computing Schur multipliers and tensor products of finite groups, *Proc. Roy. Irish Acad. Sect. A* **95** (1995), no. 2, 133–147.
26. A. Erfanian, R. Rezaei, H. Jafari, Computing the non-abelian tensor square of general linear groups, *Ital. J. Pure Appl. Math.* **24** (2008), 203–210.
27. A. Erfanian, F. Russo, N.H. Sarmin, Some considerations on the nonabelian

tensor square of crystallographic groups, Preprint.

28. The GAP Group, GAP — Groups, Algorithms, and Programming, Version 4.11, http://www.gap-system.org (2009).
29. T. Hannebauer, On nonabelian tensor squares of linear groups, *Arch. Math.* **55** (1990), no. 1, 30–34.
30. D.L. Johnson, The nonabelian tensor square of a finite split metacyclic group, *Proc. Edinburgh Math. Soc.* **30** (1987), 91–96.
31. L.-C. Kappe, Nonabelian tensor products of groups: the commutator connection, in *Groups St Andrews 1997 in Bath, Vol. II* (C. M. Campbell et al., eds.), London Math. Soc. Lecture Notes Ser. **261** (CUP, Cambridge 1999), 447–454.
32. L.-C. Kappe, M.P. Visscher, N.H. Sarmin, Two-generator two-groups of class two and their nonabelian tensor squares. *Glasg. Math. J.* **41** (1999), 417–430.
33. C. Miller, The second homology group of a group; relations among commutators, *Proc. Amer. Math. Soc.* **3** (1952), 588–595.
34. A. Magidin, R.F. Morse, Certain homological functors of 2-generator p-groups of class 2, in *Computational Group Theory and the Theory of Groups, II* (L.-C. Kappe et al., eds.), Contemporary Mathematics **511**, (Providence, Rhode Island, 2010), 127–166.
35. A. McDermott, The nonabelian tensor product of groups: Computations and structural results, Ph.D. thesis, National University of Ireland, Galway (1998).
36. P. Moravec, The exponents of nonabelian tensor products of groups, *J. Pure Appl. Algebra* **212** (2008), 1840–1848.
37. I.N. Nakaoka, Non-abelian tensor products of solvable groups, *J. Group Theory* **3** (2000), 157–167.
38. W. Nickel, nq - Nilpotent Quotients of Finitely Presented Groups (2003). A GAP package, see Ref. 28.
39. N.R. Rocco, On a construction related to the nonabelian tensor square of a group, *Bol. Soc. Brasil. Mat. (N.S.)* **22** (1991), no. 1, 63–79.
40. N.R. Rocco, A presentation for a crossed embedding of finite solvable groups, *Comm. Algebra* **22** (1994), no. 6, 1975–1998.
41. N.H. Sarmin, Infinite two-generator groups of class two and their non-abelian tensor squares. *Int. J. Math. Math. Sci.* **32** (2002), 615–625.
42. M.P. Visscher, On the nilpotency class and solvability length of nonabelian tensor products of groups, *Arch. Math.* **73** (1999), 161–171.
43. J.H.C. Whitehead, A certain exact sequence, *Ann. of Math. (2)* **52** (1950), 51–110.

CONJUGACY CLASSES OF SUBGROUPS OF FINITE p-GROUPS: THE FIRST GAP

R. BRANDL

Mathematatisches Institut, Am Hubland 12
97074 Würzburg, Germany

Let G be a finite p-group, and let $\nu(G)$ denote the number of conjugacy classes of non-normal subgroups of G. It is known that either $\nu(G) \leq 1$ or $\nu(G) \geq p$. We determine all p-groups G with $\nu(G) \leq p$.

Keywords: conjugacy classes of non-normal subgroups; finite p-groups.

1. Introduction

For a finite group G, let $\nu(G)$ denote the number of conjugacy classes of all non-normal subgroups of G. Clearly, $\nu(G) \geq 0$, and $\nu(G) = 0$ of and only if there is no non-normal subgroup of G at all. In this case, all subgroups of G are normal. These are the well-known Hamiltonian groups.

Here we are interested in the question on which values occur as $\nu(G)$. We restrict attention to finite p-groups for some fixed prime p. First, it is easy to see that the dihedral group G of order 2^n $(n \geq 3)$ satifies $\nu(G) = 2n - 4$, while for the quasidihedral group G of order 2^n $(n \geq 4)$, we have $\nu(G) = 2n - 5$. Hence for every positive integer γ, there exists a 2-group G with $\nu(G) = \gamma$ (for the case $\gamma = 1$, see Example 2.1 below).

For odd primes p, however, there is a gap (at least one!) in the set of all possible values for the number $\nu(G)$ where G varies over all p-groups:

Theorem 1.1 (R. La Haye, A. Rhemtulla[5]). *Let G be a p-group. Then $\nu(G) \leq 1$ or $\nu(G) \geq p$.*

The groups G with $\nu(G) \leq 1$ can be found in Ref. 2. In this paper, we determine all p-groups G for which $\nu(G) \leq p$. The aforementioned result of La Haye and Rhemtulla will not be used in our proof, so that we get it as an easy corollary.

Throughout this paper, all groups considered are finite. The order of an element g will be denoted by $o(g)$, and the symbol $G = [A]B$ indicates, that G is a split extension of a normal subgroup A of G by a complement B.

2. The result

Before we present our result, we introduce two prominent examples of p-groups that will turn out to be important later.

Example 2.1. Let p be a prime, and let $n \geq 3$ if $p > 2$ and $n \geq 4$ if $p = 2$. If

$$G = M(p^n) := \langle a, b \,|\, a^{p^{n-1}} = b^p = 1, a^b = a^{1+p^{n-2}} \rangle ,$$

then $\nu(G) = 1$.

The following is a special case of an example in the PhD Thesis of Leire Legarreta[6]. As the result does not seem to have been published elsewehere, we present here a short proof:

Example 2.2. Let p be an odd prime, and let $n \geq 4$ be an integer. If

$$G = L(p^n) := \langle a, b \,|\, a^{p^{n-2}} = b^{p^2} = 1, a^b = a^{1+p^{n-3}} \rangle ,$$

then $\nu(G) = p$.

Proof. First, note that $G' = \langle a^{p^{n-3}} \rangle \cong \mathbb{Z}_p$ and $\Omega_1(G) = \langle a^{p^{n-3}}, b^p \rangle \leq Z(G)$.

Let U be a non-normal subgroup of G. We claim that U must be cyclic of order p^2. If $U \cap \langle a \rangle \neq 1$, then $U \geq \langle a^{p^{n-3}} \rangle = G'$, and so U is normal in G. So $U \cap \langle a \rangle = 1$. Here, $U \cong U\langle a \rangle / \langle a \rangle \leq G/\langle a \rangle \cong \mathbb{Z}_{p^2}$, and so U is cyclic of order $\leq p^2$. If $|U| = p$, then $U \leq \Omega_1(G) \leq Z(G)$, so that U would be normal, a contradiction. So $U \cong \mathbb{Z}_{p^2}$.

Thus, all non-normal subgroups of G are contained in $\Omega_2(G) = \langle a^{p^{n-4}}, b \rangle$. This group is of order p^4 and homocyclic if $n \geq 5$. If $n = 4$, we have $\Omega_2(G) = G$.

We count the cyclic *normal* subgroups N of order p^2. If $N \leq Z(G) = \langle a^p, b^p \rangle \cong \mathbb{Z}_{p^{n-3}} \times \mathbb{Z}_p$, then $N \geq \langle a^{p^{n-3}} \rangle = G'$. If $N \not\leq Z(G)$, then $1 \neq [N, G] \leq N$, whence $N \geq G'$ also in this case. Hence the cyclic normal subgroups of G of order p^2 are precisely those cyclic subgroups of order p^2 that contain G'. The number of these is equal to p.

Now $\Omega_2(G)$ contains precisely $(p^4-p^2)/(p^2-p) = p^2+p$ cyclic subgroups of order p^2. By the above, p^2 of these are non-normal in G. Let U be one of these. Then $U \not\leq Z(G)$, and clearly $Z(G) \leq N_G(U)$. As $|G/Z(G)| = p^2$, this implies $[G : N_G(U)] = p$. Thus, the p^2 non-normal cyclic subgroups fall into p conjugacy classes of length p. This shows $\nu(G) = p$. □

The groups G with $\nu(G) \leq 1$ are known by Ref. 2. Indeed all p-groups G with $\nu(G) = 1$ are among the groups $M(p^n)$. In Mousavi[7] (see also the correction by Chen et al.[3] for non p-groups) the case $\nu(G) = 2$ has been fully classified. In the sequel, we thus may assume $p > 2$.

Examples for p-groups G with $\nu(G) = p$ have been given above, namely the group $L(p^n)$ for $n \geq 4$. We now determine *all* p-groups G with $\nu(G) \leq p$:

Theorem 2.1. *Let G be a p-group of order p^n with $\nu(G) \leq p$ $(p > 2)$. Then one of the following holds:*

(*i*) *G is abelian;*

(*ii*) *$\nu(G) = 1$ and $G \cong M(p^n)$ $(n \geq 3)$;*

(*iii*) *$\nu(G) = p$ and $G \cong L(p^n)$ $(n \geq 4)$.*

For the proof, we need to appeal to the following example:

Lemma 2.1. *Let p be a prime, let $H = M(p^{n-1})$ where $n \geq 4$ if $p > 2$ and $n \geq 5$ if $p = 2$. Moreover, let $G = H \times C$ where $C = \langle c \rangle$ is the cyclic group of order p. Then $\nu(G) = 2p$.*

Proof. First, note that $K = \langle a^p, b, c \rangle$ is an abelian normal subgroup of G of index p. Moreover, $\Omega_1(G) = \langle a^{p^{n-3}}, b, c \rangle$ is elementary abelian of order p^3, and $G' = \langle a^{p^{n-3}} \rangle \cong \mathbb{Z}_p$.

(1) If S is a non-normal subgroup of G, then S is elementary abelian.
Indeed, if S contains G', then S is normal, a contradiction. As $|G'| = p$, we get $S \cap G' = 1$, and so S embeds into G/G', whence S is abelian. Suppose $S^p \neq 1$. Then $1 \neq S^p \leq G^p = \langle a^p \rangle$ which is cyclic. So S^p contains $G' = \langle a^{p^{n-3}} \rangle$ whence S is normal, a contradiction. Thus $\exp(S) = p$.

(2) If S is non-normal, then S has precisely p conjugates in G.
Indeed, by (1), we have $S \leq \Omega_1(G) \leq K$. As K is abelian, we have $K \leq N_G(S)$, and so $[G : K] = p$ implies the statement.

(3) G has precisely p conjugacy classes of non-normal subgroups of order p.
Let S be such a subgroup. Then $S \leq \Omega_1(G)$, which contains precisely p^2+p+1 subgroups of order p. As $Z(G) = \langle a^{p^{n-3}}, c \rangle \cong \mathbb{Z}_p \times \mathbb{Z}_p$, precisely

$p+1$ of these are normal. The remaining p^2 cyclic non-normal subgroups by (2) fall into p conjugacy classes.

(4) G has precisely p conjugacy classes of non-normal subgroups $\cong \mathbb{Z}_p \times \mathbb{Z}_p$. As $\Omega_1(G)$ is elementary of order p^3, it contains precisely p^2+p+1 subgroups $\cong \mathbb{Z}_p \times \mathbb{Z}_p$. If one of them, T, say, is normal in G, then we have two possibilities: Either $T \leq Z(G)$, so $T = Z(G)$. Or, we have $1 \neq [T,G] \leq T \cap G'$, so $G' \leq T$. This is true also in the case $T = Z(G)$. Hence the normal subgroups $\cong \mathbb{Z}_p \times \mathbb{Z}_p$ are precisely those subgroups of order p^2 of $\Omega_1(G)$ which contain G'. Their number clearly is $p+1$. So we are left with p^2 non-normal subgroups of order p^2, which by (2) fall into p classes. We arrive at $\nu(G) = 2p$. □

Proof of Theorem 1.1. Proceed by induction on n. Clearly, the result is true for $n \leq 3$, so assume $n \geq 4$. Set $r = n-1$. Let N be a minimal normal subgroup of G. Clearly, we have $\nu(G/N) \leq p$, so that by induction, G/N is of one of types (i), (ii) or (iii).

By way of contradiction suppose $\nu(G/N) = p$. Let $C_1/N, \ldots, C_p/N$ be representatives for the classes of non-normal subgroups of G. As $\nu(G) \leq p$, we see that $C_1, \ldots, C_p$ are representatives for the classes of all non-normal subgroups of G. Thus, all non-normal subgroups of G contain N. In particular, all non-normal subgroups of G have nontrivial intersection. By Blackburn's result (Ref. 1, Theorem 1), we get $p = 2$, against our hypothesis.

Thus, we have $\nu(G/N) < p$, and so $G/N \not\cong L(p^r)$. Hence only the following two cases can occur: (A) Either there exists a minimal normal, subgroup N of G such that $G/N \cong M(p^r)$, or (B) every proper quotient of G is abelian.

Case A. There is a minimal normal subgroup N of G with $\bar{G} := G/N \cong M(p^r)$, where $r \geq 3$. By the presentation of $M(p^r)$, there exist elements $a, b \in G$ such that $G/N = \langle aN, bN \rangle$, where $\bar{a}$ is of order p^{r-1} and $\bar{b}$ is of order p. Moreover, we have $a^b = na^{1+p^{r-2}}$ for some $n \in N$. Clearly, $G = \langle N, a, b \rangle$.

Subcase 1. $\langle a \rangle$ is normal in G.

If $o(b) = p^2$, then $N \leq \langle b \rangle$, because $o(\bar{b}) = p$. Thus, we get $G = \langle a, b \rangle = \langle a \rangle \langle b \rangle$. If $o(a) = p^r$, then Ref. 4, p. 90f, shows $G \cong M(p^{r+1})$, and we have possibility (ii). So $o(a) = p^{r-1}$. As $|G| = p^{r+1}$, a comparison of orders yields $\langle a \rangle \cap \langle b \rangle = 1$, so that $G = [\langle a \rangle]\langle b \rangle$ is a split extension. Clearly, we may assume $a^b = a^{1+p^{r-2}}$, so that we arrive at $G \cong L(p^{r+1})$, which is possibility (iii).

Now let us assume $o(b) = p$. Let $H = \langle a, b \rangle = [\langle a \rangle]\langle b \rangle$. As before, we get $H \cong M(p^r)$. As $G = \langle n, a, b \rangle = NH$ and $|H| < |G|$, this immediately yields $G = N \times H \cong \mathbb{Z}_p \times M(p^r)$. By Lemma 2.1, we get $\nu(G) = 2p > p$, a contradiction.

SUBCASE 2. $\langle a \rangle$ is non-normal in G.
As $|G| = p^{r+1}$, we cannot have $o(a) = p^r$, so that $o(a) = p^{r-1}$. Hence $A := \langle N, a \rangle = N \times \langle a \rangle$, and A is normal in G.

Consider $T = A^{p^{r-2}}$. Then $|T| = p$, and T is normal in G. As $T \leq \langle a \rangle$, the subgroup $C = \langle a \rangle / T$ is non-normal in G/T. In particular, G/T is nonabelian, and so by the induction hypothesis, we have $G/T \cong M(p^r)$. But in this group the non-normal subgroups are all of order p. Hence $p = |C| = o(a)/p = p^{r-2}$ which implies $r = 3$ and $|G| = p^4$.

We now could conclude by appealing to the list of all groups of order p^4. However, we can give a more direct argument: First, note that $N \leq \Phi(G)$, because otherwise, $G \cong N \times M(p^3)$ and $\nu(G) = 2p$ by Lemma 2.1. Hence G is two-generator. Assume $|G/G^p| \geq p^3$. Then G has a two-generator quotient Q of exponent p and order p^3. As Q is non-abelian, we get $\nu(G) \geq \nu(Q) = p + 1$, a contradiction. Thus, $|G/G^p| = p^2$, whence G is metacyclic by Ref. 4, p. 337. If $\exp(G) \geq p^3$, then G has a cyclic subgroup of index p and so Ref. 4, p. 90f., implies $G \cong M(p^4)$. Hence $\exp(G) = p^2$, and we conclude immediately that G is a split extension of $\mathbb{Z}_{p^2}$ by $\mathbb{Z}_{p^2}$. We readily get $G \cong L(p^4)$.

Case B. All proper quotients of G are abelian.
Thus $N = G'$ is the unique minimal normal subgroup of G, and so $Z(G)$ is cyclic. Moreover, for all $x, y \in G$, we have $[x^p, y] = [x, y]^p = 1$, so $G^p \leq Z(G)$. In particular, G^p is cyclic.

Let $K = \Omega_1(G)$. As G is of class two and $p \neq 2$, we see that K is of exponent p. Let $|K| = p^f$. Assume first that $f \geq 3$. Then K contains at least p^2+p+1 subgroups of order p, and at most one of them can be normal in G. Let $U = \langle u \rangle$ be a non-normal subgroup of order p. For every $g \in G$, we have $U^g = \langle u^g \rangle = \langle u[u, g] \rangle$. As $|G'| = p$, we get that U has precisely p conjugates. Hence K contains at least $(p^2+p)/p = p+1$ conjugacy classes of non-normal subgroups. This gives us $\nu(G) \geq p + 1$ against our hypothesis.

Thus we have $f = 2$. As G is of class two, it is regular. By Ref. 4, p. 327, we have $|G/G^p| = |K| = p^2$. As G^p is cyclic, there exists $g \in G$ with $G^p = \langle g^p \rangle$. Thus, $o(g) = p^r$, and Ref. 4, p. 90f, again shows $G \cong M(p^{r+1})$. $\square$

Discussion. For $p > 2$, let G be the nonabelian group of exponent p and order p^3. It is easy to see that $\nu(G) = p + 1$, and in Lemma 2.1, we have seen p-groups G with $\nu(G) = 2p$. It would be interesting to know whether there are any gaps other than the one between 1 and p discussed here.

Added in proof (2.11.2010): It has come to the author's attention that the main result of this paper has been found independently and using different methods by G.A. Fernández-Alcober, L. Legarreta, The finite p-groups with p conjugacy classes of non-normal subgroups, Israel J. Math. 180 (2010), 189-192.

References

1. N. Blackburn, Finite groups in which the non-normal subgroups have nontrivial intersection, J. Algebra **3** (1966), 30-37.
2. R. Brandl, Groups with few non-normal subgroups, Comm. Algebra **23** (1995), 2091-2098.
3. Chen, Shun Min, Chen Gui Yun, A note on finite groups having exactly two conjugacy classes of non-normal subgroups, J. Jilin Univ. Sci. **46** (2008), 1097-1100.
4. B. Huppert, *Endliche Gruppen,* Vol. I, Springer, Berlin: 1967.
5. R. La Haye and A. Rhemtulla, Groups with a bounded number of conjugacy classes of non-normal subgroups, J. Algebra **214** (1999), 41-63.
6. L. Legarreta, Conjugacy classes of subgroups in finite p-groups, PhD Thesis, Bilbao, 2007.
7. H. Mousavi, On finite groups with few non-normal subgroups, Comm. Algebra **27** (1999), 3143-3151.

THE TUTTE POLYNOMIAL OF THE SCHREIER GRAPHS OF THE GRIGORCHUCK GROUP AND THE BASILICA GROUP

T. CECCHERINI-SILBERSTEIN

Dipartimento di Ingegneria, Università del Sannio
C.so Garibaldi 107, 82100 Benevento (Italia)
E-mail: tceccher@mat.uniroma1.it

A. DONNO* and D. IACONO

Dipartimento di Matematica "G. Castelnuovo"
Sapienza Università di Roma
P.le A. Moro 2, 00185 Roma (Italia)
**E-mail: donno@mat.uniroma1.it*
E-mail: iacono@mat.uniroma1.it

We study the Tutte polynomial of two infinite families of finite graphs. These are the Schreier graphs associated with the action of two well-known self-similar groups acting on the binary rooted tree by automorphisms: the first Grigorchuk group of intermediate growth, and the iterated monodromy group of the complex polynomial $z^2 - 1$ known as the Basilica group. For both of them, we describe the Tutte polynomial and we compute several special evaluations of it, giving further information about the combinatorial structure of these graphs

Keywords: Tutte polynomial; Schreier graph; Grigorchuk group; Basilica group; spanning subgraph; acyclic orientation; reliability polynomial; chromatic polynomial; partition function of the Ising model.

1. Introduction

The Tutte polynomial is a two-variable polynomial which can be associated with a graph, a matrix, or, more generally, with a matroid. It has many interesting applications in several areas of sciences as, for instance, Combinatorics, Probability, Statistical Mechanics, Computer Science and Biology. It was introduced by W.T. Tutte[18–20] and we will mainly refer to Refs. 4,5,11,21 as expository papers.

Given a finite graph G, its Tutte polynomial $T(G; x, y)$ satisfies a fundamental universal property with respect to the deletion-contraction reduction of the graph. Hence, any multiplicative graph invariant with respect to a deletion-contraction reduction turns out to be an evaluation of it. This polynomial is quite interesting since several combinatorial, enumerative and algebraic properties of the graph — such as the number of spanning trees, of spanning connected subgraphs, of spanning forests and of acyclic orientations of the graph — can be investigated by considering special evaluations of this polynomial. Moreover, from the Tutte polynomial one also recovers the reliability and the chromatic polynomials. It has also many interesting connections with statistical mechanical models as the Potts model[22], percolation[17], the Abelian Sandpile Model[7,15], as well as with the theory of error correcting codes[22].

In this paper, we study the Tutte polynomial of the Schreier graphs associated with the action of two well-known automorphism groups of the binary rooted tree: the Grigorchuk group and the iterated monodromy group of the complex polynomial $z^2 - 1$ known as the Basilica group. See also Ref. 10, where the Tutte polynomials of the Sierpiński graphs and the Schreier graphs of the Hanoi Towers group $H^{(3)}$ are computed.

The first Grigorchuk was introduced by R. Grigorchuk in 1980; it yields the simplest solution of the Burnside problem (an infinite, finitely generated torsion group) and the first example of a finitely generated group of intermediate (i.e. faster than polynomial but slower than exponential) growth. See Refs. 6,12 for a detailed account and further references.

The Basilica group was introduced by R. Grigorchuk and A. Żuk[13] as a group generated by a three-state automaton. It is a remarkable fact due to V. Nekrashevych[16] that this group can be described as the iterated monodromy group of the complex polynomial $z^2 - 1$; therefore, there exists a natural way to associate with it a compact limit space homeomorphic to the well-known Basilica fractal. Moreover, it is the first example of an amenable group (a highly non-trivial and deep result of L. Bartholdi and B. Virág[3]) not belonging to the class of subexponentially amenable groups, which is the smallest class containing all groups of subexponential growth and closed after taking subgroups, quotients, extensions and direct unions.

Over the last decade, Grigorchuk and a number of coauthors have developed a new exciting direction of research focusing on finitely generated groups acting by automorphisms on rooted trees, transitively on each level[2]. They proved that these groups have deep connections with the theory of profinite groups and with complex dynamics. In particular, many groups of this type satisfy a property of self-similarity (see Definition 2.5), reflected on fractalness of some limit objects associated with them[16].

In Sections 3 and 4, we study the Tutte polynomial for the Schreier graphs $\{\Gamma_n\}_{n\geq 1}$ and $\{\mathrm{B}_n\}_{n\geq 1}$ of the Grigorchuck group and the Basilica group, respectively. It follows from the recursive expression of the generators of these groups that these graphs have a cactus structure, i.e., they are union of cycles, arranged in a tree-like way. This enables us to compute the Tutte polynomial using the multiplicative property (2) (see Section 2.1). Once we have these polynomials, we compute many special evaluations of them, providing several interesting information about the combinatorial structure of these graphs and showing connections with reliability, colorability and the Ising model (see Section 2.1 for definitions and details). Note that some evaluations of the Tutte polynomial are trivial, when the graphs $\{\Gamma_n\}_{n\geq 1}$ and $\{\mathrm{B}_n\}_{n\geq 1}$ are considered with loops; for instance, the number of acyclic orientations, the chromatic polynomial and the partition function of the Ising model. Therefore, we make these computations on the graphs $\{\Gamma_n^*\}_{n\geq 1}$ and $\{\mathrm{B}_n^*\}_{n\geq 1}$, obtained from the graphs $\{\Gamma_n\}_{n\geq 1}$ and $\{\mathrm{B}_n\}_{n\geq 1}$, respectively, by deleting loops. For the considered graphs, we explicitly describe:

- *the Tutte polynomial* (Theorems 3.1 and 4.1);
- *the reliability polynomial* (Propositions 3.1 and 4.2);
- *the complexity* (Propositions 3.2 and 4.3);
- *the number of connected spanning subgraphs* (Propositions 3.3 and 4.4);
- *the number of spanning forests* (Propositions 3.4 and 4.5);
- *the number of acyclic orientations* (Propositions 3.6 and 4.7);
- *the chromatic polynomial* (Propositions 3.7 and 4.8);
- *the partition function of the Ising model* (Theorems 3.2 and 4.2).

2. Preliminaries

2.1. *The Tutte polynomial*

Throughout the paper, we deal with graphs which are connected and finite. Moreover, both multiple edges and multiple loops are allowed. As usual, $G = (V(G), E(G))$ denotes a graph with vertex set $V(G)$ and edge set $E(G)$; we will often write V and E, when there is no risk of confusion, and so $G = (V, E)$. Moreover, we denote by E_n the graph with n vertices and no edges, and by K_n the complete graph on n vertices. A subgraph $A = (V(A), E(A))$ of a graph $G = (V(G), E(G))$ is said to be *spanning* if the condition $V(A) = V(G)$ is satisfied. In particular, a *spanning subtree* of G is a spanning subgraph of G which is a tree, a *spanning forest* of G is a spanning subgraph of G which is a forest. The number of spanning trees of a graph G is called *complexity* of G and is denoted by $\tau(G)$. It is interesting to study complexity when the system grows. More precisely, given a sequence $\{G_n\}_{n\geq 1}$ of finite graphs with complexity $\tau(G_n)$, such that $|V(G_n)| \to \infty$, the limit

$$\lim_{|V(G_n)|\to\infty} \frac{\log \tau(G_n)}{|V(G_n)|},$$

when it exists, is called the *asymptotic growth constant*[14] of the spanning trees of $\{G_n\}_{n\geq 1}$.
Finally, let $k(G)$ be the number of connected components of G.

Definition 2.1. Let A be a spanning subgraph of G, then the rank $r(A)$ and the nullity $n(A)$ of A are defined as

$$r(A) = |V(A)| - k(A) = |V(G)| - k(A)$$

$$n(A) = |E(A)| - r(A) = |E(A)| - |V(A)| + k(A).$$

Definition 2.2 (Spanning subgraphs). *Let $G = (V, E)$ be a graph. The Tutte polynomial $T(G; x, y)$ of G is defined as*

$$T(G; x, y) = \sum_{A\subseteq G} (x-1)^{r(G)-r(A)}(y-1)^{n(A)}, \tag{1}$$

where the sum runs over all the spanning subgraphs A of G.

The Tutte polynomial can be also defined by a recursion process given by deleting and contracting edges. We recall that, given $G = (V, E)$, the graph $G \setminus e = (V, E - \{e\})$ is obtained from G by deleting the edge $e \in E$. The graph obtained by contracting an edge $e \in E$ is the result of the identification of the endpoints of e followed by the removal of e. We denote it by G/e. Finally, we recall that an edge in a connected graph is a *bridge* if its deletion disconnects the graph, it is a *loop* if its endpoints coincide.

Definition 2.3 (Deletion-Contraction). *Let $G = (V, E)$ be a graph. The Tutte polynomial $T(G; x, y)$ of G is defined as*

$$T(G;x,y)=\begin{cases} 1 & \textit{if } G = E_1; \\ xT(G\setminus e;x,y) & \textit{if } e \textit{ is a bridge;} \\ yT(G\setminus e;x,y) & \textit{if } e \textit{ is a loop;} \\ T(G\setminus e;x,y)+T(G/e;x,y) & \textit{if } e \textit{ is neither a bridge nor a loop.} \end{cases}$$

The recursive process to compute the Tutte polynomial in this second definition is independent of the order in which the edges are chosen: this can be proven by showing that Definitions 2.2 and 2.3 are equivalent[5].

Once we have the definition, we can state some of the main properties of the Tutte polynomial (for more details, see Refs. 4,5,11). Recall that a one point join $G * H$ of two graphs G and H is obtained by identifying a vertex v of G and a vertex w of H into a single vertex of $G * H$. The following property can be easily proven by using Definition 2.2:

$$T(G * H; x, y) = T(G; x, y)T(H; x, y). \tag{2}$$

This property will be fundamental for our computations and we will refer to it as the *multiplicative property* of the Tutte polynomial. The next lemma follows from Definition 2.3, by using the multiplicative property.

Lemma 2.1. *If C_k is a cycle of length k, with $k \geq 2$, then its Tutte polynomial $T(C_k; x, y)$ is $y + x + x^2 + \cdots + x^{k-1}$.*

Proof. The proof is by induction on the length k of the cycle. For $k = 2$, C_2 is a 2-cycle. It is trivial to verify that the Tutte polynomial of the graph

is x, using Definition 2.3. Therefore, for the graph C_2

the Tutte polynomial is $T(C_2; x, y) = y + x$, again by Definition 2.3, so that the assertion is true for $k = 2$. Now let C_{k+1} be a cycle of length $k + 1$. Let e be a fixed edge of C_{k+1}, so that by Definition 2.3,

$$T(C_{k+1}; x, y) = T(C_k; x, y) + T(C_{k+1} \setminus e; x, y).$$

Since, by the multiplicative property, $T(C_{k+1} \setminus e; x, y) = x^k$, we can apply induction and we get the required result. □

In the next sections, we will be interested in special evaluations of the Tutte polynomial, that allow us to deduce many combinatorial and algebraic properties of the graphs considered. In the following theorem, we collect many of these properties that are well-known in literature.

Theorem 2.1 (Ref. 11, Theorem 3 and 8). *Let $G = (V, E)$ be a connected graph and denote by $T(G; x, y)$ its Tutte polynomial. Then:*

(1) $T(G; 1, 1) = \tau(G)$;

(2) $T(G;1,2)$ *is the number of spanning connected subgraphs of* G*;*
(3) $T(G;2,1)$ *is the number of spanning forests of* G*;*
(4) $T(G;2,2) = 2^{|E|}$*;*
(5) $T(G;2,0)$ *is the number of acyclic orientations of* G*, i.e., orientations having no oriented cycles.*

Another fundamental interesting aspect of the Tutte polynomial is that, starting from it, one can obtain other interesting polynomials associated with the graph: the *reliability polynomial* and the *chromatic polynomial.*

More precisely, as regards the reliability polynomial $R(G,p)$, suppose that each edge of G is independently chosen to be active (or open) with probability p or inactive (closed) with probability $1-p$. Then, $R(G,p)$ is defined as the probability that in this random model there is a path of active edges between each pair of vertices of G.

As regards the colorability, we recall that a proper (or admissible) λ-coloring of the vertices of G is an assignment of λ colors to the vertices of G, in such a way that adjacent vertices have distinct colors. G is said λ-colorable if it admits a proper λ-coloring. The *chromatic number* $\chi(G)$ of G is defined as the minimal number λ such that G is λ-colorable. G is uniquely λ-colorable if $\chi(G)=\lambda$ and any λ-coloring of G induces the same partition of $V(G)$ (vertices with the same color are in the same class). The chromatic polynomial $\chi(G,\lambda)$ gives, for all values λ, the number of proper λ-colorings of G. The famous 4-color Theorem states that, if G is a planar graph, then $\chi(G,4)>0$.
The connection with the Tutte polynomial is given by the following theorem.

Theorem 2.2 (Ref. 11, Theorem 12 and 17). *Let* $G=(V,E)$ *be a graph. Then,*

(1) $R(G,p) = p^{|V(G)|-1}(1-p)^{|E(G)|-|V(G)|+1}T\left(G;1,\frac{1}{1-p}\right)$*;*
(2) $\chi(G,\lambda) = (-1)^{r(G)}\lambda^{k(G)}T(G;1-\lambda,0)$.

Finally, we want to recall the well-known connection between the Tutte polynomial of a graph and the Ising model on it, which is obtained as a special case of the Q-Potts model, for $Q=2$.

The famous Ising model of ferromagnetism consists of discrete variables called spins arranged on the vertices of the graph G. Each spin can take values ± 1 and only interacts with its nearest neighbors. Configuration of spins at two adjacent vertices i and j has energy $J>0$ if the spins have opposite values, and $-J$ if the values are the same. Let $|V(G)| = N$, and let $\vec{\sigma} = (\sigma_1, ..., \sigma_N)$ denote the configuration of spins, with $\sigma_i \in \{\pm 1\}$. The total energy of the system in configuration $\vec{\sigma}$ is then

$$E(\vec{\sigma}) = -J \sum_{i \sim j} \sigma_i \sigma_j,$$

where $i \sim j$ means that the vertices i and j are adjacent in G. The probability of a particular configuration at temperature T is given by

$$\mathbb{P}(\vec{\sigma}) = \frac{1}{Z} \exp(-\beta E(\vec{\sigma})),$$

where β is the "inverse temperature" conventionally defined as $\beta \equiv 1/(k_B T)$, and k_B denotes the Boltzmann constant. As usual in statistical physics, the normalizing constant Z, that makes the distribution above a probability measure, is called the *partition function*:

$$Z = \sum_{\vec{\sigma}} \exp(-\beta E(\vec{\sigma})).$$

It is known[22] that the partition function Z of the Ising model on G can be obtained by evaluating the Tutte polynomial $T(G; x, y)$ at the hyperbola $(x-1)(y-1) = 2$. More precisely, one has:

$$Z = 2(e^{2\beta J} - 1)^{|V(G)|-1} e^{-\beta J |E(G)|} T\left(G; \frac{e^{2\beta J} + 1}{e^{2\beta J} - 1}, e^{2\beta J}\right).$$

In the next sections we will explicitly verify this correspondence for the Schreier graphs of both the Grigorchuk and Basilica groups, using the computations of the partition functions made in Ref. 9.

2.2. *Groups of automorphisms of rooted regular trees*

We recall some basic facts about self-similar groups. Let T_q be the infinite regular rooted tree of degree q, i.e., the rooted tree in which each vertex has q children. Each vertex of the n-th level of the tree can be regarded as a word of length n in the alphabet $X = \{0, 1, \ldots, q-1\}$. Moreover, one can identify the set X^ω of infinite words in X with the set ∂T_q of infinite geodesic rays starting at the root of T_q. Next, let $S < Aut(T_q)$ be a group acting on T_q by automorphisms generated by a finite symmetric set of generators Y. Moreover, suppose that the action is transitive on each level of the tree.

Definition 2.4. The n-th *Schreier graph* Σ_n of the action of S on T_q, with respect to the generating set Y, is a graph whose vertex set coincides with the set of vertices of the n-th level of the tree, and two vertices u, v are adjacent if and only if there exists $s \in Y$ such that $s(u) = v$. If this is the case, the edge joining u and v is labelled by s.

The vertices of Σ_n are labelled by words of length n in X and the edges are labelled by elements of Y. The Schreier graph is thus a regular graph of degree $|Y|$ with q^n vertices, and it is connected, since the action of S is level-transitive.

Definition 2.5 ([16]). *A finitely generated group $S < Aut(T_q)$ is self-similar if, for all $g \in S, x \in X$, there exist $h \in S, y \in X$ such that*

$$g(xw) = yh(w),$$

for all finite words w in the alphabet X.

Self-similarity implies that S can be embedded into the wreath product $Sym(q) \wr S$, where $Sym(q)$ denotes the symmetric group on q elements, so that any automorphism $g \in S$ can be represented as

$$g = \tau(g_0, \ldots, g_{q-1}),$$

where $\tau \in Sym(q)$ describes the action of g on the first level of T_q and $g_i \in S, i = 0, ..., q-1$, is the restriction of g on the full subtree of T_q rooted at the vertex i of the first level of T_q (observe that any such subtree is isomorphic to T_q). Hence, if $x \in X$ and w is a finite word in X, we have

$$g(xw) = \tau(x)g_x(w).$$

In the next sections, the Schreier graphs of the Grigorchuk group and of the Basilica group will be described. For both of them, we recall some substitutional rules that allow to recursively construct these sequences of graphs, starting from the Schreier graph associated with the action of the group on the first level of the rooted binary tree.

3. The Tutte polynomial of the Schreier graphs of the Grigorchuk group

The Grigorchuk group admits the following description as a self-similar group of automorphisms of the rooted binary tree. It is generated by the elements

$$a = \epsilon(id, id), \qquad b = e(a, c), \qquad c = e(a, d), \qquad d = e(id, b),$$

where e and ϵ are respectively the trivial and the non-trivial permutations in $Sym(2)$. Note that each generator is an involution.

The following substitutional rules describe how to construct the graph Γ_{n+1} from Γ_n[1,12]. More precisely, the construction consists in replacing the labelled subgraphs of Γ_n on the top of the picture by new labelled graphs (on the bottom).

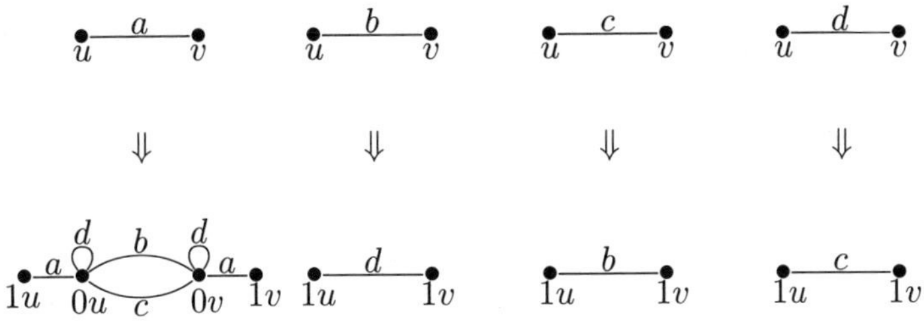

The starting point is the Schreier graph Γ_1 of the first level.

Γ_1 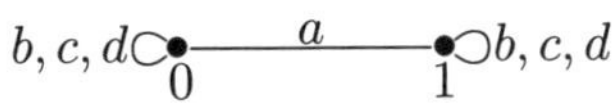

In computing the Tutte polynomial of Γ_n, we are interested in the unlabelled graph. We draw here the graphs Γ_n, for $n = 1, 2, 3$.

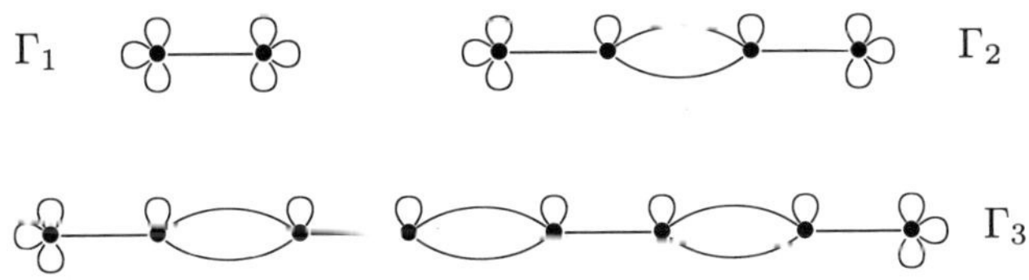

In general, one can check by using the substitutional rules that Γ_n has a linear shape, obtained by alternating bridges and 2-cycles. More precisely, it is easy to prove the following equalities:

$$|V(\Gamma_n)| = 2^n \qquad\qquad |E(\Gamma_n)| = 5 \cdot 2^{n-1} + 2.$$

As regards edges, note that there are 2^{n-1} bridges, $2^{n-1} - 1$ cycles of length 2 and $2^n + 4$ loops, since there is a loop rooted at each vertex, except for the outmost vertices, where three loops are rooted.

Since many computations are trivial for graphs with loops, it is convenient to consider Γ_n^*, defined as the graph obtained from Γ_n by erasing loops. Thus, in this case, we have

$$|V(\Gamma_n^*)| = 2^n \qquad\qquad |E(\Gamma_n^*)| = 3 \cdot 2^{n-1} - 2.$$

For every $n \geq 1$, denote by $T_n(x, y)$ the Tutte polynomial $T(\Gamma_n; x, y)$ of Γ_n and by $T_n^*(x, y)$ the Tutte polynomial $T(\Gamma_n^*; x, y)$ of Γ_n^*.

Theorem 3.1. *For each $n \geq 1$, the Tutte polynomial of the graph Γ_n is*

$$T_n(x, y) = y^{2^n+4} x^{2^{n-1}} (y + x)^{2^{n-1}-1}.$$

Proof. It suffices to apply the multiplicative property, keeping in mind that each loop, bridge or 2-cycle contributes by a factor y, x or $(y + x)$, respectively. □

Corollary 3.1. *For each $n \geq 1$, the Tutte polynomial of the graph Γ_n^* is*

$$T_n^*(x, y) = x^{2^{n-1}} (y + x)^{2^{n-1}-1}.$$

Let us start by writing the reliability polynomial $R(\Gamma_n, p)$.

Proposition 3.1. *For each $n \geq 1$, the reliability polynomial $R(\Gamma_n, p)$ is given by*

$$R(\Gamma_n, p) = p^{2^n-1} (2 - p)^{2^{n-1}-1}.$$

Proof. It suffices to apply Eq. (1) of Theorem 2.2. □

Remark 3.1. Note that the existence of loops does not change the reliability polynomial; therefore $R(\Gamma_n, p) = R(\Gamma_n^*, p)$, as one can directly check.

As regards the complexity of Γ_n, the following proposition holds.

Proposition 3.2. *The complexity of Γ_n is $2^{2^{n-1}-1}$.*

Proof. According with Formula (1) of Theorem 2.1, it suffices to compute $T_n(1, 1)$. □

Remark 3.2. The value of $\tau(\Gamma_n)$ has the following interpretation: each bridge of Γ_n must belong to any spanning subtree of Γ_n. On the other hand, a spanning subtree of Γ_n must contain exactly one edge of each 2-cycle. Then the result follows, since the number of 2-cycles of Γ_n is $2^{n-1} - 1$ and we have two choices for each 2-cycle. Also observe that loops do not contribute to $\tau(\Gamma_n)$; therefore, $T_n(1, 1) = T_n^*(1, 1)$ and so $\tau(\Gamma_n) = \tau(\Gamma_n^*)$, for each $n \geq 1$.

Corollary 3.2. *The asymptotic growth constant of the spanning trees of Γ_n is $\frac{1}{2} \log 2$.*

Proof. It suffices to compute

$$\lim_{n\to\infty} \frac{\log(\tau(\Gamma_n))}{|V(\Gamma_n)|},$$

with $|V(\Gamma_n)| = 2^n$. □

Evaluating $T_n(x, y)$ in $(1, 2)$ provides the number of connected spanning subgraphs of Γ_n. The following proposition holds.

Proposition 3.3. *The number of connected spanning subgraphs of* Γ_n *is* $2^{2^n+4} \cdot 3^{2^{n-1}-1}$.

Proof. It suffices to apply Formula (2) of Theorem 2.1. □

Remark 3.3. The value that we have found for the number of connected spanning subgraphs of Γ_n has the following interpretation: a connected spanning subgraph of Γ_n necessarily contains each bridge of the graph. On the other hand, both the edges or only one edge of each 2-cycle must belong to the subgraph (if no edge of a cycle belongs to the subgraph, then this subgraph is not connected), so that, for each of the $2^{n-1} - 1$ cycles of length 2, we have three possibilities. Finally, a connected spanning subgraph can also contain loops and so we have two possibilities for each of the $2^n + 4$ loops.

Another interesting computation concerns the number of spanning forests of Γ_n, which is given by $T_n(2, 1)$.

Proposition 3.4. *The number of spanning forests of* Γ_n *is* $2^{2^{n-1}} \cdot 3^{2^{n-1}-1}$.

Proof. It suffices to apply Formula (3) of Theorem 2.1. □

Remark 3.4. The value that we have found for the number of spanning forests of Γ_n has the following interpretation: a spanning forest of Γ_n cannot contain loops nor both the edges of a 2-cycle, since this would produce a cycle. Therefore, no edges or only one edge of each 2-cycle must belong to the forest. On the other hand, each bridge can belong to a spanning forest of Γ_n. Since the number of 2-cycles is $2^{n-1} - 1$ and the number of bridges is 2^{n-1}, we get the result.

Next, we explicitly verify that by evaluating the Tutte polynomial of Γ_n at $(2, 2)$ one gets $2^{|E(\Gamma_n)|}$ (see Formula (4) of Theorem 2.1).

Proposition 3.5. *For each* $n \geq 1$, *one has* $T_n(2, 2) = 2^{|E(\Gamma_n)|} = 2^{5\cdot 2^{n-1}+2}$.

Proof. By definition of $T_n(x,y)$, one has:

$$T_n(2,2) = 2^{2^n+4} \cdot 2^{2^{n-1}} \cdot 4^{2^{n-1}-1} = 2^{5\cdot 2^{n-1}+2}. \qquad \square$$

Finally, by evaluating the Tutte polynomial of Γ_n at $(2,0)$, we investigate the number of acyclic orientations of Γ_n. Observe that, whenever we have loops, the number of possible acyclic orientations of the graphs is 0. Therefore, we consider the graphs $\{\Gamma_n^*\}_{n\geq 1}$ without loops, whose Tutte polynomial is $T_n^*(x,y) = x^{2^{n-1}}(x+y)^{2^{n-1}-1}$.

Proposition 3.6. *The number of acyclic orientations of* Γ_n^* *is* 2^{2^n-1}.

Proof. By definition of $T_n^*(x,y)$, one has:

$$T_n^*(2,0) = 2^{2^{n-1}} \cdot 2^{2^{n-1}-1} = 2^{2^n-1}. \qquad \square$$

Remark 3.5. The value that we have found for the number of acyclic orientations of Γ_n^* has the following interpretation: we have two possible orientations for each bridge, giving the factor $2^{2^{n-1}}$. Then, each 2-cycle can receive four orientations, as shown in the following picture.

Only the first two orientations are acyclic, and so the 2-cycles give a contribution equal to $2^{2^{n-1}-1}$.

Since Γ_n has loops, it does not admit any proper coloring, so that we investigate the chromatic polynomial of Γ_n^*.

Proposition 3.7. *For each* $n \geq 1$, *the chromatic polynomial* $\chi_n(\lambda)$ *of* Γ_n^* *is*

$$\chi_n(\lambda) = -\lambda(1-\lambda)^{2^n-1}.$$

Proof. By applying Eq. (2) of Theorem 2.2, one gets

$$\chi_n(\lambda) = (-1)^{2^n-1}\lambda(1-\lambda)^{2^{n-1}}(1-\lambda)^{2^{n-1}-1} = -\lambda(1-\lambda)^{2^n-1}. \qquad \square$$

Remark 3.6. Note that $\chi_n(2) = 2$, for each $n \geq 1$, according to the fact that the graph is bipartite and so uniquely 2-colorable.

We end this section by investigating the relationship between the evaluation of the Tutte polynomial of the Schreier graph Γ_n^* at the hyperbola $(x-1)(y-1) = 2$ and the partition function of the Ising model on the same graph. In Theorem 2.1 of Ref. 9, the partition function of the Ising model on Γ_n^* has been described as

$$Z_n = 2^{2^n} \cosh(\beta J)^{3\cdot 2^{n-1}-2} \left(1 + \tanh^2(\beta J)\right)^{2^{n-1}-1}.$$

Theorem 3.2. *For each $n \geq 1$, one has*

$$2(e^{2\beta J} - 1)^{|V(\Gamma_n^*)|-1} \cdot e^{-\beta J|E(\Gamma_n^*)|} \cdot T_n^*\left(\frac{e^{2\beta J}+1}{e^{2\beta J}-1}, e^{2\beta J}\right) = Z_n. \tag{3}$$

Proof. Recall that $|E(\Gamma_n^*)| = 3 \cdot 2^{n-1} - 2$ and $|V(\Gamma_n^*)| = 2^n$. Let $e^{\beta J} = t$, so that Eq. (3) can be written as

$$\frac{2(t^2-1)^{2^n-1}}{t^{3\cdot 2^{n-1}-2}} \cdot T_n^*\left(\frac{t^2+1}{t^2-1}, t^2\right) \tag{4}$$

$$= 2^{2^n}\left(\frac{t^2+1}{2t}\right)^{3\cdot 2^{n-1}-2}\left(1+\left(\frac{t^2-1}{t^2+1}\right)^2\right)^{2^{n-1}-1}.$$

One can directly check that

$$T_n^*\left(\frac{t^2+1}{t^2-1}, t^2\right) = \frac{(t^2+1)^{2^{n-1}}(t^4+1)^{2^{n-1}-1}}{(t^2-1)^{2^n-1}}.$$

Then, it is not difficult to prove that both sides of Eq. (4) are equal to

$$\frac{2(t^2+1)^{2^{n-1}}(t^4+1)^{2^{n-1}-1}}{t^{3\cdot 2^{n-1}-2}}. \qquad \square$$

4. The Tutte polynomial of the Schreier graphs of the Basilica group

The Basilica group is a self-similar group of automorphisms of the rooted binary tree generated by the elements

$$a = e(b, id), \qquad b = \epsilon(a, id).$$

The associated Schreier graphs can be recursively constructed via the following substitutional rules:

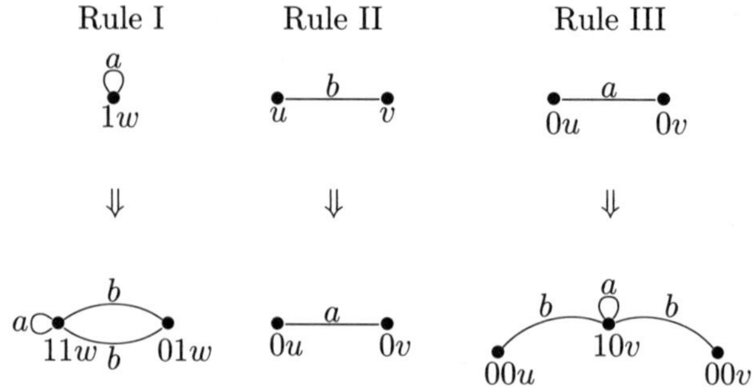

The starting point is the Schreier graph B_1 of the first level:

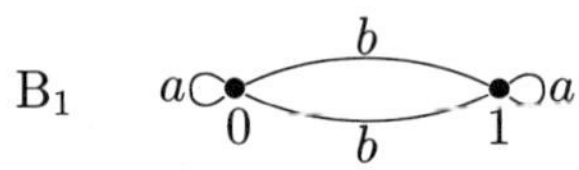

As in the case of the Grigorchuk group, we are interested in the unlabelled Schreier graphs. The following pictures of graphs B_n for $n = 1, 2, 3, 4, 5, 6$ give an idea of how Schreier graphs of the Basilica group look like. See Ref. 8 for a comprehensive analysis of finite and infinite Schreier graphs of this group. Note also that $\{B_n\}_{n\geq 1}$ is an approximating sequence for the Julia set of the polynomial $z^2 - 1$, the famous "Basilica" fractal[16].

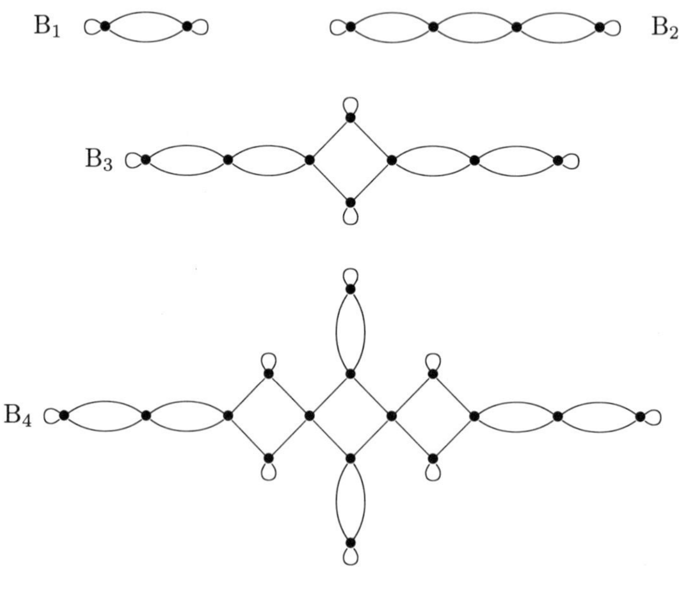

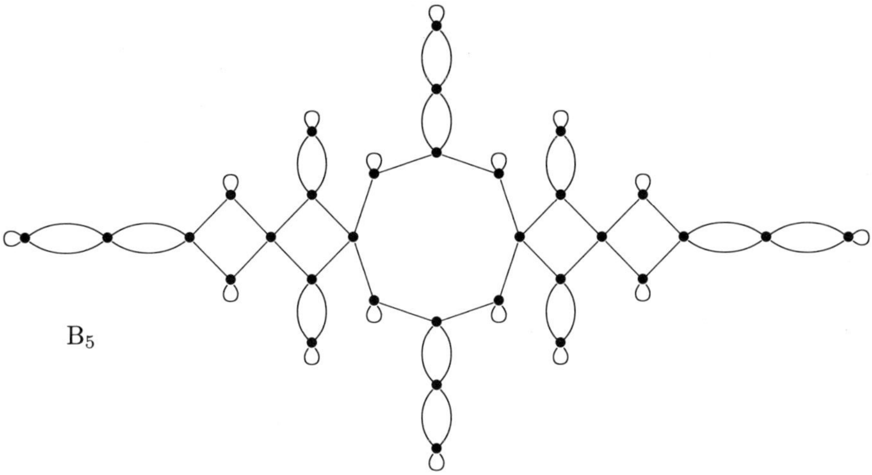

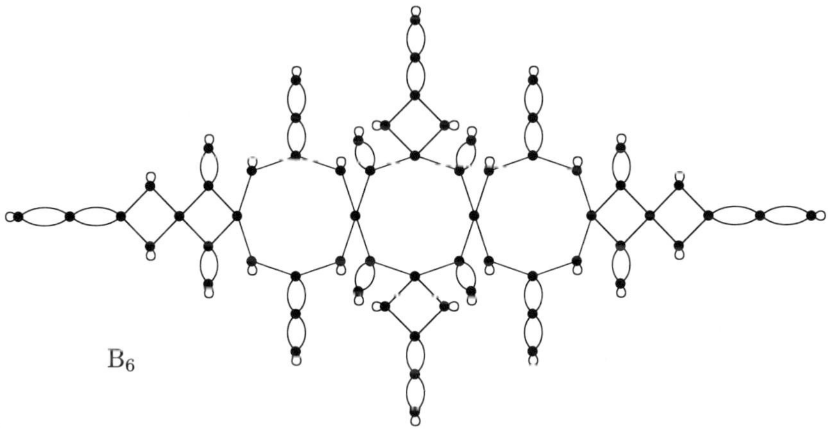

In general, it follows from the recursive definition of the generators, that each B_n is a cactus, i.e., a union of cycles (all of them are of length power of 2) arranged in a tree-like way. The maximal length of a cycle in B_n is $2^{(n+1)/2}$ if n is odd and $2^{n/2}$ if n is even. Moreover, for each $n \geq 2$, the graph B_n contains exactly 2^{n-1} loops rooted at the vertices corresponding to words in the alphabet $\{0, 1\}$ starting by 1, since the action of the generator a on these words is trivial.

Proposition 4.1. *For any $n \geq 4$, the number $c_{n,i}$ of cycles of length 2^i in* B_n *is:*

$$c_{n,i} = \begin{cases} 3 \cdot 2^{n-2i-1} & \text{for } 1 \leq i \leq \frac{n}{2} - 1 \\ 3 & \text{for } i = \frac{n}{2} \end{cases}, \qquad \text{for } n \text{ even,}$$

and

$$c_{n,i} = \begin{cases} 3 \cdot 2^{n-2i-1} & \text{for } 1 \leq i \leq \frac{n-1}{2} - 1 \\ 4 & \text{for } i = \frac{n-1}{2} \\ 1 & \text{for } i = \frac{n+1}{2} \end{cases}, \qquad \text{for } n \text{ odd.}$$

Proof. It follows from Proposition 2.2 of Ref. 9. □

To sum up, we have

$$|V(\mathrm{B}_n)| = 2^n \qquad |E(\mathrm{B}_n)| = 2^{n+1}.$$

As regards edges, note that there are 2^{n-1} loops, for $n \geq 2$, and 2 loops in B_1.

Since many computation are trivial for graphs with loops, it is convenient to define B_n^* as the graph B_n considered without loops. Thus, in this case, we have

$$|V(\mathrm{B}_n^*)| = 2^n \qquad |E(\mathrm{B}_n^*)| = 3 \cdot 2^{n-1}.$$

For every $n \geq 1$, denote by $T_n(x, y)$ the Tutte polynomial $T(\mathrm{B}_n; x, y)$ of B_n and by $T_n^*(x, y)$ the Tutte polynomial $T(\mathrm{B}_n^*; x, y)$ of B_n^*.

Theorem 4.1. *For $n \geq 4$, the Tutte polynomial of the Schreier graph B_n of the Basilica group is*

$$T_n(x,y) = y^{2^{n-1}}(y + x + \cdots + x^{2^{\frac{n-1}{2}}-1})^4 (y + x + \cdots + x^{2^{\frac{n+1}{2}}-1}) \cdot \prod_{i=1}^{\frac{n-1}{2}-1} (y + x + \cdots + x^{2^i - 1})^{3 \cdot 2^{n-2i-1}} \tag{5}$$

for n odd and

$$T_n(x,y) = y^{2^{n-1}}(y + x + \cdots + x^{2^{\frac{n}{2}}-1})^3 \prod_{i=1}^{\frac{n}{2}-1} (y + x + \cdots + x^{2^i - 1})^{3 \cdot 2^{n-2i-1}} \tag{6}$$

for n even. Moreover, one has $T_1(x, y) = y^2(x + y)$, $T_2(x, y) = y^2(x + y)^3$ and $T_3(x, y) = y^4(x + y)^4(y + x + x^2 + x^3)$.

Proof. The proof follows from Proposition 4.1 and Lemma 2.1. More precisely, by multiplicative property (2), $T_n(x,y)$ is obtained as the product of the Tutte polynomials of its cycles. □

Corollary 4.1. *For each $n \geq 4$, the Tutte polynomial of the graph B_n^* is*

$$T_n^*(x,y) = (y + x + \cdots + x^{2^{\frac{n-1}{2}}-1})^4 (y + x + \cdots + x^{2^{\frac{n+1}{2}}-1}) \cdot \prod_{i=1}^{\frac{n-1}{2}-1} (y + x + \cdots + x^{2^i-1})^{3\cdot 2^{n-2i-1}}$$

for n odd and

$$T_n^*(x,y) = (y + x + \cdots + x^{2^{\frac{n}{2}}-1})^3 \prod_{i=1}^{\frac{n}{2}-1} (y + x + \cdots + x^{2^i-1})^{3\cdot 2^{n-2i-1}}$$

for n even. Moreover, one has

$$T_1^*(x,y) = (x+y) \quad T_2^*(x,y) = (x+y)^3 \quad T_3^*(x,y) = (x+y)^4(y+x+x^2+x^3).$$

Let us start by writing the reliability polynomial $R(\mathrm{B}_n, p)$.

Proposition 4.2. *For each $n \geq 4$, the reliability polynomial $R(\mathrm{B}_n, p)$ is given by*

$$R(\mathrm{B}_n, p) = p^{2^n-1}(1-p)^{2^{n-1}+1} \left(2^{\frac{n-1}{2}} + \frac{p}{1-p}\right)^4 \left(2^{\frac{n+1}{2}} + \frac{p}{1-p}\right) \cdot \prod_{i=1}^{\frac{n-1}{2}-1} \left(\frac{p}{1-p} + 2^i\right)^{3\cdot 2^{n-2i-1}}$$

for n odd and

$$R(\mathrm{B}_n, p) = p^{2^n-1}(1-p)^{2^{n-1}+1} \left(2^{\frac{n}{2}} + \frac{p}{1-p}\right)^3 \prod_{i=1}^{\frac{n}{2}-1} \left(\frac{p}{1-p} + 2^i\right)^{3\cdot 2^{n-2i-1}}$$

for n even. Moreover, one has

$$R(\mathrm{B}_1, p) = p(2-p) \quad R(\mathrm{B}_2, p) = p^3(2-p)^3 \quad R(\mathrm{B}_3, p) = p^7(2-p)^4(4-3p).$$

Proof. It suffices to apply Eq. (1) of Theorem 2.2. □

Remark 4.1. Note that the existence of loops does not change the reliability polynomial; therefore $R(\mathrm{B}_n, p) = R(\mathrm{B}_n^*, p)$, as one can directly check.

Evaluating $T_n(x, y)$ in $(1, 1)$, we get the complexity $\tau(\mathrm{B}_n)$, i.e., the number of spanning trees of B_n.

Proposition 4.3. *The complexity of* B_n *is*

$$\tau(\mathrm{B}_n) = \begin{cases} 2^{\frac{2^{n+2}+3n-5}{6}} & \textit{for } n \textit{ odd} \\ 2^{\frac{2^{n+2}+3n-4}{6}} & \textit{for } n \textit{ even} \end{cases}.$$

Proof. It suffices to apply Formula (1) of Theorem 2.1. Indeed, one gets

$$T_n(1,1) = 2^{\frac{4(n-1)}{2}} \cdot 2^{\frac{n+1}{2}} \prod_{i=1}^{\frac{n-1}{2}-1} 2^{i \cdot 3 \cdot 2^{n-2i-1}} = 2^{\frac{2^{n+2}+3n-5}{6}}$$

for n odd and

$$T_n(1,1) = 2^{\frac{3n}{2}} \prod_{i=1}^{\frac{n}{2}-1} 2^{i \cdot 3 \cdot 2^{n-2i-1}} = 2^{\frac{2^{n+2}+3n-4}{6}}$$

for n even. For $n = 1, 2, 3$, one can directly find $T_1(1,1) = 2$, $T_2(1,1) = 2^3$, $T_3(1,1) = 2^6$. □

Remark 4.2. The previous equations can be motivated in the following way. In order to have a spanning tree of B_n, we do not have to consider loops; moreover, we have to delete exactly one edge in every cycle, so that every cycle of length 2^i contributes by a factor 2^i. Since the loops do not contribute to the number of spanning trees of B_n, note that $\tau(\mathrm{B}_n) = \tau(\mathrm{B}_n^*)$, for each $n \geq 1$.

Corollary 4.2. *The asymptotic growth constant of the spanning trees of* B_n *is* $\frac{2}{3}\log 2$.

Proof. It suffices to compute

$$\lim_{n\to\infty} \frac{\log(\tau(\mathrm{B}_n))}{|V(\mathrm{B}_n)|},$$

with $|V(\mathrm{B}_n)| = 2^n$. □

The evaluation of $T_n(x, y)$ at $(1, 2)$ provides the number of connected spanning subgraphs of B_n. More precisely, the following proposition holds.

Proposition 4.4. *The number of connected spanning subgraphs of* B_n *is*

$$T_n(1,2) = 2^{2^{n-1}} \left(1 + 2^{\frac{n-1}{2}}\right)^4 \left(1 + 2^{\frac{n+1}{2}}\right) \prod_{i=1}^{\frac{n-1}{2}-1} \left(1 + 2^i\right)^{3 \cdot 2^{n-2i-1}}$$

for n odd and

$$T_n(1,2) = 2^{2^{n-1}} \left(1 + 2^{\frac{n}{2}}\right)^3 \prod_{i=1}^{\frac{n}{2}-1} \left(1 + 2^i\right)^{3 \cdot 2^{n-2i-1}}$$

for n even. Moreover, one has

$$T_1(1,2) = 2^2 \cdot 3 \qquad T_2(1,2) = 2^2 \cdot 3^3 \qquad T_3(1,2) = 2^4 \cdot 3^4 \cdot 5.$$

Proof. It suffices to apply Formula (2) of Theorem 2.1. □

Remark 4.3. The value that we have found for the number of connected spanning subgraphs of B_n has the following interpretation: the factor $2^{2^{n-1}}$ corresponds to the possibility of choosing loops in the subgraph. On the other hand, in order to get a connected spanning subgraph, each cycle of length 2^i contributes by a factor $2^i + 1$, since we can take the whole cycle or delete exactly one edge from it.

Another interesting computation concerns the number of spanning forests of the Schreier graph B_n, which is given by $T_n(2,1)$.

Proposition 4.5. *The number of spanning forests of* B_n *is*

$$T_n(2,1) = \left(2^{2^{\frac{n-1}{2}}} - 1\right)^4 \left(2^{2^{\frac{n+1}{2}}} - 1\right) \prod_{i=1}^{\frac{n-1}{2}-1} \left(2^{2^i} - 1\right)^{3 \cdot 2^{n-2i-1}}$$

for n odd and

$$T_n(2,1) = \left(2^{2^{\frac{n}{2}}} - 1\right)^3 \prod_{i=1}^{\frac{n}{2}-1} \left(2^{2^i} - 1\right)^{3 \cdot 2^{n-2i-1}}$$

for n even. Moreover, one has

$$T_1(2,1) = 3 \qquad T_2(2,1) = 3^3 \qquad T_3(2,1) = 3^5 \cdot 5.$$

Proof. It suffices to apply Formula (3) of Theorem 2.1. □

Remark 4.4. The value that we have found for the number of spanning forests of B_n has the following interpretation: a spanning forest of B_n cannot contain loops nor the whole cycles, since this would produce a cycle. Therefore, we can choose or not each edge of any 2^i-cycle, but we cannot choose in a spanning forest all the edges of the cycle. Hence, a 2^i-cycle contributes to the number of spanning forests by a factor $2^{2^i} - 1$. Moreover, observe that $T_n(2,1) = T_n^*(2,1)$, for each $n \geq 1$.

Next, we explicitly verify that by evaluating the Tutte polynomial of B_n at $(2,2)$ one gets $2^{|E(\mathrm{B}_n)|}$ (see Formula (4) of Theorem 2.1).

Proposition 4.6. *For each $n \geq 1$, one has $T_n(2,2) = 2^{|E(\mathrm{B}_n)|} = 2^{2^{n+1}}$.*

Proof. By replacing $x = y = 2$ in Eqs. (5) and (6), it turns out that each cycle of length 2^i contributes by a factor 2^{2^i} to $T_n(2,2)$. □

Finally, by evaluating the Tutte polynomial of B_n at $(2,0)$, we investigate the number of acyclic orientations of B_n. Observe that, whenever we have loops, the number of possible acyclic orientations of the graphs is 0. Therefore, we consider the graph B_n^*.

Proposition 4.7. *The number of acyclic orientations of B_n^* is*

$$T_n^*(2,0) = \left(2^{2^{\frac{n-1}{2}}} - 2\right)^4 \left(2^{2^{\frac{n+1}{2}}} - 2\right) \prod_{i=1}^{\frac{n-1}{2}-1} \left(2^{2^i} - 2\right)^{3\cdot 2^{n-2i-1}}$$

for n odd and

$$T_n^*(2,0) = \left(2^{2^{\frac{n}{2}}} - 2\right)^3 \prod_{i=1}^{\frac{n}{2}-1} \left(2^{2^i} - 2\right)^{3\cdot 2^{n-2i-1}}$$

for n even. Moreover, one has

$$T_1^*(2,0) = 2 \qquad T_2^*(2,0) = 2^3 \qquad T_3^*(2,0) = 2^5 \cdot 7.$$

Proof. By replacing $x = 2$ and $y = 0$ in Eqs. (5) and (6), it turns out that each cycle of length 2^i contributes by a factor $2^{2^i} - 2$ to $T_n^*(2,0)$. □

Remark 4.5. The fact that each cycle of length 2^i contributes by the factor $2^{2^i} - 2$ has the following interpretation: there are two possible orientations for each edge, but we have to avoid the two cases where we create an oriented cycle (represented in the following picture in the case of a 2^3-cycle).

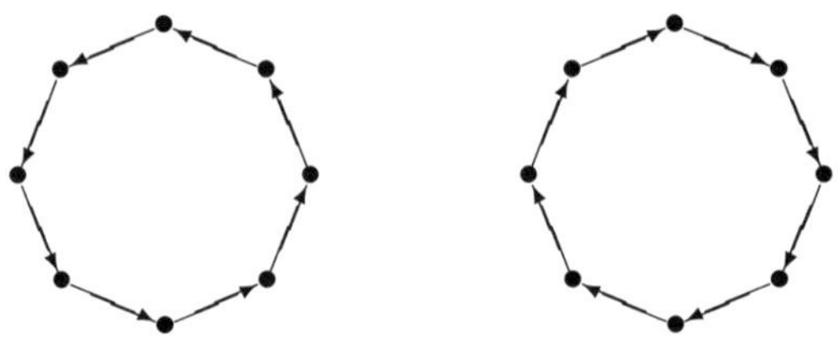

Since B_n has loops, it does not admit any proper coloring, so that we investigate the chromatic polynomial of B_n^*.

Proposition 4.8. *For each $n \geq 4$, the chromatic polynomial $\chi_n(\lambda)$ of B_n^* is*

$$\chi_n(\lambda) = -\lambda \left(\frac{1-\lambda}{\lambda} - \frac{(1-\lambda)^{2^{\frac{n-1}{2}}}}{\lambda} \right)^4 \left(\frac{1-\lambda}{\lambda} - \frac{(1-\lambda)^{2^{\frac{n+1}{2}}}}{\lambda} \right) \cdot \prod_{i=1}^{\frac{n-1}{2}-1} \left(\frac{1-\lambda}{\lambda} - \frac{(1-\lambda)^{2i}}{\lambda} \right)^{3\cdot 2^{n-2i-1}}$$

for n odd and

$$\chi_n(\lambda) = -\lambda \left(\frac{1-\lambda}{\lambda} - \frac{(1-\lambda)^{2^{\frac{n}{2}}}}{\lambda} \right)^3 \prod_{i=1}^{\frac{n}{2}-1} \left(\frac{1-\lambda}{\lambda} - \frac{(1-\lambda)^{2i}}{\lambda} \right)^{3\cdot 2^{n-2i-1}}$$

for n even. Moreover, one has

$$\chi_1(\lambda) = -\lambda(1-\lambda) \quad \chi_2(\lambda) = -\lambda(1-\lambda)^3 \quad \chi_3(\lambda) = -\lambda(1-\lambda)^5 \cdot (\lambda^2 - 3\lambda + 3).$$

Proof. It suffices to apply Eq. (2) of Theorem 2.2. □

Remark 4.6. Note that $\chi_n(2) = 2$, for each $n \geq 1$, according to the fact that the graph is bipartite and so uniquely 2-colorable.

Finally, we investigate the relationship between the evaluation of the Tutte polynomial of the Schreier graph B_n^* at the hyperbola $(x-1)(y-1) = 2$ and the partition function of the Ising model on the same graph. In Theorem 2.4 of Ref. 9, the partition function of the Ising model on B_n^* has been described as

$$Z_n = 2^{2^n} (\cosh(\beta J))^{3\cdot 2^{n-1}} \Phi_n(\tanh(\beta J)),$$

where $\Phi_n(z)$ is the generating function of closed polygons for B_n^* given by

$$\Phi_n(z) = \left(1 + z^{2^{\frac{n-1}{2}}}\right)^4 \left(1 + z^{2^{\frac{n+1}{2}}}\right) \prod_{i=1}^{\frac{n-1}{2}-1} \left(1 + z^{2^i}\right)^{3\cdot 2^{n-2i-1}}$$

for $n \geq 5$ odd and

$$\Phi_n(z) = \left(1 + z^{2^{\frac{n}{2}}}\right)^3 \prod_{i=1}^{\frac{n}{2}-1} \left(1 + z^{2^i}\right)^{3\cdot 2^{n-2i-1}}$$

for $n \geq 4$ even. Moreover,

$$Z_1 = 2^2 \cosh^2(\beta J)\left(1 + \tanh^2(\beta J)\right)$$

$$Z_2 = 2^4 \cosh^6(\beta J)\left(1 + \tanh^2(\beta J)\right)^3$$

$$Z_3 = 2^8 \cosh^{12}(\beta J)\left(1 + \tanh^2(\beta J)\right)^4\left(1 + \tanh^4(\beta J)\right).$$

Theorem 4.2. *For each $n \geq 1$, one has*

$$2(e^{2\beta J} - 1)^{|V(\mathrm{B}_n^*)|-1} \cdot e^{-\beta J|E(\mathrm{B}_n^*)|} \cdot T_n^*\left(\frac{e^{2\beta J}+1}{e^{2\beta J}-1}, e^{2\beta J}\right) = Z_n. \tag{7}$$

Proof. Here we only prove the case of n even (the computations for n odd are similar). Recall that $|E(\mathrm{B}_n^*)| = 3 \cdot 2^{n-1}$ and $|V(\mathrm{B}_n^*)| = 2^n$. Let $e^{\beta J} = t$, so that Eq. (7) can be written as

$$\frac{2(t^2-1)^{2^n-1}}{t^{3\cdot 2^{n-1}}} \cdot T_n^*\left(\frac{t^2+1}{t^2-1}, t^2\right) \tag{8}$$

$$= 2^{2^n}\left(\frac{t^2+1}{2t}\right)^{3\cdot 2^{n-1}}\left(1 + z^{2^{\frac{n}{2}}}\right)^3 \prod_{i=1}^{\frac{n}{2}-1}\left(1 + z^{2^i}\right)^{3\cdot 2^{n-2i-1}}\Bigg|_{z=\frac{t^2-1}{t^2+1}}.$$

One can directly check that

$$T_n^*\left(\frac{t^2+1}{t^2-1}, t^2\right) = \left(\frac{t^2-1}{2}\left(1 + \left(\frac{t^2+1}{t^2-1}\right)^{2^{\frac{n}{2}}}\right)\right)^3 \cdot \prod_{i=1}^{\frac{n}{2}-1}\left(\frac{t^2-1}{2}\left(1 + \left(\frac{t^2+1}{t^2-1}\right)^{2^i}\right)\right)^{3\cdot 2^{n-2i-1}}.$$

Then, it is not difficult to prove that both sides of Eq. (8) are equal to

$$\frac{(t^2+1)^{3\cdot 2^{n-1}}}{t^{3\cdot 2^{n-1}}2^{2^{n-1}}}\left(1+\left(\frac{t^2-1}{t^2+1}\right)^{2^{\frac{n}{2}}}\right)^3 \prod_{i=1}^{\frac{n}{2}-1}\left(1+\left(\frac{t^2-1}{t^2+1}\right)^{2^i}\right)^{3\cdot 2^{n-2i-1}}.$$

□

References

1. L. Bartholdi and R. Grigorchuk, On the spectrum of Hecke type operators related to some fractal groups, *Tr. Mat. Inst. Steklova* **231** (2000), Din. Sist., Avtom. i Beskon. Gruppy, 5–45; translation in *Proc. Steklov Inst. Math.* (2000), no. 4 (**231**), 1–41.

2. L. Bartholdi, R. Grigorchuk and V. Nekrashevych, From fractal groups to fractal sets, in: "Fractals in Graz"(P. Grabner and W. Woess editors), Trends in Mathematics, Birkäuser Verlag, Basel, 2003, 25–118.
3. L. Bartholdi and B. Virág, Amenability via random walks. *Duke Math Journal.* **130** (2005), no. 1, 39–56.
4. B. Bollobás, *Modern Graph Theory*, Graduate Texts in Mathematics, **184**. Springer-Verlag, New York (1998).
5. T. Brylawski and J. Oxley, The Tutte polynomial and its Applications, in: "Matroid Applications"(123–225), (N. White editor), *Encyclopedia Math. Appl.*, **40**, Cambridge Univ. Press, Cambridge, 1992.
6. T. Ceccherini-Silberstein, A. Mach and F. Scarabotti, Il gruppo di Grigorchuk di crescita intermedia. *Rend. Circ. Mat. Palermo.* **50** (2001), no. 1, 67–102.
7. R. Cori and Y. Le Borgne, The sand-pile model and Tutte polynomials. Formal power series and algebraic combinatorics (Scottsdale, AZ, 2001), *Adv. in Appl. Math.*, **30**, Issue 1–2, (2003), 44–52.
8. D. D'Angeli, A. Donno, M. Matter and T. Nagnibeda, Schreier graphs of the Basilica group, *Journal of Modern Dynamics*, **4**, no. 1, (2010), 167–205.
9. D. D'Angeli, A. Donno and T. Nagnibeda, Partition functions of the Ising model on some self-similar Schreier graphs, to appear in the Proceedings of the Euroconference "Boundaries", Graz, June 2009, (W. Woess and F. Sobieczky editors), Birkhäuser, 2010; available at `http://arxiv.org/abs/1003.0611`
10. A. Donno and D. Iacono, The Tutte polynomial of some self-similar graphs, submitted; available at `http://arxiv.org/abs/1006.5333`.
11. J. A. Ellis-Monaghan and C. Merino, *Graph Polynomials and Their Applications I: The Tutte Polynomial*, in: "Structural Analysis of Complex Networks"(Mathhias Dehmer editor), in press; available at `http://arxiv.org/abs/0803.3079v2`
12. R. Grigorchuk, Solved and unsolved problems around one group, Infinite groups: geometric, combinatorial and dynamical aspects, 117–218, *Progr. Math.*, **248**, Birkhuser, Basel, 2005.
13. R. Grigorchuk and A. Żuk, On a torsion-free weakly branch group defined by a three-state automaton. *International J. Algebra Comput.* **12** (2002), no. 1, 223–246.
14. R. Lyons, Asymptotic enumeration of spanning trees, *Combin. Probab. Comput.*, **14**, No. 4, (2005), 491–522.
15. C. Merino, Chip-firing and the Tutte polynomial, *Ann. Comb.*, **1**, No. 3, (1997), 253–259.
16. V. Nekrashevych, *Self-similar Groups*, Mathematical Surveys and Monographs, **117**. American Mathematical Society, Providence, RI, 2005.
17. J.G. Oxley and D.J.A. Welsh, The Tutte polynomial and percolation, in: "Graph theory and related topics"(Proc. Conf., Univ. Waterloo, Waterloo, Ont., 1977), 329–339, (J.A. Bondy and U.S.R. Murty editors), Academic Press, New York-London, 1979.
18. W.T. Tutte, A ring in graph theory, *Proc. Cambridge Philos. Soc.*, **43**, (1947), 26–40.

19. W.T. Tutte, A contribution to the theory of chromatic polynomials, *Canadian. J. Math.*, **6**, (1954), 80–91.
20. W.T. Tutte, On dichromatic polynomials, *J. Combinatorial Theory*, **2**, (1967), 301–320.
21. W.T. Tutte, Graph-polynomials. Special issue on the Tutte polynomial, *Adv. in Appl. Math.*, **32**, No. 1–2, (2004), 5–9.
22. D.J.A. Welsh and C. Merino, The Potts model and the Tutte polynomial. Probabilistic techniques in equilibrium and nonequilibrium statistical physics, *J. Math. Phys.*, **41**, No. 3, (2000), 1127–1152.

ON MAXIMAL SUBGROUPS OF THE ALTERNATING AND SYMMETRIC GROUPS

V. COLOMBO

Dipartimento di Matematica Pura e Applicata,
Università degli Studi di Padova,
Via Trieste, 63, 35121 Padova, Italy
E-mail: vcolombo@math.unipd.it

In this paper we consider a conjecture by A. Mann and N. Boston: they conjectured that the equality, for a fixed value of n, between the probabilistic polynomial of $\mathrm{Sym}(n)$ and the probabilistic polynomial of the direct product of $\mathrm{Alt}(n)$ with a cyclic group of order 2 reflects the non-existence of maximal subgroups of $\mathrm{Alt}(n)$ which coincide with their normalizer in $\mathrm{Sym}(n)$. We prove that this conjecture holds whenever n is a prime; but it does not hold in general (for example when $n = 21$). Even if there exists a one-to-one correspondence between maximal subgroups of $\mathrm{Alt}(n)$ and $\mathrm{Sym}(n)$ the conjecture can fail; it is the case for $n = 62$.

Keywords: alternating and symmetric groups; maximal subgroups; probabilistic polynomial.

1. Introduction

Let G be a finite group G; we may define the complex function P_G, such that $P_G(t)$ corresponds precisely to the probability of generating G with t elements, when t is a positive integer (see Section 2). Let n be a natural number; denote by $\mathrm{Sym}(n)$ and $\mathrm{Alt}(n)$ respectively the symmetric group and the alternating group of degree n. In a recent work Lucchini and Massa have proved that there exist explicit criteria to recognize whether the factor group $G/\mathrm{Frat}(G)$ is isomorphic to $\mathrm{Alt}(n)$ (with $n \geq 5$) only looking at the coefficients of $P_G(s)$, with $s \in \mathbb{C}$. Using these results they have verified that if $P_G(s) = P_{\mathrm{Sym}(n)}(s)$, with $n \geq 5$, then either $G/\mathrm{Frat}(G) \cong \mathrm{Sym}(n)$ or $G/\mathrm{Frat}(G) \cong \mathrm{Alt}(n) \times C_2$ (see Theorem 1.4 in Ref. 14). It is well known that $P_G(s) = P_{G/\mathrm{Frat}(G)}(s)$; here we want to investigate for which values of n it holds that

$$P_{\mathrm{Sym}(n)}(s) = P_{\mathrm{Alt}(n)\times C_2}(s) \tag{1}$$

The complex functions $P_{\mathrm{Sym}(n)}$ and $P_{\mathrm{Alt}(n)\times C_2}$ can be computed with GAP (Ref. 11) when n is small ($n < 12$): it turns out that Eq. (1) holds for $n = 2, 5, 6, 10$ but not for $n = 3, 4, 7, 8, 9, 11$, as was already noticed by Boston in Ref. 3. In that paper, Boston presented an interesting conjecture, formulated with Mann, to establish when Eq. (1) holds.

Conjecture 1.1 (Boston, Mann). *The validity of Eq. (1), for $n \geq 5$, reflects the non-existence of maximal subgroups of* Alt(n) *which coincide with their normalizer in* Sym(n).

Remark 1.1. In other words, Boston and Mann conjectured that, for any $n \geq 5$, $P_{\mathrm{Sym}(n)}(s) = P_{\mathrm{Alt}(n)\times C_2}(s)$ if and only if any maximal subgroup of Alt(n) does not coincide with its normalizer in Sym(n).

We study the conjecture of Boston and Mann. In Section 3 we prove that it is true when $n = p$, for any prime $p \geq 5$. But it does not hold in general; in fact we show counterexamples to the conjecture (see Section 4). Finally, in Section 5, we present an open problem concerning the behavior, with respect to Conjecture 1.1, of the class of symmetric groups Sym(n) not containing any primitive proper subgroup different from Alt(n).

2. Preliminaries

Let P be a finite poset. The *Möbius function* $\mu_P : P \times P \to \mathbb{Z}$ is defined as follows: $\mu_P(x, y) = 0$ unless $x \leq y$, when it is defined recursively by the equations:

$$\mu_P(y,y) = 1 \qquad \text{and} \qquad \sum_{x\leq z\leq y} \mu_P(z,y) = 0 \quad \text{when } x < y.$$

Two results by Crapo will be very useful in our discussion. One is the Crapo's Complement Theorem.

Definition 2.1. Let L be a finite lattice.

(a) $\hat{0}$ and $\hat{1}$ denote, respectively, the unique minimum element of L and the unique maximum element of L.
(b) A *complement* to $x \in L$ is an element $y \in L$ such that $\inf_L(x, y) = \hat{0}$ and $\sup_L(x, y) = \hat{1}$. Also $x^\perp$ denote the set of complements to x in L.

Theorem 2.1 (Crapo's Complement Theorem[6]). *Let L be a finite lattice and fix $x \in L$. Then*

$$\mu_L(\hat{0},\hat{1}) = \sum_{y,z\in x^\perp} \mu_L(\hat{0},y)\zeta_L(y,z)\mu_L(z,\hat{1}),$$

where ζ_L is the zeta function on L, with $\zeta_L(y,z)=1$ if $y\le z$, or $\zeta_L(y,z)=0$ otherwise. In particular, if $x^{\perp}$ is an antichain (that is neither $y\le z$ nor $z\le y$ holds for any pair of distinct elements $y,z\in x^{\perp}$), then

$$\mu_L(\hat{0},\hat{1})=\sum_{y\in x^{\perp}}\mu_L(\hat{0},y)\mu_L(y,\hat{1}).$$

The other is the Crapo's Closure Theorem. A *closure* on the poset P is a function $^{-}: P\to P$ satisfying the following three conditions:

(a) $x\le\bar{x}$ for all $x\in P$;
(b) if $x,y\in P$ with $x\le y$, then $\bar{x}\le\bar{y}$;
(c) $\bar{\bar{x}}=\bar{x}$ for all $x\in P$.

If $^{-}$ is a closure map on P, then $\overline{P}=\{x\in P\,|\,\bar{x}=x\}$ is a poset with order induced by the order on P. We have:

Theorem 2.2 (Crapo's Closure Theorem[7]). *Let P be a finite poset and let $^{-}: P\to P$ be a closure map. Fix $x,y\in P$ such that $y\in\overline{P}$. Then*

$$\sum_{\bar{z}=y}\mu_P(x,z)=\begin{cases}\mu_{\bar{P}}(x,y) & \text{if } x=\bar{x}\\ 0 & \text{otherwise.}\end{cases}$$

Denote by $\mathcal{L}_G$ the subgroup lattice of a finite group G; notice that if $H\le K\le G$ then $\mu_{\mathcal{L}_G}(H,K)=\mu_{\mathcal{L}_K}(H,K)$. From now on, we will write $\mu(H,K)$ instead of $\mu_{\mathcal{L}_K}(H,K)$ whenever H is a subgroup of K. Recall also that $\mu(H,K)\neq 0$ only if H is an intersection of maximal subgroups of K.

Let $n\ge 1$ be a natural number; we may define the Möbius function $\mu(n)$, that is strictly connected to the Möbius function associated to the subgroup lattice of a cyclic group.

Definition 2.2. Let $n\in\mathbb{N}$;

$$\mu(n)=\begin{cases}1 & \text{if } n=1,\\ 0 & \text{if } p^2|n \text{ for some prime } p,\\ (-1)^r & \text{if } n=p_1p_2\dots p_r \text{ where the } p_i \text{ are distinct primes.}\end{cases}$$

Remark 2.1. By the definition of $\mu(n)$, we observe that, if $n\neq 1$,

$$\sum_{d|n}\mu(d)=0.$$

For the proof of this statement see p.87 of Ref. 1.

It is now easy to prove the following proposition (see p.63 of Ref. 10).

Lemma 2.1. *Let G be a cyclic group of order n; then*

$$\mu(\langle 1\rangle, G) = \mu(n).$$

Assume that G is a transitive permutation group on a finite set Γ. A closure function on the subgroup lattice $\mathcal{L}_G$ can be defined in the following way:

Definition 2.3. Let $H \leq G$ and let $\tau = \{\Lambda_1, \ldots, \Lambda_k\}$ be the set of orbits of H in its action on Γ. We define $\overline{H} := (\mathrm{Sym}(\Lambda_1) \times \cdots \times \mathrm{Sym}(\Lambda_k)) \cap G$.

The set $\overline{\mathcal{L}_G}$ is a poset, with order induced by $\mathcal{L}_G$; for any $H \in \overline{\mathcal{L}_G}$, we will write $\bar{\mu}(H, G)$ instead of $\mu_{\overline{\mathcal{L}_G}}(H, G)$. Note that, in general, $\overline{\mathcal{L}_H} \neq \overline{\mathcal{L}_G} \cap \mathcal{L}_H$ when H is a transitive subgroup of G.

Now let $\mathcal{P}_\Gamma$ be the poset of all partitions of Γ, ordered by refinement; the maximum $\hat{1}$ of $\mathcal{P}_\Gamma$ is $\{\Gamma\}$. The orbit lattice of G is defined as

$$\mathcal{P}_\Gamma(G) := \{\tau \in \mathcal{P}_\Gamma \,|\, \text{the orbits of some } H \leq G \text{ are the parts of } \tau\}.$$

This is a poset with the order induced by the order in $\mathcal{P}_\Gamma$. If $\tau = \{\Lambda_1, \ldots, \Lambda_k\} \in \mathcal{P}_\Gamma(G)$, then we define $G(\tau) := (\mathrm{Sym}(\Lambda_1) \times \cdots \times \mathrm{Sym}(\Lambda_k)) \cap G$. Notice that $G(\tau)$ is the maximal subgroup of G whose orbits are precisely the parts of τ. We note that $H \in \overline{\mathcal{L}_G}$ if and only if there exists $\tau \in \mathcal{P}_\Gamma(G)$ with $H = G(\tau)$; hence there is a bijection between $\overline{\mathcal{L}_G}$ and $\mathcal{P}_\Gamma(G)$.

If $G = \mathrm{Sym}(\Gamma)$ then $\mathcal{P}_\Gamma(G) = \mathcal{P}_\Gamma$, and the elements of $\overline{\mathcal{L}_G}$ are the subgroups $\mathrm{Sym}(\Lambda_1) \times \cdots \times \mathrm{Sym}(\Lambda_k)$, for any partition $\{\Lambda_1, \ldots, \Lambda_k\}$ in $\mathcal{P}_\Gamma$. However if $G = \mathrm{Alt}(\Gamma)$, we have $\mathcal{P}_\Gamma(G) \neq \mathcal{P}_\Gamma$. In fact $\mathrm{Alt}(\Gamma)$ does not contain any subgroup with $|\Gamma| - 1$ orbits on Γ. The closed subgroups of $\mathrm{Alt}(\Gamma)$ are of the form $(\mathrm{Sym}(\Lambda_1) \times \cdots \times \mathrm{Sym}(\Lambda_k)) \cap \mathrm{Alt}(\Gamma)$, for any $\{\Lambda_1, \ldots, \Lambda_k\} \in \mathcal{P}_\Gamma$ with $k \neq |\Gamma| - 1$.

We now introduce the definition of probability related to a finite group. For a finite group G, the *probability* $P_G(t)$ that t randomly chosen elements of G generate G itself is given by the following formula (see Ref. 12)

$$P_G(t) = \sum_{H \leq G} \frac{\mu(H, G)}{|G : H|^t} \tag{2}$$

We may interpolate the previous integer function and define $P_G(s)$ for any complex variable s. By rearranging the addends in Eq. (2) we obtain a

Dirichlet polynomial as follows:

$$P_G(s) := \sum_{n\in\mathbb{N}} \frac{a_n}{n^s} \quad \text{where} \quad a_n := \sum_{|G:H|=n} \mu(H,G).$$

This is called the *probabilistic polynomial* of G.

Given a normal subgroup N of G we define a Dirichlet polynomial $P_{G,N}(s)$ as follows:

$$P_{G,N}(s) := \sum_{n\in\mathbb{N}} \frac{b_n}{n^s} \quad \text{where} \quad b_n := \sum_{\substack{|G:H|=n \\ HN=G}} \mu(H,G).$$

As is shown in Ref. 4, it holds

$$P_G(s) = P_{G/N}(s)P_{G,\,N}(s) \tag{3}$$

By substituting in this equation $G = \text{Sym}(n)$ and $N = \text{Alt}(n)$, for any $n \in \mathbb{N}$, we obtain:

$$\begin{aligned} P_{\text{Sym}(n)}(s) &= P_{\text{Sym}(n)/\,\text{Alt}(n)}(s) \cdot P_{\text{Sym}(n),\,\text{Alt}(n)}(s) = \\ &= P_{C_2}(s) \cdot P_{\text{Sym}(n),\,\text{Alt}(n)}(s), \end{aligned}$$

where C_2 denotes a cyclic group of order 2. Moreover, if we take $G = \text{Alt}(n) \times C_2$ and $N = \text{Alt}(n)$ in Eq. (3), we have:

$$P_{\text{Alt}(n)\times C_2}(s) = P_{C_2}(s) \cdot P_{\text{Alt}(n)}(s).$$

Then, comparing these expressions of $P_{\text{Sym}(n)}(s)$ and $P_{\text{Alt}(n)\times C_2}(s)$, we may conclude that

$$P_{\text{Sym}(n)}(s) = P_{\text{Alt}(n)\times C_2}(s) \quad \text{iff} \quad P_{\text{Sym}(n),\,\text{Alt}(n)}(s) = P_{\text{Alt}(n)}(s) \tag{4}$$

Remark 2.2. Equation (4) will be very useful in our investigation of Conjecture 1.1 by Boston and Mann: in fact instead of proving the validity (or non-validity) of Eq. (1), we will verify that $P_{\text{Sym}(n),\,\text{Alt}(n)}(s)$ is equal (or different) to $P_{\text{Alt}(n)}(s)$.

If G is an almost simple group, i.e. $S \leq G \leq \text{Aut}(S)$ with S a non abelian simple group, and G/S has prime order, then, as it is proved in p.288 of Ref. 8,

$$P_S(s) - P_{G,\,S}(s) = \sum_{H\leq S} \frac{\mu(H,S) + \mu(H,G)}{|G:H|^s}.$$

In particular it holds (see Ref. 8):

Corollary 2.1. *Suppose G an almost simple group, with* $\operatorname{soc} G = S$*, and that G/S has prime order; $P_S(s) = P_{G,S}(s)$ if $\mu(H, S) + \mu(H, G) = 0$ for any $H \leq S$.*

Remark 2.3. We will use this corollary in the next sections with G and S respectively equal to $\mathrm{Sym}(n)$ and $\mathrm{Alt}(n)$.

Now we focalize our attention on maximal subgroups of the alternating and symmetric groups. There exists a classification of these subgroups, to which many authors have contributed: see the O'Nan-Scott Theorem in Ref. 2 and the results by Liebeck, Praeger and Saxl in Ref. 13. We are going to state some basic facts on the maximal subgroups of $\mathrm{Alt}(n)$, and on their normalizers in $\mathrm{Sym}(n)$. From now on we consider $n \geq 5$.

Remark 2.4. Each maximal subgroup M of $\mathrm{Alt}(n)$ is contained in the normalizer $N_{\mathrm{Sym}(n)}(M)$, which is a proper subgroup of $\mathrm{Sym}(n)$, different from $\mathrm{Alt}(n)$. If $N_{\mathrm{Sym}(n)}(M) \neq M$, we can verify that $N_{\mathrm{Sym}(n)}(M)$ is the unique maximal subgroup of $\mathrm{Sym}(n)$, different from $\mathrm{Alt}(n)$, that contains M. In fact, we suppose that N is a maximal subgroup of $\mathrm{Sym}(n)$, different from $\mathrm{Alt}(n)$, and that $N \geq M$; then $\mathrm{Alt}(n)N = \mathrm{Sym}(n)$ and

$$\frac{N}{N \cap \mathrm{Alt}(n)} = \frac{N}{M} \cong \frac{\mathrm{Sym}(n)}{\mathrm{Alt}(n)}.$$

Therefore $|N : M| = 2$ and $M \triangleleft N \leq N_{\mathrm{Sym}(n)}(M)$. But N is maximal in $\mathrm{Sym}(n)$, and then we may conclude $N = N_{\mathrm{Sym}(n)}(M)$. In this situation, we observe: $\mu(M, \mathrm{Alt}(n)) = -1$ and $\mu(M, \mathrm{Sym}(n)) = 1$.
Otherwise, if $N_{\mathrm{Sym}(n)}(M) = M$, we obtain $\mu(M, \mathrm{Alt}(n)) = -1$ and $\mu(M, \mathrm{Sym}(n)) = 0$.

Remark 2.5. If M is an intransitive or imprimitive maximal subgroup of $\mathrm{Alt}(n)$, then $M \neq N_{\mathrm{Sym}(n)}(M)$. Any intransitive maximal subgroup of $\mathrm{Alt}(n)$ is of the form

$$(\mathrm{Sym}(a) \times \mathrm{Sym}(b)) \cap \mathrm{Alt}(n)$$

with $\mathrm{Sym}(a) \times \mathrm{Sym}(b)$ an intransitive maximal subgroup of $\mathrm{Sym}(n)$. We observe that any subgroup isomorphic to $\mathrm{Sym}(a) \times \mathrm{Sym}(b)$ in $\mathrm{Sym}(n)$, with $a + b = n$, is not contained in $\mathrm{Alt}(n)$. Hence the map $M \mapsto N_{\mathrm{Sym}(n)}(M)$ induces a bijection between intransitive maximal subgroups of $\mathrm{Alt}(n)$ and $\mathrm{Sym}(n)$. Any imprimitive maximal subgroup of $\mathrm{Alt}(n)$ is of the form

$$(\mathrm{Sym}(a) \wr \mathrm{Sym}(b)) \cap \mathrm{Alt}(n)$$

where $\mathrm{Sym}(a)\wr\mathrm{Sym}(b)$ is an imprimitive maximal subgroup of $\mathrm{Sym}(n)$. It is easy to prove that any subgroup of $\mathrm{Sym}(n)$ isomorphic to $\mathrm{Sym}(a)\wr\mathrm{Sym}(b)$, with $ab = n$, is not contained in $\mathrm{Alt}(n)$. Then there exists a bijection between maximal imprimitive subgroups of $\mathrm{Alt}(n)$ and $\mathrm{Sym}(n)$: to each maximal imprimitive subgroup of $\mathrm{Alt}(n)$ corresponds its normalizer in $\mathrm{Sym}(n)$.

3. The prime case

We suppose p a prime, with $p \geq 5$. In this section we will prove that $\mathrm{Sym}(p)$ satisfies Conjecture 1.1 by Boston and Mann (see Theorems 3.1 and 3.2). First of all we make some considerations on the form of affine maximal subgroups of $\mathrm{Sym}(p)$ and $\mathrm{Alt}(p)$.

Remark 3.1. Each affine maximal subgroup of $\mathrm{Sym}(p)$ is isomorphic to $C_p \rtimes C_{p-1}$; then its intersection with $\mathrm{Alt}(p)$ is isomorphic to $C_p \rtimes C_{(p-1)/2}$. Any affine maximal subgroup M of $\mathrm{Alt}(p)$ is a subgroup of the form $C_p \rtimes C_{(p-1)/2}$, and it does not coincide with its normalizer in $\mathrm{Sym}(p)$. Moreover we note that any affine subgroup of $\mathrm{Alt}(p)$ is of the form $C_p \rtimes C_t$, with C_t a subgroup of $C_{(p-1)/2}$; we will deal with these subgroups in the next proofs.

By using Theorem 2 and Table I of Ref. 15, we may deduce that if $p \neq 11, 23$ and $p \neq (q^d - 1)/(q - 1)$, for any couple of natural numbers (q, d), with q a prime power and $q > 4$ if $d = 2$, then $\mathrm{Sym}(p)$ contains proper transitive subgroups, different from $\mathrm{Alt}(p)$, only of affine type. On the other hand, if $p = 11$, $p = 23$ or $p = (q^d - 1)/(q - 1) > 5$, then $\mathrm{Sym}(p)$ contains some almost simple proper transitive subgroups, different from $\mathrm{Alt}(p)$, but, as is stated in Corollary 3 of Ref. 15, these subgroups are all contained in $\mathrm{Alt}(p)$.

Then, for any value of p, it results that $\mathrm{Sym}(p)$ contains only maximal transitive subgroups, different from $\mathrm{Alt}(p)$, of affine type. Denote by $\mathcal{M}_{\mathrm{Sym}(p)}$ the set of all the intransitive and affine maximal subgroups of $\mathrm{Sym}(p)$, and by $\mathcal{M}_{\mathrm{Alt}(p)}$ the set of the intransitive and affine maximal subgroups of $\mathrm{Alt}(p)$. The set $\mathcal{M}_{\mathrm{Sym}(p)}$ coincides with the set of all maximal subgroups of $\mathrm{Sym}(p)$, different from $\mathrm{Alt}(p)$, for any p prime. Moreover, if M is an affine or an intransitive maximal subgroup of $\mathrm{Alt}(p)$, then, by Remark 2.5 and Remark 3.1, $M \neq N_{\mathrm{Sym}(p)}(M)$. Hence there exists a bijection from $\mathcal{M}_{\mathrm{Alt}(p)}$ to $\mathcal{M}_{\mathrm{Sym}(p)}$, namely $M \to N_{\mathrm{Sym}(p)}(M)$.

Lemma 3.1. *Let $H < \mathrm{Alt}(p)$, $H \neq \langle 1 \rangle$, such that if N is a maximal subgroup of $\mathrm{Alt}(p)$ with $H \leq N$, then $N \in \mathcal{M}_{\mathrm{Alt}(p)}$. It holds*

$$\mu(H, \mathrm{Sym}(p)) = -\mu(H, \mathrm{Alt}(p)).$$

Proof. If H is not an intersection of elements of $\mathcal{M}_{\mathrm{Alt}(p)}$ then, by the bijection between $\mathcal{M}_{\mathrm{Sym}(p)}$ and $\mathcal{M}_{\mathrm{Alt}(p)}$, it results that H is also not an intersection of elements of $\mathcal{M}_{\mathrm{Sym}(p)}$; hence it follows $\mu(H, \mathrm{Alt}(p)) = \mu(H, \mathrm{Sym}(p)) = 0$. Then we have to calculate $\mu(H, \mathrm{Alt}(p))$ and $\mu(H, \mathrm{Sym}(p))$ when H is an intersection of elements of $\mathcal{M}_{\mathrm{Alt}(p)}$.

We consider $H \in \mathcal{M}_{\mathrm{Alt}(p)}$ a maximal subgroup of $\mathrm{Alt}(p)$; H is obtained as the intersection between $N_{\mathrm{Sym}(p)}(H)$ and $\mathrm{Alt}(p)$. As we have already observed in Remark 2.4, we obtain $\mu(H, \mathrm{Sym}(p)) = -\mu(H, \mathrm{Alt}(p))$. We suppose now H is an intersection of maximal subgroups in $\mathcal{M}_{\mathrm{Alt}(p)}$, but H is not maximal in $\mathrm{Alt}(p)$. In other words H is an intersection of some maximal affine or intransitive subgroups of $\mathrm{Sym}(p)$ with $\mathrm{Alt}(p)$. We observe that the intersection of two maximal affine subgroups of $\mathrm{Sym}(p)$ is intransitive, and then it follows that H is intransitive. Applying the Crapo's Closure Theorem to $\mathcal{L}_{\mathrm{Sym}(p)}$, we obtain

$$\sum_{\substack{T \leq \mathrm{Sym}(p) \\ \bar{T} = \mathrm{Sym}(p)}} \mu(H, T) = \begin{cases} \bar{\mu}(H, \mathrm{Sym}(p)) & \text{if } H \text{ is closed in } \mathcal{L}_{\mathrm{Sym}(p)} \\ 0 & \text{otherwise.} \end{cases} \tag{5}$$

If there does not exist any transitive subgroup $T \notin \{\mathrm{Alt}(p), \mathrm{Sym}(p)\}$ such that $H \leq T$, then H has the form $(\mathrm{Sym}(\Omega_1) \times \cdots \times \mathrm{Sym}(\Omega_k)) \cap \mathrm{Alt}(p)$ for some partition $\{\Omega_1, \cdots, \Omega_k\}$ of $\{1, \ldots, p\}$, with $2 < k < p-1$ (in fact $H \neq \langle 1 \rangle$). We observe that H is not closed in $\mathcal{L}_{\mathrm{Sym}(p)}$; from Eq. (5)

$$\mu(H, \mathrm{Sym}(p)) + \mu(H, \mathrm{Alt}(p)) = 0.$$

Then we may suppose that there exists at least one transitive subgroup $T \notin \{\mathrm{Alt}(p), \mathrm{Sym}(p)\}$ such that $H \leq T$. We have $T \leq A$, with A a maximal affine subgroup of $\mathrm{Sym}(p)$; so H is an intransitive subgroup of A. It holds $A = P \rtimes K = N_{\mathrm{Sym}(p)}(P)$, for some $P \cong C_p$ and $K \cong C_{p-1}$; then H is a cyclic subgroup of order dividing $p-1$. By hypothesis, $H \neq \langle 1 \rangle$, and then H is not closed in $\mathcal{L}_{\mathrm{Sym}(p)}$. From Eq. (5)

$$\mu(H, \mathrm{Sym}(p)) = -\mu(H, \mathrm{Alt}(p)) - \sum_{\substack{T \text{ trans.} \\ H \leq T < \mathrm{Sym}(p) \\ T \neq \mathrm{Alt}(p)}} \mu(H, T) \tag{6}$$

The subgroup P has p complements in A all pairwise disjoint. Then H is contained only in one complement of P; without loss of generality we may assume $H \leq K$.

Denote by t the number of the maximal affine subgroups of $\mathrm{Sym}(p)$ containing H. We recall that the intersection of two maximal affine subgroups is intransitive; then any transitive subgroup T that appears in Eq. (6), is contained in only one maximal affine subgroup of $\mathrm{Sym}(p)$. It follows that

$$\sum_{\substack{T \text{ trans.} \\ H \leq T < \mathrm{Sym}(p) \\ T \neq \mathrm{Alt}(p)}} \mu(H,T) = t \cdot \sum_{\substack{T \text{ trans.} \\ H \leq T \leq A}} \mu(H,T) = t \cdot \sum_{H \leq R \leq K} \mu(H, PR).$$

Fix R; we proceed to calculate $\mu(H, PR)$. We consider the lattice L of subgroups of PR containing H: in this lattice $\hat{0} = H$ and $\hat{1} = PR$. We notice that the set $(PH)^{\perp} = \{X \in L \,|\, X \cap PH = H \text{ and } \langle X, PH\rangle = PR\}$ of the complements to PH in L is reduced to the element R; applying the Crapo's Complement Theorem we obtain:

$$\mu(H,PR) = \sum_{\bar{R} \in (PH)^{\perp}} \mu(H,\bar{R}) \cdot \mu(\bar{R}, PR) = \mu(H,R) \cdot \mu(R, PR).$$

We are supposing $H \neq \langle 1 \rangle$; then $R \neq \langle 1 \rangle$ and R is a maximal subgroup of PR. Hence $\mu(R, PR) = -1$ and

$$\mu(H, PR) = -\mu(H, R).$$

$H \lhd R$, so $\mu(H,R) = \mu(\langle 1 \rangle, R/H)$. Let $|H| = h$ and $|R| = r$; R/H is cyclic, and then, by Lemma 2.1, it follows that

$$\mu(H, PR) = -\mu(\langle 1 \rangle, R/H) = -\mu(r/h).$$

We obtain that

$$\sum_{\substack{T \text{ trans.} \\ H \leq T < \mathrm{Sym}(p) \\ T \neq \mathrm{Alt}(p)}} \mu(H,T) = -t \cdot \sum_{h|r \text{ and } r|p-1} \mu(r/h).$$

Set $\bar{r} = r/h$, for any r such that $h|r$ and $r|p-1$; then we observe that $\sum_{h|r \text{ and } r|p-1} \mu(r/h) = \sum_{\bar{r}|((p-1)/h)} \mu(\bar{r})$. Moreover $(p-1)/h \neq 1$, because $H \leq \mathrm{Alt}(p)$, and so, as observed in Remark 2.1, it follows that

$$\sum_{\bar{r}|((p-1)/h)} \mu(\bar{r}) = 0.$$

Then, from Eq. (6), we may conclude $\mu(H, \mathrm{Sym}(p)) + \mu(H, \mathrm{Alt}(p)) = 0$. □

Suppose $p \neq 11, 23$ and $p \neq (q^d - 1)/(q-1)$, with q a prime power and $q > 4$ if $d = 2$. Then $\mathrm{Alt}(p)$ does not contain any almost simple proper transitive subgroup; each maximal subgroup M of $\mathrm{Alt}(p)$ is intransitive or affine (see Ref. 15). Then $\mathcal{M}_{\mathrm{Alt}(p)}$ is the set of all maximal subgroups of $\mathrm{Alt}(p)$, and any maximal subgroup of $\mathrm{Alt}(p)$ does not coincide with its normalizer in $\mathrm{Sym}(p)$. Hence to verify the validity of Conjecture 1.1, we have to prove Eq. (1), that is equivalent to verify the following statement (see Remark 2.2):

Theorem 3.1. *Let $p \geq 5$ prime, with $p \neq 11, 23$ and $p \neq (q^d-1)/(q-1)$, for any couple of natural numbers (q, d), with q a prime power and $q > 4$ if $d = 2$. Then*

$$P_{\mathrm{Sym}(p),\,\mathrm{Alt}(p)}(s) = P_{\mathrm{Alt}(p)}(s).$$

Proof. Applying Corollary 2.1, we observe that it suffices to prove the equality $\mu(H, \mathrm{Sym}(p)) = -\mu(H, \mathrm{Alt}(p))$ for any $H \leq \mathrm{Alt}(p)$. We recall that, with our hypothesis, $\mathrm{Alt}(p)$ has only affine and intransitive maximal subgroups; then, using Lemma 3.1, it remains to prove this equality when $H = \mathrm{Alt}(p)$ and $H = \langle 1 \rangle$.

If $H = \mathrm{Alt}(p)$, then $\mu(\mathrm{Alt}(p), \mathrm{Sym}(p)) = -\mu(\mathrm{Alt}(p), \mathrm{Alt}(p))$: in fact $\mathrm{Alt}(p)$ is a maximal subgroup of $\mathrm{Sym}(p)$ and $\mu(\mathrm{Alt}(p), \mathrm{Sym}(p)) = -1$, from the definition of μ.

Let now $H = \langle 1 \rangle$. H is closed in $\mathcal{L}_{\mathrm{Sym}(p)}$; then from Eq. (5) we have

$$\sum_{\substack{T \leq \mathrm{Sym}(p) \\ \bar{T} = \mathrm{Sym}(p)}} \mu(\langle 1 \rangle, T) = \bar{\mu}(\langle 1 \rangle, \mathrm{Sym}(p)).$$

It is a known result that $\bar{\mu}(\langle 1 \rangle, \mathrm{Sym}(p)) = (-1)^{p-1}(p-1)! = (p-1)!$ (see for example Ref. 17, p.128). Then

$$\sum_{T \text{ trans., } T \leq \mathrm{Sym}(p)} \mu(\langle 1 \rangle, T) = (p-1)!,$$

and so

$$\mu(\langle 1 \rangle, \mathrm{Sym}(p)) + \mu(\langle 1 \rangle, \mathrm{Alt}(p)) = -\sum_{\substack{T \text{ trans., } T \leq \mathrm{Sym}(p) \\ T \neq \mathrm{Sym}(p) \\ T \neq \mathrm{Alt}(p)}} \mu(\langle 1 \rangle, T) + (p-1)!.$$

T is a transitive subgroup of $\mathrm{Sym}(p)$ and then, by hypothesis, T is contained in a maximal affine subgroup. The number of the maximal affine subgroups of $\mathrm{Sym}(p)$ is equal to the number of the subgroups of $\mathrm{Sym}(p)$ with order p; it is $(p-2)!$. These maximal subgroups are all conjugated in $\mathrm{Sym}(p)$. Then, without loss of generality, we consider one of these maximal subgroups, that we denote by A, and we obtain:

$$\sum_{\substack{T \text{ trans.}, T\leq \mathrm{Sym}(p)\\ T\neq \mathrm{Sym}(p)\\ T\neq \mathrm{Alt}(p)}} \mu(\langle 1\rangle, T) = (p-2)! \cdot \sum_{\substack{T \text{ trans.}\\ T\leq A}} \mu(\langle 1\rangle, T) \tag{7}$$

Let $A = P \rtimes K$, with $P \cong C_p$ and $K \cong C_{p-1}$; hence we may consider $T = P \rtimes R$, with $R \leq K$. Then

$$\sum_{\substack{T \text{ trans.}\\ T\leq A}} \mu(\langle 1\rangle, T) = \sum_{PR\leq A} \mu(\langle 1\rangle, PR) = \sum_{R\leq K} \mu(\langle 1\rangle, PR).$$

If $R = \langle 1\rangle$, obviously $\mu(\langle 1\rangle, P) = -1$. Fix now $R \neq \langle 1\rangle$; we observe that the p complements to P in $\mathcal{L}_{PR}$ are pairwise disjoint. We apply the Crapo's Complement Theorem to the lattice $\mathcal{L}_{PR}$:

$$\begin{aligned}\mu(\langle 1\rangle, PR) &= \sum_{\bar{R}\in P^{\perp}} \mu(\langle 1\rangle, \bar{R}) \cdot \mu(\bar{R}, PR) = \\ &= p \cdot \mu(\langle 1\rangle, R) \cdot (-1) = -p \cdot \mu(\langle 1\rangle, R).\end{aligned}$$

Let $|R| = r$; R is cyclic, then, by Lemma 2.1, it follows that $\mu(\langle 1\rangle, R) = \mu(r)$. Hence, if $R \neq \langle 1\rangle$, $\mu(\langle 1\rangle, PR) = -p \cdot \mu(r)$.
Then we obtain:

$$\sum_{R\leq K} \mu(\langle 1\rangle, PR) = -1 - p \cdot \sum_{\substack{r|p-1\\ r\neq 1}} \mu(r).$$

It holds that $\sum_{\substack{r|p-1\\ r\neq 1}} \mu(r) = \underbrace{\sum_{r|p-1} \mu(r)}_{=0} - \mu(1) = -\mu(1) = -1$, and then

$$\sum_{R\leq K} \mu(\langle 1\rangle, PR) = -1 + p.$$

From Eq. (7)

$$\sum_{\substack{T \text{ trans., } T \leq \mathrm{Sym}(p) \\ T \neq \mathrm{Sym}(p) \\ T \neq \mathrm{Alt}(p)}} \mu(\langle 1 \rangle, T) = (p-2)! \cdot (p-1) = (p-1)!$$

and so we may conclude $\mu(\langle 1 \rangle, \mathrm{Sym}(p)) + \mu(\langle 1 \rangle, \mathrm{Alt}(p)) = 0$. □

So we have proved that, if $p \neq 11, 23$ and $p \neq (q^d - 1)/(q-1)$ (with $q > 4$ if $d = 2$), $\mathrm{Sym}(p)$ satisfies Conjecture 1.1 by Boston and Mann. We are going to analyze the remaining cases. If $p = 11$, $p = 23$ or $p = (q^d - 1)/(q-1) > 5$, with q a prime power, then $\mathrm{Alt}(p)$ contains some almost simple maximal subgroups which coincide with their normalizers in $\mathrm{Sym}(p)$ (see Corollary 3 and Table I in Ref. 15). We verify that Conjecture 1.1 holds in all these cases, that is Eq. (1) is not true; by Remark 2.2 we know that it is equivalent to prove the following statement.

Theorem 3.2. *If $p = 11$, $p = 23$ or $p = (q^d - 1)/(q - 1) > 5$ (with q a prime power), then*

$$P_{\mathrm{Sym}(p),\, \mathrm{Alt}(p)}(s) \neq P_{\mathrm{Alt}(p)}(s).$$

Proof. We recall (see Section 2) that

$$P_{\mathrm{Alt}(p)}(s) - P_{\mathrm{Sym}(p),\, \mathrm{Alt}(p)}(s) = \sum_{H \leq \mathrm{Alt}(p)} \frac{\mu(H, \mathrm{Alt}(p)) + \mu(H, \mathrm{Sym}(p))}{|\,\mathrm{Sym}(p) : H|^s} \tag{8}$$

Let $p \in \{11, 23\}$. By Corollary 3 in Ref. 15, we know that, if $p = 11$ or $p = 23$, the only almost simple maximal subgroups of $\mathrm{Alt}(p)$ are those isomorphic to the Mathieu group M_p; these subgroups coincide with their normalizer in $\mathrm{Sym}(p)$. We consider all the subgroups H of $\mathrm{Alt}(p)$ having the same order as M_p; we want to calculate the following term of the sum in Eq. (8)

$$\sum_{\substack{H \leq \mathrm{Alt}(p) \\ |H| = |\,\mathrm{M}_p\,|}} \frac{\mu(H, \mathrm{Alt}(p)) + \mu(H, \mathrm{Sym}(p))}{(|\,\mathrm{Sym}(p)|/|\,\mathrm{M}_p\,|)^s} \tag{9}$$

Suppose H is such that if N is a maximal subgroup of $\mathrm{Alt}(p)$ and $H \leq N$, then $N \in \mathcal{M}_{\mathrm{Alt}(p)}$; in other words, H is contained only in affine or intransitive maximal subgroups of $\mathrm{Alt}(p)$. Hence, by Lemma 3.1, we may conclude that $\mu(H,\mathrm{Alt}(p))+\mu(H,\mathrm{Sym}(p))=0$.

Then we may assume that H is contained in at least one almost simple maximal subgroup of $\mathrm{Alt}(p)$; hence $H\le M$, with M isomorphic to M_p. But $|H|=|\mathrm{M}_p|=|M|$, and then H has to coincide with the maximal subgroup M. We deduce that $H=N_{\mathrm{Sym}(p)}(H)$ and, as observed in Remark 2.4, it follows that $\mu(H,\mathrm{Alt}(p))+\mu(H,\mathrm{Sym}(p))=-1$. Denote by m the number of maximal subgroups of $\mathrm{Alt}(p)$ isomorphic to M_p; from Eq. (9) we have

$$\sum_{\substack{H\le \mathrm{Alt}(p)\\ |H|=|\mathrm{M}_p|}} \frac{\mu(H,\mathrm{Alt}(p))+\mu(H,\mathrm{Sym}(p))}{(|\mathrm{Sym}(p)|/|\mathrm{M}_p|)^s} = -\frac{m}{(|\mathrm{Sym}(p)|/|\mathrm{M}_p|)^s}\neq 0.$$

Hence the term in Eq. (8) with denominator equal to $(|\mathrm{Sym}(p)|/|\mathrm{M}_p|)^s$ is different from 0; it follows that

$$P_{\mathrm{Alt}(p)}(s)-P_{\mathrm{Sym}(p),\,\mathrm{Alt}(p)}(s)\neq 0.$$

We suppose now that $p=(q^d-1)/(q-1)>5$ for some $q,d\in\mathbb{N}$, with q a prime power; we consider the set $\mathcal{Q}$ of couples (d,q) with $(q^d-1)/(q-1)=p$. For any $(d,q)\in\mathcal{Q}$, $\mathrm{Alt}(p)$ contains some maximal transitive subgroups isomorphic to $\mathrm{P\Gamma L}(d,q)$ that coincide with their normalizers in $\mathrm{Sym}(p)$. The set $\mathcal{Q}$ is finite, hence we can consider $(\hat{d},\hat{q})\in\mathcal{Q}$ such that $|\mathrm{P\Gamma L}(\hat{d},\hat{q})|\ge |\mathrm{P\Gamma L}(d,q)|$, for any $(q,d)\in\mathcal{Q}$. We proceed in an analogous way to the cases $p=11,23$ to prove $P_{\mathrm{Alt}(p)}(s)\neq P_{\mathrm{Sym}(p),\,\mathrm{Alt}(p)}(s)$. We consider all the subgroups H of $\mathrm{Alt}(p)$ having the same order as $\mathrm{P\Gamma L}(\hat{d},\hat{q})$; we want to calculate the following term of the sum in Eq. (8):

$$\sum_{\substack{H\le \mathrm{Alt}(p)\\ |H|=|\mathrm{P\Gamma L}(\hat{d},\hat{q})|}} \frac{\mu(H,\mathrm{Alt}(p))+\mu(H,\mathrm{Sym}(p))}{(|\mathrm{Sym}(p)|/|\mathrm{P\Gamma L}(\hat{d},\hat{q})|)^s} \tag{10}$$

Let H be such that if N is a maximal subgroup of $\mathrm{Alt}(p)$ and $H\le N$, then $N\in\mathcal{M}_{\mathrm{Alt}(p)}$; by Lemma 3.1, we conclude $\mu(H,\mathrm{Alt}(p))+\mu(H,\mathrm{Sym}(p))=0$.

Then we may assume that H is contained in at least one almost simple maximal subgroup of $\mathrm{Alt}(p)$; hence $H\le\mathrm{P\Gamma L}(d,q)$, for some $(d,q)\in\mathcal{Q}$. But we are supposing $|H|=|\mathrm{P\Gamma L}(\hat{d},\hat{q})|$, and $|\mathrm{P\Gamma L}(\hat{d},\hat{q})|$ is maximal between the orders of the maximal almost simple subgroups of $\mathrm{Alt}(p)$; therefore, if $H\le\mathrm{P\Gamma L}(d,q)$, it follows that $|\mathrm{P\Gamma L}(d,q)|=|\mathrm{P\Gamma L}(\hat{d},\hat{q})|$, and H has to coincide with $\mathrm{P\Gamma L}(d,q)$. So we obtain that $H=N_{\mathrm{Sym}(p)}(H)$ and, by Remark 2.4, $\mu(H,\mathrm{Alt}(p))+\mu(H,\mathrm{Sym}(p))=-1$. Denote by γ the number of maximal subgroups of $\mathrm{Alt}(p)$ with order equal to $|\,\mathrm{P\Gamma L}(\hat{d},\hat{q})|$; from Eq. (10) we have

$$\sum_{\substack{H\leq \mathrm{Alt}(p)\\ |H|=|\,\mathrm{P\Gamma L}(\hat{d},\hat{q})|}} \frac{\mu(H,\mathrm{Alt}(p))+\mu(H,\mathrm{Sym}(p))}{(|\,\mathrm{Sym}(p)|/|\,\mathrm{P\Gamma L}(\hat{d},\hat{q})|)^s} = -\frac{\gamma}{(|\,\mathrm{Sym}(p)|/|\,\mathrm{P\Gamma L}(\hat{d},\hat{q})|)^s} \neq 0.$$

It follows that $P_{\mathrm{Alt}(p)}(s) - P_{\mathrm{Sym}(p),\,\mathrm{Alt}(p)}(s) \neq 0$. □

4. Some counterexamples

We have shown that Conjecture 1.1 is true when n is a prime. In this section we will show that this conjecture does not hold in general for $n \geq 5$.

It is very difficult to establish whether $\mathrm{Sym}(n)$, for a generic n not prime, satisfies the conjecture or not, because of the heterogeneity in the behavior of maximal subgroups of $\mathrm{Alt}(n)$ and $\mathrm{Sym}(n)$. We are going to explain more in details how these difficulties arise.

Suppose that any maximal subgroup of $\mathrm{Alt}(n)$ does not coincide with its normalizer in $\mathrm{Sym}(n)$; this condition is necessary but not sufficient to guarantee the existence of a bijection between the set of the maximal subgroups of $\mathrm{Alt}(n)$ and the set of the maximal subgroups of $\mathrm{Sym}(n)$, different from $\mathrm{Alt}(n)$. In fact we know that there exists a one-to-one correspondence between the maximal subgroups of $\mathrm{Alt}(n)$ and $\mathrm{Sym}(n)$ of types intransitive and imprimitive (see Remark 2.5), but $\mathrm{Sym}(n)$ can contain a maximal primitive subgroup, different from $\mathrm{Alt}(n)$, that intersects $\mathrm{Alt}(n)$ in a subgroup which is not maximal in $\mathrm{Alt}(n)$. In this situation the Möbius numbers in $\mathrm{Alt}(n)$ and in $\mathrm{Sym}(n)$ of this subgroup can be not opposite, and it could follow that $P_{\mathrm{Alt}(n)}(s) \neq P_{\mathrm{Sym}(n),\,\mathrm{Alt}(n)}(s)$.

Moreover we note that, even if n is chosen such that the map $M \mapsto M \cap \mathrm{Alt}(n)$ induces a bijection between the set $\mathcal{S}$ of maximal supplements of $\mathrm{Alt}(n)$ in $\mathrm{Sym}(n)$ and the set $\mathcal{A}$ of maximal subgroups of $\mathrm{Alt}(n)$, it is not clear whether and why this could imply $P_{\mathrm{Alt}(n)}(s) = P_{\mathrm{Sym}(n),\mathrm{Alt}(n)}(s)$. The problem is that there is no hope that this bijection can be extended to a bijection between the subgroups that can be obtained as intersections of elements of $\mathcal{A}$ and those that are intersections of elements of $\mathcal{S}$.

In particular it is very difficult to study what happens when we consider the imprimitive subgroups. Indeed, it is possible that the intersection of few (even two) imprimitive maximal subgroups of $\mathrm{Sym}(n)$ is already contained in $\mathrm{Alt}(n)$.

Remark 4.1. The case $n = 21$ is very interesting, because it realizes the situations just explained: in fact we will prove that there is not a bijection between the maximal subgroups of Alt(21) and the maximal supplements of Alt(21) in Sym(21), and two maximal imprimitive subgroups of Sym(21) can have intersection in Alt(21). In particular we are going to verify that $n = 21$ is a counterexample to Conjecture 1.1.

4.1. $n = 21$

We denote by G a primitive subgroup of Sym(21) isomorphic to PGL(2, 7): for example let $G = \langle(2,4)(5,6)(7,9)(10,11)(12,16)(14,20)(15,19)(17,18),$ $(1,14,17,9,13,3)(2,5,10,19,16,4)(6,7,21,8,20,12)(11,15,18)\rangle$. All primitive subgroups of Sym(21) are known (see Ref. 9); from this list we may deduce that G is a maximal subgroup of Sym(21). Let D be the derived subgroup of G; hence $D \cong \mathrm{PSL}(2,7)$, and D is transitive on $\{1,\ldots,21\}$. Notice that $D = G \cap \mathrm{Alt}(21)$; we will prove that D is not maximal in Alt(21) and it corresponds to the intersection of two maximal imprimitive subgroups of Sym(21). Moreover we will verify that

$$\mu(D, \mathrm{Alt}(21)) \neq -\mu(D, \mathrm{Sym}(21)) \tag{11}$$

and that this fact produces the failure of Conjecture 1.1.

D has two distinct systems of imprimitivity B_1 and B_2 on $\{1,\ldots,21\}$; each of these has 7 blocks, with any block having cardinality 3. We can easily compute B_1 and B_2 with GAP; we obtain:

$$B_1 = \{[1,14,20],[2,10,16],[3,13,21],[4,11,12],[5,7,18],[6,9,17],[8,15,19]\},$$
$$B_2 = \{[1,3,8],[2,6,15],[4,5,19],[7,9,13],[10,11,21],[12,14,17],[16,18,20]\}.$$

Then D is contained in two maximal imprimitive subgroups of Sym(21) isomorphic to $\mathrm{Sym}(3) \wr \mathrm{Sym}(7)$, that we denote by I and J; they are the only maximal imprimitive subgroups of Sym(21) containing D. Moreover we obtain that $D = I \cap J = I \cap J \cap \mathrm{Alt}(21)$.

We have now to determine the maximal primitive subgroups of Alt(21) and Sym(21) containing D. It is easy to prove that D is contained in two maximal subgroups of Sym(21) isomorphic to Sym(7); in particular it corresponds to the intersection of these two subgroups. Obviously it follows that D is also contained in two maximal subgroups of Alt(21) isomorphic to Alt(7). We have already observed that D is contained in one maximal subgroup isomorphic to PGL(2, 7), that is G. So we are able to represent the lattice of subgroups of Sym(21) containing D:

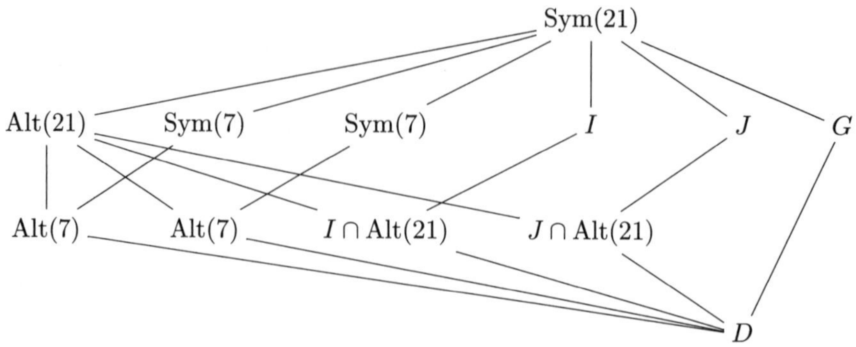

Note that there is not a bijection between maximal subgroups of Alt(21) and maximal supplements of Alt(21) in Sym(21). From the diagram we may calculate the Möbius numbers of D in Alt(21) and in Sym(21):

$$\mu(D, \mathrm{Alt}(21)) = 3 \quad \text{and} \quad \mu(D, \mathrm{Sym}(21)) = 1.$$

Then the inequality in Eq. (11) holds. We recall that

$$P_{\mathrm{Alt}(21)}(s) - P_{\mathrm{Sym}(21),\,\mathrm{Alt}(21)}(s) = \sum_{H \leq \mathrm{Alt}(21)} \frac{\mu(H, \mathrm{Alt}(21)) + \mu(H, \mathrm{Sym}(21))}{|\,\mathrm{Sym}(21) : H|^s} \tag{12}$$

By considering all subgroups H of Alt(21) with the same order as D, that is 168, we obtain that:

$$\sum_{\substack{H \leq \mathrm{Alt}(21) \\ |H| = |D|}} \frac{\mu(H, \mathrm{Alt}(21)) + \mu(H, \mathrm{Sym}(21))}{(|\,\mathrm{Sym}(21)|/|D|)^s} \neq 0$$

Hence the term in Eq. (12) with denominator equal to $(|\,\mathrm{Sym}(21)|/|D|)^s$ is different from 0; it follows that

$$P_{\mathrm{Alt}(21)}(s) \neq P_{\mathrm{Sym}(21),\,\mathrm{Alt}(21)}(s).$$

But we may verify that any maximal subgroup of Alt(21) does not coincide with its normalizer in Sym(21); then we conclude that Conjecture 1.1 by Boston and Mann does not hold if $n = 21$.

Remark 4.2. In the case $n = 21$ there does not exist a one-to-one correspondence between maximal subgroups of Alt(21) and maximal supplements of Alt(21) in Sym(21); this fact causes the failure of Conjecture 1.1.

We show now an example ($n = 62$) in which this bijective correspondence exists; but also in this case the conjecture fails.

4.2. $n = 62$

There exists a bijection between the maximal subgroups of Alt(62) and the maximal subgroups of Sym(62) different from Alt(62); this bijection maps each maximal subgroup M of Sym(62) to $M \cap \text{Alt}(62)$. This implies, in particular, that any maximal subgroup of Alt(62) does not coincide with its normalizer in Sym(62).

Using a result of Shareshian (see Corollary 4.14 in Ref. 16), we know that

$$\mu(\langle 1\rangle, \text{Sym}(62)) + \mu(\langle 1\rangle, \text{Alt}(62)) = -62!.$$

We recall that

$$P_{\text{Alt}(62)}(s) - P_{\text{Sym}(62),\,\text{Alt}(62)}(s) = \sum_{H \leq \text{Alt}(62)} \frac{\mu(H, \text{Alt}(62)) + \mu(H, \text{Sym}(62))}{|\,\text{Sym}(62) : H|^s}.$$

Therefore the term in this sum with denominator equal to $|\,\text{Sym}(62)|^s$ is different from zero, and then

$$P_{\text{Alt}(62)}(s) \neq P_{\text{Sym}(62),\,\text{Alt}(62)}(s).$$

Hence Conjecture 1.1 is not true.

5. Open problem

Note that Sym(62) contains some primitive proper subgroups different from Alt(62). So the natural next step in our investigation is to consider symmetric groups that do not contain primitive subgroups different from Alt(n). By Cameron (see Ref. 5), we know that the infinite family $\overline{\mathcal{F}}$ of these groups has density 1 in the set of all symmetric groups. Moreover we notice that for each $\text{Sym}(n) \in \overline{\mathcal{F}}$ there exists a bijective correspondence between maximal subgroups of Alt(n) and maximal supplements of Alt(n) in Sym(n).

In this case we ask if Sym(n) satisfies the conjecture of Boston and Mann or not, and if all the symmetric groups in $\mathcal{F}$ have the same behavior. We have not yet obtained an answer to this question.

We think that the possibility of expressing subgroups of the alternating group as intersections of two (or few) imprimitive maximal subgroups of the symmetric group, could imply the existence of a counterexample to Conjecture 1.1, even among the groups in $\mathcal{F}$. On the other hand, different considerations and partial results seem to indicate that if such a counterexample exists, then the degree must be quite large.

References

1. G.E. Andrews, *Number theory*, Saunders, Philadelphia, Pa., 1971.
2. M. Aschbacher and L.L. Scott, Maximal subgroups of finite groups, *Journal of Algebra* **92** (1985), 44-80.
3. N. Boston, A probabilistic generalization of the Riemann zeta function, *Analytic Number Theory vol. 1 (Allerton Park, IL, 1995)***138** (1996), 155-162.
4. K.S. Brown, The coset poset and probabilistic zeta function of a finite group, *Journal of Algebra* **225** (2000), no. 2, 989-1012.
5. P.J. Cameron, *Permutation groups*, London Mathematical Society Student Texts 45, Cambridge University Press, Cambridge, 1999.
6. H. Crapo, The Möbius function of a lattice, *J. Combin. Theory* **1** (1966), 126-134.
7. H. Crapo, Möbius inversion in lattices, *Arch. Math. (Basel)* **19** (1968), 595-607.
8. F. DallaVolta, A. Lucchini and F. Morini, Some remarks on the probability of generating an almost simple group, *Glasgow Math. J.***45** (2003), 281-291.
9. J.B. Dixon and B. Mortimer, *Permutation groups*, Graduate Texts in Mathematics vol.163, Springer, New York, 1996.
10. M. Downs, The Möbius function of PSL$(2, q)$, with application to the maximal normal subgroups of the modular group, *J. London Math. Soc.* **43** (1991), no. 2, 61-75.
11. The GAP Group, *GAP-Groups, Algorithms, and Programming*, Version 4.4.10 (2007).
12. P. Hall, The Eulerian functions of a group, *Quart. J. Math.* **7** (1936), 134-151.
13. M.W. Liebeck, C.E. Praeger and J.Saxl, A classification of the maximal subgroups of the finite alternating and symmetric groups., *Journal of Algebra* **111** (1987), 365-383.
14. A. Lucchini and M. Massa, The probabilistic zeta function of alternating and symmetric groups, *Algebra Colloquium* **16** (2009), no. 2, 195-216.
15. P.P. Pálfy, On Feit's examples of intervals in subgroup lattices, *Journal of Algebra* **116** (1988), 471-479.
16. J. Shareshian, On the Möbius number of the subgroup lattice of the symmetric group, *Journal of Combinatorial Theory, Series A* **78** (1997), 236-267.
17. R.P. Stanley, *Enumerative combinatorics. Vol.1*, Cambridge Studies in Advanced Mathematics vol.49, Cambridge University Press, Cambridge, 1997.

MARKOV'S PROBLEMS THROUGH THE LOOKING GLASS OF ZARISKI AND MARKOV TOPOLOGIES

D. DIKRANJAN* and D. TOLLER**

Dipartimento di Matematica e Informatica, Università di Udine,
Via delle Scienze 206, 33100 Udine, Italy
**E-mail: dikranja@dimi.uniud.it*
***E-mail: daniele.toller@uniud.it*

Dedicated to the 50-th anniversary of Professor Dmitri Shakhmatov

Markov posed a series of challenging problems on infinite groups. We discuss a unified approach to four of them, based on the extensive use of several topologies that allows for their better interpretation, and provides an efficient tool for their solution in many cases. The backbone of this approach is a chain $\mathfrak{Z}_G \subseteq \mathfrak{M}_G \subseteq \mathfrak{P}_G$ of three topologies (the Zariski topology, the Markov topology and the precompact Markov topology), that every group G carries. Here we propose also a chain of subsequent refinements $P_\kappa \mathfrak{Z}_G$ of the Zariski topology, which allows for an easier presentation and understanding of Hesse's solution of the first Markov problem (concerning the equality $\mathfrak{Z}_G = \mathfrak{M}_G$). In the case of an abelian group G, the topology $\mathfrak{P}_G$ is used to "approximate" the Zariski topology $\mathfrak{Z}_G$ in the proof of one of the main theorems, establishing the coincidence of all three topologies for abelian groups. On the other hand, the equality $\mathfrak{M}_G = \mathfrak{P}_G$ turns out to be a criterion for a soluble divisible group G to be abelian.

Keywords: Markov topology, Zariski topology; Markov's problems; (Elementary) Algebraic subset; Potentially dense subset; (Non-) Topologizable group; CZ-group; Infinite permutation group; van der Waerden group; Heisenberg group.

1. Introduction

1.1. *The Zariski rod of a group*

Markov introduced five special families of subsets of a group G:

Definition 1.1. (Ref. 45) A subset X of a group G is called:

(a) *elementary algebraic* if there exist an integer $n > 0$, elements

$g_1, \ldots, g_n \in G$ and $\varepsilon_1, \ldots, \varepsilon_n \in \mathbb{Z}$, such that

$$X = \{x \in G : g_1 x^{\varepsilon_1} g_2 x^{\varepsilon_2} \cdots g_n x^{\varepsilon_n} = e_G\};$$

(b) *additively algebraic* if X is a finite union of elementary algebraic subsets of G;
(c) *algebraic* if X is an intersection of additively algebraic subsets of G;
(d) *unconditionally closed* if X is closed in *every* Hausdorff group topology on G;
(e) *potentially dense* if G admits *some* Hausdorff group topology $\mathcal{T}$ such that X is dense in $(G, \mathcal{T})$.

The family of algebraic subsets of a group G coincides with the family of closed subsets of a T_1 topology $\mathfrak{Z}_G$ on G, called the *Zariski topology* (Refs. 20,21). Markov did not explicitly introduce this topology, although it was implicitly present in Refs. 45–47 (through the *algebraic closure* of a subset X, i.e., the smallest algebraic subset of the group G that contains X). It was explicitly introduced only in 1977 by Bryant[7] under the name *verbal topology.*

More recently, in a series of papers starting in 1999 with Ref. 4, Baumslag, Myasnikov and Remeslennikov developed the study of algebraic geometry over an abstract group G: in analogy to the well-known Zariski topology from algebraic geometry, the authors consider finite powers G^n of a group G, and introduce the *Zariski topology on* G^n using solution sets of n-variable equations. In the case $n = 1$, this topology is Bryant's verbal topology. Here we keep the name Zariski topology and the notation $\mathfrak{Z}_G$ for this topology following this use since 2003 when the first drafts of Refs. 22,24 were ready (their main results were reported later in Refs. 17,20,21).

In *linear* groups the term *Zariski topology* is used for a different standardly defined topology. Namely, for a field K consider the topology $\mathcal{A}_{K^n}$ of the affine space K^n having as a subbase of the closed sets the family of zero-sets of n-variable polynomials over K. The full linear group $\mathrm{GL}_n(K)$ (as well as its subgroups) carry the topology induced by $\mathcal{A}_{K^{n^2}}$ (via the embedding in K^{n^2}). Usually *this* topology of the linear groups is called Zariski topology. In order to avoid confusion, we use the term *affine topology*, when we refer to this topology for a linear group G and denote it by $\mathcal{A}_G$. We discuss linear groups only in §§5, 7 (in general, $\mathcal{A}_G \neq \mathfrak{Z}_G$ for a linear group G, see Corollaries 7.1 and 7.2).

Let $\mathbb{E}_G$ be the family of all elementary algebraic subsets of a group G. Then the cardinality

$$\mathrm{bd}(G) = \min\{|S| : S \subseteq \mathbb{E}_G,\ \bigcup S = G \setminus \{e_G\}\}$$

is called the *bound* of G (Ref. 10); G is called *κ-bound* for a (possibly finite) cardinal κ if $\mathrm{bd}(G) \leq \kappa$ (see Ref. 36, Definition 1). Clearly, $\mathrm{bd}(G) \leq |G|$; while $\mathrm{bd}(G) < \infty$ if and only if $\mathfrak{Z}_G$ is discrete. This definition was inspired by Podewski[54], where κ-bound groups are called *κ-gebunden* (see also Ref. 35).

To connect the bound to the Zariski topology we need to recall some notions from topology. If $(X, \mathcal{T})$ is a topological space and κ is an infinite cardinal, a union of $\leq \kappa$-many closed subsets is called an *F_κ-set.* The topology $\mathcal{T}$ is called a *P_κ-topology,* if every F_κ-set is closed. Every topology $\mathcal{T}$ on X admits a coarsest P_κ-topology $P_\kappa\mathcal{T}$ containing $\mathcal{T}$, called the *P_κ-modification* of $\mathcal{T}$ (namely the topology having as a base of its closed sets all F_κ-sets of $(X, \mathcal{T})$).

Let G be a group and let δ_G denote the discrete topology of G. For an infinite cardinal κ, the P_κ-modification $P_\kappa\mathfrak{Z}_G$ of the Zariski topology of G will be called *κ-Zariski topology* of G. Clearly, $P_\kappa\mathfrak{Z}_G$ is discrete for all $\kappa \geq |G|$, but $P_\kappa\mathfrak{Z}_G$ may be non-discrete for some infinite $\kappa < |G|$. We call the chain of topologies

$$\mathfrak{Z}_G \leq P_\omega\mathfrak{Z}_G \leq \ldots \leq P_\kappa\mathfrak{Z}_G \leq \ldots \leq P_{|G|}\mathfrak{Z}_G = \delta_G$$

the *Zariski rod* of G. If $\mathfrak{Z}_G$ is not discrete, then $\mathrm{bd}(G)$ is infinite and the least cardinal κ with $P_\kappa\mathfrak{Z}_G = \delta_G$ coincides with $\mathrm{bd}(G)$. This allows us to consider $\mathrm{bd}(G)$ also as a *measure of the failure of $\mathfrak{Z}_G$ to be discrete.* The importance of the groups G with discrete $\mathfrak{Z}_G$, called *$\mathfrak{Z}$-discrete* in the sequel, becomes clear in §1.2.

1.2. *The Markov topologies on a group and Markov's problems*

As noticed in Ref. 21, the family of unconditionally closed subsets of G coincides with the family of closed subsets of a T_1 topology $\mathfrak{M}_G$ on G, called *the Markov topology of* G. It coincides with the infimum (taken in the lattice of all topologies on G) of all Hausdorff group topologies on G.

A Hausdorff group is said to be *precompact,* if G is topologically isomorphic to a subgroup of a compact Hausdorff group. In analogy with $\mathfrak{M}_G$, let $\mathfrak{P}_G$ be the infimum of all precompact Hausdorff group topologies on G. If G admits no such topologies, then $\mathfrak{P}_G = \delta_G$ is discrete. Call $\mathfrak{P}_G$ *the precompact Markov topology of G* (Ref. 22). Note that $(G, \mathfrak{Z}_G)$, $(G, \mathfrak{M}_G)$ and $(G, \mathfrak{P}_G)$ are quasi-topological groups in the sense of Ref. 2, i.e., the inversion and shifts are continuous. In general, $\mathfrak{Z}_G \subseteq \mathfrak{M}_G \subseteq \mathfrak{P}_G$ for every group G. If G is abelian, then $\mathfrak{Z}_G = \mathfrak{M}_G = \mathfrak{P}_G$ (see Theorem 6.6), although $\mathfrak{M}_G \neq \mathfrak{P}_G$ may occur in some nilpotent groups of class 2 (Example 7.1).

Markov posed the following problem (without the explicit use of topologies):

Markov's First Problem: Does $\mathfrak{Z}_G = \mathfrak{M}_G$ hold true for an arbitrary group G?

He proved that these two topologies coincide in the countable case[45]. The further progress in the study of the class $\mathcal{MZ}$ of groups for which Markov and Zariski topologies coincide will be discussed in §3.

If the group G is infinite, the topology $\mathfrak{M}_G$ is discrete if and only if G is *non-topologizable*, i.e., it does not admit a non-discrete Hausdorff group topology (see Definition 4.1). In 1945, Markov (Ref. 46, Problem 4) asked:

Markov's Second Problem: Does there exist an infinite non-topologizable group?

This problem remained unsolved for many years, until Shelah[57] constructed under CH a non-topologizable group of size ω_1. Actually, Hesse[35] eliminated CH from Shelah's construction, thus presenting the first example of a non-topologizable group in ZFC of size ω_1. Finally, in 1980, Ol′shanskij[51] built up a countable non-topologizable group in ZFC (for more details see §4).

An important connection between this problem of Markov and the bound was found by Podewski[54]. He proved that a group G with $\mathrm{bd}(G) = |G|$ (he called such a group *ungebunden*, see also Ref. 35) is always topologizable. (More precisely, it admits the maximum number $2^{2^{|G|}}$ of Hausdorff group topologies, see Theorem 4.7.) Hesse[35] showed that this condition is not necessary: for any uncountable cardinal λ he found a topologizable group G of size λ with $\mathrm{bd}(G) = \omega$ (so $\mathrm{bd}(G) < |G|$).

Markov was interested also to describe the groups admitting a *connected* Hausdorff group topology. If τ is such a topology on a group G, then every proper τ-closed subgroup H of G has index at least $\mathfrak{c}$. (Indeed, if τ is a connected Hausdorff group topology on G, then the non-trivial completely regular quotient space G/H is connected, hence $|G/H| = [G : H] \geq \mathfrak{c}$.) In particular, all proper $\mathfrak{M}_G$-closed subgroups of G must have index at least

$\mathfrak{c}$. This is why Markov asked:

Markov's Third Problem: If every proper $\mathfrak{M}_G$-closed subgroup of a group G has index $\geq \mathfrak{c}$, does G admit a connected Hausdorff group topology?

The last section of Markov's paper[46] is exclusively dedicated to the following problem:

Markov's Fourth Problem: Which subsets of a group G are potentially dense in G?

Markov[46] succeeded in proving that a subset of a countable group is potentially dense if and only if it is Zariski-dense. In particular, every infinite subset A of $\mathbb{Z}$ is potentially dense in $\mathbb{Z}$ (this easily follows also from the uniform distribution theorem of Weyl[70]: if $A = (a_n)$, then by Weyl's theorem there exists an irrational number α such that the sequence $(a_n\alpha)$ is dense (even uniformly distributed) modulo 1, so the embedding $\mathbb{Z} \to \mathbb{T}$ defined by $n \mapsto n\alpha$ induces a Hausdorff group topology τ on $\mathbb{Z}$ such that A is τ-dense). More on this problem will be given in Theorem 6.8.

As mentioned in the abstract, a prominent role in our exposition is played by the Zariski and Markov topologies. This determined also our choice to include only these four problems of Markov in the survey. The reader can find more on the problems we didn't mention here in the survey Ref. 12, §3.5.

This survey is organized as follows. In §2 we fix notation and terminology, and present some general results used frequently in the sequel. To Markov's first problem is dedicated §3. Here we shortly discuss Hesse's impressive solution (Theorem 3.1) and present a review of some more recent results, including a general reflection principle from Ref. 21 (Corollary 3.4) which yields a first proof of Perel′man's theorem about the equality $\mathfrak{Z}_G = \mathfrak{M}_G$ for abelian groups G (Corollary 3.5). Then in §4 we consider various ways to topologize a group and discuss groups G such that $\mathfrak{M}_G$ (resp. $\mathfrak{Z}_G$, or $\mathfrak{P}_G$) is discrete, these groups will be called $\mathfrak{M}$-discrete (resp. $\mathfrak{Z}$-discrete, $\mathfrak{P}$-discrete). Section 5 abandons for a while Markov's problems, it is dedicated to Kaplanski's CZ-groups, and in particular to those groups (called $\mathfrak{Z}$-Noetherian) that are CZ-groups with respect to their Zariski topology, i.e. groups such that their Zariski topology is Noetherian. Our choice is justified by the fact that starting from here all Zariski spaces

$(G, \mathfrak{Z}_G)$ will be Noetherian. This will be heavily used in §6, where we treat the case of abelian groups. Here properties of the Noetherian spaces are needed for the explicit computation of the Zariski-closure (Theorem 6.4). The next step is to realize the Zariski-closure of a given subset X of an abelian group G as a closure with respect to some Hausdorff group topology on G (Theorem 6.5, its proof in Ref. 22 uses this explicit description of the Zariski-closure). In Theorem 6.8 this realization theorem is used to describe the potentially dense subset of the abelian groups of size $\leq 2^{\mathfrak{c}}$. At the end of this section (§6.2) we discuss applications towards partial solutions of Markov's first (Theorem 6.6) and fourth problem (Theorem 6.8, Corollary 6.3). In Theorem 6.7 we prove that a soluble divisible group G is abelian if and only if $\mathfrak{M}_G = \mathfrak{P}_G$. The Zariski topology of some linear groups is described in §7. In the final §8 we collect all open questions.

It is a pleasure to dedicate this survey to the fiftieth anniversary of our friend and colleague Dima Shakhmatov, whose contribution to this topic is not limited only to his more than generous and crucial participation in Refs. 20–25.

2. General results

If X is a set, then $[X]^{<\omega}$ and $[X]^{\leq\omega}$ denote the set of all finite subsets of X and all (at most) countable subsets of X, respectively. We denote by $\mathbb{N}$ the set of all natural numbers, while $\mathbb{Z}$ denotes the group of integers, $\mathbb{R}$ the group of reals and $\mathbb{T} = \mathbb{R}/\mathbb{Z}$ the circle group.

Let $\mathcal{G} = \{G_i \mid i \in I\}$ be a family of groups. Adopting terminology and notation from abelian group theory, we denote by $G = \prod_{i\in I} G_i$ the group having the cartesian product of $\mathcal{G}$ as underlying set, with componentwise defined operation, and we call G *direct product* of $\mathcal{G}$. The subgroup S of G consisting of the elements which have only finitely many non-trivial coordinates will be called *direct sum* of $\mathcal{G}$.

A topological space X is called:

(a) *Noetherian* if X satisfies the descending chain condition on closed sets;
(b) *irreducible*, if $X = F \cup G$, with closed subsets F, G, implies $X = F$ or $X = G$.

A subset Y of a topological space X is *irreducible*, if Y is irreducible as a topological subspace of X. It is easy to see that a space X is Noetherian if and only if all subspaces of X are compact.

Let e_G denote the neutral element of a group G. An elementary algebraic set $X = \{x \in G : g_1 x^{\varepsilon_1} g_2 x^{\varepsilon_2} \cdots g_n x^{\varepsilon_n} = e_G\}$ as in Definition 1.1 will be denoted by E_w, where w is an abbreviation for the defining word $w(x) = g_1 x^{\varepsilon_1} g_2 x^{\varepsilon_2} \cdots g_n x^{\varepsilon_n}$ considered as an element of the free product $G * \langle x \rangle$.

Note that, for every element $g \in G$, the centralizer $C_G(g) = \{x \in G \mid xgx^{-1}g^{-1} = e_G\}$ is an elementary algebraic subset of G, so the centralizer $C_G(S)$ of every subset S of G is an algebraic subset.

The following useful technique for building Noetherian spaces was proposed by Bryant[7]:

Proposition 2.1 (Ref. 7). *Let $\mathscr{E}$ be a family of subsets of a set X closed under finite intersections and satisfying the descending chain condition. Assume also that $X \in \mathscr{E}$. Then the family $\mathscr{E}^{\cup}$ consisting of finite unions of the members of $\mathscr{E}$ forms the family of closed sets of a unique topology $\mathcal{T}_{\mathscr{E}}$ on X, such that the space $(X, \mathcal{T}_{\mathscr{E}})$ is a Noetherian space.*

Remark 2.1. We give here some comments about the use of this proposition.

(a) It easily yields the result that finite products of Noetherian spaces are Noetherian (Ref. 7, Lemma 3.4).
(b) The typical situation where one can use Proposition 2.1 is when $\mathbb{E}_G$ satisfies the descending chain condition (i.e., every intersection of a family $\mathfrak{S} \subseteq \mathbb{E}_G$ coincides with the intersection of some finite subfamily $\mathfrak{S}_1$ of $\mathfrak{S}$). Then the family $\mathscr{E}$ of all finite intersections of members of $\mathbb{E}_G$ is closed under finite intersections and satisfies the descending chain condition, so Proposition 2.1 can be applied. This situation occurs in free groups and in abelian groups, thus yielding that the Zariski topology is Noetherian if the group is either free (Corollary 5.2) or abelian (Theorem 6.1).

Let $\mathcal{T}$ be a T_1-topology on a group G. If the "inverse" operation $x \mapsto x^{-1}$ is $\mathcal{T}$-continuous, and the binary operation $(x, y) \mapsto xy$ is separately $\mathcal{T}$-continuous, then Kaplansky calls the pair $(G, \mathcal{T})$ a *T_1-group*; see Ref. 38, Chap. IV, p. 27. The same pair is often called a T_1 *quasi-topological group*; see Ref. 2.

Then Kaplanski (Ref. 38, Chap. IV, p. 28) also defines a C-group to be a T_1-group such that for every $g \in G$ the mapping $x \mapsto x^{-1}gx$ is continuous.

Proposition 2.2. *$(G, \mathfrak{Z}_G)$ is a C-group.*

Proposition 2.3 (Ref. 27). *Let N be a normal subgroup of a group G, and let $\bar{G} = G/N$. Then the following conditions are equivalent:*

(1) *N is $\mathfrak{Z}_G$-closed;*
(2) *the quotient topology $\bar{\mathfrak{Z}}_G$ is a T_1 topology;*
(3) *$\mathfrak{Z}_{\bar{G}} \subseteq \bar{\mathfrak{Z}}_G$;*
(4) *the canonical map $\pi\colon (G, \mathfrak{Z}_G) \to (\bar{G}, \mathfrak{Z}_{\bar{G}})$ is continuous.*

The equivalence between (1) and (2) is proved as in the case of topological groups. The equivalence between (3) and (4) is obvious because the quotient topology $\bar{\mathfrak{Z}}_G$ is the final topology of the map $\pi\colon (G, \mathfrak{Z}_G) \to \bar{G}$, and (3) implies (2) as $\mathfrak{Z}_{\bar{G}}$ is T_1. Finally, to prove that (1) implies (4), one sees that the pre-image of an elementary algebraic subset of $\bar{G}$ is an algebraic subset of G.

Theorem 2.1 (Ref. 27). *Let $\{G_i \mid i \in I\}$ be a family of groups, and $G = \prod_{i\in I} G_i$. Then the Zariski topology $\mathfrak{Z}_G$ of the direct product is coarser than the product topology $\prod_{i\in I} \mathfrak{Z}_{G_i}$.*

Those two topologies on a direct product need not coincide even in very simple cases. For example, the Zariski topology of $G = \mathbb{Z}\times\mathbb{Z}$ is the co-finite topology (see Example 6.1), so neither $\mathbb{Z}\times\{0\}$ nor $\{0\}\times\mathbb{Z}$ are Zariski closed in G, whereas they are certainly closed in the product topology.

In Ref. 27 the authors prove also the obvious analogues of Theorem 2.1 for direct sums.

3. Groups for which Markov and Zariski topologies coincide

Here we discuss the first of Markov's problems, namely when $\mathfrak{Z}_G = \mathfrak{M}_G$ occurs for a group G.

Although Hesse gave a strong and very impressive negative solution of Markov's first problem by 1979, it is fair to say that his solution remained completely unknown to the large audience for long time. Indeed, in 1992, this problem is qualified as "still open" in the survey Ref. 12. Similarly, the quite recent survey Ref. 41 does not mention Hesse's solution and quotes the recent consistent counterexample announced in Ref. 58, where an appropriate modification of Shelah's construction (Ref. 57) under CH is shown to produce an example of a group G such that $\mathfrak{Z}_G \neq \mathfrak{M}_G$. A possible explanation can be the somewhat surprising fact that Hesse (maybe unaware of that problem) never claimes a solution, and he never published the solution obtained in his PhD thesis[35].

Here comes Hesse's powerful theorem:

Theorem 3.1 (Ref. 35). *Let λ, κ be infinite cardinal numbers with $cf(\lambda) > \kappa = cf(\kappa)$. Then there exists a group G such that :*

(a) $|G| = \lambda$;
(b) every Hausdorff semigroup topology on G is discrete (in particular, $\mathfrak{M}_G$ is discrete);
(c) $bd(G) > \kappa$.

Since $\mathrm{bd}(G) > \kappa$ means that $P_\kappa \mathfrak{Z}_G$ (hence $\mathfrak{Z}_G$ too) is non-discrete, while $\mathfrak{M}_G$ is discrete by item (b), one can deduce from the above theorem that there is a large gap between $\mathfrak{M}_G$ and $\mathfrak{Z}_G$, so $G \notin \mathcal{MZ}$. For the sake of convenience we formulate the following corollary giving an explicit negative solution to Markov's first problem.

Corollary 3.1. *For every cardinal λ of uncountable cofinality there exists a group G of size λ such that $\mathfrak{M}_G$ is discrete while $\mathfrak{Z}_G$ is not discrete, so $G \notin \mathcal{MZ}$.*

In particular, there exist groups $G \notin \mathcal{MZ}$ of size ω_1 and $\mathfrak{c}$.

In the rest of this section we recall some more recent results towards the description of the class $\mathcal{MZ}$ of groups with coinciding Zariski and Markov topologies.

Now we isolate various types of subgroup embeddings that "behave well" with respect to the Markov and Zariski topologies.

Definition 3.1 (Ref. 21, Definitions 1.1, 2.1). *A subgroup H of a group G is called:*

(a) super-normal in G *if for every $g \in G$ there exists $h \in H$ such that $gxg^{-1} = hxh^{-1}$ for every $x \in H$;*
(b) Zariski embedded in G *if the injection $(H, \mathfrak{Z}_H) \hookrightarrow (G, \mathfrak{Z}_G)$ is continuous;*
(c) Markov embedded in G *if the injection $(H, \mathfrak{M}_H) \hookrightarrow (G, \mathfrak{M}_G)$ is continuous;*
(d) Hausdorff embedded in G *if every Hausdorff group topology on H can be extended to a Hausdorff group topology on G.*

All three conditions (b), (c), (d) are trivially satisfied by finite subgroups.

The next theorem characterizing Hausdorff embedded normal subgroups is taken from Ref. 21.

Theorem 3.2. *Let N be a normal subgroup of a group G. Then N is Hausdorff embedded in G if and only if the automorphisms of N induced by conjugation by elements of G are continuous for any Hausdorff group topology on N.*

Easy examples show that even an index two (hence, normal) subgroup may strongly fail to be Hausdorff embedded (Ref. 25).

Obviously, a subgroup H of a group G is super-normal if and only if $G = HC_G(H)$. In particular, if H is abelian, then H is super-normal in G if and only if it is central. The next lemma relates the various types of embeddings:

Lemma 3.1 (Ref. 21). *Let H be a subgroup of a group G. Then:*

(a) if H is super-normal in G, then H is Hausdorff embedded and Zariski embedded in G.
(b) if H is Hausdorff embedded then H is Markov embedded.

Concerning item (a) of the previous lemma, in Ref. 25 the authors show that a normal Hausdorff embedded subgroup H of a group G need not be super-normal in G.

If a countable subgroup H of a group G fails to be Hausdorff embedded, then this failure can always be witnessed by some *metric* group topology on H (Ref. 21).

Theorem 3.3 (Ref. 21). *Let H be a countable subgroup of a group G. If every metric group topology on H can be extended to a (not necessarily metric) group topology on G, then H is Hausdorff embedded in G.*

Remark 3.1. If $\mathcal{A}$ is a countable family of additively algebraic sets of a group G, then for each countable subset X of G one can take the (countable) subgroup H of G generated by X and by all (countably many) elements of G coming from the countably many words defining the members of $\mathcal{A}$. Then $H \cap A$ is an additively algebraic subset of H, for each $A \in \mathcal{A}$.

The following useful concept from set theory will be needed in the sequel.

Definition 3.2. A family $\mathcal{C} \subseteq [X]^{\leq\omega}$ is called a *club in* $[X]^{\leq\omega}$, if

(a) $\bigcup\{C_n : n \in \mathbb{N}\} \in \mathcal{C}$ for every chain $C_0 \subseteq C_1 \subseteq \cdots \subseteq C_n \subseteq \dots$ in $\mathcal{C}$,

(b) for every $Y \in [X]^{\leq\omega}$ there exists $C \in \mathcal{C}$ containing Y.

One of the main reasons why clubs are so useful is that countable intersection of clubs in $[X]^{\leq\omega}$ is still a club in $[X]^{\leq\omega}$.

For a group G, a typical example of a club in $[G]^{\leq\omega}$ is

$$\mathcal{S}(G) = \{H \in [G]^{\leq\omega} : H \text{ is a subgroup of } G\}.$$

The following general reflection principle for the Zariski closure was proved in Ref. 21:

Theorem 3.4. *If G is a group and $\mathcal{A}$ is a countable family of subsets of G, then the family*

$$\{H \in \mathcal{S}(G) : \mathrm{Cl}_{\mathfrak{Z}_H}(H \cap A) = H \cap \mathrm{Cl}_{\mathfrak{Z}_G} A \text{ for all } A \in \mathcal{A}\}$$

contains a club in $[G]^{\leq\omega}$.

Corollary 3.2 (Ref. 21). *Let G be a group and $\mathcal{A}$ a countable family of $\mathfrak{Z}_G$-closed subsets of G. Then the family*

$$\{H \in \mathcal{S}(G) : H \cap A \text{ is } \mathfrak{Z}_H\text{-closed for every } A \in \mathcal{A}\}$$

contains a club in $[G]^{\leq\omega}$.

The next corollary should be compared with Remark 3.1.

Corollary 3.3 (Ref. 21). *Assume that G is a group, $\mathcal{A}$ a countable family of $\mathfrak{Z}_G$-closed subsets of G and X is a countable subset of G. Then there exists a countable subgroup H of G containing X such that $H \cap A$ is $\mathfrak{Z}_H$-closed for each $A \in \mathcal{A}$.*

It turns out that the version of reflection for $\mathfrak{M}_G$ similar to the one for $\mathfrak{Z}_G$ obtained in Theorem 3.4 *characterizes* the groups $G \in \mathcal{MZ}$.

Theorem 3.5 (Ref. 21). *For a group G the following conditions are equivalent:*

(i) $G \in \mathcal{MZ}$;

(ii) For every subset $A \subseteq G$, the family

$$\mathcal{M}_A = \{H \in \mathcal{S}(G) : \mathrm{Cl}_{\mathfrak{M}_H}(H \cap A) = H \cap \mathrm{Cl}_{\mathfrak{M}_G} A\}$$

contains a club in $[G]^{\leq\omega}$;

(iii) For every $\mathfrak{M}_G$-closed subset $A \subseteq G$, the family

$$\mathcal{E}_A = \{H \in \mathcal{S}(G) : H \cap A \text{ is } \mathfrak{M}_H\text{-closed}\}$$

contains a club in $[G]^{\leq\omega}$.

This theorem shows that $G \in \mathcal{MZ}$ is completely determined by countable subgroups of G.

As an application of Theorem 3.5, we get a new class of groups G for which $\mathfrak{Z}_G = \mathfrak{M}_G$:

Corollary 3.4 (Ref. 21). *$\mathcal{MZ}$ contains all groups G such that the family*

$$\mathcal{N}_G = \{N \in \mathcal{S}(G) : N \text{ is Markov embedded in } G\}$$

contains some club in $[G]^{\leq\omega}$.

Indeed, given a $\mathfrak{M}_G$-closed subset $A \subseteq G$, we have $\mathcal{N}_G \subseteq \mathcal{E}_A$, and the conclusion follows from the implication (iii)→(i) of Theorem 3.5.

Corollary 3.5 (Ref. 21). *$\mathcal{MZ}$ contains all abelian groups.*

Indeed, let H be a countable subgroup of an abelian group G. Then H is central, hence super-normal in G. Hence, by Lemma 3.1, H is Markov embedded in G. We have proved that $\mathcal{S}(G) \subseteq \mathcal{N}_G$. Now the conclusion follows from Corollary 3.4.

The fact that Markov and Zariski topologies on abelian groups coincide was attributed to Perel′man by Markov, although such a proof has never appeared. For a much stronger version of this result see §6.

Follows another application of Corollary 3.4.

Corollary 3.6 (Ref. 21). *If $G = A \times \left(\bigoplus_{i\in I} G_i\right)$, where A is an abelian group and each group G_i is countable, then $G \in \mathcal{MZ}$.*

Indeed, for $J \in [I]^{\leq\omega}$ let $G_J = \bigoplus_{j\in J} G_j$. In order to be able to apply Corollary 3.4 it suffices to check that the family

$$\mathcal{C} = \left\{N \times G_J : J \in [I]^{\leq\omega} \text{ and } N \in \mathcal{S}(A)\right\}$$

is a club in $[G]^{\leq\omega}$ contained in $\mathcal{N}_G$. Obviously $\mathcal{C}$ is a club in $[G]^{\leq\omega}$. For $J \in [I]^{\leq\omega}$, the subgroup G_J of $\bigoplus_{i\in I} G_i$ is super-normal, being a direct summand. Since A is abelian, this implies that $N \times G_J$ is super-normal in G for every $N \in \mathcal{S}(A)$. Lemma 3.1 implies that $N \times G_J$ is Markov embedded in G.

As a corollary of her main theorem (Ref. 60, Theorem 7), Sipacheva obtains (Ref. 60, Corollary 11), that is a τ-option of a weaker version of Corollary 3.4 (with "Markov embedded" replaced by "Hausdorff embedded" in the definition of $\mathcal{N}_G$). From this corollary she deduces the above Corollary 3.6.

4. Topologizing a group

Definition 4.1. Recall that a group G is called *topologizable* if it admits a non-discrete Hausdorff group topology (so G is *non-topologizable* if the only Hausdorff group topology of G is the discrete one). As the Markov topology of a non-topologizable group G is the discrete topology, such a group G will also be called $\mathfrak{M}$-*discrete.* Similarly, a group G such that the Zariski topology $\mathfrak{Z}_G$ is discrete will be called $\mathfrak{Z}$-*discrete.*

Since every abelian group embeds into some power of the circle group $\mathbb{T} = \mathbb{R}/\mathbb{Z}$, infinite abelian groups are topologizable. Hence a group G with infinite center cannot be $\mathfrak{M}$-discrete, as $Z(G)$ is always super-normal in G, hence Markov embedded, and $\mathfrak{M}_{Z(G)}$ is non-discrete. This is why we shall be interested mainly in groups with finite or just trivial center in this section.

4.1. *Topologizing a group of permutations*

For an infinite set X, let $S(X)$ denote the group of all permutations of X. It has a standard topology obtained from the embedding of $S(X)$ in the product X^X equipped with the product topology where X is taken discrete (it coincides with the pointwise convergence topology of X^X). Note that a local base of neighborhoods of id_X in $S(X)$ is given by the the family of subgroups S_F, where S_F is the pointwise stabilizer of the finite subset F of X. This is a Hausdorff group topology (actually totally disconnected, as each subgroup S_F is clopen). We denote by $\mathcal{T}_p(X)$ (or simply $\mathcal{T}_p$ when no confusion is possible) this topology of $S(X)$. It is non-discrete and non-precompact when X is infinite.

The following remarkable theorem of Gaughan[31] gives a suprising connection with the Markov topology:

Theorem 4.1. *Let X be an infinite set. Then $\mathfrak{M}_{S(X)} = \mathcal{T}_p$ is the topology of pointwise convergence.*

Note that in this case the Markov topology is a group topology, so it is Hausdorff.

Lemma 4.1. *If F has at least three elements, then $S_F(X)$ is $\mathfrak{Z}_{S(X)}$-closed.*

Proof. Let $G = S(X)$, and let $S(F)$ be the subgroup of all permutations of F in G. As $|F| > 2$, the center of $S(F)$ is trivial, so the stabilizer $S_F(X)$ is precisely the centralizer $C_G(S(F))$. Therefore, $S_F(X)$ is $\mathfrak{Z}_G$-closed. □

Pestov[53] answered negatively the third Markov's problem by a rather complex counter-example. Later, Remus[55] noticed that a quite easy example can be obtained using the symmetric groups $S(X)$. Indeed, by Theorem 4.1 every Hausdorff group topology on $G = S(X)$ is totally disconnected (as $\mathcal{T}_p$ is totally disconnected and this property is preserved by taking finer topologies). Hence one must only ensure to have all proper $\mathfrak{M}_G$-closed subgroups of index at least $\mathfrak{c}$. Since $\mathfrak{M}_G = \mathcal{T}_p$, every proper $\mathfrak{M}_G$-closed subgroup is the intersection of proper $\mathfrak{M}_G$-open subgroups. So it suffices to see that every proper $\mathfrak{M}_G$-open subgroup has index at least $\mathfrak{c}$. Since every such subgroup is contained in some stabilizer $S_x(X)$ of a point $x \in X$ (these are maximal subgroups), it sufficies to note that $S_x(X)$ has index $|X|$. So choosing X with $|X| \geq \mathfrak{c}$ one obtains a counter-example to Markov's third problem.

Since $\mathrm{Aut}(G)$ is a subgroup of $S(G)$, it inherits from $(S(G), \mathcal{T}_p))$ a Hausdorff group topology. This is exactly the pointwise convergence topology of $\mathrm{Aut}(G)$ when G is taken discrete. We will keep the same notation for this topology to avoid proliferation of notation. A local base is formed by the subgroups

$$Stab_F = \{\gamma \in \mathrm{Aut}(G) : \gamma(f) = f \text{ for all } f \in F\}, \quad F \in [G]^{<\omega}.$$

So, the automorphism groups admit a naturally defined Hausdorff group topology. Yet it may happen to be discrete. This motivated the following question posed by A. D. Taimanov (see §4.2 for the solution):

Question 4.1. (Ref. 63, Question 1.1) Does there exist a group whose automorphism group is non-topologizable?

In Ref. 62 Taĭmanov introduced the group topology T_G on a group G: let $F \in [G]^{<\omega}$ be a finite subset of G and

$$C_G(F) = \bigcap_{x \in F} C_G(x)$$

the centralizer of F in G. Then $\mathcal{C} = \{C_G(F) : F \in [G]^{<\omega}\}$ is a family of subgroups of G closed under finite intersections, and T_G is the topology such that $\mathcal{C}$ is a local base at e_G of T_G.

If G is center-free, then the embedding $(G, T_G) \hookrightarrow (\mathrm{Aut}(G), \mathcal{T}_p)$ is also topological.

Here are some easy to establish properties of this topology (Ref. 18, Lemma 4.1):

(a) the T_G-closure of $\{e_G\}$ coincides with $Z(G)$; so T_G is Hausdorff if and only if G is center-free;
(b) G is abelian if and only if T_G coincides with the indiscrete topology of G;
(c) in case G is finitely generated, $Z(G)$ is T_G-open; in particular T_G is discrete if and only if G is center-free.

One can easily see that T_G has nice functorial properties, e.g., every surjective homomorphism $f : (G, T_G) \to (H, T_H)$ is continuous (see Ref. 18, Proposition 4.2). Nevertheless, even the canonical projection $\pi : (G, T_G) \to (G/Z(G), T_{G/Z(G)})$ need not be open (see Ref. 18, Example 4.3).

Lemma 4.2. *Let $G = \prod_{i\in I} G_i$. Then $\prod_{i\in I} T_{G_i} \leq T_G$. If $I = \{1, \ldots, n\}$ is finite, then $T_{G_1} \times \ldots \times T_{G_n} = T_G$.*

Definition 4.2 (Ref. 18). *A group G is* Taĭmanov *if T_G is discrete.*

We denote by $\mathfrak{T}$ the class of Taĭmanov groups. Clearly, every Taĭmanov group is center-free (in particular, $\mathfrak{T}$ contains no non-trivial abelian group). For a center-free group G and a finite normal subgroup N, $G/N \in \mathfrak{T}$ implies $G \in \mathfrak{T}$ (see Ref. 18, Lemma 4.13).

It follows from Lemma 4.2 that the class $\mathfrak{T}$ is closed under taking finite products. On the other hand, easy examples show that $\mathfrak{T}$ is not closed under taking quotients and subgroups (Ref. 18, Example 4.7).

Remark 4.1. (Ref. 18 Non-abelian free groups are Taĭmanov, so there exist arbitrarily large Taĭmanov groups. Moreover, every group is the quotient of a Taĭmanov group.

Theorem 4.2 (Ref. 18,Theorem 4.11, Lemma 4.10). *For every abelian group G there exists a group $H \in \mathfrak{T}$ containing G as a subgroup and such that $|H| = \omega \cdot |G|$. If G has no elements of order 2, then one can arrange to have $[H : G] = 2$.*

Theorem 4.3 (Ref. 18, Theorem 4.18). *Let X be a set with $|X| > 2$.*

(a) If $|X| \leq \omega$, then $S(X) \in \mathfrak{T}$.
(b) If $|X| > \mathfrak{c}$, then $S(X) \notin \mathfrak{T}$.

It turns out that the groups $S(X)$ are helpful also in questions related to the connected topologization.

4.2. $\mathfrak{Z}$-discrete and $\mathfrak{M}$-discrete groups

We start with a discussion on the $\mathfrak{Z}$-discrete groups, i.e., groups G with finite $\mathrm{bd}(G)$. Obviously, an infinite $\mathfrak{Z}$-*discrete* group is also $\mathfrak{M}$-*discrete*, so provides a solution of Markov's second problem. Since the popularity of this problem has been much wider than that of the first one, this explains the major interest in $\mathfrak{Z}$-discrete groups.

One can easily check that a group G is $\mathfrak{Z}$-discrete if and only if there exist elementary algebraic subsets $E_1, \ldots, E_n$ such that $E_1 \cup \ldots \cup E_n = G \setminus \{e_G\}$ (in the terminology of Ref. 36, G is n-bound, i.e., $\mathrm{bd}(G) \leq n$).

The problem to construct a countable $\mathfrak{Z}$-discrete (hence, non-topologizable) group is equivalent to Markov's second problem for countable groups. It was resolved by Ol′shanskij[51] (see also Ref. 52, Theorem 31.5).

Example 4.1. Ol′shanskij produced a countable $\mathfrak{Z}$-discrete group using an appropriate quotient G of the (countable) Adian group $A(n, m)$. More precisely, he establised that actually $\mathrm{bd}(G) \leq 2m$ holds true.

If a group G is $\mathfrak{Z}$-discrete, then $\mathfrak{Z}_G = \mathfrak{M}_G = \mathfrak{P}_G$ holds because all three topologies become discrete. Such groups are extremely hard to come by, but a variety of examples were constructed since the pioneer work of Ol′shanskij:

Example 4.2.

(a) Klyachko and Trofimov[40] constructed a finitely generated torsion-free $\mathfrak{Z}$-discrete group.
(b) Trofimov[64] proved that every group admits an embedding into a $\mathfrak{Z}$-discrete group.
(c) Morris and Obraztsov[48] modified Ol′shanskij's example to build, for any sufficiently large prime p, a continuum of pairwise non-isomorphic infinite $\mathfrak{Z}$-discrete groups of exponent p^2, all of whose proper subgroups are cyclic.

Using Ol′shanskij's technique, Trofimov[65] answered positively Question 4.1, constructing an infinite, finitely generated, center-free, $\mathfrak{Z}$-discrete group G such that $G \cong \mathrm{Aut}\, G$.

The next theorem gives an "external" characterization of $\mathfrak{Z}$-discrete groups in terms of Zariski embeddings.

Theorem 4.4 (Ref. 21, Theorem 6.14). *For a group H the following conditions are equivalent:*

(i) *H is simultaneously Hausdorff embedded, Markov embedded and Zariski embedded in every group G that contains H as a subgroup,*
(ii) *H is Zariski embedded in every group G that contains H as a subgroup,*
(iii) $\mathfrak{Z}_H$ *is discrete.*

Example 4.3. This theorem gives easy examples of non-Zariski-embedded subgroups. For example, the infinite abelian group $H = \mathbb{Z}$ is not $\mathfrak{Z}$-discrete, hence there exists a group G containing H as a subgroup such that H is not Zariski embedded in G.

The groups with finite bound seem to be of interest also in the finite case (see Ref. 36, Definition 1, for the description of all finite 2-bound groups, where an unpublished manuscript of G. Cherlin is quoted, describing the finite 1-bound groups).

The delicate example found by Hesse[35] distinguishes between $\mathfrak{Z}$-*discrete* and $\mathfrak{M}$-*discrete* groups (indeed the groups provided by Corollary 3.1 are $\mathfrak{M}$-discrete but not $\mathfrak{Z}$-discrete).

Let us recall that a group G with $|G| = \omega_1$ is called a *Kurosh group*, if all proper subgroups of G are countable. Let us introduce the following stronger notion:

Definition 4.3. Let $m \in \mathbb{N}$. An uncountable group G is said to be a *m-Kurosh group*, if $A^m = G$ for every subset A of G with $|A| = |G|$ and $A^{-1} = A$.

Clearly every m-Kurosh group of size ω_1 is a Kurosh group.

For an uncountable group G consider the following condition:

(S) for every subgroup H of G with $|H| < |G|$ there exists $F \in [G]^{<\omega}$ such that $\bigcap_{x \in F} x^{-1}Hx$ is finite.

Clearly, a group G satisfying (S) cannot have countably infinite normal subgroups (actually, (S) implies a stronger property — the core H_G of any subgroup H of G with $|H| < |G|$ is finite and $H_G = \bigcap_{x \in F} x^{-1}Hx$ for some $F \in [G]^{<\omega}$).

The utility of these notions becomes clear in the following original sufficient condition ensuring $\mathfrak{M}$-discreteness elaborated by Shelah[57].

Proposition 4.1. *If* $m \in \mathbb{N}$, *then every m-Kurosh group G satisfying (S) is* $\mathfrak{M}$-*discrete.*

Indeed, for every Hausdorff group topology $\mathcal{T}$ on G, there exists a $\mathcal{T}$-neighbourhood V of e_G with $V^{-1} = V$ and $V^m \neq G$. Then our hypothesis on G to be m-Kurosh yields $|V| < |G|$, so also the subgroup $H = \langle V \rangle$ satisfies $|H| < |G|$. Then $O = \bigcap_{x \in F} x^{-1}Hx$ is finite for some $F \in [G]^{<\omega}$, according to (S). As each $x^{-1}Hx \in \mathcal{T}$, this proves that $e_G \in O \in \mathcal{T}$. Since $\mathcal{T}$ is Hausdorff and O is finite, it follows that $\mathcal{T}$ is discrete.

Remark 4.2. The above criterion was used by Shelah[57] to produce the first consistent example of a non-topologizable group. He worked under the assumption of CH and produced a 10000-Kurosh group G of size ω_1 satisfying (S) with $|F| = 2$ for every subgroup H of G with $|H| < |G|$ (actually, he managed to have each of these subgroups H malnormal in G, i.e. $H \cap x^{-1}Hx = e_G$ for every $x \in G \setminus H$).

Here comes the counterpart of Theorem 4.4 providing an "external" characterization of discreteness of $\mathfrak{M}_H$ in terms of Markov and Hausdorff embeddings.

Theorem 4.5 (Ref. 21, Theorem 6.9). *For a group H the following conditions are equivalent:*

- *(i) H is Hausdorff embedded in every group G that contains H as a subgroup,*
- *(ii) H is Markov embedded in every group G that contains H as a subgroup,*
- *(iii) $\mathfrak{M}_H$ is discrete.*

Following Lukàcs[44], call a group G *hereditarily $\mathfrak{M}$-discrete*, if for every subgroup H of G all quotients of H are $\mathfrak{M}$-discrete (the term *hereditarily non-topologizable group* is used in Ref. 44). This is the largest class of $\mathfrak{M}$-discrete groups closed with respect to taking subgroups and quotients. Let us say immediately, that no examples of infinite hereditarily $\mathfrak{M}$-discrete groups are known so far (Ref. 20, Question 929).

The origin of this class comes from the categorically compact groups introduced in Ref. 29. A topological group (G, τ) is called *categorically compact* if, for every topological group H, the projection $G \times H \to H$ sends closed subgroups of $G \times H$ onto closed subgroups of H. By Kuratowski's closed projection theorem, every compact group is categorically compact. This implication is invertible for soluble groups[29]. It is not known whether it holds true even in the case of discrete groups, i.e., whether every discrete categorically compact group is finite. It is proved in Ref. 29 that a countable discrete group is categorically compact if and only if it is hereditarily $\mathfrak{M}$-discrete.

We conclude with a couple of results on topologizable groups.

Belyaev[5] proved the following.

Theorem 4.6 (Ref. 5). *No countably infinite locally finite group is $\mathfrak{M}$-discrete.*

Now we state Podewski's result about the topologizability of groups with $\mathrm{bd}(G) = |G|$.

Theorem 4.7 (Ref. 54). *Let G be a group such that $bd(G) = |G|$. Then G admits the maximum number $2^{2^{|G|}}$ of Hausdorff group topologies.*

It turns out that the above theorem is reversible for countable groups:

Corollary 4.1 (Ref. 54). *For a countable group G, the following conditions are equivalent:*

- *G is not $\mathfrak{M}$-discrete (i.e., it admits at least one non-discrete Hausdorff group topology);*
- *$bd(G) = |G|$;*
- *G admits the maximum number $2^{2^{|G|}}$ of Hausdorff group topologies.*

A long series of countable topologizable groups can be found in Hesse[34], where the stronger property $\mathrm{bd}(G) = |G|$ was established for the group G; that entails, according to Corollary 4.1, that G admits actually $2^{2^{|G|}}$ Hausdorff group topologies:

Theorem 4.8 (Ref. 34). *If a countable group G has one of the following properties, then G is topologizable:*

(a) G has an infinite normal subgroup N such that $[N : c_N(F)] < \omega$ for every finite subset F of N;
(b) G has an infinite normal subgroup N that is an FC-group;
(c) G is an FC-nilpotent group;
(d) G contains an infinite solvable normal subgroup;
(e) G is locally nilpotent.

Obviously, (d) implies that every countable solvable group admits a nondiscrete Hausdorff group topology. This partially answers positively a question due to Sharma[56].

4.3. $\mathfrak{P}$-discrete and $\mathfrak{P}$-Hausdorff groups

Observe that a group G is $\mathfrak{P}$-discrete (i.e., $\mathfrak{P}_G$ is discrete) if and only if G is not *maximally almost periodic* (according to von Neumann's[49] terminology), i.e., it does not admit any precompact Hausdorff group topology.

Example 4.4. We recall here two well known examples of $\mathfrak{P}$-discrete groups that are not $\mathfrak{M}$-discrete, so that $\mathfrak{M}_G \neq \mathfrak{P}_G$.

(a) A classical example of a $\mathfrak{P}$-discrete group is the group $SL(2,\mathbb{C})$ (Ref. 50). Since $SL(2,\mathbb{C})$ is topologizable (by its usual topology induced by $\mathbb{C}^4$), it is not $\mathfrak{M}$-discrete.
(b) It easily follows from Theorem 4.1, that for any infinite set X the group $S(X)$ is $\mathfrak{P}$-discrete, but not $\mathfrak{M}$-discrete.

The next theorem, proved in Ref. 28, uses the fact that compact divisible groups are connected, so that the structure theory of compact connected groups can be applied[37].

Theorem 4.9. *Every divisible precompact solvable group is abelian.*

According to Theorem 4.9, a divisible solvable non-abelian group admits no precompact group topology. This gives:

Corollary 4.2. *Every divisible solvable non-abelian group is $\mathfrak{P}$-discrete.*

Examples of divisible solvable non-abelian groups are given in Example 7.1.

The discrete topology is notoriously a group topology, this explains our next step to pay attention to those groups G where $\mathfrak{Z}_G$ (resp., $\mathfrak{M}_G$, $\mathfrak{P}_G$) is a group topology. Since these topologies are always T_1, such topological groups are necessarily Hausdorff, i.e., the topological group $(G,\mathfrak{Z}_G)$ (resp., $(G,\mathfrak{M}_G)$, $(G,\mathfrak{P}_G)$) is Hausdorff. The final part of this subsection will be dedicated to the non-$\mathfrak{P}$-discrete (hence, non-$\mathfrak{M}$-discrete) groups G such that both $\mathfrak{M}_G$ and $\mathfrak{P}_G$ are group topologies, yet non-discrete ones.

Theorem 4.10 (Ref. 27). *Let $\{F_i \mid i \in I\}$ be a non-empty family of finite center-free groups, $G = \prod_{i\in I} F_i$, and $S = \bigoplus_{i\in I} F_i$. Then:*

(a) $\mathfrak{Z}_G = \mathfrak{M}_G = \mathfrak{P}_G = \prod_{i\in I} \mathfrak{Z}_{F_i}$.
(b) $\mathfrak{Z}_S = \mathfrak{M}_S = \mathfrak{P}_S = \mathfrak{Z}_G \restriction_S$.

In particular, both $\mathfrak{Z}_G$ and $\mathfrak{Z}_S$ are Hausdorff group topologies, $\mathfrak{Z}_G$ is compact, and $\mathfrak{Z}_S$ is precompact. Moreover, they both have a local base of clopen subgroups of finite index, so they are zero-dimensional.

The proof makes use of the fact that for each $i \in I$ the subgroup $\{e_{F_i}\} \times \prod_{j\in I\setminus\{i\}} F_j$ is a centralizer, as the group F_i is center-free, so it is $\mathfrak{Z}_G$-closed.

Item (b) of the above theorem implies that the group $S = \bigoplus_{i\in I} F_i$ admits no Hausdorff group topologies strictly coarser than $\mathfrak{Z}_S$. This weaker property of $(S, (\prod_{i\in I} \mathfrak{Z}_{F_i}) \restriction_S)$ was established already in Ref. 14.

Take the direct product G of any infinite family of finite, center-free groups. Its Zariski topology $\mathfrak{Z}_G$ is compact Hausdorff by Theorem 4.10, so it is not Noetherian. This negatively answers Question 1, and positively answers Question 2, from Ref. 22.

Example 4.5. Following Ref. 15, Remark 3.13, we say that a topological group G is a *van der Waerden group* if G is a compact Hausdorff group on which every homomorphism to a compact Hausdorff group is continuous.

(a) Clearly, the class $\mathcal{W}$ of van der Waerden groups is closed under taking Hausdorff quotients and finite products and contains all finite groups.

(b) It is a theorem of van der Waerden[68], that $\mathcal{W}$ contains all compact, connected, semi-simple Lie groups (for the group $SO(3,\mathbb{R})$ of rotations of $\mathbb{R}^3$ see also Ref. 16).

Theorem 4.11. *For a compact Hausdorff group (G,τ) the following are equivalent:*

(a) $(G,\tau) \in \mathcal{W}$;

(b) τ *is the only (hence, the finest) precompact group topology on* G;

(c) $\mathfrak{P}_G$ *is Hausdorff;*

(d) $\mathfrak{P}_G = \tau$.

The equivalence of (a) and (b) is proved in Ref. 13, p. 195. The equivalence of (b), (c) and (d) follows from the definitions. This theorem shows that a van der Waerden group (G,τ) is completely determined by the underlying abstract group G, as $\tau = \mathfrak{P}_G$. Hence, one may speak of van der Waerden groups without specifying the topology. Namely, $\mathcal{W}$ can be viewed as the class of abstract groups G, for which $\mathfrak{P}_G$ is compact Hausdorff.

Item (c) of the above theorem and Theorem 4.10 imply that every product $G = \prod_{i\in I} F_i$ of finite center-free groups is a van der Waerden group.

5. $\mathfrak{Z}$-Noetherian groups

5.1. *Kaplanski's CZ-groups*

The next definitions are due to Kaplanski (Ref. 38, p.28, p.54):

Definition 5.1. Let (G, τ) be a T_1 quasi-topological group. Then (G, τ) is

- a *Z-group* if it is Noetherian as a topological space,
- a *CZ-group* if it is both a C-group and a Z-group.

In other words, (G, τ) is a CZ-group if it is a quasi-topological group, a T_1 Noetherian space, and for every $g \in G$ the mapping $x \mapsto x^{-1}gx$ is continuous. Kaplanski introduced the CZ-groups in studying linear groups, he noticed that these are CZ-groups if equipped with their affine topology. See Ref. 38, Chapters 4 and 8, for the elementary properties of CZ-groups, and Ref. 67, Chapter 5, for further results.

Let us see that infinite CZ-groups are never topological groups. In effect, this property holds in a slight more general setting:

Remark 5.1. Let G be an infinite group equipped with a Noetherian T_1 topology. Then the following conditions are equivalent:

(a) G is Hausdorff,
(b) G is finite,
(c) G is a topological group.

Indeed, (a) $\rightarrow$ (b) holds true in any Noetherian topological space. The implications (b) $\rightarrow$ (c) $\rightarrow$ (a) obviously follow from the fact that G is a T_1 space.

In general, the fact that (G, τ) is a CZ-group does not imply that the quotient $(G/N, \bar{\tau})$ is a CZ-group for every normal subgroup N (see Example 5.3). Nevertheless the following result holds.

Proposition 5.1 (Ref. 67, Lemma 5.14). *Let G be a CZ-group, and N be a closed normal subgroup of a G. Then the quotient group G/N is a CZ-group with respect to the quotient topology.*

Nice properties on the structure of CZ-groups were established in Refs. 8,9,71:

Theorem 5.1 (Ref. 9, Theorem 1, Theorem 2). *Let G be a locally finite CZ-group. Then, for every prime p, the maximal p-subgroups are conjugate. If G is also locally soluble, then G is soluble.*

Theorem 5.1 inspired the following question posed by Wehrfritz (Ref. 67, Question 10).

Question 5.1. Is a locally soluble CZ-group necessarily soluble (or perhaps only hyperabelian)?

The answer is positive for linear groups.

Theorem 5.2 (Ref. 8). *Every periodic locally soluble CZ-group G is nilpotent by abelian-of-finite-rank by finite. If in addition G is locally nilpotent, then G is nilpotent-by-finite.*

Answering affermatively another question posed by Wehrfritz (Ref. 67, Question 11) Wilson[71] proved that every locally nilpotent CZ-group is hypercentral.

5.2. $\mathfrak{Z}$-*Noetherian groups and* $\mathfrak{Z}$-*compact groups*

In the sequel we will consider groups G such that the Zariski topology $\mathfrak{Z}_G$ is Noetherian, i.e. $(G, \mathfrak{Z}_G)$ is a Z-group.

Definition 5.2. A group G is called $\mathfrak{Z}$-*Noetherian* (resp., $\mathfrak{Z}$-*compact*), if the Zariski topology $\mathfrak{Z}_G$ is Noetherian (resp., compact).

As $(G, \mathfrak{Z}_G)$ is always a C-group, a group G is a $\mathfrak{Z}$-Noetherian group if and only if the pair $(G, \mathfrak{Z}_G)$ is a CZ-group.

As compactness in this section is not intended to include automatically "Hausdorff", $\mathfrak{Z}$-Noetherian groups are $\mathfrak{Z}$-compact. On the other hand, by Theorem 4.10, there exist infinite $\mathfrak{Z}$-compact groups that cannot be $\mathfrak{Z}$-Noetherian, as the next example shows.

Example 5.1. Let G be a $\mathfrak{Z}$-Noetherian group (for example, an abelian group). Then $(G, \mathfrak{Z}_G)$ is a CZ-group, so by Remark 5.1 we can conclude that $\mathfrak{Z}_G$ is Hausdorff if and only if G is finite if and only if $(G, \mathfrak{Z}_G)$ is a topological group. The first equivalence may fail for some non-commutative groups. Indeed, any infinite group G with discrete $\mathfrak{Z}_G$ (see Example 4.2) provides an example where $(G, \mathfrak{Z}_G)$ is a Hausdorff topological group.

As $\mathfrak{Z}_H \subseteq \mathfrak{Z}_G \restriction_H$ for every subgroup H of a group G, one concludes that the class of $\mathfrak{Z}$-Noetherian groups is stable under taking subgroups:

Lemma 5.1 (Ref. 7, Lemma 3.3). *Subgroups of a $\mathfrak{Z}$-Noetherian group are $\mathfrak{Z}$-Noetherian.*

Moreover, one can easily see that the class of $\mathfrak{Z}$-compact groups is stable only under taking Zariski-closed subgroups. In other words, a $\mathfrak{Z}_G$-closed subgroup of a $\mathfrak{Z}$-compact group is $\mathfrak{Z}$-compact, and the condition to be $\mathfrak{Z}_G$-closed is not redundant (see Ref. 27).

The following two theorems are also due to Bryant:

Theorem 5.3 (Ref. 7, Corollary 3.7, Theorem 3.8). *A group G is $\mathfrak{Z}$-Noetherian if one of the following holds:*

- *G is abelian-by-finite;*
- *G is finitely generated abelian-by-nilpotent-by-finite.*

Theorem 5.4 (Ref. 7, Theorem 3.5). *Let G be a linear group. Then its Zariski topology $\mathfrak{Z}_G$ is coarser than its affine topology $\mathcal{A}_G$. In particular, G is $\mathfrak{Z}$-Noetherian.*

Since countable free groups are isomorphic to subgroups of the linear group $SO_3(\mathbb{R})$ (Ref. 3), the above theorem along with Lemma 5.1 implies the following corollary.

Corollary 5.1. *Every countable free group is $\mathfrak{Z}$-Noetherian.*

The option in parentheses of the following result follows from Proposition 5.1.

Proposition 5.2. *If N is a $\mathfrak{Z}_G$-closed normal subgroup of a $\mathfrak{Z}$-compact (resp. $\mathfrak{Z}$-Noetherian) group G, then also the quotient group G/N is $\mathfrak{Z}$-compact (resp. $\mathfrak{Z}$-Noetherian).*

So both classes of $\mathfrak{Z}$-Noetherian and $\mathfrak{Z}$-compact groups are stable under taking quotients with respect to Zariski-closed normal subgroups.

Making use of Remark 3.1, one can prove the following:

Theorem 5.5 (Ref. 27). *A group G is $\mathfrak{Z}$-Noetherian if and only if every countable subgroup of G is $\mathfrak{Z}$-Noetherian.*

Since by Nielson-Schreier's Theorem subgroups of free groups are free, from this theorem and Corollary 5.1 one can deduce that all free groups are $\mathfrak{Z}$-Noetherian (Ref. 27). We shall see below that it follows also from an older result of Guba.

Example 5.2. Let us see that for an infinite set X the group $S(X)$ is not $\mathfrak{Z}$-Noetherian. Indeed, it follows immediately from Lemma 5.1 and Olshanskij's example 4.1 of a countable $\mathfrak{Z}$-discrete group G, that $S(X)$ is not $\mathfrak{Z}$-Noetherian.

Let us give also an immediate proof, making no use of this deep example. Let F_1 be a subset of X with at least three elements, and let $F_1 \subsetneq F_2 \subsetneq F_3 \subsetneq \dots \subsetneq X$ be an infinite chain of subsets of X. Then one has the following strictly descending infinite chain of subgroups of $G = S(X)$:

$$S(X) \supsetneq S_{F_1}(X) \supsetneq S_{F_2}(X) \supsetneq S_{F_3}(X) \supsetneq \dots$$

By Lemma 4.1, each subgroup $S_{F_i}(X)$ is a $\mathfrak{Z}_G$-closed subset of G, so the group G is not $\mathfrak{Z}$-Noetherian.

The next example shows that the quotient of a $\mathfrak{Z}$-Noetherian group need not be $\mathfrak{Z}$-Noetherian.

Example 5.3. Take X an infinite set, F a free group and N a normal subgroup of F such that $F/N = S(X)$. As noted above, F is $\mathfrak{Z}$-Noetherian (see also Corollary 5.2), but the quotient group F/N is not $\mathfrak{Z}$-Noetherian, according to Example 5.2.

Let us see now that the larger class of $\mathfrak{Z}$-compact groups is closed under arbitrary products:

Theorem 5.6 (Ref. 27). *Let $\{G_i \mid i \in I\}$ be a family of groups, and $G = \prod_{i\in I} G_i$. Then the Zariski topology $\mathfrak{Z}_G$ of the direct product is compact if each $(G_i, \mathfrak{Z}_{G_i})$ is compact.*

Proof. It is enough to note that by Theorem 2.1, $\mathfrak{Z}_G$ is coarser than the product topology $\prod_{i\in I} \mathfrak{Z}_{G_i}$. □

Finite products of $\mathfrak{Z}$-Noetherian groups are $\mathfrak{Z}$-Noetherian by Remark 2.1 (a) and Lemma 5.1. In the following theorem we describe when an arbitrary direct product or sum is a $\mathfrak{Z}$-Noetherian group.

Theorem 5.7 (Ref. 27). *Let $\{G_i \mid i \in I\}$ be a non-empty family of groups, $G = \prod_{i\in I} G_i$ and $S = \bigoplus_{i\in I} G_i$. Then the following conditions are equivalent:*

(i) every G_i is $\mathfrak{Z}$-Noetherian and all but finitely many of the groups G_i are abelian;
(ii) G is $\mathfrak{Z}$-Noetherian;
(iii) S is $\mathfrak{Z}$-Noetherian.

Let D_8 denote the 8 element-dyhedral group, and consider the product group $G = D_8^{\mathbb{N}}$. Then G is nilpotent of nilpotency class 2, but it is not $\mathfrak{Z}$-Noetherian by Theorem 5.7. As the class of $\mathfrak{Z}$-Noetherian groups contains all abelian groups by Theorem 5.3, in particular we obtain that it is not stable under taking central extensions.

5.3. *Equations in a free group*

In Ref. 33, Guba considers equations with more than one variable in a group, that are expressions of the following form:

$$g_1 x_{i_1}^{m_1} g_2 x_{i_2}^{m_2} \cdots g_n x_{i_n}^{m_n} = e_G,$$

where $g_1, g_2, \ldots, g_n$ are elements of G, $m_1, m_2, \ldots, m_n$ are integers, and $x_{i_1}, x_{i_2}, \ldots, x_{i_n}$ are taken from a fixed finite set of variables $\{x_1, x_2, \ldots, x_N\}$. Then he proves the following result.

Theorem 5.8 (Ref. 33). *Every system of equations in a free group from a finite number of variables over a free group is equivalent to a finite subsystem.*

Considering only equations in one variable yields that the family of finite intersections of elementary algebraic subsets of a free group is closed under taking arbitrary intersections. Then one can apply Remark 2.1 (b) to get the following corollary of Theorem 5.8.

Corollary 5.2. *Every free group is $\mathfrak{Z}$-Noetherian.*

In Ref. 27 the authors provide a different proof of Corollary 5.2 based on Theorem 5.5 (see the comment in the previous subsection).

Let F be a free non-abelian group, $\mathcal{B}$ be the family of subsets of F consisting of singletons or sets of the type $fC_F(g)h$, for $f, g, h \in F$, and finally let $\mathcal{B}^{\cup}$ the family of finite unions of elements of $\mathcal{B}$. One can see that every element of $\mathcal{B}$ is an elementary algebraic subset of F, and so every $C \in \mathcal{B}^{\cup}$ is an additively algebraic subset of F.

The following result considerably sharpens Theorem 5.8 in describing the solution-set of a system of equations in one variable in a free group (i.e. the intersection of elementary algebraic subsets).

Theorem 5.9 (Ref. 11). *Let F be a free non-abelian group and let S be a system of equations in one variable with coefficients from F. Let V be the set of all solutions of S in F. Then $V \in \mathcal{B}^{\cup}$.*

Applying Remark 2.1 (b), we finally obtain that $\mathcal{B}^{\cup}$ is exactly the family of $\mathfrak{Z}_F$-closed subsets. In particular, the elements of $\mathcal{B}$ are the $\mathfrak{Z}_F$-closed irreducible subsets of F.

6. The Zariski space $(G, \mathfrak{Z}_G)$ of an abelian group G

This section is dedicated to the abelian case, so the groups will always be written additively. According to Ref. 61, an abelian group G is called *almost torsion-free* if $G[n] = \{g \in G : ng = 0\}$ is finite for every $n > 1$.

The commutativity of the operation in an abelian group G allows us to consider only words of the form $w = nx + g$, for $n \in \mathbb{Z}$, and $g \in G$. The corresponding elementary algebraic subset is

$$E_{nx+g} = \begin{cases} \emptyset & \text{if } nx + g = 0 \text{ has no solution in } G, \\ G[n] + x_0 & \text{if } x_0 \text{ is a solution of } nx + g = 0. \end{cases}$$

On the other hand, $G[n] + g = E_{nx-ng}$, so the non-empty elementary algebraic subsets of G are exactly the cosets of the subgroups of G of the form $G[n]$. Since every descending chain of this kind of subgroups stabilizes, we can deduce that every descending chain of elementary algebraic subsets stabilizes, so an application of Proposition 2.1 (see Remark 2.1 (b)) gives:

Theorem 6.1 (Ref. 7). *Every abelian group G is $\mathfrak{Z}$-Noetherian. Moreover, every algebraic set in G is additively algebraic.*

Now we can conclude with Remark 5.1 that $\mathfrak{Z}_G$ is Hausdorff for no infinite abelian group G. Nevertheless, $\mathfrak{Z}_G$ is always T_1. In the next example we carry out the first step in understanding the Zariski topology of an abelian group by describing the abelian groups G for which $\mathfrak{Z}_G$ is the coarsest T_1 topology, i.e., the cofinite topology.

Example 6.1. (Ref. 61, Theorem 5.1) For an abelian group G, the following conditions are equivalent:

(i) every proper algebraic subset of G is finite (i.e., $\mathfrak{Z}_G$ coincides with the cofinite topology of G),
(ii) every proper elementary algebraic subset of G is finite,
(iii) either G is almost torsion-free, or G has exponent p for some prime p.

The next lemma shows that the Zariski topology behaves well under taking subgroups. This is a typical property in the abelian case (for counterexamples in the non-abelian case, see Example 4.3). The proof easily follows either from the explicit description of the elementary algebraic subsets in the abelian case, or from Lemma 3.1 (a).

Lemma 6.1. *For every subgroup H of an abelian group G one has $\mathfrak{Z}_G \restriction_H = \mathfrak{Z}_H$.*

We need an invariant closely related to the exponent:

Definition 6.1. Let G be an abelian group.

(i) (Refs. 22,32) If G is bounded, then the *essential order* $eo(G)$ of G is the smallest positive integer n such that nG is finite.
(ii) If G is not bounded, we define $eo(G) = 0$.

If G is a bounded abelian group with $n = eo(G)$ and mG is finite for some $m \in \mathbb{N} \setminus \{0\}$, then n divides m. In particular, the essential order $eo(G)$ of a bounded group G divides its exponent.

It is well known fact that in a Noetherian space every subspace admits a unique decomposition in *irreducible components*. Now we see that the essential order describes the connected component and the irreducible component of the identity of the space $(G, \mathfrak{Z}_G)$ for an abelian group G.

Theorem 6.2 (Ref. 22, Theorem 4.6). *Let G be an infinite abelian group and $n = eo(G)$. Then:*

(i) $G[n]$ has finite index in G, so it is a $\mathfrak{Z}_G$-clopen subgroup of G;
(ii) $G[n]$ is both the connected component and the irreducible component of the identity of the space $(G, \mathfrak{Z}_G)$.

Consequently, the Zariski space $(G, \mathfrak{Z}_G)$ of an abelian group G is connected if and only if it is irreducible. This occurs precisely when $G = G[n]$, where $n = eo(G)$ (Ref. 22, Corollary 4.7).

6.1. *Description of the Zariski topology*

The next two notions are the core of the Zariski topology of an abelian group.

Definition 6.2 (Ref. 19). *Let G be an abelian group and X be a countably infinite subset of G. We say that:*

(i) X is almost n-torsion *in G for a given $n \in \mathbb{N}$, if $X \subseteq G[n]$ and the set $\{x \in X : dx = h\}$ is finite for each $h \in G$ and every proper divisor d of n.*
(ii) X is a $\mathfrak{Z}_G$-atom *is the subspace $(X, \mathfrak{Z}_G \restriction_X)$ of $(G, \mathfrak{Z}_G)$ carries the cofinite topology.*

Obviously, an infinite subset of a $\mathfrak{Z}_G$-atom is still a $\mathfrak{Z}_G$-atom, the same applies to almost n-torsion sets as well.

For $n \in \mathbb{N}$, a countably infinite subset X of an abelian group G is almost n-torsion if and only if $G[n]$ has exponent n, $\mathrm{cl}_{\mathfrak{Z}_G}(X) = G[n]$ and X is a $\mathfrak{Z}_G$-atom.

Proposition 6.1 (Ref. 19). *Suppose that $n \in \mathbb{N}$ and G is an abelian group. Then for every infinite set $Y \subseteq G[n]$, the following conditions are equivalent:*

(i) Y contains an almost n-torsion set,
(ii) $G[n]$ is an irreducible subgroup of G of exponent n, and $\mathrm{cl}_{\mathfrak{Z}_G}(Y) = G[n]$,
(iii) $eo(G[n]) = n$ and $\mathrm{cl}_{\mathfrak{Z}_G}(Y) = G[n]$.

Using this proposition one obtains the following connection between $\mathfrak{Z}_G$-atoms and almost n-torsion sets.

Theorem 6.3 (Ref. 19). *A countably infinite subset X of an abelian group G is a $\mathfrak{Z}_G$-atom if and only if $X - a$ is almost n-torsion for some $a \in G$ and $n \in \mathbb{N}$.*

One can show that $n \in \mathbb{N}$ in the above theorem depends only on X, not on $a \in G$. We denote by $\mathfrak{n}(X)$ this uniquely determined number n; it coincides with $eo(G[n])$, where $G[n]$ is the Zariski closure of the almost n-torsion set $X - a$ (see Proposition 6.1).

Using Theorem 6.3 one can see that *the irreducible subsets of $(G, \mathfrak{Z}_G)$ are precisely the sets containing a dense $\mathfrak{Z}_G$-atom.* In the case of closed irreducible subsets one obtains the following nice description:

Corollary 6.1. *For a subset F of an abelian group G, the following conditions are equivalent:*

(i) F is a closed, irreducible subset of $(G, \mathfrak{Z}_G)$,
(ii) F is a $\mathfrak{Z}_G$-irreducible elementary algebraic set,
(iii) F is a $\mathfrak{Z}_G$-connected elementary algebraic set,
(iv) F coincides with the $\mathfrak{Z}_G$-closure of some $\mathfrak{Z}_G$-atom.

The next theorem is a variation of Ref. 22, Theorem 8.1. Its proof follows the main line of the proof of Ref. 22, Theorem 8.1, using the decomposition in irreducible components and the description of the (closed) irreducible subset of $(G, \mathfrak{Z}_G)$.

Theorem 6.4. *Let X be an infinite subset of an abelian group G, let D be the finite set of $\mathfrak{Z}_G$-isolated points of X, and let $X \setminus D = \bigcup_{i=1}^{k} X_i$ be* *the (unique) decomposition of* $X \setminus D$ *into irreducible components. Then for every* $i \leq k$ *there exists a* $\mathfrak{Z}_G$*-atom* A_i*,* $\mathfrak{Z}_G$*-densely contained in* X_i*, such that*

$$\mathrm{cl}_{\mathfrak{Z}_G}(X) = D \cup \bigcup_{i=1}^{k} \mathrm{cl}_{\mathfrak{Z}_G}(A_i) = D \cup \bigcup_{i=1}^{k} (a_i + G[n_i]),$$

where $n_i = \mathfrak{n}(A_i)$ *and* $a_i \in A_i$*.*

This theorem shows that the $\mathfrak{Z}_G$-atoms of a Zariski space $(G, \mathfrak{Z}_G)$ play the same fundamental role as the usual converging sequences in a metric space. This fact is used (see Ref. 22, Corollaries 8.8 and 8.9) to deduce the next corollary from Theorem 6.4:

Corollary 6.2. *For an abelian group* G*, the space* $(G, \mathfrak{Z}_G)$ *is:*

(a) hereditarily separable; that is, every subset X *of* $(G, \mathfrak{Z}_G)$ *has a countable dense subset.*
(b) Fréchet-Urysohn; that is, if X *is a subset of* G *and* $x \in \mathrm{cl}_{\mathfrak{Z}_G}(X)$*, then there exists a sequence of points* $\{x_j : j \in \mathbb{N}\}$ *of* X *converging to* x*.*

There exist easy examples of Noetherian T_1-spaces that are neither separable nor Fréchet-Urysohn (see Ref. 22), so both properties in the above corollary are specific for the Zariski space $(G, \mathfrak{Z}_G)$ of an abelian group G.

6.2. *Realization theorem and applications to Markov's problems 1 and 4*

Let G be a (non-necessarily abelian) group and $\mathcal{T}$ be any Hausdorff group topology on G. Then $\mathfrak{Z}_G \subseteq \mathfrak{M}_G \subseteq \mathcal{T}$, and therefore the $\mathcal{T}$-closure of a subset of G must be contained in its $\mathfrak{Z}_G$-closure. In other words, the $\mathfrak{Z}_G$-closure of a given set $S \subseteq G$ is the *biggest* subset of G that one could attain by taking the closure of S in *any* Hausdorff group topology on G. This leads to the question whether the $\mathfrak{Z}_G$-closure of S can actually be *realized* by taking the closure of S in *some* Hausdorff group topology on G. It turns out that this is possible when $|G| \leq \mathfrak{c}$ and the Hausdorff group topology can be chosen to be precompact and metric.

Theorem 6.5 (Ref. 22). *Let* G *be an abelian group with* $|G| \leq \mathfrak{c}$*, and let* X *be a subset of* G*. Then there exists a precompact metric group topology* $\mathcal{T}$ *on* G *such that* $\mathrm{cl}_{\mathcal{T}}(X) = \mathrm{cl}_{\mathfrak{Z}_G}(X)$*.*

A stronger form of this theorem was proved in Ref. 22, with the single set X replaced by a countable family $\mathscr{X}$ of subsets of G. We prefer to give this weaker version which allows for an easier first approach.

As a first application of Theorem 6.5, let us see that all three topologies coincide on abelian groups:

Theorem 6.6 (Ref. 22). *If G is an abelian group, then $\mathfrak{Z}_G = \mathfrak{M}_G = \mathfrak{P}_G$.*

Let us give a sketch of the proof of this theorem. Since $\mathfrak{Z}_G \subseteq \mathfrak{M}_G \subseteq \mathfrak{P}_G$, it suffices to prove only that $\mathfrak{P}_G \subseteq \mathfrak{Z}_G$. Let us pick a $\mathfrak{P}_G$-closed subset F of G and show that F is also $\mathfrak{Z}_G$-closed. According to Corollary 6.2 (a), there exists a countable subset X of F such that $\mathrm{cl}_{\mathfrak{Z}_G}(X) = \mathrm{cl}_{\mathfrak{Z}_G}(F)$. Fix an arbitrary $x \in \mathrm{cl}_{\mathfrak{Z}_G}(F) = \mathrm{cl}_{\mathfrak{Z}_G}(X)$, and let H be the subgroup of G generated by X and x. By Lemma 6.1, $x \in H \cap \mathrm{cl}_{\mathfrak{Z}_G}(X) = \mathrm{cl}_{\mathfrak{Z}_G \restriction H}(X) = \mathrm{cl}_{\mathfrak{Z}_H}(X)$. According to Theorem 6.5, there exists a precompact group topology $\mathcal{T}$ on H with $\mathrm{cl}_{\mathfrak{Z}_H}(X) = \mathrm{cl}_{\mathcal{T}}(X)$. Using extensions of $\mathcal{T}$-continuous characters of H to characters of G and a sufficiently rich family of characters of G vanishing on H to separate the points of G/H, extend $\mathcal{T}$ to a precompact group topology $\mathcal{T}^*$ on G. Then

$$x \in \mathrm{cl}_{\mathfrak{Z}_H}(X) = \mathrm{cl}_{\mathcal{T}}(X) \subseteq \mathrm{cl}_{\mathcal{T}^*}(X) \subseteq \mathrm{cl}_{\mathfrak{P}_G}(X) \subseteq \mathrm{cl}_{\mathfrak{P}_G}(F) = F.$$

This proves that F is $\mathfrak{Z}_G$-closed.

In the particular case when G is almost torsion-free the equality $\mathfrak{Z}_G = \mathfrak{M}_G$ was earlier proved in Ref. 61.

As a by-product we obtain the following unexpected topological 'commutativity criterion' for soluble divisible groups:

Theorem 6.7. *For a soluble divisible group G the following are equivalent:*

(a) *G is abelian;*
(b) *$\mathfrak{M}_G = \mathfrak{P}_G$;*
(c) *$\mathfrak{P}_G$ is non-discrete.*

Proof. The implication (a) $\to$ (b) follows from Theorem 6.6. According to Theorem 4.8, $\mathfrak{M}_G$ is non-discrete, so (b) $\to$ (c). By Corollary 4.2, (c) implies that G is abelian, so (c) $\to$ (a). □

Here we discuss another application of Zariski-density to Markov's fourth problem on potentially density.

Tkachenko and Yaschenko[61] proved the following theorem: if an abelian group G of size at most $\mathfrak{c}$ is either almost torsion-free or has exponent p

for some prime p, then every infinite subset of G is potentially dense in G (in this case the infinite subsets are precisely the Zariski-dense subsets according to Example 6.1). Then they asked whether the restriction $|G| \leq \mathfrak{c}$ can be relaxed to $|G| \leq 2^{\mathfrak{c}}$. This inequality is motivated by the well known fact that a Hausdorff space with a dense subset D has size at most $2^{2^{|D|}}$; in particular, a separable Hausdorff space has size at most $2^{\mathfrak{c}}$.

In Ref. 24 the following natural characterization of potentially dense subsets X of abelian groups of size at most $2^{\mathfrak{c}}$ is obtained: X is potentially dense in G if and only if $|G| \leq 2^{\mathfrak{c}}$ and X is Zariski-dense in G. More precisely:

Theorem 6.8. *Let X be a subset of an abelian group G such that $|G| \leq 2^{\mathfrak{c}}$. Then the following conditions are equivalent:*

(i) X is potentially dense in G,
(ii) X is $\mathcal{T}$-dense in G for some precompact group topology on G,
(iii) $\mathrm{cl}_{\mathfrak{Z}_G}(X) = G$.

If $|G| \leq \mathfrak{c}$, then the precompact topology $\mathcal{T}$ can be chosen to be also metric.

As a precompact metric group has size at most $\mathfrak{c}$, the restraint $|G| \leq \mathfrak{c}$ in the final part of the theorem is necessary.

In Ref. 23 the remaining case of uncountable sets is investigated. The following sufficient condition that guarantees potential density of a subset S of an abelian group G is proposed:

Definition 6.3 (Ref. 23). *Let κ be an infinite cardinal. We say that a subset S of an abelian group G is κ-wide if $nS \not\subseteq \langle S' \rangle$ for every $n \in \mathbb{N}$ and each $S' \subseteq S$ with $|S'| < \kappa$.*

Theorem 6.9 (Ref. 23). *Let κ be an infinite cardinal and S be a κ-wide subset of an abelian group G such that $|G| \leq 2^{2^{\kappa}}$. Then there exists a precompact Hausdorff group topology τ on G such that S is dense in (G, τ).*

In other words, κ-widness implies potential density for subsets of groups of size $\leq 2^{2^{\kappa}}$. Moreover, the topology τ on G witnessing potential density can be chosen to be precompact.

When S is uncountable and G belongs to a large class of abelian groups, this sufficient condition turns out to be also necessary for the potential density of S in G.

If S is a potentially dense subset in G, then there exists a Hausdorff group topology τ on G such that S is dense in (G, τ). But then for every

$n \in N$ the subset nS will be dense in nG as an image of S under the continuous map $s \mapsto ns$. Hence, $|nG| \leq 2^{2^{|nS|}}$.

Corollary 6.3 (Ref. 23). *If G is an abelian group with $|nG| = |G|$ for every $n \in \mathbb{N}$, then the following conditions are equivalent for an uncountable subset S of G:*

(a) *S is potentially dense in G;*
(b) *S is τ-dense in some precompact Hausdorff group topology τ on G;*
(c) *$|G| \leq 2^{2^{|nS|}}$ holds for every $n \in N$.*

The hypothesis of this theorem is satisfied, for example, for (almost) torsion-free or divisible abelian groups.

7. The Zariski topology of some linear groups

7.1. *The group* $(K^*, \cdot) \ltimes (K^n, +)$

Let K be a field, n be a positive integer, and consider the action of the group $(K^*, \cdot)$ on the additive group $(K^n, +)$ defined by scalar multiplication. Denote by G_K, or simply G, the semidirect product $K^* \ltimes K^n$, and by N its normal subgroup $N_K = \{1\} \times K^n$.

Let $K(T)$ be the field of rational functions over K. For $r \in K(T)$ with finite set Z of roots in K of its denominator, let $\operatorname{dom} r$ denote the domain $K \setminus Z$ of the rational evaluation function $K \setminus Z \to K$, $x \mapsto r(x)$, associated to r.

In the sequel we use functions $K \to K^n$ having each component a rational function, and that we consider as elements of $K(T)^n$. Hence, for such a function $\vec{r} = (r_1, \ldots, r_n) \in K(T)^n$, we let $\operatorname{dom} \vec{r} = \bigcap_i \operatorname{dom} r_i$, this is a co-finite subset of K.

For $\vec{r} \in K(T)^n$, let

$$F_{\vec{r}} = \{(x, \vec{r}(x)) \in G \mid x \in K^* \cap \operatorname{dom} \vec{r} = \operatorname{dom} \vec{r} \setminus \{0_K\}\}.$$

One can identify the subset $F_{\vec{r}}$ of $K^* \times K^n$ with the 'graph' of the n-tuple of rational functions $\vec{r}$: $K^* \cap \operatorname{dom} \vec{r} \to K^n$.

Let $\mathcal{B}$ be the family consisting of the following subsets of G:

- singletons of G, and the whole G;
- cosets of the normal subgroup N of G;
- graphs $F_{\vec{r}}$, with $\vec{r} \in K(T)^n$.

Theorem 7.1 (Ref. 28). *G is a $\mathfrak{Z}$-Noetherian group, and every algebraic subset of G is a finite union of members of $\mathcal{B}$.*

Note that G is not nilpotent as $Z(G) = 1$, but it is solvable of class 2, as $G' = N$ is abelian. Note also that G is a linear group, as it is isomorphic to the matrix group $\begin{pmatrix} K^* & K^n \\ \vec{0}^t & I_n \end{pmatrix}$, a subgroup of $\mathrm{GL}_{n+1}(K)$. In particular, we can consider the affine topology $\mathcal{A}_G$ of the group G. From Theorem 7.1 one obtains:

Corollary 7.1 (Ref. 28). *The Zariski topology $\mathfrak{Z}_G$ is properly contained in the affine topology $\mathcal{A}_G$.*

The linear space $V = K^n$ in the definition of the group G can be replaced by a K-linear space V of arbitrary dimension, not necessarily finite-dimensional. In Ref. 28, the authors prove similar results for the semidirect product obtained in this way.

The next theorem was proved in Ref. 28 using the fact that if K is a field such that

$$\mathrm{char}\, K = 0 \text{ or } \mathrm{char}\, K = p \text{ and the extension } K/\mathbb{F}_p \text{ is not algebraic,} \quad (\dagger)$$

then there exist a Hausdorff group topology on $(K, +)$, and an element $a \in K^*$ such that the group automorphism $x \mapsto ax$ of K is not continuous.

Theorem 7.2. *N_K is not Hausdorff embedded in G_K if $(\dagger)$ holds.*

As $N \cong (K, +)^n$ is abelian, its Zariski-Markov topology $\mathfrak{Z}_N$ is the cofinite topology of N by Example 6.1, so it follows from Theorem 7.1 that N is a $\mathfrak{Z}_G$-closed, Zariski embedded normal subgroup of G.

If K is countable, then G is countable as well, so $\mathfrak{Z}_G = \mathfrak{M}_G$ by Markov's Theorem. As a consequence, Theorem 7.1 describes also the Markov topology of G, and for any subgroup the conditions of being Markov embedded and Zariski embedded coincide. In particular N is Markov embedded in G.

From item (a) and (b) of Lemma 3.1, we conclude that a super-normal subgroup is both Zariski embedded and Markov embedded. The converse does not hold true; actually, a Zariski and Markov embedded normal subgroup need not be even Hausdorff embedded. In fact, if $K = \mathbb{Q}$, then $G_\mathbb{Q}$ is countable, so $N_\mathbb{Q}$ is Zariski and Markov embedded, but not Hausdorff embedded, in $G_\mathbb{Q}$ (by Theorem 7.2). Moreover $N_\mathbb{Q}$ is normal, $\mathfrak{Z}_{G_\mathbb{Q}}$-closed, $\mathfrak{M}_{G_\mathbb{Q}}$-closed.

7.2. *The Heisenberg group*

If K is a field, the *Heisenberg group with coefficients in* K is the group H_K of unitriangular upper matrixes

$$H_K = \begin{pmatrix} 1 & K & K \\ 0 & 1 & K \\ 0 & 0 & 1 \end{pmatrix}$$

which we will denote simply H from now on.

Consider now rational functions $r = \frac{\beta+\gamma T}{\alpha+\delta T} \in K(T)$, with $\alpha, \beta, \gamma, \delta \in K$ such that α or δ is not zero, and $\delta = 0$ if $\operatorname{char} K \neq 2$. We define the following subset of H

$$A_r = \left\{ \begin{pmatrix} 1 & x & K \\ & 1 & r(x) \\ & & 1 \end{pmatrix} \in H \mid x \in \operatorname{dom} r \right\}.$$

Denote L and M the following subgroups of H:

$$L = L_K = \begin{pmatrix} 1 & 0 & K \\ & 1 & K \\ & & 1 \end{pmatrix}, \quad M = M_K = \begin{pmatrix} 1 & K & K \\ & 1 & 0 \\ & & 1 \end{pmatrix}.$$

Let $\mathcal{C}$ denote the family consisting of the following subsets of H:

- singletons, the whole H;
- cosets either of $Z(H)$ or L;
- the subsets A_r.

Theorem 7.3. *H is a $\mathfrak{Z}$-Noetherian group and the $\mathfrak{Z}_H$-closed sets are the finite unions of elements of $\mathcal{C}$.*

Note that the zero rational function $r = 0 \in K(T)$ gives $A_r = M$, so both L and M are $\mathfrak{Z}_H$-closed subsets of H by Theorem 7.3.

We shall see below (Corollary 7.2) that L is a non-Zariski embedded subgroup of H, and the same result holds for M; in particular, neither L nor M are super-normal in H. However, their intersection $L \cap M = Z(H)$ is super-normal in H.

Identifying L with the group K^2 via the isomorphism

$$\begin{aligned} K^2 &\to L \\ (y, z) &\mapsto \begin{pmatrix} 1 & 0 & y \\ & 1 & z \\ & & 1 \end{pmatrix}, \end{aligned}$$

one can see that the $\mathfrak{Z}_H \restriction_L$-closed sets are the unions of a finite subset of K^2 and a set $K \times F$, for a finite subset F of K. The following corollary compares the three topologies $\mathfrak{Z}_L$, $\mathfrak{Z}_H \restriction_L$ and $\mathcal{A}_L$ on L.

Corollary 7.2. *For the group L, the following holds: $\mathfrak{Z}_L \subsetneq \mathfrak{Z}_H \restriction_L \subsetneq \mathcal{A}_L$. In particular, the Zariski topology $\mathfrak{Z}_H$ and the affine topology $\mathcal{A}_H$ differ.*

Indeed, one can easily see that $\mathfrak{Z}_L \subsetneq \mathfrak{Z}_H \restriction_L$, as $\mathfrak{Z}_L$ is the cofinite topology by Example 6.1. To prove the second inclusion to be proper, observe that the subset

$$\{(y, z) \in K^2 \mid z = y\} = \{(y, y) \in K^2 \mid y \in K\}$$

is $\mathcal{A}_{K^2}$-closed; so

$$\left\{ \begin{pmatrix} 1 & 0 & y \\ & 1 & y \\ & & 1 \end{pmatrix} \in H \mid y \in K \right\}$$

is $\mathcal{A}_L$-closed, but it is not $\mathfrak{Z}_H \restriction_L$-closed. Now recall that $\mathcal{A}_L$ is $\mathcal{A}_H \restriction_L$ by definition, so $\mathfrak{Z}_H \restriction_L \subsetneq \mathcal{A}_H \restriction_L$ yields $\mathfrak{Z}_H \subsetneq \mathcal{A}_H$.

Here we provide examples of $\mathfrak{P}$-discrete groups.

Example 7.1. (Ref. 28) Let K be a field of characteristic 0.

(i) Then the group H_K is $\mathfrak{P}$-discrete.
(ii) If in addition K is also algebraically closed, then also G_K is $\mathfrak{P}$-discrete.

Indeed, since the group H_K is a nilpotent (so solvable) non-abelian group, to see (i) it suffices to apply Corollary 4.2. At this end we need to check that H_K is divisible. Since $Z(H_K) \cong (K, +)$ is divisible, it suffices to check that $H_K/Z(H_K)$ is divisible. This follows from $H_K/Z(H_K) \cong (K, +) \times (K, +)$. For (ii) see Ref. 28.

Theorem 7.3 and Example 7.1 yields that $\mathfrak{Z}_{H_K} \neq \mathfrak{P}_{H_K}$ for the nilpotent group H_K, the Heisenberg group over a field K with $\operatorname{char} K = 0$. This partially answers Question 12.1 from Ref. 22. See Question 8.13 for its still unsolved parts.

8. Open questions

We start with some open questions taken from Refs. 20,21 about the class $\mathcal{MZ}$.

Question 8.1. (Ref. 20) For which uncountable cardinals τ does the free group of rank τ belong to $\mathcal{M}Z$?

Question 8.2. (Ref. 20) For which infinite cardinals τ does the permutation group $S(\tau)$ of a set of size τ belong to $\mathcal{M}\mathcal{Z}$?

Question 8.3. (Ref. 21) Let H be an abelian subgroup of G.

(i) If the index of H in G is finite, does G belong to $\mathcal{M}\mathcal{Z}$?
(ii) If the index of H in G is countable, does G belong to $\mathcal{M}\mathcal{Z}$?

What is the answer to both items (i) and (ii) if one additionally assumes that H is a normal subgroup of G?

Question 8.4. (Refs. 20,21)

(i) Is $\mathcal{M}Z$ closed under taking subgroups?
(ii) Is $\mathcal{M}Z$ closed under taking (finite) direct products?
(iii) Is $\mathcal{M}Z$ closed under taking (finite) direct sums?

By Corollary 3.6, the direct sum of countable groups belongs to $\mathcal{M}\mathcal{Z}$.

Sipacheva[59] announced that $\mathcal{M}\mathcal{Z}$ contains all subgroup of a direct sum of countable groups, but the argument has a gap (see Ref. 21). Apparently, the question remains still open, so we formulate it for the sake of completeness:

Question 8.5. (Ref. 21) Let G be a subgroup of a direct sum of countable groups. Does G belong to $\mathcal{M}\mathcal{Z}$?

A negative answer to this question will provide also a negative answer to Question 8.4 (i).

The next three questions are related to Markov's fourth problem.

Question 8.6. (Ref. 20) Let G be a group of size at most $2^{\mathfrak{c}}$ and $\mathcal{E}$ a countable family of subsets of G. Can one find a Hausdorff group topology $\mathcal{T}_{\mathcal{E}}$ on G such that the $\mathcal{T}_{\mathcal{E}}$-closure of every $E \in \mathcal{E}$ coincides with its $\mathfrak{M}_G$-closure?

For an abelian group G the answer is positive, and in fact the topology $\mathcal{T}_{\mathcal{E}}$ in this case can be chosen to be precompact by Theorem 6.8.

The counterpart of Question 8.6 for $\mathfrak{Z}_G$ instead of $\mathfrak{M}_G$ has a negative answer. Indeed, let G be an infinite $\mathfrak{M}$-discrete, not $\mathfrak{Z}$-discrete group, and let e be the unit element of G. Then $G\setminus\{e\}$ is not $\mathfrak{Z}_G$-closed, so the $\mathfrak{Z}_G$-closure of $G \setminus \{e\}$ must coincide with G. On the other hand, the only Hausdorff group topology $\mathcal{T}$ on G is the discrete topology, and so the $\mathcal{T}$-closure of $G \setminus \{e\}$ is $G \setminus \{e\}$.

Let us consider now the counterpart of Question 8.6 for $\mathfrak{P}_G$ instead of $\mathfrak{M}_G$.

Question 8.7. (Ref. 20) Let G be a group of size at most $2^{\mathfrak{c}}$ having at least one precompact Hausdorff group topology, and let $\mathcal{E}$ be a countable family of subsets of G. Can one find a precompact Hausdorff group topology $\mathcal{T}_{\mathcal{E}}$ on G such that the $\mathcal{T}_{\mathcal{E}}$-closure of every $E \in \mathcal{E}$ coincides with its $\mathfrak{P}_G$-closure?

Again, for an abelian group G the answer is positive by Theorem 6.8.

For an infinite abelian group G, $\mathfrak{Z}_G$ is neither Hausdorff, nor a group topology on G, but still has various nice properties, e.g., the space $(G, \mathfrak{Z}_G)$ is Noetherian (Theorem 6.1), hereditarily separable and Fréchet-Urysohn (Corollary 6.2), moreover it has only finitely many connected components, and each component is an irreducible space[22].

Question 8.8. (Ref. 20) Let G be an abelian group with $|G| \leq 2^{\mathfrak{c}}$ and $\mathcal{E}$ a family of subsets of G with $|\mathcal{E}| < 2^{|G|}$. Does there exist a precompact Hausdorff group topology $\mathcal{T}_{\mathcal{E}}$ on G such that the $\mathcal{T}_{\mathcal{E}}$-closure of each $E \in \mathcal{E}$ coincides with its $\mathfrak{Z}_G$-closure?

The answer is positive for countable families $\mathcal{E}$ (Ref. 24). Moreover, it was shown in Ref. 22 that if $|G| \leq \mathfrak{c}$, one can choose the approximating topology $\mathcal{T}_{\mathcal{E}}$ to be even metric (see Theorem 6.5 for the case of $\mathcal{E}$ consisting of a single subset X of G).

The restriction on the cardinality of the group G in Questions 8.6, 8.7 and Question 8.8 is obviously necessary since the closure of a countable set in a Hausdorff topology cannot exceed $2^{\mathfrak{c}}$.

The next five problems concern Zariski, Markov, and Hausdorff embeddings of subgroups.

Question 8.9. Let H be a (normal) subgroup of a group G. If H is Zariski embedded in G, must it also be Markov embedded in G?

Our next problem should be compared with Theorem 4.5.

Problem 8.1. *(Ref. 21) Describe the class of groups G such that G is Markov (respectively, Hausdorff) embedded in every group that contains G as a normal subgroup.*

We note that the counterpart of Problem 8.1 with "Markov" (respectively, "Hausdorff") replaced by "Zariski" has been completely solved in Ref. 25.

Let $\{G_i : i \in I\}$ be a family of groups, and for every $i \in I$, let H_i be a Hausdorff embedded subgroup of G_i. Then the Cartesian product $\prod_{i\in I} H_i$ is a Hausdorff embedded subgroup of the Cartesian product $\prod_{i\in I} G_i$. Similarly, the direct sum $\bigoplus_{i\in I} H_i$ is a Hausdorff embedded subgroup of the direct sum $\bigoplus_{i\in I} G_i$. In other words, the class of Hausdorff embeddings is closed under taking both Cartesian products and direct sums. It is not known whether the remaining two classes of Markov embeddings and of Zariski embeddings are closed under taking Cartesian products and direct sums. This is an open question even in the weakest possible form:

Question 8.10. (Ref. 21) Assume H_1 is Markov (Zariski) embedded in G_1. Is it true that for every group G_2 the subgroup $H_1 \times G_2$ of $G = G_1 \times G_2$ is Markov (Zariski) embedded in G?

Recall that in Theorem 7.2 we proved that N_K is not Hausdorff embedded in G_K whenever (†) holds. We do not know what happens when (†) fails, this leaves open the following:

Question 8.11. Is N_K Hausdorff embedded in G_K, provided that the field K is an infinite algebraic extension of $\mathbb{F}_p$?

In case (†) holds, one can still ask:

Question 8.12. If K is an uncountable field with (†), is then N_K Markov embedded in G_K? What about the general case (i.e., when (†) fails)?

A positive answer would give that $N = N_K$ is both Zariski and Markov embedded, but not Hausdorff embedded in $G = G_K$, as for the countable case.

A negative answer would give that N is not Markov embedded in G, i.e. $\mathfrak{M}_G \restriction_N \not\subseteq \mathfrak{M}_N$; but N is Zariski embedded in G, i.e. $\mathfrak{Z}_G \restriction_N = \mathfrak{Z}_N$. As the T_1 topology $\mathfrak{M}_G \restriction_N$ contains the cofinite topology $\mathfrak{Z}_N = \mathfrak{M}_N$, we would have

$$\mathfrak{Z}_G \restriction_N = \mathfrak{Z}_N = \mathfrak{M}_N \subsetneq \mathfrak{M}_G \restriction_N$$

and in particular the Zariski and Markov topologies of G would differ.

Recall that if K is a field with $\operatorname{char} K = 0$, the nilpotent group H_K is $\mathfrak{P}$-discrete, but not $\mathfrak{Z}$-discrete.

Question 8.13. (Ref. 22) If G is a nilpotent group, does $\mathfrak{Z}_G = \mathfrak{M}_G$ necessarily hold?

Question 8.14. Let G be an infinite group. Is $\mathfrak{Z}_G$ (resp. $\mathfrak{M}_G$, $\mathfrak{P}_G$) a group topology, if it is Hausdorff?

Question 8.15. (Ref. 5) Is it true that an infinite locally finite group is not $\mathfrak{M}$-discrete?

The above question has a partial, positive answer in the countable case by Theorem 4.6.

The next two questions are motivated by Remark 4.1 and Theorem 4.3, respectively:

Question 8.16. Is every group a subgroup of a Taĭmanov group?

Question 8.17. If $|X| = \omega_1$, is $S(X) \in \mathfrak{T}$? And in the case $|X| = \mathfrak{c}$? And if $\omega < |X| \leq \mathfrak{c}$?

Let

$$\mathfrak{N} = \{G : \mathfrak{Z}_G \text{ is Noetherian}\}, \quad \mathfrak{C} = \{G : \mathfrak{Z}_G \text{ is compact}\},$$
$$\text{and } \mathfrak{H} = \{G : \mathfrak{Z}_G \text{ is Hausdorff}\}.$$

Then $\mathfrak{N}$ is stable under taking finite products and subgroups, and under taking quotients with respect to Zariski closed normal subgroups (see Proposition 5.2, see Theorem 5.7 for the description of the arbitrary products that belong to $\mathfrak{N}$).

While $\mathfrak{N}$ contains all abelian groups, all free groups, and is stable under taking finite products, it is not stable under taking central extensions, in particular a nilpotent group of nilpotency class 2 need not be in $\mathfrak{N}$.

On the other hand, $\mathfrak{C}$ is stable under taking arbitrary products (Theorem 5.6), Zariski closed subgroups, and quotients with respect to Zariski closed normal subgroups (Proposition 5.2). From Theorem 4.10 one obtains examples witnessing non-stability of $\mathfrak{C}$ with respect to taking subgroups.

Problem 8.2. *Establish further permanence properties of the classes* $\mathfrak{N}$, $\mathfrak{C}$, $\mathfrak{H}$.

It is a pleasure to thank the referee for the very careful reading and the numerous useful suggestions that improved the paper; thanks are also due to Giovanni Bacci for the technical support.

References

1. K. Appel, *One-variable equations in free groups*, Proc. Amer. Math. Soc. 19 (1968), pp. 912–918.
2. A. Arhangel′skiĭ, M. Tkachenko, *Topological groups and related structures*, Atlantis Studies in Mathematics, 1. Atlantis Press, Paris; World Scientific Publishing Co. Pte. Ltd., Hackensack, NJ, 2008, pp. xiv+781.
3. S. Balcerzyk, J. Mycielski, *On the existence of free subgroups of topological groups*, Fund. Math. 44 (1957), pp. 303–308.
4. G. Baumslag, A. Myasnikov, V. Remeslennikov, *Algebraic geometry over groups. I: Algebraic sets and ideal theory*, J. Algebra 219 (1999), pp. 16–79.
5. V.V. Belyaev, *Topologization of countable locally finite groups*, Algebra and Logic 34 (1996), pp. 339–342.
6. D. Bormotov, R. Gilman, A. Myasnikov, *Solving one-variable equations in free groups*, J. Group Theory 12 (2009), pp. 317–330.
7. R. Bryant, *The verbal topology of a group*, J. Algebra 48 (1977), no. 2, pp. 340–346.
8. R. Bryant, J.S. Wilson, *On locally finite CZ-groups*, Quart. J. Math. Oxford 30 (1979), no. 2, pp. 265–270.
9. R. Bryant, T. Yen, *Sylow subgroups of locally finite CZ-groups*, Arch. Math. (Basel) 31 (1978/79), no. 2, pp. 117–119.
10. G. Cherlin, MR review # MR0723069 (85j:20028) to Ref. 36, 1985.
11. I.M. Chiswell, V. Remeslennikov, *Equations in free groups with one variable. I*, J. Group Theory 3 (2000), pp. 445–466.
12. W.W. Comfort, K.H Hofmann, D. Remus, *Topological groups and semigroups*, Recent progress in general topology (Prague, 1991), North-Holland, Amsterdam (1992), pp. 57–144.
13. W.W. Comfort, *Compact groups: subgroups and extensions*, in: Á. Császár, ed., Colloquia Mathematica Societas János Bolyai 41 (Topology Theory and Applications), Proc. Colloquium on Topology, Eger (Hungary), Bolyai János Matematikai Társulat, Budapest 1983), pp. 183–198.
14. W.W. Comfort, D. Grant, *Cardinal invariants, pseudocompactness and minimality: Some recent advances in the topological theory of topological groups*, Topology Proc. **6** (1978), pp. 227–265.
15. W. Comfort, D. Remus, *Longs chains of topological group topolgies – a continuation*, Topology Appl. **75** (1997), pp. 51–79.
16. W. Comfort, L.C. Robertson *Images and quotients of $SO(3,\mathbb{R})$: Remarks on a theorem of van der Waerden*, Rocky Mountain Jour. Math. **17** n. 1 (1987), pp. 1–13.
17. D. Dikranjan, *The Markov–Zariski topology of an abelian group*, talk given at the Prague Topological Symposium, August 2006.
18. D. Dikranjan, A. Giordano Bruno, *Arnautov's problems on semitopological isomorphisms*, Applied General Topology 10 (2009), no. 1, pp. 85-119.
19. D. Dikranjan, D. Shakhmatov, *Forcing hereditarily separable countably compact group topologies on Abelian groups*, Topology Appl. 151 (2005), pp. 2–54.
20. D. Dikranjan, D. Shakhmatov, *Selected topics from the structure theory of topological groups*, in: Open problems in topology II, E. Perl. Ed., Elsevier

(2007), pp. 389–406.
21. D. Dikranjan, D. Shakhmatov, *Reflection principle characterizing groups in which unconditionally closed sets are algebraic*, J. Group Theory 11 (2008), no. 3, pp. 421–442.
22. D. Dikranjan, D. Shakhmatov, *The Markov–Zariski topology of an abelian group*, J. Algebra 324 (2010), pp. 1125–1158.
23. D. Dikranjan, D. Shakhmatov, *Hewitt-Marczewski-Pondiczery type theorem for abelian groups and Markov's potential density*, Proc. Amer. Math Soc. 138, no. 8 (2010), pp. 2979–2990.
24. D. Dikranjan, D. Shakhmatov, *A Kronecker-Weyl theorem for subsets of Abelian groups*, Advances in Mathematics, 226 (2011) 4776–4795.
25. D. Dikranjan, D. Shakhmatov, *Extension of group topologies and Zariski embeddings*, Preprint 2007.
26. D. Dikranjan, M. Tkačenko, *Algebraic structure of the small countably compact abelian groups*, Forum Math. **15** (2003) pp. 811–837.
27. D. Dikranjan, D. Toller, *The Markov-Zariski topology of some non-abelian groups*, work in progress.
28. D. Dikranjan, D. Toller, *The Markov-Zariski topology of some linear groups*, work in progress.
29. D. Dikranjan, V. Uspenskij, *Categorically compact topological groups*, J. Pure Appl. Algebra 126 (1998), no. 1-3, pp. 149-168.
30. L. Fuchs, *Infinite abelian groups*, Academic Press, New York, (1973).
31. E.D. Gaughan, *Topological group structures of infinite symmetric groups*, Proc. Nat. Acad. Sci. U.S.A. 58 (1967), pp. 907–910.
32. B. Givens, K. Kunen, *Chromatic Numbers and Bohr Topologies*, Topology Appl. 131 (2003), pp. 189–202.
33. V. Guba, *Equivalence of infinite systems of equations in free groups and semigroups to finite subsystems*, Mat. Zametki 40, No. 3 (1986), pp. 321–324.
34. G. Hesse, *Non-discrete topologizability of some countable generalized abelian groups*, Preprint Nr. 35, Bibliotek der Insitutes für Mathematik der technischen Universitẗ Hannover, 1975.
35. G. Hesse, *Zur Topologisierbarkeit von Gruppen*, PHD thesis, Univ. Hannover, Hannover (1979).
36. G. Hesse, J. Reineke, *2-bound finite groups*, J. Algebra 85 (1983), no. 1, 86–103.
37. K.-H. Hofmann and S. A. Morris, *The structure of compact groups. A primer for the student—a handbook for the expert*, de Gruyter Studies in Mathematics, 25 (Walter de Gruyter & Co., Berlin, 1998).
38. I. Kaplansky, *An introduction to differential algebra.* Second edition. Actualites Scientifiques et Industrielles, No. 1251. Publications de l'Institut de Mathematique de l'Universite de Nancago, No. V. Hermann, Paris, 1976. 64 pp.
39. O. Kharlampovich, A. Myasnikov, *Algebraic Geometry over Free Groups: Lifting Solutions into Generic Points*, Contemp. Math. AMS, Algorithms, Languages, Logic (Borovik, ed.), CONM/ 378, 2005, pp. 213–318.
40. A.A. Klyachko, A.V. Trofimov, *The number of non-solutions of an equation*

in a group, J. Group Theory 8 (2005), pp. 747–754.

41. K. Kozlov, E. Reznichenko, O. Sipacheva, *Moscow questions on topological algebra*, Open problems in topology II, Elsevier Science, 2007.
42. J. Lenox, D. Robinson, *The theory of infinite soluble groups*, Oxford Univ. Press 2004.
43. F. Levin, *One Variable Equation Over Groups*, Arch. Math. 15 (1964), No. 1, pp. 179–188.
44. G. Lukács, *Hereditarily non-topologizable groups*, Topology Proc. 33 (2009), pp. 269-275.
45. A.A. Markov, *On unconditionally closed sets*, Comptes Rendus Doklady AN SSSR (N.S.) 44 (1944), pp. 180–181 (in Russian).
46. A.A. Markov, *On free topological groups*, Izv. Akad. Nauk SSSR, Ser. Mat., vol. 9 (1945), no. 1, pp. 3–64 (in Russian). English translation in: A.A. Markov, *Three papers on topological groups: I. On the existence of periodic connected topological groups, II. On free topological groups, III. On unconditionally closed sets*, Amer. Math. Soc. Transl. 1950, (1950) no. 30, 120 pp.
47. A.A. Markov, *On unconditionally closed sets*, Mat. Sbornik 18 (1946), pp. 3–28 (in Russian). English translation in: A.A. Markov, *Three papers on topological groups: I. On the existence of periodic connected topological groups, II. On free topological groups, III. On unconditionally closed sets*, Amer. Math. Soc. Transl. 1950, (1950) no. 30, 120 pp.; another English translation in: *Topology and Topological Algebra*, Transl. Ser. 1, vol. 8, Amer. Math. Soc., 1962, pp. 273-304.
48. S. Morris, V. Obraztsov, *Nondiscrete topological groups with many discrete subgroups*, Topology Appl. 84 (1998), pp. 105–120.
49. J. von Neumann, *Almost periodic functions in a group I*, Trans. Amer. Math. Soc. **36**, (1934), pp. 445–492.
50. J. von Neuman, E. Wigner, *Minimally almost periodic groups*, Annals of Math. (Ser. 2) 41 (1940), pp. 746–750.
51. A. Yu. Ol′shanskij, *A note on countable non-topologizable groups*, Vestnik Mosk. Gos. Univ. Mat. Mekh. 3 (1980), p. 103.
52. A. Yu. Ol′shanskij, *Geometry of defining relations in groups*, Translated from the 1989 Russian original by Yu.A. Bakhturin. Mathematics and its Applications (Soviet Series), 70. Kluwer Academic Publishers Group, Dordrecht, 1991.
53. V. Pestov, *Absolutely closed sets and a hypothesis of A. A. Markov*, Siberian Math. J. 29 (1988), pp. 260–266; original (in Russian): Sibirsk. Mat. Zh. 29 (1988), pp. 124–132.
54. K.-P. Podewski, *Topologisierung algebraischer Strukturen,* Rev. Roumaine Math. Pures Appl. 22 (9), 1283–1290 (1977).
55. D. Remus, *A short solution of Markov's problem on connected group topologies*, Proc. Amer. Math. Soc. 110 (1990), pp. 1109–1110.
56. P. Sharma, *Hausdorff topologies on groups*, The Proceedings of the 1981 Topology Conference (Blacksburg, Va., 1981). Topology Proc. 6 (1981), no. 1, 77–98.
57. S. Shelah, *On a problem of Kurosh, Jonsson groups, and applications*, in

Word problems II, North-Holland (1980), pp. 373–394.
58. O.V. Sipacheva, *Consistent solution of Markov's problem*, arXiv:math/0605558v2.
59. O.V. Sipacheva, *A class of groups in which all unconditionally closed sets are algebraic*, Fundam. Prikl. Mat. 12 (2006), no. 8, pp. 217–222 (in Russian). English translation in: J. Math. Sci. (N. Y.) 152 (2008), no. 2, pp. 288–291.
60. O.V. Sipacheva, *Unconditionally τ-closed and τ-algebraic sets in groups*, Topology Appl. 155 (2008), no. 4, pp. 335–341.
61. M. G. Tkachenko, I. Yaschenko, *Independent group topologies on abelian groups*, Topology Appl. 122 (2002), pp. 435–451.
62. A.D. Taĭmanov, *Topologizable groups. II.* Sibirsk. Mat. Zh. **19** no. 5 (1978), pp. 1201-1203 (in Russian). English translation in: Siberian Math. J. **19** no. 5 (1978), pp. 848–850.
63. A. D. Taĭmanov, *Unsolved problems of topological algebra*, Shtiintsa, Kishinev (1985) (in Russian).
64. A. V. Trofimov, *A theorem on embedding into a nontopologizable group*, Vestnik Moskov. Univ. Ser. I Mat. Mekh. no. 3 (2005), pp. 60–62 (in Russian). English translation in: Moscow Univ. Math. Bull. 60, no. 3 (2005), pp. 42–44.
65. A. V. Trofimov, *A perfect nontopological group*, Vestnik Moskov. Univ. Ser. I Mat. Mekh. 71, no. 1 (2007), pp. 7–13 (in Russian). English translation in: Moscow Univ. Math. Bull. 62, no. 1 (2007), pp. 5–11.
66. B.A.F. Wehrfritz, *A Note on Periodic Locally Soluble Groups*, Arch. Math. 18 (1967), No. 6, pp. 577–579.
67. B.A.F. Wehrfritz, *Infinite linear groups*, Springer-Verlag, Berlin, Heidelberg, New York (1973).
68. B.L. van der Waerden, *Stetigkeitssätze für halbeinfache Liesche Gruppen*, Math. Z. **36** (1933) pp. 780–786.
69. T.S. Weigel, *Residual properties of free groups, III*, Israel J. Math. 77, No. 1-2 (1992), pp. 65–81.
70. H. Weyl, *Über die Gleichverteilung von Zahlen mod. Eins*, Math. Ann. 77 (1916), pp. 313-352 (in German).
71. J.S. Wilson, *Locally Nilpotent CZ-Groups are Hypercentral*, J. London Math. Society 19 (1979), No. 2, pp. 62–64.

LINEAR GROUPS WITH FINITE DIMENSIONAL ORBITS

M. R. DIXON

Department of Mathematics
University of Alabama
Tuscaloosa, AL 35487-0350, U.S.A.
Email:mdixon@gp.as.ua.edu

L. A. KURDACHENKO

Department of Algebra, Facultet of mathematic and mechanik
National University of Dnepropetrovsk
Gagarin prospect 72, Dnepropetrovsk 10, 49010, Ukraine
Email:lkurdachenko@i.ua

J. OTAL

Department of Mathematics - IUMA
University of Zaragoza, Pedro Cerbuna 12 50009 Zaragoza, Spain
Email:otal@unizar.es

If G is a subgroup of the group $GL(F, A)$ of all automorphisms of a vector space A on a field F we say that G *is a linear group with finite dimensional G–orbits*, if $\dim_F(aFG)$ is finite for each element $a \in A$. In this paper the structure of two types of linear groups with finite dimensional orbits, namely *groups with boundedly finite dimensional orbits on* A and groups with *finite G–orbits of subspaces* is considered.

Keywords: linear group; bounded finite dimensional orbit; finite orbit of subspaces.

Introduction

Let A be a vector space over a field F. We let $GL(F, A)$ denote the group of all non-singular F–linear transformations of A under composition. If G is a subgroup of $GL(F, A)$ then G acts (faithfully) on A via

$$(a, g) \longrightarrow ag, \ a \in A, \ g \in G.$$

A subspace $B \leq A$ is said to be G*–invariant* if $bg \in B$ for every $b \in B$ and $g \in G$. These subspaces of A give rise to other substructures of A that we now review. We let

$$FG = \{\sum_{g\in G} t_g g \mid g \in G,\ t_g \in F,\ t_g = 0 \text{ for all but finitely many } g \in G\}$$

denote the group ring of G over F. Any element of FG can be thought as an F–linear transformation of A so that this is naturally an FG–module. Under this meaning G–invariant subspaces of A are just FG–submodules of A. Very often, the study of the structure of the FG–submodules of A gives the possibility of finding a basis in which the elements of G have a very simple form. In the first part of this paper we want to consider the influence of local submodules of the form

$$aFG = \{a\alpha \mid \alpha \in FG\},\ a \in A,$$

on the structure of the FG–module A. As an illustration we consider the following example. Suppose that $\dim_F(aFG) = 1$ for each element $a \in A$. It is not hard to see that in this case the group G is abelian; moreover, every subspace of A is an FG–submodule and the group G becomes isomorphic to a subgroup of the multiplicative group $U(F)$ of the field F. A natural question arises as to what can be said! concerning those vector spaces A on which G acts with *bounded finite dimensional orbits*. By this we mean that there exists a positive integer $o \geq 1$ such that $\dim_F(aFG) \leq o$ for each element $a \in A$.

Given $a \in A$, the set

$$aG = \{ag \mid g \in G\}$$

is said to be *the G–orbit of a*. We note that if there exists some $o \geq 1$ such that $|aG| \leq o$ for each $a \in A$, then $\dim_F(aFG) \leq o$ for each element $a \in A$. However the converse it is not true. Indeed, if A is a finite dimensional vector space and $\dim_F(A) = k$, trivially we have that $\dim_F(aFG) \leq k$ for each element $a \in A$. We also note that this hypothesis on finite dimensional vector spaces is very strong. In fact, suppose that $|aG| \leq o$ for each $a \in A$, and let $\{a_1, \cdots, a_k\}$ be a basis of A. Since $|aG| = |G : C_G(a)|$ is finite for every $a \in A$, every centralizer $C_G(a_i)$ has finite index in G and so has $C = C_G(a_1) \cap \cdots \cap C_G(a_k)$. But every element of C acts trivially on A, and therefore $C = \langle 1 \rangle$. Thus G is finite. In Ref. 1, vector spaces A such that the G–orbit of their elements $a \in A$ have size at most o were studied. In ! this paper we pursue this further. Suppose that A has an FG–submodule D such that $\dim_F(D) = k$ is finite and every subspace of A/D is G–invariant. Obviously we have $aFG \leq aF + D$, so that $\dim_F(aFG) \leq k + 1$ is bounded in terms of k and a natural question is whether the converse also holds. Thus if $\dim_F(aFG) \leq o$ is bounded for each element $a \in A$, *must A contain an FG–submodule D satisfying the following conditions:*

(1) $\dim_F(D)$ is bounded by some function of o; and
(2) every subspace of A/D is G–invariant?

The first main result of this paper (Theorem A below) shows that the answer to this question is very close to this, but in general the answer is negative as the following specific example shows. Let ωFG be the augmentation ideal of the group ring FG, that is the two-sided ideal of FG generated by the set $\{g - 1 \mid g \in G\}$. Let A be a vector space with basis $\{a_n \mid n \in \mathbb{N}\}$ over a finite field F of characteristic $p > 0$. Then the mapping g defined by

$$\begin{cases} a_{2n}g = a_{2n} \\ a_{2n+1}g = a_{2n} + a_{2n+1} \end{cases}$$

is an F–automorphism of A. Let $G = \langle g \rangle$. Clearly $A(\omega FG) = \bigoplus_{n \in \mathbb{N}} a_{2n}F$ has infinite dimension. Since G acts trivially on $A/A(\omega FG)$, every subspace of $A/A(\omega FG)$ is G–invariant. Also $|\langle g \rangle| = p$, so $|aG| \leq p$ and hence $\dim_F(aFG) \leq p$ for each $a \in A$. If now D is an FG-submodule satisfying conditions (1) and (2) then there exists $n \in \mathbb{N}$ such that the subspace spanned by a_{2n} and a_{2n+1} has trivial intersection with D. Then $(a_{2n+1} + D) \cdot g = a_{2n} + a_{2n+1} + D$ and it is easy to see that the subspace spanned by $a_{2n+1} + D$ is not G-invariant. Thus in this example no such submodule D exists.

Theorem A. *Let G be a subgroup of $GL(F, A)$. Suppose that there is a positive integer $o \geq 1$ such that $\dim_F(aFG) \leq o$ for each element $a \in A$. Then*

(1) A has an FG–submodule D such that $\dim_F(D) \leq f_1(o)$, for some positive integer valued function f_1, and every subspace of A/D is K–invariant where $K = C_G(D)$;
(2) K is normal in G and has an abelian G–invariant subgroup T such that $A(\omega FT) \leq D$ and K/T is isomorphic to a subgroup of $U(F)$; and
(3) T is an elementary abelian p–group if char $F = p > 0$ and a torsion-free abelian group otherwise.

If A has finite dimension and G is periodic, a well-known theorem of I. Schur ensures that G is locally finite (see Corollary 4.8, Ref. 7). The next corollary provides a generalization of Schur's theorem.

Corollary A1. *Let G be a subgroup of $GL(F, A)$. Suppose that there is a positive integer $o \geq 1$ such that $dim_F(aFG) \leq o$ for each element $a \in A$. If G is periodic, then G is locally finite.*

A group G is said to be *generalized radical* if G has an ascending series

$$\langle 1 \rangle = G_0 \triangleleft G_1 \triangleleft \cdots \triangleleft G_\alpha \triangleleft G_{\alpha+1} \triangleleft \cdots \triangleleft G_\gamma = G$$

whose factors $G_{\alpha+1}/G_\alpha$ are locally nilpotent or locally finite.

Another consequence of Theorem A is the next result.

Corollary A2. *Let $G \leq GL(F, A)$ be a subgroup of the group of all F–automorphisms of a vector space A over a field F. Suppose that there is a positive integer $o \geq 1$ such that $dim_F(aFG) \leq o$ for each element $a \in A$. If G is locally generalized radical, then G is soluble–by–locally finite.*

If B is a subspace of A, then we put

$$cl_G(B) = \{Bg \mid g \in G\}.$$

The set $cl_G(B)$ is said to be *the G–orbit of the subspace B*. We note that $|cl_G(B)| = |G : N_G(B)|$, where $N_G(B) = \{g \in G \mid Bg = B\}$. Obviously if the subspace B is G–invariant, then $|cl_G(B)| = 1$. The situation when $cl_G(B)$ is finite for each subspace B of A is therefore of some interest and the second part of this paper is devoted to the study of those groups G for which $cl_G(B)$ is finite for each subspace B. Our main result in this regard is the following theorem.

Theorem B. *Let G be a subgroup of $GL(F, A)$. Suppose that the G–orbit of every subspace of A is finite. Then*

(1) G has a normal subgroup K of finite index such that every subspace of A is K–invariant;
(2) G is central–by–finite; and
(3) there exists a positive integer k such that $|cl_G(B)| \leq k$ for every subspace B of A.

1. Vector spaces whose cyclic *FG*–submodules have bounded finite dimension

Let A be a vector space on a field F. Throughout this Section $G \leq GL(F, A)$ *will be a subgroup of the group of all F–automorphisms of A.*

If B is a subset of A the *centralizer of B in G* is the subgroup of G consisting of all elements fixing B elementwise, that is,

$$C_G(B) = \{g \in G \mid bg = b \text{ for all } b \in B\}.$$

Actually $C_G(B)$ is a normal subgroup of G provided B is an FG–submodule of A. If B is an F–subspace of A, then *the normalizer of B in G* is the subgroup consisting of all elements of G fixing B globally, that is,

$$N_G(B) = \{g \in G \mid Bg \subseteq B\}.$$

Thus if B is an FG-module then $G = N_G(B)$. Also *the norm of B in G* is the subgroup

$$\mathrm{Norm}_G(B) = \bigcap_{b \in B} N_G(bF).$$

We remark that $\mathrm{Norm}_G(B)$ is the intersection of the normalizers of all F–submodules of B and that $G = \mathrm{Norm}_G(A)$ if and only if every submodule of A is G-invariant. If C is an FG–submodule of B, then $\mathrm{Norm}_G(B) \subset N_G(C)$.

We start with some elementary properties that we use very often. The first of them is rather obvious.

Lemma 1.1. *Suppose that there is a positive integer o such that $dim_F(aFG) \leq o$ for each element $a \in A$.*

(1) If H is a subgroup of G, then $dim_F(aFH) \leq o$ for each element $a \in A$;
(2) If B is an FG–submodule of A, then $dim_F((a + B)FG) \leq o$ for each element $a \in A$;
(3) If H is a subgroup of G and B is an FH–submodule of A, we have that $dim_F(bFH) \leq o$ for each element $b \in B$; and
(4) If H is a subgroup of G and $C \leq B$ are FH–submodules of A, then $dim_F((b + C)FH) \leq o$ for each element $b \in B$.

Lemma 1.2. *Suppose that $G = Norm_G(A)$. If $g \in G$, then there exists some $\alpha_g \in F$ such that $ag = \alpha_g a$ for each element $a \in A$.*

Proof. Let $\{a_\lambda \mid \lambda \in \Lambda\}$ be a basis of B. Since $G = \mathrm{Norm}_G(A)$, every subspace of A is G–invariant. In particular, $a_\lambda F$ is G–invariant for every $\lambda \in \Lambda$. It follows that $a_\lambda g = \alpha_\lambda a_\lambda$, for suitable $\alpha_\lambda \in F$. We claim that the scalars only depend on g and not on the index λ. For, let $b = a_\lambda + a_\mu$ for certain indices $\lambda, \mu \in \Lambda$. Then

$$bg = (a_\lambda + a_\mu)g = a_\lambda g + a_\mu g = \alpha_\lambda a_\lambda + \alpha_\mu a_\mu.$$

On the other hand, bF is also G–invariant, so that

$$bg = \gamma b = \gamma(a_\lambda + a_\mu) = \gamma a_\lambda + \gamma a_\mu,$$

for some $\gamma \in F$. Since a_λ and a_μ are linearly independent, we have $\alpha_\mu = \gamma = \alpha_\lambda$ and the result now follows.

This has some important consequences.

Corollary 1.3. *Suppose that $G = Norm_G(A)$. Then G is isomorphic to a subgroup of the multiplicative group $U(F)$ of F.*

Proof. It follows from Lemma 1.2 that for all $a \in A$ and $g \in G$ there exists a unique $\alpha_g \in F$ such that $ag = \alpha_g a$. Then the mapping $f : g \to \alpha_g$, $g \in G$, is a homomorphism from G into $U(F)$. Clearly Ker $f = C_G(A) = \langle 1 \rangle$ and so f is a monomorphism.

Corollary 1.4. *The subgroup $Norm_G(A)$ lies in $\zeta(G)$, the center of G.*

Proof. Let $g \in \mathrm{Norm}_G(A)$. By Lemma 1.2, there exists some $\alpha_g \in F$ such that $ag = \alpha_g a$ for every $a \in A$. If $x \in G$, then we have

$$a(gx) = (\alpha_g a)x = \alpha_g(ax) \text{ and } a(xg) = (ax)g = \alpha_g(ax).$$

This implies that $a(gxg^{-1}x^{-1}) = a$ for every $a \in A$ and hence $gx = xg$ since $C_G(A) = \langle 1 \rangle$, as required.

We now establish a result which enables us to use inductive arguments in the sequel.

Lemma 1.5. *Suppose that there is a positive integer o such that $dim_F(aFG) \leq o$ for each element $a \in A$ and assume that $d \in A$ is chosen such that $dim_F(dFG) = o$. Let $D = dFG$ and $H = C_G(dFG)$. Then $dim_F((a + D)FH) \leq o - 1$ for every $a \in A$.*

Proof. We first remark that if $u, v \in A$ satisfy $uFG \cap vFG = 0$ then $C_G(u + vFG) = C_G(u)$. Indeed, $C_G(u) \leq C_G(u + vFG)$ and if $x \in C_G(u + vFG)$ then $ux + vFG = u + vFG$ so that $u(x-1) \in vFG \cap uFG = 0$. Consequently $x \in C_G(u)$ and our claim follows.

We now suppose to the contrary that our result is false, and choose $c \in A$ such that $\dim_F((c + D)FH) = o$. Since

$$o = \dim_F((c + D)FH) \leq \dim_F((c + D)FG) \leq \dim_F cFG \leq o,$$

using Lemma 1.1, it follows that

$$\dim_F((c + D)FG) = \dim_F(cFG) = o.$$

Let $C = cFG$. Now $(c+D)FG = (C+D)/D$ and $(C+D)/D \cong C/C \cap D$ has dimension o. Also C has dimension o and hence the F-subspace $C \cap D$ is $\{0\}$. Consequently, by our initial remark, we have $C_G(d+C) = C_G(d)$.

Let $e = c+d$ and $E = eFG$. Then $e+C = d+C$ and hence $E+C = D+C$. Since $D \cap C = \{0\}$ it follows that $E/E \cap C \cong (E+C)/C \cong D$. However D is o-dimensional and $E \cap C$ is an F-subspace of E, which is also o-dimensional. Hence $E \cap C = \{0\}$ and by our initial remark we see that $C_G(e+C) = C_G(e)$. It now follows, since $e+C = d+C$, that $C_G(d) = C_G(e)$. In similar fashion we also have $C_G(c) = C_G(e)$ and hence $C_G(c) = C_G(d)$. Hence $C_G(D) \leq C_G(c)$ and $cC_G(D) = cH = c$ so that $(c+D)H = cH + D = \{c+D\}$, which contradicts the fact that $\dim_F((c+D)FH) = o$, and this is the contradiction sought.

Proof of Theorem A. If $o = 1$, then $\dim_F(aFG) = 1$ for all $a \in A$ and hence $G = \mathrm{Norm}_G(A)$. By Corollary 1.3, G is isomorphic to a subgroup of $U(F)$ and so the result follows, upon taking $D = \{0\}, K = G$ and $T = 1$.

Therefore we suppose that $o > 1$. Choose $d_1 \in A$ such that $\dim_F(d_1FG) = o$, let $D_1 = d_1FG$ and $L = C_G(D_1)$. By Lemma 1.5, $\dim_F((a+D_1)FL) \leq o-1$ for each element $a \in A$. Choose $d_2 \in A$ such that $\dim_F((d_2+D_1)FL)$ is maximal and let $\dim_F((d_2+D_1)FL) = k \leq o-1$. Let $B/D_1 = (d_2+D_1)FL$. Then $\dim_F(B/D_1) = k \leq o-1$ is finite and $\dim_F(B/D_1) = k \leq o-1$.

We let $\{e_1+D_1, \cdots, e_k+D_1\}$ denote a basis of B/D_1 and set $D_2/D_1 = (B/D_1)FG$. Then

$$D_2/D_1 = (e_1+D_1)FG + \cdots + (e_k+D_1)FG.$$

By Lemma 1.1, $\dim_F((e_j+D_1)FG) \leq o$ for every j, so that $\dim_F(D_2/D_1) \leq ko \leq (o-1)o$. It follows that $\dim_F(D_2) \leq o+(o-1)o$. Let $H = C_L(B/D_1)$. By Lemma 1.5, $\dim_F((a+B)FH) \leq o-2$ for each element $a \in A$. Since $B \leq D_2$, $\dim_F((a+D_2)FH) \leq o-2$ for each element $a \in A$. Let $R = C_G(D_2)$. By construction D_1 lies in D_2, so that $R = C_G(D_2) \leq C_G(D_1) = L$. Since $B \leq D_2$, $R \leq C_L(B/D_1) = H$. It follows that $\dim_F((a+D_2)FR) \leq o-2$ for each element $a \in A$. Proceeding in this way, after finitely many steps we construct an FG–submodule D such that if $K = C_G(D)$ then $\dim_F((a+D)FK) = 1$, for each $a \in A$, and $\dim_F(D) \leq f_1(o)$, where f_1 is a suitable positive integer valued function depending only on o.

Thus $K \leq \mathrm{Norm}_G(A/D)$ and the above proof suggests that we may choose f_1 to be defined by

$$f_1(o) = o + (o-1)o + (o-2)o + \cdots + 2o =$$

$$= o((o-1)+(o-2)+\cdots+2+1) = o^2(o-1)/2.$$

Let $T = C_K(A/D)$. Then T stabilizes the chain $\{0\} \leq D \leq A$ and hence T is abelian (see 1.C.1 of Ref. 3, for example). Furthermore, if char $F = p > 0$, then the additive group of A is an elementary abelian p–group and so is T. If char $F = 0$, then the additive group of A is torsion-free abelian and so is T. Finally $K/T \leq GL(A/D, F)$ and hence is isomorphic to a subgroup of $U(F)$, by Corollary 1.3.

`Proof of Corollary A1`. With K as in Theorem A we have that K is metabelian and hence locally finite. Moreover, G/K is isomorphic to a subgroup of the general linear group $GL_m(F)$, where $m = \dim_F(D)$. Thus it suffices to note that the periodic subgroups of $GL_m(F)$ are locally finite (see Corollary 4.8 of Ref. 7, for example).

`Proof of Corollary A2`. As above K is metabelian and G/K is isomorphic to a subgroup of $GL_m(F)$, where $m = \dim_F(D)$. Suppose that G/C has a free subgroup X of rank 2. Since X is finitely generated, X is generalized radical. However generalized radical groups cannot be free. This contradiction shows that G/C has no free subgroups. It suffices to apply Corollary 10.17, Ref. 7.

We continue examining the structure described in Theorem A. Following A. I. Maltsev[5], a group G is said to have *finite special rank* r if every finitely generated subgroup of G can be generated by r elements and r is the least positive integer with this property. All the consequences will be deduced from the next result.

Lemma 1.6. *Let F, A and $G \leq GL(F, A)$ as above. Suppose that A includes an FG–submodule D such that $dim_F(D) = d$ is finite. Put $K = C_G(D)$ and let U/K be an abelian p'–subgroup of G/K, where $p = char\ F$. Then U/K has finite special rank at most d.*

`Proof`. Let

$$\{0\} = D_0 \leq D_1 \leq \cdots \leq D_n = D$$

be a U–chief series of D. Clearly, $n \leq d$. In other words, every factor D_j/D_{j-1} is a simple FU–module and it follows that the factor-group $U/C_U(D_j/D_{j-1})$ is locally cyclic (see Corollary 2.4, Ref. 4). In particular $U/C_U(D_j/D_{j-1})$ has special rank 1. If $V = \cap_{1\leq j\leq n} C_U(D_j/D_{j-1})$, by Remak's theorem we obtain an embedding

$$U/V \hookrightarrow U/C_U(D_1) \times \cdots \times U/C_U(D_n/D_{n-1}),$$

which shows that the factor-group U/V has special rank at most n. Note that every element of V acts trivially on each factor of the series

$$\{0\} = D_0 \leq D_1 \leq \cdots \leq D_n = D,$$

from which it is not hard to see that $V/C_G(D)$ is a p - group. Thus $V/C_G(D)$ is trivial so $V = C_G(D)$. Therefore U/K has special rank at most $n \leq d$.

Corollary 1.7. *Let F, A and $G \leq GL(F, A)$ as above, and suppose that G is a locally p'–group, where $p = char\ F$. Suppose also that A includes an FG–submodule D such that $dim_F(D) = d$ is finite. Then there exists an integer valued function r such that $G/C_G(D)$ has finite special rank at most $r(d)$.*

Proof. Put $K = C_G(D)$ and let q be a prime other than p. We choose an arbitrary Sylow q–subgroup Q/K of G/K. By Lemma 1.6, every abelian subgroup of Q/K has finite special rank at most d. Applying a result by Yu. I. Merzlyakov[6], we have that Q/K has finite special rank at most $\frac{d(5d+1)}{2}$. Let now L/K be an arbitrary finite subgroup of G/K. Applying the above conclusion, we have that every Sylow q–subgroup of L/K has at most $\frac{d(5d+1)}{2}$ generators. By Theorem A, Ref. 2, L/K has at most

$$r(d) = 1 + \frac{d(5d+1)}{2}$$

generators. It follows that G/K has finite special rank at most $r(d)$.

Corollary 1.8. *Let F, A and $G \leq GL(F, A)$ as above, and suppose that G is a locally p'–group, where $p = char\ F$. Suppose also that A includes an FG–submodule D such that $dim_F(D) = d$ is finite. Then $G/C_G(D)$ is abelian–by–finite.*

Proof. Put $K = C_G(D)$. By Corollary 1.7, G/K has finite special rank at most $r(d)$ for a suitable integer valued function. Since G/K is isomorphic to a subgroup of $GL_{r(d)}(F)$, it suffices to apply Theorem 10.9, Ref. 7.

Corollary 1.9. *Let F, A and $G \leq GL(F, A)$ as above, and suppose that G is a periodic p'–subgroup, where $p = char\ F$. Suppose also that there exists a positive integer o such that $dim_F(aFG) \leq o$ for each element $a \in A$. Then there exists a locally cyclic central subgroup K such that G/K is soluble–by–finite.*

Proof. Let D be the FG–submodule of A and K, T be the subgroups which appeared in Theorem A. By Corollary 1.3, K is isomorphic to a subgroup of $U(F)$. Being periodic, K is locally cyclic. By Corollary 1.4, $K \leq \zeta(G)$. Since $\dim_F(D)$ is finite, it suffices to apply Corollary 1.8.

2. Vector spaces whose subspaces have finite G–orbits

We keep the notation throughout this paper, namely A is a vector space over a field F and $G \leq GL(A, F)$ is a linear group. The aim of this section is to study the situation when $cl_G(B) = \{Bg \mid g \in G\}$ is finite for each F–subspace B of A.

Suppose that all the orbits $cl_G(B)$ are finite for each F–subspace B of A. In particular the orbits $cl_G(aF)$ are finite for every element $a \in A$. If T is a transversal of $N_G(aF)$ in G, then we have

$$aFG = \sum_{g \in T} (aF)g,$$

and the finiteness of the set T implies that $\dim_F(aFG) \leq |T|$. It readily follows that $\dim_F(aFG)$ is finite. If we further assume that the orbits are *boundedly finite*, that is there exists some positive integer o such that $|cl_G(B)| \leq o$ for every subspace B of A, then the same arguments give that $\dim_F(aFG) \leq o$ for each element $a \in A$, a case described within Theorem A. This is the reason why the results of this section are the natural sequel of those of the first one. It turns out that the assumption that all the G–orbits are finite implies that they are in fact boundedly finite.

We begin our study with the following result.

Lemma 2.1. *Let F, A and $G \leq GL(F, A)$ be as above. If $0 \neq a \in A$, then the factor-group $N_G(aF)/C_G(aF)$ is isomorphic to a subgroup of the multiplicative group $U(F)$ of F.*

Proof. If $g \in N_G(aF)$, then there is a unique $0 \neq \alpha_g \in F$ such that $ag = \alpha_g a$. We define the mapping $f : g \to \alpha_g,\ g \in N_G(aF)$. It is not hard to see that f is a homomorphism from G into $U(F)$. Since Ker $f = C_G(a) = C_G(aF)$, we have that $N_G(aF)/C_G(aF)$ is isomorphic to some subgroup of $U(F)$, as required.

Lemma 2.2. *Let F, A and $G \leq GL(A, F)$ be as above. If $a \in A$ and the G–orbit of aF is finite, then the dimension of the FG–submodule aFG is finite and the factor-group $G/C_G(aFG)$ is abelian–by–finite.*

Proof. Put $B = aF$. Then the set $\{Bg \mid g \in G\}$ is finite. Put

$$\{Bg \mid g \in G\} = \{B_1, \cdots, B_m\}.$$

Then $aFG = B_1 + \cdots + B_m$ has finite dimension over F. Clearly

$$L = \bigcap_{g \in G} N_G(Bg) = N_G(B_1) \cap \cdots \cap N_G(B_m)$$

is a normal subgroup of G of finite index. For every $1 \leq j \leq m$ we put $K_j = L \cap C_G(B_j)$ so that

$$L/K_j = L/(L \cap C_G(B_j)) \cong LC_G(B_j)/C_G(Bj) \leq N_G(B_j)/C_G(B_j).$$

By Lemma 2.1, L/K_j is abelian. Since

$$aFG = B_1 \cap \cdots \cap B_m,$$

we have that

$$C_G(aFG) = C_G(B_1) \cap \cdots \cap C_G(B_m).$$

By Remak's theorem, we have an embedding

$$L/(L \cap C_G(aFG)) \hookrightarrow L/(L \cap C_G(B_1)) \times \cdots \times L/(L \cap C_G(B_m)),$$

which shows that the factor-group $L/(L \cap C_G(aFG))$ is abelian. Since the index $|G : L|$ is finite, the result follows.

Lemma 2.3. *Let F, A and $G \leq GL(F, A)$ be as above. Suppose that the G–orbit of every subspace of A is finite. Then A has an FG–submodule B such that $dim_F(A/B)$ and $|G : Norm_G(B)|$ are finite.*

Proof. If $G = \mathrm{Norm}_G(A)$ we set $B = A$ and we are done. Therefore we can assume that there is some $u_1 \in A$ such that the subspace $U_1 = u_1F$ is not G–invariant. This means that there is an element $g_1 \in G$ such that $U_1g_1 \neq U_1$. Since $cl_G(U_1)$ is finite, the index $|G : N_G(U_1)|$ is finite. Let $G_1 = N_G(U_1)$ and $V_1 = U_1G = u_1FG$. By Lemma 2.2, V_1 has finite dimension and, by construction, $U_1g_1 \leq V_1$. There exists an F–subspace W_1 such that $A = V_1 \oplus W_1$, and we remark that $\dim_F(A/W_1)$ is finite. Again we have that $cl_G(W_1)$ is finite. Suppose that

$$\{W_1g \mid g \in G\} = \{B_1, \cdots, B_m\}.$$

Since $\dim_F(A/W_1) = \dim_F(A/W_1g)$ for each $g \in G$, $\dim_F(A/B_j)$ is finite for all $1 \leq j \leq m$. Let

$$Y_1 = B_1 \cap \cdots \cap B_m$$

so that $\dim_F(A/Y_1)$ is finite, since there is an embedding

$$A/Y_1 \hookrightarrow A/B_1 \oplus \cdots \oplus A/B_m,$$

by Remak's theorem. If $G_1 = \mathrm{Norm}_G(Y_1)$ then we set $B = Y_1$ and the result follows. Therefore we may suppose there is some $u_2 \in Y_1$ such that $U_2 = u_2F$ is not G_1–invariant, which means that there is an element $g_2 \in G_1$ such that $U_2g_2 \neq U_2$. By construction $U_2g_2 \leq Y_1$. Furthermore $U_1g_2 = U_1$. Since $cl_G(U_1 \oplus U_2)$ is finite, $|G : N_G(U_1 \oplus U_2)|$ is finite. Put $G_2 = N_G(U_1 \oplus U_2)$ and $V_2 = U_2G = U_2FG = u_2FG$. By Lemma 2.3, V_2 has finite dimension. By construction $U_2g_2 \leq V_2$. Proceeding as above, we find an FG–submodule Y_2 such that $Y_2 \cap (V_1 \oplus V_2) = \{0\}$ and $\dim_F(A/Y_2)$ is finite. If $G_2 = \mathrm{Norm}_G(Y_2)$ we set $B = Y_2$. Otherwise we repeat this process. If this process does not terminate then we can construct elements $\{u_n \mid n \in \mathbb{N}\}$ of A and $\{g_n \mid n \in \mathbb{N}\}$ of G such that for all $n \in \mathbb{N}$ the following conditions are satisfied:

(i) $U_ng_n \neq U_n$, where $U_n = u_nF$;
(ii) $U_ng_{n+k} = U_n$ for all $k \in \mathbb{N}$; and
(iii) $U_{n+1}FG \cap (U_1FG + \cdots + U_nFG) = \{0\}$.

Applying (iii) we have that

$$\sum_{n\in\mathbb{N}} U_nFG = \bigoplus_{n\in\mathbb{N}} U_nFG$$

and hence,

$$\sum_{n\in\mathbb{N}} U_n = \bigoplus_{n\in\mathbb{N}} U_n.$$

Also

$$(\bigoplus_{n\in\mathbb{N}} U_n)g_j = \bigoplus_{n\in\mathbb{N}} U_ng_j \text{ for all } j \in \mathbb{N}.$$

By (i) $U_jg_j \neq U_j$ and by (ii), if $k > j$, $U_jg_k = U_j$. It follows from (iii) that

$$(\bigoplus_{n\in\mathbb{N}} U_n)g_k \neq (\bigoplus_{n\in\mathbb{N}} U_n)g_j \text{for } k \neq j.$$

Thus $cl_G(\oplus_{n\in\mathbb{N}}U_n)$ is infinite, a contradiction which shows that the above process terminates in finitely many steps, from which the result follows.

We can now gather quite a lot of information concerning A and G in this case.

Proposition 2.4. *Let F, A and $G \leq GL(F, A)$ be as above. Suppose that the G–orbit of every subspace of A is finite. Then there exist a subgroup $H \leq G$, an FG–submodule $B \leq A$ and a subset $\{a_1, \cdots, a_n\} \subseteq A$ such that*

(1) *The index* $|G:H|$ *is finite;*
(2) $A = a_1F \oplus \cdots \oplus a_nF \oplus B$;
(3) *if* $x \in H$ *then there is* $\beta_x \in F$ *such that* $bx = \beta_x b$ *for each element* $b \in B$;
(4) *if* $x \in H$ *then for all* $1 \leq j \leq n$ *there are* $\alpha_{j,x} \in F$ *such that* $a_j x = \alpha_{j,x} a_j$; *and*
(5) *there is a positive integer* m *such that* $\alpha_{j,x} = \varepsilon_{j,x}\beta_x$ *and* $\varepsilon_{j,x}^m = 1$ *for all* $x \in H$.

Proof. By Lemma 2.3, A has an FG–submodule B such that $\dim_F(A/B)$ and $|G : \mathrm{Norm}_G(B)|$ are finite. Let $\{a_1 + B, \cdots, a_n + B\}$ be a basis of the factor-space A/B. It is clear that

$$a_1F \oplus \cdots \oplus a_nF \oplus B = A.$$

Let

$$H = \mathrm{Norm}_G(B) \cap N_G(a_1F) \cap \cdots \cap N_G(a_nF).$$

Since the G–orbits of subspaces of A are finite, $|G : H|$ is finite. Pick $x \in H$ and note that, since $H \leq \mathrm{Norm}_G(B)$, Lemma 1.2 shows that there exists $\beta_x \in F$ such that $bx = \beta_x b$ for each element $b \in B$. By the choice of H, the subspaces a_1F, ..., a_nF are H–invariant and it follows that there are elements $\alpha_{j,x} \in F$ such that $a_jx = \alpha_{j,x}a_j$ for each $1 \leq j \leq n$. Pick $0 \neq b \in B$ and put $a = a_1 + \cdots + a_n + b$. We have

$$ax = (a_1 + \cdots + a_n + b)x = \alpha_{1,x}a_1 + \cdots + \alpha_{n,x}a_n + xb.$$

Since the G–orbits of subspaces of A are finite, $m := |G : N_G(aF)|$ is finite. Then $y := x^m \in N_G(aF)$ and hence aF is $\langle y \rangle$–invariant. Thus

$$ay = \eta a = \eta a_1 + \cdots + \eta a_n + \eta b,$$

where $\eta \in F$. On the other hand,

$$ay = ax^m = \alpha_{1,x}^m a_1 + \cdots + \alpha_{n,x}^m a_n + \beta_x^m b.$$

Since the elements $a_1, \cdots, a_n, b$ are linearly independent, we deduce that

$$\alpha_{1,x}^m = \cdots = \alpha_{n,x}^m = \beta_x^m = \eta.$$

In particular, $(\alpha_{j,x}\beta_x^{-1})^m = 1$, for every $1 \leq j \leq n$. Put $\varepsilon_{j,x} = \alpha_{j,x}\beta_x^{-1}$ so that $\varepsilon_{j,x}^m = 1$ and $\alpha_{j,x} = \varepsilon_{j,x}\beta_x$, as required.

We note that the converse of this result also holds.

Corollary 2.5. *Let F, A and $G \leq GL(A, F)$ be as above. Suppose that the G–orbit of every subspace of A is finite. Then G has a subgroup H of finite index and there are positive integers n, m and an embedding*

$$\upsilon : H \longrightarrow U := \underbrace{U(F) \times \cdots \times U(F) \times U(F)}_{n+1}$$

such that

$$Im\ \upsilon = (\underbrace{C_m \times \cdots \times C_m}_{n})D,$$

where $C_m = \{\kappa \in U(F) \mid \kappa^m = 1\}$ and D is the diagonal subgroup of U.

`Proof`. We keep the notation of Proposition 2.4. Let H be the subgroup of G defined there. There are an FG–submodule $B \leq A$ and a subset $\{a_1, \cdots, a_n\} \subseteq A$ such that

$$A = a_1 F \cdots + a_n F + B.$$

If $x \in H$, define the mapping

$$\upsilon : x \longrightarrow (\varepsilon_{1,x}\beta_x, \cdots, \varepsilon_{n,x}\beta_x, \beta_x),$$

where $bx = \beta_x b$ and $a_j x = \varepsilon_{j,x}\beta_x a_j$ $(1 \leq j \leq n)$. It is not hard to see that υ is a homomorphism of H into $U = U(F)^{n+1}$. Clearly Ker $\upsilon = C_H(A) = \langle 1 \rangle$ and Im $\upsilon = (C_m \times \cdots \times C_m)D$, where $C_m = \{\kappa \in U(F) \mid \kappa^m = 1\}$ and D is the diagonal subgroup of U.

`Proof of Theorem B`. Let H be the subgroup of G defined in Corollary 2.5, and keep the notation there. It follows that the index $|G : H|$ is finite. Since every subgroup C_m is cyclic of order m,

$$|(\underbrace{C_m \times \overset{n}{\cdots} \times C_m})D : D| \leq m^n.$$

Clearly the preimage of D in H lies in a subgroup of $\mathrm{Norm}_G(A) \cap H$. This subgroup has finite index in G and the existence of K is then guaranteed. Corollary 1.4 implies that G is center-by-finite. Finally, let $|G : K| = k$. If B is a subspace of A then B is K-invariant so $K \leq N_G(B)$. Then $|G : N_G(B)| \leq |G : K| = k$ and the result follows, since $|cl_G(B)| = |G : N_G(B)|$.

Acknowledgements

The second and third authors were supported by Proyecto MTM2010-19938-C03-03 of Dirección General de Investigación del MICINN (Spain).

References

1. M. R. Dixon, L. A. Kurdachenko, J. Otal. *Linear groups with bounded action.* Alg. Colloquium, to appear.
2. R. M. Guralnick. *On the number of generators of a finite group.* Arch. Math. **53** (1989), 521–523.
3. O. H. Kegel and B. A. F. Wehrfritz, *Locally Finite Groups*, North-Holland, Amsterdam, London, 1973, Volume 3.
4. L. A. Kurdachenko, J. Otal and I. Ya. Subbotin, *Groups with prescribed quotient groups and associated module theory*, World Scientific, New Jersey 2002.
5. A. I. Maltsev. *On groups of finite rank.* Mat. Sbornik **22** (1948), 351–352.
6. Yu. I. Merzlyakov. *On locally soluble groups of finite rank.* Algebra i logika **3** (1964), 5–16.
7. B. A. F. Wehrfritz. *Infinite Linear Groups*, Springer-Verlag, Berlin, 1973.

THREE-DIMENSIONAL LOOPS AS SECTIONS IN A FOUR-DIMENSIONAL SOLVABLE LIE GROUP

Á. FIGULA

Institute of Mathematics, University of Debrecen,
Debrecen, H-4010, P.O.Box: 12, Hungary
E-mail: figula@math.klte.hu

We classify all three-dimensional connected topological loops such that the group topologically generated by their left translations is the four-dimensional connected Lie group G which has trivial center and precisely two one-dimensional normal subgroups. We show that G is not the multiplication group of connected topological proper loops.

Keywords: topological loops; sharply transitive sections in groups; multiplication group of loops; solvable Lie groups.

1. Introduction

A loop $(L, \cdot)$ is a quasigroup with identity element $e \in L$. The left translations $\lambda_a : x \mapsto ax$ and the right translations $\rho_a : x \mapsto xa$, $a \in L$, generate a permutation group on the set L. This group is called the multiplication group $Mult(L)$ of L (Ref. 2). The subgroup G of $Mult(L)$ generated by all left translations of L is the group of left translations of L. The stabilizer $Inn(L)$ of $e \in L$ in the group $Mult(L)$ is called the inner mapping group of L. The multiplication group and the inner mapping group of L are essential tools for investigating the structure of the loop L. An important problem is to be analyzed under which circumstances a group is the multiplication group of a loop. The answer to this question was given in 11 which says that we can use certain conditions for transversals. These conditions were applied to study the structures of the multiplication groups and the inner mapping groups of finite loops (Refs. 3, 9, 10, 12, 16).

Another relevant problem is the classification of loops L having a group G as the group of the left translations of L. This problem is equivalent to the following problem in group theory: to find the subgroups H of G such that the core of H in G is trivial and after this to determine all sections

$\sigma : G/H \to G$ such that $\sigma(H) = 1 \in G$, $\sigma(G/H)$ generates G and acts sharply transitively on the left cosets xH, $x \in G$. The last property means that for given cosets g_1H, g_2H there exists precisely one $z \in \sigma(G/H)$ which satisfies the equation $zg_1H = g_2H$. In particular if G is a connected Lie group and the section σ is continuous with the above property (it is called continuous sharply transitive section), then we obtain all connected topological loops L such that G is the group of left translations of L (Ref. 13, Section 1). In this paper we investigate connected topological loops.

Concrete classifications of 2-dimensional connected topological loops L having a Lie group of dimension 3 as the group of left translations of L are given in 13, Section 22 and 23, and in 5. The Lie groups which are multiplication groups of 2-dimensional topological loops are determined in 4. In 6 we classify all 3-dimensional simply connected topological loops L having a 4-dimensional nilpotent Lie group as the group of left translations of L. Moreover, we prove that these groups are not multiplication groups of 3-dimensional topological loops.

In recent research we deal with the class of solvable Lie groups and study which groups in this class occur as the group of left translations respectively as the multiplication group of 3-dimensional topological loops. The 4-dimensional indecoposable Lie algebras are determined (Ref. 15). The Lie algebra in the first class given in 15 is nilpotent. Here we treat the Lie algebra in the second class. This Lie algebra depends on a real parameter $a \neq 0$, has trivial centre and precisely two 1-dimensional ideals. It has two kinds of paracomplex structures (Ref. 1, Section 2) and admits only for $a = 1$ an invariant complex structure (Ref. 14, Theorem 5, pp. 24-25).

In this paper we classify the 3-dimensional connected topological loops L which are sections in the 4-dimensional connected solvable Lie group G corresponding to this Lie algebra. These loops L are homeomorphic to $\mathbb{R}^3$ and their multiplications depend on one continuous real function of two or three variables (cf. Theorem 3.1). We give an easy proof of the fact that G is not the multiplication group of connected topological proper loops (cf. Theorem 3.2). Later we intend to extend our research to other 4-dimensional Lie groups.

2. Preliminaries

A binary system $(L, \cdot)$ is called a loop if there exists an element $e \in L$ such that $x = e \cdot x = x \cdot e$ holds for all $x \in L$ and the equations $a \cdot y = b$ and $x \cdot a = b$ have precisely one solution, which we denote by $y = a\backslash b$ and $x = b/a$. A loop L is proper if it is not a group.

The left and right translations $\lambda_a : y \mapsto a \cdot y : L \times L \to L$ and $\rho_a : y \mapsto y \cdot a : L \times L \to L$, $a \in L$, are bijections of L. The permutation group $Mult(L)$ generated by all left and right translations of the loop L is called the multiplication group of L and the stabilizer of $e \in L$ in the group $Mult(L)$ is called the inner mapping group $Inn(L)$ of L.

Let K be a group, let $S \leq K$. Denote by $C_K(S)$ the core of S in K (the largest normal subgroup of K contained in S). A group K is isomorphic to the multiplication group of a loop if and only if there exists a subgroup S with $C_K(S) = 1$ and two left transversals A and B to S in K such that $a^{-1}b^{-1}ab \in S$ for every $a \in A$ and $b \in B$ and $K = \langle A, B \rangle$ (cf. Theorem 4.1 in Ref. 11, p. 118). Proposition 2.7 in Ref. 11, p. 114, yields the following

Lemma 2.1. *Let L be a loop with multiplication group $Mult(L)$ and inner mapping group $Inn(L)$. Then the normalizer $N_{Mult(L)}(Inn(L))$ is the direct product $Inn(L) \times Z(Mult(L))$, where $Z(Mult(L))$ is the center of the group $Mult(L)$.*

A loop L is called topological if L is a topological space and the binary operations $(x,y) \mapsto x \cdot y$, $(x,y) \mapsto x \backslash y, (x,y) \mapsto y/x : L \times L \to L$ are continuous. Let G be a connected Lie group, H be a subgroup of G. Let $\sigma : G/H \to G$ be a continuous section with respect to the natural projection $G \to G/H$. This is called continuous sharply transitive section, if the set $\sigma(G/H)$ operates sharply transitively on G/H, which means that to any g_1H and g_2H there exists precisely one $z \in \sigma(G/H)$ with $zxH = yH$. Every connected topological loop having a Lie group G as the group topologically generated by its left translations is isomorphic to a loop L realized on the factor space G/H, where H is a closed subgroup of G with $C_G(H) = 1$ and $\sigma : G/H \to G$ is a continuous sharply transitive section with $\sigma(H) = 1 \in G$ such that the subset $\sigma(G/H)$ generates G. The multiplication of L on the manifold G/H is defined by $xH * yH = \sigma(xH)yH$. Moreover, the subgroup H is the stabilizer of the identity element $e \in L$ in the group G.
Since there does not exist a multiplication with identity on the sphere S^2 a simply connected 3-dimensional topological loop L having a Lie group G as the group topologically generated by its left translations is homeomorphic either to $\mathbb{R}^3$ or to S^3 (see Ref. 8, p. 210). If the group G is solvable, then the simply connected loop L must be homeomorphic to $\mathbb{R}^3$ because of the sphere S^3 is not a solvmanifold (cf. Theorem 3.2 in Ref. 8, p. 208).
The kernel of a homomorphism $\alpha : (L, \cdot) \to (L', *)$ of a loop L into a loop L' is a normal subloop N of L, i.e. a subloop of L such that

$$x \cdot N = N \cdot x, \quad (x \cdot N) \cdot y = x \cdot (N \cdot y), \quad x \cdot (y \cdot N) = (x \cdot y) \cdot N.$$

A loop L is solvable if it has a series $1 = L_0 \leq L_1 \leq \cdots \leq L_n = L$, where L_{i-1} is normal in L_i and L_i/L_{i-1} is an abelian group, $i = 1, \cdots, n$.

3. Three-dimensional topological loops as sections in a 4-dimensional solvable Lie group

A list of the 4-dimensional indecomposable Lie algebras can be found in Ref. 15, Table I, p. 988. All these Lie algebras are solvable. Now we consider the Lie algebra $A_{4,2}^a$, $a \neq 0$, in Table I in Ref. 15. This Lie algebra has a codimension 1 abelian ideal and it has a representation as subalgebra of $\mathbf{gl}_4(\mathbb{R})$. In this section we classify the 3-dimensional connected topological loops such that the Lie algebra of the group G topologically generated by their left translations is $A_{4,2}^a$, $a \neq 0$. The multiplication of the loops in this class depends on a continuous real function of two or three variables. We prove that the group G cannot be the multiplication group of connected topological proper loops. We often use the following lemma.

Lemma 3.1. *Let $f : \mathbb{R} \to \mathbb{R}$ be a continuous function such that for all $z_1, z_2 \in \mathbb{R}$ we have*

$$f(z_2) + e^{-z_2} f(z_1) = f(z_1 + z_2). \tag{1}$$

Then $f(z) = K(1 - e^{-z})$, where K is a real constant.

Proof. Interchanging z_1 and z_2 in (1) we get $f(z_1)+e^{-z_1} f(z_2) = f(z_1+z_2)$. The right hand side of the last equation is equal to the right hand side of equation (1). Hence for $z_1 z_2 \neq 0$ one has $\frac{f(z_1)}{1-e^{-z_1}} = \frac{f(z_2)}{1-e^{-z_2}} = K$ for a suitable constant $K \in \mathbb{R}$ and the assertion follows. $\sqcap$

Theorem 3.1. *Let G be the four-dimensional connected solvable Lie group with trivial center having precisely two one-dimensional normal subgroups and choose for G the representation*

$$G = \left\{ g(x_1, x_2, x_3, x_4) = \begin{pmatrix} e^{ax_4} & 0 & 0 & x_1 \\ 0 & e^{x_4} & x_4 e^{x_4} & x_2 \\ 0 & 0 & e^{x_4} & x_3 \\ 0 & 0 & 0 & 1 \end{pmatrix}, x_i \in \mathbb{R}, i = 1, 2, 3, 4 \right\}$$

with fixed number $a \in \mathbb{R} \setminus \{0\}$. Let H be a one-dimensional subgroup of G which is not normal in G. If H is not contained in the commutator subgroup G' of G, then using automorphism of G we may choose $H = \{g(0, 0, 0, x_4); x_4 \in \mathbb{R}\}$. There does not exist connected topological loop L having G as the group topologically generated by the left translations of L

and H as the stabilizer of $e \in L$.

If H is contained in the commutator subgroup of G and $a \neq 1$, then using automorphisms of G we may choose H as one of the following subgroups:

$$H_1 = \{g(0,0,x_3,0); x_3 \in \mathbb{R}\},\ H_2 = \{g(x_1,0,x_1,0); x_1 \in \mathbb{R}\},$$

$$H_3 = \{g(x_1,x_1,0,0); x_1 \in \mathbb{R}\}.$$

For $a = 1$ using automorphisms of G we may assume that $H = H_1$.

a) Every continuous sharply transitive section $\sigma : G/H_1 \to G$ with the properties that $\sigma(G/H_1)$ generates G and $\sigma(H_1) = 1$ is given by the map $\sigma_f : g(x,y,0,z)H_1 \mapsto g(x, y + ze^z f(x,z), e^z f(x,z), z)$, where $f : \mathbb{R}^2 \to \mathbb{R}$ is a continuous function with $f(0,0) = 0$ such that the function f does not fulfill the identities $f(x,0) = 0$ and $f(0,z) = K(1 - e^{-z})$, $K \in \mathbb{R}$, simultaneously. The multiplication of the loop L_f corresponding to σ_f can be written as

$$(x_1,y_1,z_1) * (x_2,y_2,z_2) = (x_1 + e^{az_1}x_2, y_1 + y_2e^{z_1} - z_2e^{z_1}f(x_1,z_1), z_1 + z_2). \tag{2}$$

Every loop L_f defined by (2) is solvable.

b) Each continuous sharply transitive section $\sigma : G/H_2 \to G$ such that $\sigma(G/H_2)$ generates G and $\sigma(H_2) = 1$ has the form

$$\sigma_h : g(x,y,0,z)H_2 \mapsto g(x + e^{az}h(x,y,z), y + ze^z h(x,y,z), e^z h(x,y,z), z),$$

$a \neq 1$, where $h : \mathbb{R}^3 \to \mathbb{R}$ is a continuous function with $h(0,0,0) = 0$ such that h does not satisfy the identities $h(x,y,0) = 0$ and $h(0,0,z) = K(1 - e^{-z})$, $K \in \mathbb{R}$, simultaneously and such that for all triples (x_1,y_1,z_1) and $(x_2,y_2,z_2) \in \mathbb{R}^3$ the equations

$$y = y_2 - e^{z_2 - z_1}y_1 + e^{z_2 - z_1}z_1 h(x,y,z_2 - z_1), \tag{3}$$

$$x = x_2 - x_1e^{a(z_2 - z_1)} + e^{az_2}(e^{-z_1} - e^{-az_1})h(x,y,z_2 - z_1) \tag{4}$$

have a unique solution $(x,y) \in \mathbb{R}^2$. The multiplication of the loop L_h corresponding to σ_h is determined by

$$(x_1,y_1,z_1) * (x_2,y_2,z_2) =$$

$$\left(x_1 + e^{az_1}(x_2 + h(x_1,y_1,z_1)[1 - e^{(a-1)z_2}]), y_1 + e^{z_1}(y_2 - z_2h(x_1,y_1,z_1)), z_1 + z_2\right). \tag{5}$$

c) Every continuous sharply transitive section $\sigma : G/H_3 \to G$ with the properties $\sigma(G/H_3)$ generates G and $\sigma(H_3) = 1$ is given by the map

$$\sigma_f : g(x,0,y,z)H_3 \mapsto g(x + e^{az}f(x,y,z), e^z f(x,y,z), y, z),\ a \neq 1,$$

where $f : \mathbb{R}^3 \to \mathbb{R}$ *is a continuous function with* $f(0,0,0) = 0$ *such that* f *does not fulfill the identities* $f(x,y,0) = -x$ *and* $f(0,0,z) = c(1-e^{-az})$, $c \in \mathbb{R}$, *simultaneously and such that for all triples* (x_1,y_1,z_1), $(x_2,y_2,z_2) \in \mathbb{R}^3$ *there is a unique* $x \in \mathbb{R}$ *satisfying the equation*

$$x = x_2 - x_1 e^{a(z_2-z_1)} +$$

$$e^{az_2-z_1}[y_1(z_2-z_1) + (1-e^{(1-a)z_1}) f(x, y_2 - e^{z_2-z_1} y_1, z_2 - z_1)]. \quad (6)$$

The multiplication of the loop L_f *corresponding to* σ_f *is defined by*

$$(x_1,y_1,z_1) * (x_2,y_2,z_2) =$$

$$\big(x_1 + e^{az_1}(x_2 - y_2 z_1 e^{(a-1)z_2} + f(x_1,y_1,z_1)[1-e^{(a-1)z_2}]), y_1 + e^{z_1} y_2, z_1 + z_2\big). \quad (7)$$

Proof. The linear representation of the Lie group G is given in Ref. 7, p. 164. The Lie algebra $\mathbf{g}$ of G is given by the basis $\{e_1,e_2,e_3,e_4\}$ with $[e_1,e_4] = ae_1$, $[e_2,e_4] = e_2$, $[e_3,e_4] = e_2 + e_3$, $a \neq 0$. As $\mathbb{R}e_1$ and $\mathbb{R}e_2$ are ideals of $\mathbf{g}$ the subalgebra $\mathbf{h}$ of the 1-dimensional non-normal subgroup H of G does not contain e_1, e_2. First we assume that $\mathbf{h}$ is not contained in the commutator subalgebra $\mathbf{g}' = \langle e_1,e_2,e_3 \rangle$ of $\mathbf{g}$. Hence one has $\mathbf{h} = \mathbb{R}(ae_1 + be_2 + ce_3 + e_4)$ with $a,b,c \in \mathbb{R}$. Using the automorphism $e_1 \mapsto e_1$, $e_2 \mapsto e_2$, $e_3 \mapsto e_3$ and $e_4 \mapsto e_4 + ae_1 + be_2 + ce_3$ of $\mathbf{g}$ we can assume $\mathbf{h} = \mathbb{R}e_4$. Now we assume that there is a connected topological loop L having the group G as the group topologically generated by its left translations and $H = \{\exp te_4; t \in \mathbb{R}\}$ as the stabilizer of $e \in L$. As every 1-dimensional subalgebra which is not contained in the commutator subalgebra of the Lie algebra $\mathbf{g}$ is conjugate to $\mathbf{h}$ every element of G not contained in the commutator subgroup G' of G has a fixed point on G/H. As the set $\sigma(G/H)$ is the set of the left translations of the loop L and the left translations have no fixed point $\sigma(G/H)$ should be contained in G'. This is a contradiction to the fact that the left translations of L generate G and the first assertion is proved.

If the subalgebra $\mathbf{h}$ of H is contained in the commutator subalgebra $\mathbf{g}'$, then H has the form $H = \exp\ t(b_1e_3 + b_2e_1 + b_3e_2)$, $t \in \mathbb{R}$, with $b_1 \neq 0$ or $b_2b_3 \neq 0$. If $a \neq 1$, then each automorphism φ of $\mathbf{g}$ is given by $\varphi(e_1) = ke_1$, $\varphi(e_2) = le_2$, $\varphi(e_3) = ne_2 + le_3$, $\varphi(e_4) = f_1e_1 + f_2e_2 + f_3e_3 + e_4$ with $kl \neq 0$, $k,l,n,f_1,f_2,f_3 \in \mathbb{R}$. If $a = 1$, then the automorphism group of $\mathbf{g}$ consists of the linear mappings $\alpha(e_1) = k_1e_1 + k_2e_2$, $\alpha(e_2) = le_2$, $\alpha(e_3) = n_1e_1 + n_2e_2 + le_3$, $\alpha(e_4) = f_1e_1 + f_2e_2 + f_3e_3 + e_4$ with $k_1l \neq 0$,

$k_1, k_2, l, n_1, n_2, f_1, f_2, f_3 \in \mathbb{R}$. If $b_1 \neq 0$ and $a \neq 1$, then we can change H by an automorphism φ of G such that H has one of the following forms

$$H_1 = \{\exp te_3; t \in \mathbb{R}\}, \quad H_2 = \{\exp t(e_3 + e_1); t \in \mathbb{R}\}.$$

If $b_1 \neq 0$ and $a = 1$, then using an automorphism α of G we may assume that $H = H_1$. If $b_1 = 0$ and $a \neq 1$, then we can change H by an automorphism φ of G such that H is the subgroup $H_3 = \{\exp t(e_1 + e_2); t \in \mathbb{R}\}$. If $b_1 = 0$ and $a = 1$, then for all $b_2, b_3 \in \mathbb{R}$ the subgroup $H = \{\exp\ t(b_2e_1 + b_3e_2); t \in \mathbb{R}\}$ is normal in G which is a contradiction.

First we deal with the case that $H = H_1 = \{g(0,0,k,0);\ k \in \mathbb{R}\}$. Since all elements of G have a unique decomposition as $g(x, y, 0, z)g(0, 0, k, 0)$, any continuous function $f : \mathbb{R}^3 \to \mathbb{R}; (x, y, z) \mapsto f(x, y, z)$ determines a continuous section $\sigma : G/H_1 \to G$ given by

$$\sigma : g(x, y, 0, z)H_1 \mapsto g(x, y, 0, z)g(0, 0, f(x, y, z), 0) =$$

$$g(x, y + ze^z f(x, y, z), e^z f(x, y, z), z).$$

The section σ is sharply transitive if and only if for every triples (x_1, y_1, z_1), $(x_2, y_2, z_2) \in \mathbb{R}^3$ there exists precisely one triple $(x, y, z) \in \mathbb{R}^3$ such that

$$g(x, y{+}ze^z f(x, y, z), e^z f(x, y, z), z)g(x_1, y_1, 0, z_1) = g(x_2, y_2, 0, z_2)g(0, 0, t, 0)$$

holds with a suitable $t \in \mathbb{R}$. This gives the equations

$$z = z_2 - z_1,\ x = x_2 - x_1e^{a(z_2-z_1)},\ t = e^{-z_1} f(x_2 - x_1e^{a(z_2-z_1)}, y, z_2 - z_1),$$

$$0 = y - y_2 + y_1e^{z_2-z_1} - z_1e^{z_2-z_1} f(x_2 - x_1e^{a(z_2-z_1)}, y, z_2 - z_1).$$

These are equivalent to the condition that for every $z_0 = z_2 - z_1$, $x_0 = x_2 - x_1e^{az_0}$ and $z_1 \in \mathbb{R}$ the function $g : y \mapsto y - z_1e^{z_0} f(x_0, y, z_0) : \mathbb{R} \to \mathbb{R}$ is a bijective mapping. Let be $\psi_1 < \psi_2 \in \mathbb{R}$ then $g(\psi_1) \neq g(\psi_2)$, e.g. $g(\psi_1) < g(\psi_2)$. We consider

$$0 < g(\psi_2) - g(\psi_1) = \psi_2 - \psi_1 - z_1e^{z_0}[f(x_0, \psi_2, z_0) - f(x_0, \psi_1, z_0)]$$

as a linear function of $z_1 \in \mathbb{R}$. If $f(x_0, \psi_2, z_0) \neq f(x_0, \psi_1, z_0)$, then there exists a $z_1 \in \mathbb{R}$ such that $g(\psi_2) - g(\psi_1) = 0$, which is a contradiction. Hence the function $f(x, y, z) = f(x, z)$ does not depend on y. In this case g is a monotone function and every continuous function $f(x, z)$ with $f(0, 0) = 0$ determines a loop multiplication.

This loop is proper precisely if the set

$$\sigma(G/H_1) = \{g(x, y + ze^z f(x, z), e^z f(x, z), z); x, y, z \in \mathbb{R}\}$$

generates the whole group G. The set $\sigma(G/H_1)$ contains the subgroup

$$G_1 = \{g(x, y, f(x, 0), 0);\ x, y \in \mathbb{R}\} < G' = [G, G],$$

and the subset $F_2 = \{g(0, ze^z f(0, z), e^z f(0, z), z);\ z \in \mathbb{R}\}$. As $G_1 \cap F_2 = \{1\}$ the set $\sigma(G/H_1)$ generates G if the group G_1 has dimension 3. This is the case if and only if the subgroup $G_2 = \{g(x, 0, f(x, 0), 0); x \in \mathbb{R}\}$ is not a one-parameter subgroup. But G_2 is a one-parameter subgroup precisely if $f(x, 0) = \lambda x$, $\lambda \in \mathbb{R}$. In this case the set $\sigma(G/H_1)$ generates G if there exists an element $h \in F_2$ with $h^{-1} G_1 h \neq G_1$. For $h = g(0, ze^z f(0, z), e^z f(0, z), z) \in F_2$, where $z \neq 0$, we have

$$h^{-1} g(x, y, \lambda x, 0) h = g(xe^{-az}, ye^{-z} - ze^{-z}\lambda x, e^{-z}\lambda x, 0).$$

Hence $h^{-1} G_1 h = G_1$ if and only if for $a \neq 1$ one has $f(x, 0) = 0$ and for $a = 1$ we have $f(x, 0) = \lambda x$, $\lambda \in \mathbb{R}$. Using this for $a \neq 1$ the group G_1 reduces to $\widetilde{G_1} = \{g(x, y, 0, 0); x, y \in \mathbb{R}\}$ and for $a = 1$ to $G_1^* = \{g(x, y, \lambda x, 0); x, y \in \mathbb{R}\}$, $\lambda \in \mathbb{R}$. The group $\widetilde{G_1}$ and for $a = 1$ the group G_1^* are normal subgroups of G. The set $\sigma(G/H)$ does not generate G precisely if for $a \neq 1$ the set $F_2\widetilde{G_1}/\widetilde{G_1}$, respectively for $a = 1$ the set $F_2 G_1^*/G_1^*$ is a one-parameter subgroup of $G/\widetilde{G_1}$, respectively G/G_1^*. Since for $a \neq 1$ one has

$$g(\mathbb{R}, \mathbb{R}, e^{z_1} f(0, z_1), z_1) g(\mathbb{R}, \mathbb{R}, e^{z_2} f(0, z_2), z_2) =$$

$$g(\mathbb{R}, \mathbb{R}, e^{z_1+z_2} f(0, z_2) + e^{z_1} f(0, z_1), z_1 + z_2)$$

and for $a = 1$ we have

$$g(\mathbb{R}, \mathbb{R}, e^{z_1}(\lambda x_1 + f(0, z_1)), z_1) g(\mathbb{R}, \mathbb{R}, e^{z_2}(\lambda x_2 + f(0, z_2)), z_2) =$$

$$g(\mathbb{R}, \mathbb{R}, e^{z_1}[e^{z_2}(\lambda x_2 + f(0, z_2)) + \lambda x_1 + f(0, z_1)], z_1 + z_2)$$

the set $\sigma(G/H_1)$ does not generate G if and only if in both cases $f(x, 0) = 0$ and for all $z_1, z_2 \in \mathbb{R}$ one has $f(0, z_2) + e^{-z_2} f(0, z_1) = f(0, z_1 + z_2)$. Using Lemma 3.1 for the real function $f(0, z)$ we obtain $f(0, z) = K(1 - e^{-z})$, where $K \in \mathbb{R}$. Hence the set $\sigma(G/H_1)$ does not generate G if for all $x, z \in \mathbb{R}$ one has $f(x, 0) = 0$ and $f(0, z) = K(1 - e^{-z})$, $K \in \mathbb{R}$.

Now we represent the loop L_f in the coordinate system $(x, y, z) \mapsto g(x, y, 0, z)H_1$. Then the product $(x_1, y_1, z_1) * (x_2, y_2, z_2)$ will be determined if we apply $\sigma(g(x_1, y_1, 0, z_1)H_1) = g(x_1, y_1 + z_1 e^{z_1} f(x_1, z_1), e^{z_1} f(x_1, z_1), z_1)$ to the left coset $g(x_2, y_2, 0, z_2)H_1$ and find in the image coset the element of G which lies in the set $\{g(x, y, 0, z)H_1;\ x, y, z \in \mathbb{R}\}$. A direct computation gives multiplication (2) in assertion a).

The set $N = \{(x, y, 0); x, y \in \mathbb{R}\}$ is a normal subgroup isomorphic to $\mathbb{R}^2$ of the loop L_f and one has $(0, 0, z_1)N * (0, 0, z_2)N = (0, 0, z_1 + z_2)N$. Hence the factor loop L_f/N is isomorphic to the Lie group $\mathbb{R}$ and therefore the loop L_f is solvable.

Now we assume that $H = H_2 = \{g(k, 0, k, 0);\ k \in \mathbb{R}\}$. As all elements of G can be written in a unique way as $g(x, y, 0, z)g(k, 0, k, 0)$, every continuous function $h : \mathbb{R}^3 \to \mathbb{R}; (x, y, z) \mapsto h(x, y, z)$ determines a continuous section $\sigma : G/H_2 \to G$ defined by

$$\sigma_h : g(x, y, 0, z)H_2 \mapsto g(x, y, 0, z)g(h(x, y, z), 0, h(x, y, z), 0) =$$

$$g(x + e^{az}h(x, y, z), y + ze^{z}h(x, y, z), e^{z}h(x, y, z), z). \tag{8}$$

The section σ is sharply transitive if and only if for each triples (x_1, y_1, z_1), $(x_2, y_2, z_2) \in \mathbb{R}^3$ there exists precisely one triple $(x, y, z) \in \mathbb{R}^3$ such that

$$g(x + e^{az}h(x, y, z), y + ze^{z}h(x, y, z), e^{z}h(x, y, z), z)g(x_1, y_1, 0, z_1) =$$

$$g(x_2, y_2, 0, z_2)g(0, 0, t, 0) \tag{9}$$

for a suitable $t \in \mathbb{R}$. Equation (9) gives $z = z_2 - z_1$, $t = e^{-z_1}h(x, y, z_2 - z_1)$ and that equations (3), (4) in assertion b) must have a unique solution $(x, y) \in \mathbb{R}^2$.

Now we investigate under which circumstances the set $\sigma(G/H_2)$ generates the group G. The set $\sigma(G/H_2)$ contains the subgroup

$$G_1 = \{g(x + h(x, y, 0), y, h(x, y, 0), 0);\ x, y \in \mathbb{R}\} < G' = [G, G],$$

and the subset $F_2 = \{g(e^{az}h(0, 0, z), ze^{z}h(0, 0, z), e^{z}h(0, 0, z), z);\ z \in \mathbb{R}\}$. One has $G_1 \cap F_2 = \{1\}$. The set $\sigma(G/H_2)$ generates G if dim $G_1 = 3$. This happens if the group $G_2 = \{g(h(0, y, 0), y, h(0, y, 0), 0); y \in \mathbb{R}\}$ or the group $G_3 = \{g(x + h(x, 0, 0), 0, h(x, 0, 0), 0); x \in \mathbb{R}\}$ has dimension 2. The group G_2, respectively G_3 is a one-parameter subgroup precisely if $h(0, y, 0) = cy$, $c \in \mathbb{R}$, respectively $h(x, 0, 0) = bx$, $b \in \mathbb{R}$. As

$$g(x + h(x, 0, 0), 0, h(x, 0, 0), 0)g(h(0, y, 0), y, h(0, y, 0), 0) =$$

$$g(x + h(x, 0, 0) + h(0, y, 0), y, h(x, 0, 0) + h(0, y, 0), 0)$$

the group G_1 has dimension 3 if and only if the function $h(x, y, 0)$ is different from $bx + cy$, $b, c \in \mathbb{R}$. Assuming this the set $\sigma(G/H_2)$ generates G if there exists an element $h \in F_2$ such that $h^{-1}G_1h \neq G_1$ holds. For $h = g(e^{az}h(0, 0, z), ze^{z}h(0, 0, z), e^{z}h(0, 0, z), z) \in F_2$ with $z \neq 0$ one has

$$h^{-1}g(x + bx + cy, y, bx + cy, 0)h =$$

$$g([(b+1)x+cy]e^{-az}, ye^{-z}-z(bx+cy)e^{-z}, (bx+cy)e^{-z}, 0).$$

We obtain $h^{-1}G_1 h = G_1$ if and only if $b = c = 0$. Then the group $G_1 = \{g(x, y, 0, 0);\ x, y \in \mathbb{R}\}$ is a normal subgroup of G. The set $\sigma(G/H_2)$ generates G, if the set $(F_2 G_1)/G_1$ is not a one-parameter subgroup of G/G_1. Because of

$$g(\mathbb{R}, \mathbb{R}, e^{z_1}h(0,0,z_1), z_1)g(\mathbb{R}, \mathbb{R}, e^{z_2}h(0,0,z_2), z_2) =$$

$$g(\mathbb{R}, \mathbb{R}, e^{z_1+z_2}h(0,0,z_2) + e^{z_1}h(0,0,z_1), z_1+z_2),$$

the set $\sigma(G/H_2)$ does not generate G precisely if for all $z_1, z_2 \in \mathbb{R}$ the equality $h(0,0,z_2) + e^{-z_2}h(0,0,z_1) = h(0,0,z_1+z_2)$ holds. Using Lemma 3.1 for the function $h(0,0,z)$ we get $h(0,0,z) = K(1-e^{-z})$ with $K \in \mathbb{R}$. Therefore the loop L_h is proper if and only if the function $h : \mathbb{R}^3 \to \mathbb{R}$ does not satisfy the identities $h(x,y,0) = 0$ and $h(0,0,z) = K(1-e^{-z})$ simultaneously.

The multiplication of the loop L_h in the coordinate system $(x,y,z) \mapsto g(x,y,0,z)H_2$ is determined if we apply $\sigma_h(g(x_1,y_1,0,z_1)H_2)$ given by (8) to the left coset $g(x_2,y_2,0,z_2)H_2$ and find in the image coset the element of G which lies in the set $\{g(x,y,0,z)H_2;\ x,y,z \in \mathbb{R}\}$. A direct computation yields multiplication (5) of assertion b).

Now we consider the case that $H = H_3 = \{g(k,k,0,0);\ k \in \mathbb{R}\}$. As all elements of G have a unique decomposition as $g(x,0,y,z)g(k,k,0,0)$, any continuous function $f : \mathbb{R}^3 \to \mathbb{R}; (x,y,z) \mapsto f(x,y,z)$ determines a continuous section $\sigma : G/H_3 \to G$ given by

$$\sigma : g(x,0,y,z)H_3 \mapsto g(x,0,y,z)g(f(x,y,z), f(x,y,z), 0, 0) =$$

$$g(x + e^{az}f(x,y,z), e^{z}f(x,y,z), y, z).$$

The set $\sigma(G/H_3)$ acts sharply transitively on the factor space G/H_3 if and only if for every triples (x_1,y_1,z_1), $(x_2,y_2,z_2) \in \mathbb{R}^3$ there exists precisely one triple $(x,y,z) \in \mathbb{R}^3$ such that

$$g(x + e^{az}f(x,y,z), e^{z}f(x,y,z), y, z)g(x_1, 0, y_1, z_1) =$$

$$g(x_2, 0, y_2, z_2)g(t,t,0,0) \tag{10}$$

for a suitable $t \in \mathbb{R}$. Equation (10) yields that $z = z_2 - z_1$, $y = y_2 - e^{z_2-z_1}y_1$, $t = y_1(z_2-z_1)e^{-z_1} + e^{-z_1}h(x, y_2 - e^{z_2-z_1}y_1, z_2-z_1)$ and equation (6) in assertion c) must have a unique solution $x \in \mathbb{R}$.

The set $\sigma(G/H_3)$ contains the subgroup

$$G_1 = \{g(x + f(x,y,0), f(x,y,0), y, 0);\ x, y \in \mathbb{R}\} < G' = [G,G],$$

and the subset $F_2 = \{g(e^{az}f(0,0,z), e^z f(0,0,z), 0, z);\ z \in \mathbb{R}\}$. We have $G_1 \cap F_2 = \{1\}$. The set $\sigma(G/H_3)$ generates G if G_1 is a 3-dimensional group. This is the case if the subgroup

$$G_2 = \{g(x + f(x,0,0), f(x,0,0), 0, 0); x \in \mathbb{R}\}$$

or the subgroup

$$G_3 = \{g(f(0,y,0), f(0,y,0), y, 0); y \in \mathbb{R}\}$$

has dimension 2. The group G_2, respectively G_3 is a one-parameter subgroup precisely if $f(x,0,0) = cx$, $c \in \mathbb{R}$, respectively $f(0,y,0) = dy$, $d \in \mathbb{R}$. As

$$g(x + f(x,0,0), f(x,0,0), 0, 0)g(f(0,y,0), f(0,y,0), y, 0) =$$

$$g(x + f(x,0,0) + f(0,y,0), f(x,0,0) + f(0,y,0), y, 0)$$

we have $\dim G_1 = 3$ if and only if the function $f(x,y,0)$ is different from $cx + dy$, $c, d \in \mathbb{R}$. In this case the set $\sigma(G/H_3)$ generates G if there exists an element $h \in F_2$ such that $h^{-1}G_1 h \neq G_1$ holds. For $h = g(e^{az}f(0,0,z), e^z f(0,0,z), 0, z) \in F_2$ with $z \neq 0$ we have

$$h^{-1}g(x + cx + dy, cx + dy, y, 0)h =$$

$$g([(c+1)x + dy]e^{-az}, (-zy + cx + dy))e^{-z}, e^{-z}y, 0).$$

Hence $h^{-1}G_1 h = G_1$ if and only if one has $c = -1$ and $d = 0$. Then the group $G_1 = \{g(0, -x, y, 0);\ x, y \in \mathbb{R}\}$ is a normal subgroup of G and the set $\sigma(G/H_3)$ generates G, if the set $(F_2 G_1)/G_1$ is not a one-parameter subgroup of G/G_1. Because of

$$g(e^{az_1}f(0,0,z_1), \mathbb{R}, \mathbb{R}, z_1)g(e^{az_2}f(0,0,z_2), \mathbb{R}, \mathbb{R}, z_2) =$$

$$g(e^{a(z_1+z_2)}f(0,0,z_2) + e^{az_1}f(0,0,z_1), \mathbb{R}, \mathbb{R}, z_1 + z_2),$$

the set $\sigma(G/H_3)$ does not generate G precisely if for all $z_1, z_2 \in \mathbb{R}$ the identity $f(0,0,z_2) + e^{-az_2}f(0,0,z_1) = f(0,0,z_1+z_2)$ holds. Using Lemma 3.1 for the function $f(0,0,z)$ we obtain $f(0,0,z) = c(1 - e^{-az})$ with a real constant $c \in \mathbb{R}$. A direct computation yields that the multiplication of the loop L_f corresponding to the section σ_f in the coordinate system $(x,y,z) \mapsto g(x,0,y,z)H_3$ is given by (7) and the assertion c) is proved. □

Theorem 3.2. *The 4-dimensional connected solvable Lie group G which has trivial center and precisely two one-dimensional normal subgroups is not the multiplication group of connected topological proper loops.*

Proof. Every 1-dimensional connected topological loop L having a Lie group as its multiplication group is already a Lie group (cf. Theorem 18.18 in Ref. 13, p. 248). If the multiplication group of a 2-dimensional connected topological proper loop is a Lie group, then it is nilpotent (cf. Theorem 1 in Ref. 4). Since the group G has trivial center all 3-dimensional connected topological loop L having the group G as the group topologically generated by the left translations of L are homeomorphic to $\mathbb{R}^3$ and their multiplications are given by (2), (5) and (7) in Theorem 3.1. Denote by H_i, $i = 1, 2, 3$, the subgroups of G given in Theorem 3.1. If the group G is also the group generated by all left and right translations of 3-dimensional topological loops, then the inner mapping group of L is the group H_1 if L is defined by (2), the group H_2 if L is defined by (5) and the group H_3 if L is given by (7). Since the commutator subgroup G' of G is a 3-dimensional abelian normal subgroup of G and each subgroup H_i, $i = 1, 2, 3$, is contained in G' the normalizer of H_i in G coincides with the group G'. As the Lie algebra **g** of the group G has trivial center we have a contradiction to Lemma 2.1 and the assertion follows. □

Acknowledgments

Sincere thanks to the referee for the careful reading of the manuscript and for many suggestions. This paper was supported by the Hungarian Scientific Research Fund (OTKA) Grant PD 77392 and by the EEA and Norway Grants (Zoltán Magyary Higher Education Public Foundation).

References

1. A. Andrada, M. L. Barberis, I. G. Dotti and G. P. Ovando, Product structures on four dimensional solvable Lie algebras. *Homology Homotopy Appl.* **7** (1), 2005, 9-37.
2. R. H. Bruck, Contributions to the Theory of Loops. *Trans. Amer. Math. Soc.* **60**, 1946, 245-354.
3. P. Csörgö and M. Niemenmaa, On connected transversals to nonabelian subgroups. *European J. Combin.* **23**, 2002, 179-185.
4. Á. Figula, The multiplication groups of 2-dimensional topological loops. *J. Group Theory* **12**, 2009, 419-429.
5. Á. Figula, Topological loops with three-dimensional solvable left translation group. *Aequat. Math.* **79**, 2010, 83-97.
6. Á. Figula, On the multiplication groups of 3-dimensional topological loops. accepted for publication *J. Lie theory*, 2010, pp. 31.
7. R. Ghanam, I. Strugar and G. Thompson, Matrix Representation for Low Dimensional Lie Algebras. *Extracta Math.* **20**, 2005, no. 2, 151-184.

8. V. V. Gorbatsevich, On three-dimensional homogeneous spaces. *Sib. Math. Zh.* **18**, 1977, 280-293. English transl.: *Sib. Math. J.* **18**, 200-210.
9. M. Niemenmaa, Finite loops with dihedral inner mapping groups are solvable. *J. Algebra* **273**, 2004, 288-294.
10. M. Niemenmaa, On connected transversals to subgroups whose order is a product of two primes. *European J. Combin.* **18**, 1997, 915-919.
11. M. Niemenmaa and T. Kepka, On Multiplication Groups of Loops. *J. Algebra* **135**, 1990, 112-122.
12. M. Niemenmaa and A. Vesanen, On connected transversals in the projective special linear group $PSL(2,7)$. *J. Algebra* **166**, 1994, 455-460.
13. P. T. Nagy and K. Strambach, *Loops in Group Theory and Lie Theory*, Walter de Gruyter 35. Berlin, New York, 2002.
14. G. Ovando, Invariant complex structures on solvable real Lie groups. *Manuscr. Math.* **103**, 2000, 19-30.
15. J. Patera, R. T. Sharp, P. Winternitz and H. Zassenhaus, Invariants of real low dimension Lie algebras. *J. Math. Phys.* **17** (6), 1976, 986-994.
16. A. Vesanen, Finite classical groups and multiplication groups of loops. *Math. Proc. Camb. Philos. Soc.* **117** 1995, 425-429.

A NOTE ON FINITE p-GROUPS WITH A MAXIMAL ELEMENTARY SUBGROUP OF RANK 2

G. GLAUBERMAN

Department of Mathematics
The University of Chicago
5734 S. University Avenue
Chicago, IL 60637, USA
E-mail: gg@math.uchicago.edu

Suppose p is an odd prime and S is a finite p-group containing a maximal elementary abelian subgroup of order p^2. We show that every elementary abelian subgroup of S has order at most p^p, and we discuss related open questions and applications to simple groups and representation theory. (This is joint work with Nadia Mazza.)

Keywords: p-groups; elementary abelian.

1. Introduction and examples

Let p be a prime and S be a finite p-group. An elementary abelian subgroup E of S has *rank* k if $|E| = p^k$. We say that E is a *maximal* elementary abelian subgroup of S if we never have $E < F$ for some elementary abelian subgroup F of S. In this expository note, based on a recent joint article with N. Mazza[1], we discuss some consequences of the following hypothesis:

(H) there exists a maximal elementary abelian subgroup E of rank 2 in S.

For details of proofs and references, see the joint article[1].

We thank the organizers for a very enjoyable conference.

In mathematics, usually

bigger is *more complicated.*

So one's first question may be: why do we study groups with such a narrow restriction?

Curiously, groups satisfying (H) can be harder, rather than easier, to study than larger groups. This happens, for example, in the proof of the Feit-Thompson Odd Order Theorem[2] (pp. 845, 903), and of many other theorems on classifying simple groups (as in Ref. 3 pp. 67–69); and recently in the study of endotrivial modules in representation theory (Section 3 below).

Now assume (H). It is easy to show that S is "small" in the sense that it has no *normal* elementary abelian subgroup of rank at least $p+1$ (see the proof of Proposition 2.9 of Mazza's article[4]). Our main result shows that this is true for *non-normal* subgroups as well, if p is odd.

Theorem 1.1 ([1]). *Assume p is odd and (H) is satisfied. Then every elementary abelian subgroup of S has rank at most p.*

We give examples to show that the bound is best possible for p odd and that a bound of 4 is best possible for $p = 2$.

Here, p may be any prime. Let $\mathbb{F}$ be the Galois field $\mathrm{GF}(p^p)$ and σ be the field automorphism given by $\sigma(x) = x^p$.

Example 1. For each element a in $\mathbb{F}$, let $t(a)$ be the translation that maps x to $x + a$ for each x in $\mathbb{F}$. Let

$$T = \{t(a) | a \in \mathbb{F}\} = \text{ translation group of } \mathbb{F}$$

Then σ and T are contained in the symmetric group on $\mathbb{F}$, and σ normalizes T. Let $S = T\langle\sigma\rangle$.

In the group S, $\langle\sigma, t(1)\rangle$ is a maximal elementary abelian subgroup of rank 2, and T is an elementary abelian subgroup of rank p. Thus, S satisfies (H), and the bound of rank p in the theorem is best possible.

Example 2. Suppose $p = 2$, so that $\mathbb{F} = \mathrm{GF}(4)$. Let $\overline{x} = \sigma(x) = x^2$ for all x in $\mathbb{F}$. For a, b, c in $\mathbb{F}$, let

$$M(a,b,c) = \begin{pmatrix} 1 & a & c \\ 0 & 1 & b \\ 0 & 0 & 1 \end{pmatrix}.$$

Let U be the group of all matrices $M(a,b,c)$ for a, b, c in $\mathbb{F}$. Define a mapping t on U by

$$t(M(a,b,c)) = M(\overline{b}, \overline{a}, \overline{a}\overline{b} + \overline{c}).$$

Then t is an automorphism of U and $t^2 = 1$. Let S be the semidirect product $U \rtimes \langle t \rangle$.

Then $\langle t, M(0,0,1)\rangle$ is a maximal elementary abelian subgroup of rank 2 in S, and the set $\{M(0,b,c)|b,c \in \mathbb{F}\}$ is an elementary abelian subgroup of rank 4 in S.

A theorem of Anne MacWilliams Patterson[5] and a short argument in the joint article[1] (p. 1734) show that a 2-group satisfying (H) cannot have an elementary abelian subgroup of rank 5; thus 4 is the maximum possible rank.

2. Ideas in proof and open questions

Assume that p is odd and S is a counterexample to the theorem. We describe some steps toward a contradiction.

Here, S satisfies (H) and contains an elementary abelian subgroup A of rank at least $p+1$. Let E be a maximal elementary abelian subgroup of rank 2 in S.

N. Blackburn and L. Héthelyi have studied extensively the consequences of (H). Our proof[1] relies heavily on their work.

The first main step in the proof is the following:

Proposition 2.1. *Suppose E normalizes some subgroup T of exponent p in S. Then $|T| \leq p^p$.*

Summary of proof. The result is easy if T is elementary abelian; see the proof of Proposition 2.9 in Mazza's article[4]. In the general case, the proof uses theorems of Blackburn, M. Lazard, and M. Suzuki about p-groups of maximal class and p-groups of class at most $p-1$. □

Given the proposition, take A as above. Since A contains a subgroup of rank $p+1$, we may assume for convenience that A itself has rank $p+1$.

If $p = 3$, a theorem of D. Jonah and M. Konvisser[6] asserts that we may choose A normal in S, contrary to the proposition. Thus, $p \geq 5$.

Now a theorem of J. Alperin and the author[7] says that we may choose A such that $A \lhd N$ for some normal subgroup N of S. Then a theorem of P. Hall eventually gives a contradiction to the proposition.

The theorem suggests the following generalizations as open problems:

Problems. Assume (H), and suppose p is odd and T is a subgroup of S.

(1) Can T be generated by p elements? (See Thompson's result in Ref. 8, Theorem X.3.4.)
(2) Suppose T has exponent p. Is $|T| \leq p^p$?

The theorem gives affirmative answers if T is elementary abelian.

3. Applications

In this section, we assume that S contains some elementary abelian subgroup of rank two, but we do not assume (H).

Let $\mathcal{E}(S)$ be the set of all elementary abelian subgroups of S of rank at least two. We form an undirected graph in which the vertices are the elements of $\mathcal{E}(S)$ and the edges consist of all unordered pairs $\{A, B\}$ such that $A < B$.

As in the joint article[1], pp. 1730–1731, a great deal is known about the graph of $\mathcal{E}(S)$. In particular, if S is a Sylow p-subgroup of a group G, then the study of G is much easier if $\mathcal{E}(S)$ is connected; while if $\mathcal{E}(S)$ is not connected, then S satisfies (H). Thus, the theorem yields:

Corollary 3.1. *If p is odd and S contains an elementary abelian subgroup of rank at least $p+1$, then $\mathcal{E}(S)$ is connected.*

This yields an application to representation theory. Suppose S is a Sylow p-subgroup of a group G and K is a field of characteristic p. We may regard K as a trivial KG-module. Then a KG-module M is called *endotrivial* if

$$\mathrm{Hom}_K(M, M) \cong K \oplus P \text{ for some projective } KG\text{-module } P.$$

For example, if $M^{\otimes m} \cong K \oplus P'$ for some positive integer m and some projective KG-module P', then M is called a *torsion endotrivial* module, and M is known to be endotrivial.

There is a standard way (described in the joint article[1], p. 1735) to construct endotrivial modules from torsion endotrivial modules and a minimal projective resolution of K as a KG-module, and there are no other endotrivial KG-modules if $\mathcal{E}(S)$ is connected. Therefore, the theorem yields another consequence:

Corollary 3.2. *Suppose p is odd, S is a Sylow p-subgroup of a finite group G, K is a field of characteristic p, and S contains an elementary abelian subgroup of rank at least $p+1$. Then every endotrivial KG-module arises as described in Ref. 1.*

References

1. G. Glauberman and N. Mazza, *J. Algebra* **323**, 1729 (2010).
2. W. Feit and J. G. Thompson, *Pacific J. Math.* **13**, 775 (1963).

3. D. Gorenstein, R. Lyons and R. Solomon, *The classification of the finite simple groups. Number 2*, Mathematical Surveys and Monographs, Vol. 40 (American Mathematical Society, Providence, RI, 1996).
4. N. Mazza, *J. Algebra* **320**, 4242 (2008).
5. A. R. MacWilliams, *Trans. Amer. Math. Soc.* **150**, 345 (1970).
6. M. Konvisser and D. Jonah, *J. Algebra* **34**, 309 (1975).
7. J. L. Alperin and G. Glauberman, *J. Algebra* **203**, 533 (1998).
8. B. Huppert and N. Blackburn, *Finite groups. III*, Grundlehren der Mathematischen Wissenschaften [Fundamental Principles of Mathematical Sciences], Vol. 243 (Springer-Verlag, Berlin, 1982).

FINITELY GENERATED FREE BY C_{p^n} PRO-p GROUPS

W. HERFORT

University of Technology at Vienna,
Vienna, Austria
E-mail: w.herfort@tuwien.ac.at

P. A. ZALESSKII

University of Brasilia,
Brasilia-DF, Brazil,
E-mail: pz@mat.unb.br

We prove that a topologically finitely generated pro-p group G is the extension of a free pro-p group by a cyclic p group if and only if it contains a finitely generated dense subgroup which is the extension of a free group by a cyclic p group. Moreover G is isomorphic to the pro-p completion of this dense subgroup.

Keywords: Pro-p group; Free by cyclic; Permutation extension.

1. The structure theorem

Our main result is Theorem 1.1 below, which one can derive from the more general result of free pro-p-by-cyclic pro-p groups in Ref. 3. However, we find it worth to provide a short and more direct proof using essentially only the result of C. Scheiderer[8] for finitely generated free pro-p by C_p extensions: Let G contain a finitely generated normal free pro-p group with quotient a cyclic p group. First we embed G into a *permutational extension* H and, after having established that H is a free pro-p product $N_H(t) \amalg E_0$ for t of order p and E_0 free pro-p, we apply a version of the Kurosh subgroup theorem found independently by L .Ribes–W. Herfort[2], D. Haran[1], and O.V. Mel'nikov[5] to the finitely generated subgroup G of H. Such simple proof is not available if G is not finitely generated[3].

As has been noted $H = E \rtimes K$ with F finitely generated free pro-p and K a cyclic p group with all torsion elements conjugate into K must be a permutational extension. When K is not cyclic this need not be true.

Notation is standard as in the References 6,7 and 9.

Theorem 1.1. *Let G be a finitely generated free pro-p-by-cyclic pro-p group. Then G contains a dense subgroup G_0 which is finitely generated (as an abstract group) and is itself an extension of a free group by a cyclic finite p group. Moreover, G is isomorphic to the pro-p completion of G_0.*

For giving a proof the following Lemma is needed:

Lemma 1.1. *Let $G = F \rtimes K$ with F free pro-p and K a cyclic p group. Suppose that every finite subgroup of G is F-conjugate into K. Then, for any $T \leq K$,*

(i) $N_G(T) = C_F(T) \rtimes K$;
(ii) Every finite subgroup of $N_G(T)$ is $C_F(T)$-conjugate to a subgroup of K;

Proof.
i) observe that $g \subset N_G(T)$ can be written as $g = fk$ with $f \in F$ and $k \in K$. Then $T = T^g = T^{fk}$ reads modulo F as $T = T^k$ so that $k \in K$ and hence $f \in C_F(T)$ follows.

ii) Let R be a finite subgroup of $N_G(T)$ and, replacing R by RT if necessary, we can assume that it contains T. By the hypothesis there exists $f \in F$ with $R^f \leq K$; hence $T^f \leq K$. Therefore $TT^f \leq K$ and, since $F \triangleleft G$, for every element $t \in T$ one has $t^{-1}t^f \in K \cap F$. As $K \cap F = \{1\}$ it follows that $f \in C_F(T)$. □

Proof of Theorem 1.1. We perform a few steps.

Step 1: Let $G = F \rtimes K$ with F finitely generated free pro-p, K a cyclic p-group of order p^n and $n \geq 1$. Embed G into a finitely generated free pro-p-by cyclic pro-p group $H = E \rtimes K$ with E again free pro-p and so that every torsion element is conjugate into K; a description of the method can be found in Sec.3[4].

Now, in $H = E \rtimes K$, every torsion element is conjugate to an element in K. During step 3 we show that H is a *permutational extension*, i.e., that there is a finite set I of indices, subgroups A_i of K and subsets Z_i of F with $H = HNN(K, A_i, Z_i, I)$. The latter denotes the pro-p completion of the discrete version of a permutational extension given by generators and relations as follows:

$$H_0 := \langle K, Z_i \in I \mid (\forall i \in I) \ \ [A_i, Z_i] = 1 \rangle.$$

Fix an element t of order p in K. When $K \cong C_p$ then $K = \langle t \rangle$ and H is a permutational extension $H = HNN(K, Z, Z_0)$, as can be seen by using the main result from Ref. 8 yields

$$H = (C_E(t) \times K) \amalg E_0 \tag{1}$$

for some free pro-p factor E_0; then let Z and Z_0 be bases of the free pro-p groups $C_E(t)$ and E_0 respectively.

Letting "bar" denote passing to the quotient modulo $(C_E(t))_E$ from Eq. (1) we can deduce that $C_{\overline{E}}(t) = \{1\}$.

In the sequel it will be convenient to omit the "bars" for the images in $\overline{G}$ of elements in K.

Step 2: We claim that every finite subgroup $R = \langle r \rangle$ of $\overline{H}$ is E-conjugate into K. Decompose $r = \overline{e}c$ with $e \in E$ and $c \in K$. When $|R| < |K|$ then by the minimality assumption on K the proper subgroup $E \rtimes \langle c \rangle$ of H is a permutational extension. But then, as a consequence of the relations in the permutational extension $E \rtimes \langle c \rangle$ we find that every finite subgroup of $\bar{E} \rtimes \langle c \rangle$ is conjugate into K. On the other hand, when $R \cong K$, by the minimality assumption on $|K|$ we can $\overline{E}$-conjugate $\langle r^p \rangle$ into K. Thus $r^p = c^p$ and so $R \leq N_{\overline{H}}(c) = K$, since $C_{\overline{E}}(c) = K$. Therefore Prop. 14 (see Ref. 4) implies that $\overline{H} \cong K \amalg E_0$.

Step 3: We come to proving that H is a permutational extension. Lemma 1.1 and the minimality assumption on the rank of F together imply that either $t \in Z(H)$ or that $N_H(t) = HNN(K, B_j, Y_j, j \in J)$ is a permutational extension. When $t \in Z(H)$ then in $H/\langle t \rangle$ every finite subgroup is F-conjugate into $K/\langle t \rangle$ and by the minimality assumption on $|K|$ we find that $H/\langle t \rangle = HNN(K/\langle t \rangle, A_i, Z_i, i \in I)$ is a permutational extension. Therefore $H = HNN(K, \langle A_i, t \rangle, Z_i, i \in I)$ is a permutational extension.

Let us turn to the case $Z(H) = \{1\}$. We devise a "model"

$$\tilde{H} = HNN(K, B_j, Y_j, j \in J) \amalg E_0,$$

define a map from the subset $K \cup \bigcup_{j \in J} Y_j \cup E_0$ of $\tilde{H}$ to the corresponding subset of H and extend it to become an epimorphism $\omega : \tilde{H} \to H$ by means of the universal property of the permutational extension. Now the restriction of ω to $N_{\tilde{H}}(t) = HNN(K, B_j, Y_j, j \in J)$ is an isomorphism onto $N_H(t)$. In particular, setting $\tilde{E} := \omega^{-1}(E)$, we can see that $\tilde{E} = C_{\tilde{E}}(t) \amalg \coprod_{l=0}^{p-1} E_0^{t^l}$ must have the same rank as E. Hence ω is an isomorphism and so H is a permutational extension.

Step 4: We want to prove that

$$G = \left(\coprod_{\dot{t}\in T_G/G} N_G(\sigma(\dot{t})) \right) \amalg F_0, \tag{2}$$

where T_G stands for the set of subgroups of G of order p, and σ is a section from T_G/G to T_G.

Observe first that $H = N_H(t) \amalg E_0$. Next, by a version of the Kurosh subgroup theorem (Theorem 4.4, Ref. 2) for finitely generated subgroups of a free pro-p product of pro-p groups, we find that the finitely generated subgroup G of H is a free pro-p product of the form

$$G = \left(\coprod_{i\in I} N_G(t^{r_i}) \right) \amalg F_1$$

with F_1 a free pro-p factor. Collecting the factors $N_G(t^{r_i})$ with $t^{r_i} \in G$ and letting F_0 denote the free pro-p product of F_1 and all the remaining normalizers $N_G(t^{r_i})$ that are themselves free pro-p, we find Eq. (2); the image of σ consists of all the $\langle t\rangle^{r_i}$ that are contained in G.

Step 5: By induction we can assume that for every $\dot{t} \in T_G/G$ there is a finitely generated $G_{\dot{t}0}$ which is free by cyclic dense in $N_G(\sigma(\dot{t}))$ and whose pro-p completion is $N_G(\sigma(\dot{t}))$. Moreover, there is a discrete free group Φ_0 dense in F_0 whose pro-p completion is isomorphic to F_0. Since G is finitely generated pro-p completion commutes with forming the free product; hence $G_0 := \left(*_{\dot{t}\in T_G/G} G_{\dot{t}0} \right) * \Phi_0$ is dense in G and its pro-p completion is isomorphic to G. □

Here is a consequence of our result.

Corollary 1.1. *Let the finitely generated pro-p group G act on a pro-p tree with finite cyclic stabilizers. Then G is virtually free pro-p.*

Proof. Since G is finitely generated there is a descending series of clopen normal subgroups U_k with $\bigcap U_k = \{1\}$. Then, setting $\tilde{N}_k := \langle G_v \cap U_k \mid v \in T \rangle$, Lemma 3.7 of Ref. 7 implies that $G = \varprojlim G_k$ with each $G_k := G/\tilde{N}_k$ virtually free pro-p. Now Eq.(2) applies to G_k and so gives a bound $|T_{G_k}/G_k| \leq d(G_k) \leq d(G)$; therefore there is k_0 so that $T_{G_{k+1}}$ maps onto T_{G_k} whenever $k \geq k_0$. Now a standard projective limit argument shows that $T_G = \varprojlim T_{G_k}$ and so $|T_G/G| \leq d(G)$ holds. Therefore T_G is clopen and so there is an open normal subgroup N of G with $T_G \cap N = \emptyset$. So N is torsion free and so, acting freely upon a pro-p tree, by Serre's result[7] [Theorem 3.4], G is virtually free pro-p. □

2. An example

We come to the promised example of a semidirect product $G = F \rtimes K$ with all torsion elements, up to conjugation, contained in K but G is not a permutational extension. This will follow from the fact that there is a finite subgroup of G not conjugate into K. That must hold for permutational extensions.

Proposition 2.1. *Let G be a pro-2 group with pro-2-presentation $G = \langle a, b, t, s \mid a^2 = b^2 = [a, b] = [b, a^t] = a^t b(ab)^{-s}\rangle$ and $F := (s, t)_G$. Then, setting $K := \langle a, b\rangle$, G enjoys the following properties:*

(i) $G = F \rtimes K$.

(ii) *F is free pro-2 on the set $\{t, t^a, t^b, s, s^a\}$. More precisely, K can be realized as a group of automorphisms for the free pro-2 group $F(t, y, z, s, u)$ such that*

$$t^b = y, z^b = z, s^b = uz^{-1}, u^b = sz, t^a = tz, y^a = yz, z^a = z^{-1}, s^a = u$$

and so that the holomorph $F(t, y, z, s, u) \rtimes K$ is isomorphic to G. The isomorphism is induced naturally by sending K and $\{t, s\}$ identically to K and $\{t, s\}$.

(iii) *All torsion elements in G are, up to conjugation, contained in K. However $\langle b, a^t\rangle \cong K$ is a maximal finite subgroup of G not conjugate to K.*

Proof. (i) is evident. (ii) can be verified by explicit computation – the relation $s^b = s^a t^{-a} t$ is helpful. For checking (iii) it suffices to remark that a^t and $ba^t = a^t b = (ab)^s$ are both conjugate into K. Suppose that there is $f \in F$ with $\langle a, b\rangle^f = \langle b, a^t\rangle$. Then we must have $a^f = a^t$ and $b^f = b$. Therefore $f \in \langle z\rangle$. On the other hand $C_F(a) = \langle ty^{-1}\rangle$ and so $f \in \langle z\rangle \cap \langle t, y\rangle = \{1\}$. So $f = 1$ and hence $\langle b, a^t\rangle = K$, a contradiction. □

Acknowledgements

The first author thanks for support by the University of Technology in Vienna and the Department of Mathematics at Bicocca Milano for the kind invitation in February 2010. The second author would like to thank for support by the University of Technology Vienna during a stay in September 2010.

References

1. D. Haran, *On closed subgroups of free products of profinite groups.* Proc. London Math. Soc. (3) **55** (1987), no. 2, 266–298.
2. W. Herfort and L. Ribes, *Subgroups of free pro-p groups.* Math. Proc. Camb. Philos. Soc. **101** (1986), 197– 206.
3. W. Herfort and P. A. Zalesskii, *Cyclic Extensions of free pro-p groups,* Journal of Algebra, **216** (1999) 511–547.
4. W. Herfort and P. A. Zalesskii, *Virtually free pro-p groups whose torsion elements have finite centralizers,* Bull. London Math. Soc. 2008 **40** (2008) 929–936.
5. O. V. Mel'nikov, *Subgroups and Homology of Free Products of Profinite Groups*, Math. USSR Izvestiya, **34**, 1, (1990), 97-119.
6. L. Ribes and P. A. Zalesskii, *Profinite Groups*, Springer 2000.
7. L. Ribes and P. A. Zalesskii, *Pro-p Trees and Applications,* (2000), Chapter, Ser. Progress in Mathematics, Birkhäuser Boston (2000), Ed. A. Shalev, D. Segal
8. C. Scheiderer, *The structure of some virtually free pro-p groups,* Proc. Amer. Math. Soc. **127** (1999) 695-700.
9. J.S. Wilson, *Profinite Groups,* London Math.Soc. Monographs (Clarendon Press, Oxford, 1998)

FINITE NONABELIAN 2-GROUPS ALL OF WHOSE MINIMAL NONABELIAN SUBGROUPS ARE ISOMORPHIC TO $M_{2^{n+1}}$

Z. JANKO

Mathematical Institute, University of Heidelberg,
69120 Heidelberg, Germany
E-mail: janko@mathi.uni-heidelberg.de

We give a complete classification of the title groups which generalizes a special case for $n = 3$.

Keywords: minimal nonabelian 2-groups; maximal normal abelian subgroups; quaternion subgroups.

Here we classify completely the title groups, where the special case for $n = 3$ was done in Ref. 1, Theorem 4.2. Here the modular 2-group $M_{2^{n+1}}$ is defined with

$$M_{2^{n+1}} = \langle a, t \,|\, a^{2^n} = t^2 = 1,\ a^t = a^{1+2^{n-1}} \rangle, n \geq 3.$$

This classification for an arbitrary $n \geq 3$ (fixed) is very involved and shows that our technique to determine 2-groups with given minimal nonabelian subgroups is quite efficient (see also Refs. 2–4. Hence this particular case can be considered as a part of a general project to determine the structure of 2-groups, when the structure of minimal nonabelian subgroups is known.

Our notation is standard. In particular, for a p-group G and $n \geq 1$ we define:

$$\Omega_n(G) = \langle x \in G \mid x^{p^n} = 1 \rangle,\ \mho_n(G) = \langle x^{p^n} \mid x \in G \rangle,$$

and the generalized quaternion group is defined with

$$Q_{2^n} = \langle a, b \mid a^{2^{n-1}} = 1,\ b^2 = a^{2^{n-2}},\ a^b = a^{-1} \rangle,\ n \geq 3.$$

Theorem. *Let G be a finite nonabelian 2-group all of whose minimal nonabelian subgroups are isomorphic to $\mathrm{M}_{2^{n+1}}$, $n \geq 3$ fixed. Then $E = \Omega_1(G)$ is elementary abelian and let A be a maximal normal abelian subgroup of G containing E. Then $A = \mathrm{C}_G(E)$, all elements in $G \setminus A$ are of order 2^n, G/A is elementary abelian of order ≤ 4, and if $x \in G \setminus A$, then $|E : \mathrm{C}_E(x)| = 2$, $A_0 = \mathrm{C}_A(x) = \mathrm{C}_E(x)\langle x^2 \rangle$, A/A_0 is cyclic, $\langle A, x\rangle'$ is cyclic of order $|A/A_0|$, $\langle A, x\rangle' \cap A_0 = \Omega_1(\langle x \rangle)$, x acts invertingly on A/A_0 and on $\mho_{n-2}(A)$. We have two possibilities:*

(a) $\exp(A) = 2^m \geq 2^n$ *in which case* $|G : A| = 2$ *and* A *is of type* $(2^m, 2^{n-2}, 2, ..., 2)$;

(b) $\exp(A) \leq 2^{n-1}$ *in which case* A *is of type* $(2^{n-1}, 2^r, 2, ..., 2)$ *with* $1 \leq r \leq n-2$ *and either* $|G : A| = 2$ *or* $|G : A| = 4$*, where in the last case* $G/\Omega_{n-2}(A) \cong \mathrm{Q}_8$.

Conversely, if G is a 2-group of type (a) or (b), then each minimal nonabelian subgroup of G is isomorphic to $\mathrm{M}_{2^{n+1}}$, $n \geq 3$ fixed.

Proof. Let G be a title group with $n \geq 3$ fixed. Since G has no dihedral subgroups, $E = \Omega_1(G)$ is elementary abelian. Set $A = \mathrm{C}_G(E)$ so that A is normal in G. If A is nonabelian, then consider a minimal nonabelian subgroup $X \cong \mathrm{M}_{2^{n+1}}$ in A. But then all three involutions in X are central in X which contradicts the structure of $\mathrm{M}_{2^{n+1}}$. Hence A is abelian and so $A = \mathrm{C}_G(E)$ is a unique maximal abelian subgroup of G containing E.

Let x be any element in $G - A$ so that $o(x) \geq 4$ and set $\langle z \rangle = E \cap \langle x \rangle$, where z is an involution. Set $T = E\langle x \rangle$ and $E_0 = \mathrm{C}_E(x)$ so that $E > E_0$ and $T_0 = \mathrm{C}_T(x) = E_0\langle x \rangle$ is abelian. Let $T_1 \leq T$ be such that $T_1 > T_0$ and $|T_1 : T_0| = 2$. Set $E_1 = T_1 \cap E$ so that $T_1 = E_1\langle x \rangle$, $|E_1 : E_0| = 2$ and let e be an involution in $E_1 \setminus E_0$. Then $1 \neq [e, x] \in T_0 \cap E = E_0$ and so $[e, x]$ is an involution commuting with e and x and therefore $\langle e, x\rangle' = \langle [e, x] \rangle$ which implies that $\langle e, x \rangle$ is minimal nonabelian (Proposition 1.3 in Ref. 2) and so $\langle e, x \rangle \cong \mathrm{M}_{2^{n+1}}$. In particular, $\langle [e, x] \rangle = \Omega_1(\Phi(\langle e, x \rangle)) = \langle z \rangle$, $e^x = ez$ and $\langle e \rangle$ normalizes $\langle x \rangle$ which gives $o(x) = 2^n$. We have proved that all elements in $G \setminus A$ are of order 2^n and so G/A is elementary abelian. In particular, $x^2 \in A$ and so x induces an involutory automorphism on E. Hence x centralizes E/E_0 and so T_0 is normal in T with $T/T_0 \cong E/E_0$.

Suppose that $|E/E_0| > 2$ and let $f \in E - E_1$. Then considering $T_2 = \langle f \rangle T_0$ we get in the same way as above that $\langle f, x \rangle \cong \mathrm{M}_{2^{n+1}}$ and so $f^x = fz$. But then $ef \notin E_0$ and $(ef)^x = (ez)(fz) = ef$, a contradiction. We have proved that $|E : E_0| = 2$ and $[E, x] = \langle z \rangle = \Omega_1(\langle x \rangle)$.

Let $t \in E - E_0$ so that $t^x = tz$ and $\langle t, x \rangle \cong \mathrm{M}_{2^{n+1}}$. Let s be an arbitrary element in A. We may set $s^x = s^{-1}w$ for some $w \in A$. Note that $\mathrm{o}(x) = 2^n$, $\mathrm{o}(sx) = 2^n$ and $t^{sx} = t^x = tz$ so that $[t, sx] = z$ with $z \in \mathrm{Z}(A\langle x \rangle)$ and so $\langle t, sx \rangle \cong \mathrm{M}_{2^{n+1}}$. In particular, $\langle sx \rangle > \langle z \rangle$ and so $(sx)^{2^{n-1}} = (x)^{2^{n-1}} = z$. We compute (noting that $x^{-2} \in A$ and $(sx)^2 \in A$)

$$w = ss^x = s(x^{-1}sx) = sx^{-2}xsx = x^{-2}(sx)(sx) = x^{-2}(sx)^2.$$

Therefore $w^{2^{n-2}} = x^{-2^{n-1}}(sx)^{2^{n-1}} = zz = 1$ and so $s^x = s^{-1}w$ implies that x acts invertingly on $A/\Omega_{n-2}(A)$. Since

$$(s^{2^{n-2}})^x = (s^x)^{2^{n-2}} = (s^{-1}w)^{2^{n-2}} = s^{-2^{n-2}} = (s^{2^{n-2}})^{-1},$$

we see that x also inverts each element in $\mho_{n-2}(A)$.

We shall determine the structure of $A_0 = \mathrm{C}_A(x)$, where $A_0 \geq E_0\langle x^2 \rangle$. Suppose that there is an element v of order 4 in A_0 such that $v^2 = z' \neq z$. Then taking an involution $t \in E - E_0$ we get $(vt)^2 = z'$ and $[vt, x] = [v, x]^t[t, x] = [t, x] = z$ and so $\langle vt, x \rangle$ is minimal nonabelian. But $\Phi(\langle vt, x \rangle) \geq \langle z, z' \rangle \cong E_4$ which contradicts the structure of $\mathrm{M}_{2^{n+1}}$. Hence A_0 is of type $(2^{s_0}, 2, ..., 2)$ with $s_0 \geq n-1$ and $\Omega_1(A_0) = E_0$. Suppose that A_0 contains an element y of order 2^n. Then $\mho_{n-2}(\langle y \rangle) \cong \mathrm{C}_4$ is inverted by x, a contradiction. Hence $s_0 = n-1$ and so $A_0 = E_0\langle x^2 \rangle$. Also $A_1 = EA_0$ is of type $(2^{n-1}, 2, ..., 2)$ with $\mho_{n-2}(A_1) = \langle z \rangle$. For each $l \in A_1 - A_0$, $l^2 \in \langle x^2 \rangle$, $[l, x] = z$ and so $\langle l, x \rangle \cong \mathrm{M}_{2^{n+1}}$.

For each $y \in Ax$ $(x \in G - A)$, $\mathrm{C}_A(y) = A_0$ and so $y^2 \in A_0$. This implies that x acts invertingly on A/A_0. Let A^*/A_0 be a subgroup of order 2 in A/A_0 and let a^* be an element in $A^* - A_0$. We have $(a^*)^x = a^*s$ with $1 \neq s \in A_0$ since x acts invertingly on A/A_0. Then $a^* = (a^*)^{x^2} = (a^*s)^x = a^*s^2$ and so $s^2 = 1$ and s is an involution in $E_0 = \Omega_1(A_0)$. The involution $s = [a^*, x]$ commutes with a^* and x and so $\langle a^*, x \rangle$ is minimal nonabelian. Hence $\langle a^*, x \rangle \cong \mathrm{M}_{2^{n+1}}$ with $\Phi(\langle a^*, x \rangle) = \langle x^2 \rangle$ and so $[a^*, x] = s = z$. If $\langle a^*, x^2 \rangle$ is cyclic, then $\mathrm{o}(a^*) = 2^n$ and x inverts $\mho_{n-2}(\langle a^* \rangle) \cong \mathrm{C}_4$ with $\mho_{n-2}(\langle a^* \rangle) \leq A_0$, a contradiction. Hence $\langle a^*, x^2 \rangle$ is noncyclic and so $i = a^*y$ is an involution for a suitable $y \in \langle x^2 \rangle$, where $i \in E - E_0$. Thus $E \leq A^*$ and so $A^* = E\langle x^2 \rangle = A_1$. We have proved that $E\langle x^2 \rangle/A_0$ is a unique subgroup of order 2 in A/A_0 and so A/A_0 is cyclic. The mapping $a \mapsto a^{-1}a^x$ $(a \in A)$ is a homomorphism from A onto $\langle A, x \rangle'$ with the kernel $A_0 = \mathrm{Z}(\langle A, x \rangle)$.

Thus $A/A_0 \cong \langle A, x\rangle'$. Since x acts invertingly on the cyclic group A/A_0, we have $|A : (\langle A, x\rangle' A_0)| = 2$ and therefore $\langle A, x\rangle' \cap A_0 = \langle z\rangle$ noting that $[A_1, x] = \langle z\rangle$.

(i) Suppose that A possesses an element a of order 2^n. Then each element $x \in G - A$ inverts $\mho_{n-2}(\langle a\rangle) = \langle v\rangle \cong C_4$ which implies $|G : A| = 2$, $v \notin A_0 = C_A(x)$ and $v^2 \in E_0$. Since $A_1 = E\langle x^2\rangle$ is of type $(2^{n-1}, 2, ..., 2)$, we have $v^2 = z$ and $v \in A_1 - A_0$. This implies that $|A/A_0| \geq 2^{n-1}$. Let y be an element in $A \setminus A_0$ such that $\langle y\rangle$ covers A/A_0 and so $o(y) \geq 2^{n-1}$. Suppose that $o(y) = 2^{n-1}$ in which case $|A/A_0| = 2^{n-1}$, A splits over A_0 and $y^{2^{n-2}} = e \in E \setminus E_0$. But then x inverts $\mho_{n-2}(\langle y\rangle) = \langle e\rangle$ and so x centralizes e, contrary to $C_E(x) = E_0$. It follows that $o(y) = 2^m \geq 2^n$. By the above, $\langle y\rangle \cap A_0 = \langle z\rangle$. Since $\langle y\rangle$ covers A/A_0 and $\langle x^2\rangle$ covers A_0/E_0, we get $A = E\langle y\rangle\langle x^2\rangle$, where $\langle y\rangle \cap \langle x^2\rangle = \langle z\rangle$. Let $y_0 \in \langle y\rangle$ be of order 2^{n-1}. Then $o(y_0x^2) = 2^{n-2}$, $\Omega_1(\langle y_0x^2\rangle) \setminus \{1\} \in E \setminus E_0$ and $\langle y, x^2\rangle = \langle y\rangle \times \langle y_0x^2\rangle$ so that A is of type $(2^m, 2^{n-2}, 2, ..., 2)$. We have obtained groups of type (a) of our theorem.

(ii) Suppose that $\exp(A) = 2^{n-1}$. In that case $\langle x^2\rangle$ is a cyclic subgroup in A of the maximal possible order. Let C be a complement of $\langle x^2\rangle$ in A so that $A = C \times \langle x^2\rangle$. Set $C_0 = A_0 \cap C$, $C_1 = A_1 \cap C$ so that C_1 is elementary abelian since $A_1/\langle x^2\rangle$ is elementary abelian. Hence we have $A_1 = C_1 \times \langle x^2\rangle$, $A_0 = C_0 \times \langle x^2\rangle$, $|C_1 : C_0| = 2$, $E = C_1 \times \langle z\rangle$ and $E_0 = C_0 \times \langle z\rangle$. We know that $A/A_0 \cong C/C_0$ is cyclic of order 2^r, $r \geq 1$. Let $c \in C$ be such that $\langle c\rangle$ covers C/C_0. Then $c^{2^{r-1}} = e$ is an involution in $C_1 \setminus C_0$ so that $o(c) = 2^r$, $r \leq n-1$. Suppose that $r = n-1$. Since x inverts $\mho_{n-2}(\langle c\rangle) = \langle e\rangle$, so x centralizes $e \notin E_0$, a contradiction. We have proved that $r \leq n-2$ and so A is of type $(2^{n-1}, 2^r, 2, ..., 2)$, where $1 \leq r \leq n-2$. It follows that $\Omega_{n-2}(A) = \langle c, E, x^4\rangle$ is of index 2 in A, $\Omega_{n-2}(A)$ is normal in G and all elements in $A \setminus \Omega_{n-2}(A)$ are of order 2^{n-1}.

Assume in addition that G/A is elementary abelian of order ≥ 4. Let $B/\Omega_{n-2}(A)$ be a subgroup of order 2 in $G/\Omega_{n-2}(A)$ distinct from $A/\Omega_{n-2}(A)$. But then each element $d \in B - \Omega_{n-2}(A)$ is of order 2^n but $d^2 \in \Omega_{n-2}(A)$ is of order $\leq 2^{n-2}$, a contradiction. Hence $A/\Omega_{n-2}(A)$ is a unique subgroup of order 2 in $G/\Omega_{n-2}(A)$. Since $G/\Omega_{n-2}(A)$ cannot be cyclic (noting that G/A is elementary abelian of order ≥ 4), it follows that $G/\Omega_{n-2}(A)$ is generalized quaternion. But then $d(G/\Omega_{n-2}(A)) = 2$ and so $|G/A| = 4$ and $G/\Omega_{n-2}(A) \cong Q_8$. We have obtained groups of type (b) of our theorem.

It remains to be shown that all obtained groups G of type (a) or (b) have the property that each minimal nonabelian subgroup M of G is isomorphic to $M_{2^{n+1}}$, $n \geq 3$ fixed. First suppose that $|M : (M \cap A)| = 2$ and let $x \in M \setminus A$. Then $M \cap A \not\leq A_0 = C_A(x)$. Since A/A_0 is cyclic, it follows that $(M \cap A) \setminus A_0$ contains an element $l \in A_1 \setminus A_0$, where $A_1 = E\langle x^2 \rangle$. But then we know that $\langle l, x \rangle \cong M_{2^{n+1}}$ and so $\langle l, x \rangle = M$. Now suppose that $|M : (M \cap A)| > 2$ in which case $|G/A| = 4$ and $G/\Omega_{n-2}(A) \cong Q_8$. In this case M covers Q/A. But for any $m \in M \setminus A$, $o(m^2) = 2^{n-1}$ and so $m^2 \in A \setminus \Omega_{n-2}(A)$. Thus M covers also $G/\Omega_{n-2}(A)$ and so $M/(M \cap \Omega_{n-2}(A)) \cong Q_8$. Hence M' covers $(M \cap A)/(M \cap \Omega_{n-2}(A))$. But $|M'| = 2$ which gives a contradiction since all elements in $A \setminus \Omega_{n-2}(A)$ are of order $2^{n-1} \geq 4$. Our theorem is proved. □

References

1. Y. Berkovich and Z. Janko, On subgroups of finite p-groups, *Israel J. Math.*, **171** (2009), 29–49.
2. Z. Janko, On finite nonabelian 2-groups all of whose minimal nonabelian subgroups are of exponent 4, *J.Algebra* **315** (2007), 801–808.
3. Z. Janko, On minimal nonabelian subgroups in finite p-groups, *J.Group Theory* **12** (2009), 289–303.
4. Z. Janko, Finite nonabelian 2-groups all of whose minimal nonabelian subgroups are metacyclic and have exponent 4, *J.Algebra* **submitted** (2009).

TWISTED CONJUGACY IN CERTAIN ARTIN GROUPS

A. JUHÁSZ

Department of Mathematics
Technion, Israel Institute of Technology
Haifa 32000, ISRAEL

In memoriam Karl Gruenberg

We show that the twisted conjugacy problem for extra-large Artin groups and for the class of CLTTF Artin groups introduced by J. Crisp in Ref. 18 is solvable, relative to length preserving automorphisms. We also show that these groups have infinite number of such twisted conjugacy classes and twisted conjugacy of elements of a standard parabolic subgroup which is invariant under the automorphism in the group , implies twisted conjugacy already in the parabolic subgroup.

Keywords: Artin groups; Twisted conjugacy; Reidemeister number; van Kampen diagrams; Small cancellation Theory

1. Introduction

Suppose we are given a pair (G, σ), where G is a group and σ is an automorphism (endomorphism) of G. Elements a and b in G are *σ-twisted conjugate*, or briefly *σ-conjugate* if there exists an element g in G such that

$$b = ga(\sigma(g))^{-1} \tag{1}$$

Thus, usual conjugacy is σ-conjugacy, with $\sigma = I_G$, the identity automorphism. Like conjugacy, σ-conjugacy is an equivalence relation and the corresponding equivalence classes are termed *σ-twisted conjugacy classes* or if σ is fixed then they are termed *twisted conjugacy classes.* This notion arises naturally in representation theory (see Ref. 18 and references therein) and in topology (see Ref. 7 and references therein). It seems that the first appearance of this notion in representation theory was in F. Gantmakher (see Ref. 9) and in topology in K. Reidemeister in Ref. 16. Recently there is some interest in twisted conjugacy also in cryptography (see Ref. 17).

For more details on twisted conjugacy and in particular on the first two problems considered below see Refs. 6–8,10,11,18,19 and references therein.

In this paper we are interested in the following general problems in certain types of Artin groups.

1. Twisted Conjugacy Problem: Given $a, b \in G$. *Decide whether there exists an element* $g \in G$ *such that* $b = ga(\sigma(g))^{-1}$.

2. Property R(∞). *Given* G *and* σ. *Does* G *have infinite number of* σ*-conjugacy classes?* If the answer is "yes" then we say that G *has property* $R(\infty)$ *relative to* σ. If G has property $R(\infty)$ relative to every automorphism then we say that G *has property* $R(\infty)$.

3. σ-Frattini embedding of σ-invariant subgroups: We say that a σ-invariant subgroup H of G is *σ-Frattini embedded in* G if for every $a, b \in H$ which are σ-conjugate in G are already σ-conjugate in H. In other words if there exists an element g in G such that $b = ga(\sigma(g))^{-1}$, then there exists an element h in H such that $b = ha(\sigma(h))^{-1}$.

These problems have been dealt with in several classes of groups. See the references mentioned above. In the present work we focus on a class of Artin groups. It seems that the only subclass of Artin groups for which the twisted conjugacy problem is solved is the subclass of braid groups. (See Ref. 10) These groups are known to be of "finite type", i.e., the corresponding Coxeter groups obtained by adding $x^2 = 1$ for every standard generator x is finite. The groups we deal with are of infinite type and the proof is completely different.

Recall from Ref. 1 that for every finite graph Γ with vertex set V and edge set E in which the edges are labelled by natural numbers $2, 3, 4, \ldots$, there corresponds a uniquely defined Artin group which is generated by $\{x_v | v \in V\}$ and related by $\{R_e | e \in E\}$ such that for each edge e of Γ with endpoints $v_i, v_j \in V$ and label n_e, R_e is the relation $U = V$ where U is the initial subword of $(x_i x_j)^{n_e}$ of length n_e and V is the initial subword of $(x_j x_i)^{n_e}$ of length n_e. Here we wrote x_i for x_{v_i}. Denote $n(R_e) = n_e$ and $r(R_e) = \{x_i, x_j\}$ (the endpoints of e).

We shall restrict ourselves to length preserving automorphisms of the following classes of Artin groups, the first of which was introduced by John Crisp in Ref. 4 . Thus, let Γ be a labelled graph as above with at least 3 vertices and let $G = G_\Gamma$ be the Artin group defined by Γ. G is a *CLTTF–Artin group* if Γ satisfies the following requirements:

(i) Γ is connected;
(ii) Γ contains no triangles;
(iii) $n_e \geq 3$ for every edge $e \in E$.

We also consider *extra-large Artin groups.* These are Artin groups for which $n_e \geq 4$, for every edge e of the defining graph Γ. These groups were introduced and studied by K. Appel and P. Schupp in Ref. 1. Their automorphism groups are not yet known.

Here are the main results.

Theorem A. *Let G be a CLTTF–Artin group or an extra-large Artin group and let σ be a length preserving automorphism of G. Then the σ-twisted conjugacy problem is solvable for G.*

Theorem B. *Let G and σ be as in theorem A. Then G has the property $R(\infty)$ relative to σ.*

Theorem C. *Let G and σ be as in Theorem A. Let Γ_0 be a σ-invariant subgraph of Γ and let H be the corresponding subgroup of G. Then H is σ-Frattini embedded in G.*

Example. Let Γ be a graph of a CLTTF - Artin group. It follows from Ref. 4 that if Γ contains neither separating vertices nor separating edges then every outer automorphism of G is length preserving. Hence for these groups theorems A, B and C hold for every automorphism. In particular, G has property $R(\infty)$.

The proofs of these results are based on the small cancellation theory of annular diagrams developed in Ref. 14 for $V(6)$ presentations, together with the results in Refs. 12,13.

The work is organised as follows: In Section 1 we recall some basic results concerning twisted conjugacy. We also recall briefly results on $V(6)$-diagrams from Refs. 12–14 and explain their relevance to our work.

In Section 2 we show that the diagrams obtained via results of Section 1 can be assumed to be thin. This allows in Section 3, to prove Theorems A, B, and C.

Theorems A and C hold for the corresponding Coxeter groups. (See Ref. 1)

2. Preliminary Results

2.1. *Length preserving automorphisms of G and extensions by an automorphism*

Definition 2.1. Let G be a group given by a presentation $\langle X|\mathcal{R}\rangle$ and let F be the free group freely generated by X. An automorphism σ of G induced by a length-preserving automorphism of F is termed *length-preserving.*

Let $\hat{G}^{(\sigma)}$ denote the HNN-extension with base group and associated groups all coinciding with G, homomorphism σ and stable letter t. This is the *ascending HNN-extension of G by σ* which we shall refer to as the *extension of G by σ*. When σ is fixed, we write $\hat{G}$ instead of $\hat{G}^{(\sigma)}$. Setting $T := \{txt^{-1}\sigma(x)^{-1} \mid x \in X\}$ we note that $\hat{G} = \langle X \cup \{t\} \mid \hat{\mathcal{R}}\rangle$ is the canonical presentation with $\hat{\mathcal{R}} := \mathcal{R} \cup T$. By the normal form theorem (Ref. 15, p.182) there is a canonical embedding θ of the base group G into $\hat{G}$. Although the proof of the following result is straightforward we include it for the sake of completeness.

Lemma 2.1. *Let G_0 and σ be as above and let $a, b \in G$. Then there exists a $g \in G$ such that $ga\sigma(g)^{-1} = b$ if and only if $g(at)g^{-1} = bt$ in $\hat{G}$.*

Proof. We have by definition $b = ga\sigma(g^{-1})$ if and only if $b = ga(tg^{-1}t^{-1})$ in $\hat{G}_0$, i.e., $g(at)g^{-1} = bt$. □

Our Lemma allows to reformulate Theorems $A - C$ about G into corresponding ones for $\hat{G}$. Notice however, that twisted conjugacy in G is not just conjugacy in $\hat{G}$ but rather it is a conjugacy of elements at and bt of $\hat{G}$, $a, b \in G$, by elements of G. (G – conjugacy).
These are the theorems we shall prove.

Theorem A' *Let G be a CLTTF or an extra-large Artin group and let σ be a length preserving automorphism of G. Let $\hat{G}$ be the extension of G by σ. Then for given elements a and b of G it is decidable whether at and bt are conjugate in $\hat{G}$ by an element g of G.*

Theorem B' *Let notation and assumptions be as in Theorem A'. Then $\hat{G}$ contains an infinite number of non σ-conjugate elements of the form at, with $a \in G$.*

Theorem C' *Let notation and assumptions be as in Theorem A and let H be as in Theorem C. If $a, b \in H$ and there is an element $g \in G$ such* *that* $g(at)g^{-1} = bt$ *in* $\hat{G}$*, then there exists an element* $h \in H$ *such that* $h(at)h^{-1} = bt$.

The method we are going to employ in the proofs of Theorems A', B', and C' is small cancellation theory with van Kampen and annular diagrams. For basic results on diagrams see Ref. 15, Ch. V.

2.2. *σ-cyclical reduction and conjugation*

Definition 2.2. When σ is a length preserving automorphism of F, and A, B, C and H are reduced words in F and $G = F/\mathcal{R}^F$ then we shall find it useful to write

(i) $A \sim_C B$ if A is σ-conjugate to B in G, i.e., if $A = CB(\sigma(C))^{-1}$.
(ii) $A \rightarrow_C B$ if $A = CB(\sigma(C))^{-1}$, reduced as written holds in F. We say that B is the result of σ-cyclical reduction (using C).
(iii) In the same vein we write $A \rightharpoonup_H A^*$ if $A = HA'$ is reduced as written and $A^* = \sigma(A')H$. We say that B is the result of σ-cyclical conjugation of A (using H).

Straightforward computation yields the following observation.

Lemma 2.2.

(a) $A \rightarrow_C B$ *implies* $A \sim_C B$ *and* $|B| < |A|$ *if* $C \neq 1$.
(b) Suppose that $A \rightharpoonup_H A^*$ *and* $B \rightharpoonup_K B^*$ *in* F*. Then, in* G*,* $A \sim_C B$ *if and only if* $A^* \sim_{H\sigma(C)K^{-1}} B^*$*. Moreover,* $r(A^*) \subseteq r(A) \cup r(\sigma(A))$ *and the set* $L(A) := \{A^* \in F \mid \exists H \in F : \quad A \rightharpoonup_H A^*\}$ *is finite, where* $r(A)$ *is the set of letters (up to sign) which occur in* A.
(c) $A \sim B$ *in* F *if and only if* A *and* B *can* σ*-cyclically reduced to the same normal form.*

In particular every sequence of σ-cyclical reductions terminates after a finite number of steps so that a unique normal form $\hat{A}$ *of* A *can be achieved.*

Therefore, given $A, B \in F$, and having to decide whether or not $A \sim B$ in G, we may first apply the operations "$\rightarrow$" and "$\rightharpoonup$" in order to get representatives A^* and B^* of smallest length in its respective σ-conjugacy classes in F. Then $A \sim B$ if and only if $A^* \sim B^*$.

We assume the reader to be familiar with annular diagrams and its connection to the conjugacy problem. See Fig. 1.

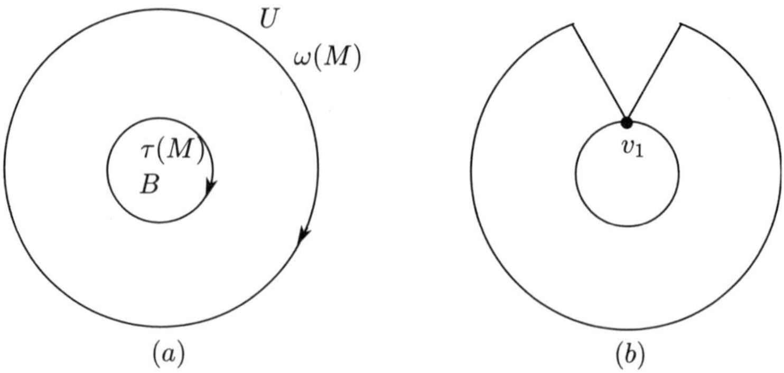

Fig. 1

Let $\hat{G}$ be the extension of G by σ, as above and assume that $g(at)g^{-1} = bt$ in $\hat{G}$, for $a, b, g \in G$. Denote $k = tgt^{-1}$. Then $k = \sigma(g) \in G$ and $1 = g(at)g^{-1}(t^{-1}b^{-1}) = ga(tgt^{-1})^{-1}b^{-1} = gak^{-1}b^{-1}$, i.e., $gak^{-1}b^{-1} = 1$. Choose reduced words A, B, C and K in F such that $\pi(A) = a$, $\pi(B) = b$, $\pi(C) = g$ and $\pi(K) = k$ where π is the natural projection of F on G. Then by Theorem 1.1 in Ref. 15, p. 237 there is a van Kampen $\mathcal{R}$-diagram M_0 over F with boundary cycle $\gamma_1\alpha\kappa_1^{-1}\beta^{-1}$, where $\Phi(\gamma_1) = C$, $\Phi(\alpha) = A$, $\Phi(\kappa_1) = K$ and $\Phi(\beta) = B$. Now $tgt^{-1}k^{-1} = 1$ in $\hat{G}$, therefore by Theorem 1.1 in Ref. 15, p. 237 there is an $\hat{\mathcal{R}}$-diagram M_1 with boundary cycle $\xi_1\gamma_2^{-1}\xi_2^{-1}\kappa_2$ with $\Phi(\xi_1) = t$, $\Phi(\gamma_2) = C$, $\Phi(\xi_2) = t$ and $\Phi(\kappa_2) = K$. Since $\Phi(\kappa_1) = K = \Phi(\kappa_2)$ and $\Phi(\gamma_1) = C = \Phi(\gamma_2)$, we can amalgamate the two van Kampen diagrams along (γ_1, γ_2) and (κ_1, κ_2) to get an annular $\hat{\mathcal{R}}$-diagram M_0 with outer boundary label At and inner boundary label Bt: $M_0 = M \bigcup_{\substack{\kappa_1 = \kappa_2 \\ \gamma_1 = \gamma_2}} M_1$. See Figs. 1(c) and 1(d).

The diagram M_0 will be our main subject of study. In general, it is not of much help, unless it has nice combinatorial and geometrical structure. In the remaining subsections we show that this is the case for annular $\mathcal{R}$-diagrams, where $\mathcal{R}$ is as in the Main theorems. More precisely, we recall the small cancellation condition $V(6)$, which implies nice geometrical and combinatorial structure and show that annular $\mathcal{R}$- diagrams can be modified such that they satisfy this condition and hence the general theory of $V(6)$ diagrams is applicable. The main technical result is that this holds true also for the $\hat{\mathcal{R}}$-diagram M_0 defined above. This is the content of Section 2.

Before dealing with annular diagrams we have to study simply connected diagrams. This we do in the next subsection.

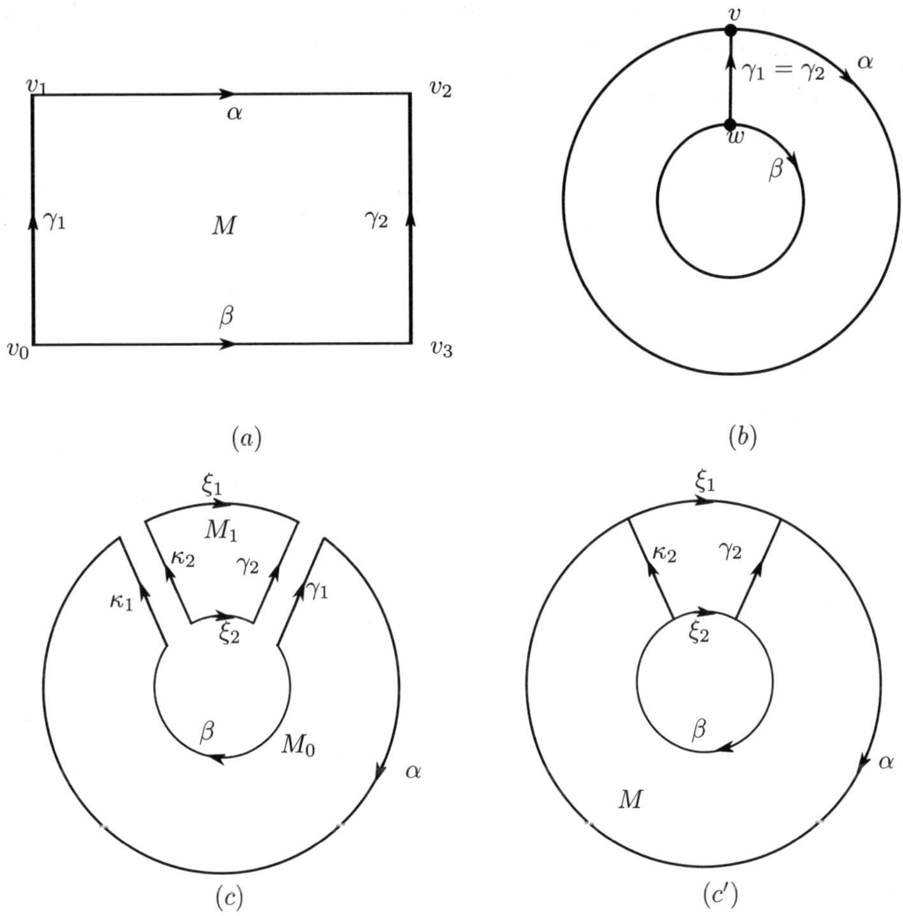

Fig. 2

2.3. *The small cancellation condition* $V(6)$

We recall from Refs. 12–14 some basic results on $V(6)$-diagrams. Usually we follow the notation of Ref. 15.

Definition 2.3. Let M be a map. (Not necessarily simply connected or annular.) M satisfies the *condition* $V(6)$ if for each inner region D, $d(D) \geq 4$ and if ∂D contains a vertex with valency three then $d(D) \geq 6$.

In particular, the condition $V(6)$ implies the conditions $C(4)\&T(4)$ and the condition $C(6)$ and every $V(6)$ diagram is a $W(6)$ diagram, as defined in Ref. 12, hence the results in Ref. 12 are applicable here. Thus, let M be a $W(6)$-diagram. The main results of Ref. 12 are Theorem 2.2 on page 59, which gives the analogue of Greendlinger's Lemma (See Ref. 15, p.250) and Theorem 2.7 with Lemma 2.7, which gives a detailed description of the layer structure of M. In particular, Theorem 2.7 in Ref. 12 states that every region D in M induces a decomposition of M into a disjoint union of annular one-layer diagrams with boundaries homotopic to ∂D.

These two results have the following corollary.

Recall the notions of *k-corner regions*, *layers* and *layer structure* from Ref. 12 – Definitions 2.1 and 2.2 on pages 59-60 and Definition 2 on page 64.

Proposition 2.1 (Ref. 12). *Let M be a connected, simply connected map which contains more than one region. If M satisfies the condition $V(6)$, then*

(a) M has at least two distinct corner regions. It has exactly two corner regions if and only if it is a one-layer map.
(b) if $\Lambda = (\mathcal{L}_0, \ldots, \mathcal{L}_p)$ is any layer decomposition of M and $|\mathcal{L}_1| = q$ then M has at least q corner regions. If the condition $T(4)$ is satisfied then we may assume that these corner regions are k-corner regions with $k \leq 2$.

Proof. Part (a) follows from Ref. 12, Corollary 1 and part (b) follows from Lemma 2.10 in Ref. 12, p. 86 and the layer structure of M. □

Following Ref. 12 we shall denote by $\mathcal{R}eg(M)$ the set of all the regions of a diagram M. Also, denote by $\mathcal{M}$ the collection of all the connected, simply connected $\mathcal{R}$-diagrams with connected interior, where $\mathcal{R}$ is a fixed finite set of Artin relators, as in Theorems A, B and C. As in Ref. 13, Theorem 3.1, we correspond to each $M \in \mathcal{M}$ a diagram M', which we call the *derived diagram*, of M the regions of which are *the derived regions*.

The next theorem shows the relevance of the $V(6)$ condition to Artin groups that occur in Theorems A, B and C.

Theorem 2.1 (Ref. 13). *For $M \in \mathcal{M}$ let M' be the derived diagrams as described in Ref. 13, section 3. Then*

(a) Each region $\Delta \in \mathcal{R}eg(M')$ is simply connected with connected interior, homeomorphic to an open disc.

(*b*) *For* $\Delta_1, \ \Delta_2 \in \mathcal{R}eg(M')$, $\|\partial\Delta_1 \cap \partial\Delta_2\| \leq 1$ *and every inner vertex of* $\partial\Delta_1 \cap \partial\Delta_2$ *has valency 3, in both* Δ_1 *and* Δ_2.

(*c*) M' *satisfies the condition* $V(6)$.

Proof. Only part (b) includes a statement not already from Ref. 13, Theorem 3.1. Now, the inner vertices of a boundary path of M labelled by consequtive power of a fixed $x \in X \cup X^{-1}$, so its valency can only be 3. □

Definition 2.4.

(a) A path μ in M is *shortest in* M if $|\mu| \leq |\nu|$, for every path ν in M, the endpoints of which are those of μ.

(b) Let $W \epsilon F$ be a reduced word. The *syllable length* of W, denoted $||W||$, is the syllable length of W in the free product $\langle x_1 \rangle * \langle x_2 \rangle * \cdots * \langle x_n \rangle$. For example, $||x_1^5|| = 1$, $||x_1^3 x_2^7|| = 2$ and $||x_1^2 x_2^4 x_1^7|| = 3$.

(c) For a region D in M with a boundary label $R \subset \mathcal{R}$ denote $n(D) = n(R)$ and $r(D) = r(R)$.

Lemma 2.3. *Let* M *be a van Kampen diagram over a CLTTF or extra-large Artin group. Let* μ *be a boundary path of* M *and let* Λ *be a layer decomposition of its derived diagram* M'. *If* μ *is shortest in* M *then* μ *contains no corner regions of* M'.

Lemma 2.3 follows in a standard way from Theorem 2.1(b) and the result of Appel and Schupp below. We omit details.

Lemma 2.4 (Appel-Schupp's Lemma, Ref. 1, Lemmas 6 and 7). *Let* $M \in \mathcal{M}$ *and let* $\Delta \in \mathrm{Reg}(M')$. *Assume that* $\partial\Delta$ *has a decomposition in* M *by* $\partial\Delta = \mu\bar{\nu}$. *Then each of the following holds:*

(*a*) *if* $\| \mu \| \leq n(\Delta)$ *then* $|\mu| \leq |\nu|$;

(*b*) *if* $\| \mu \| < n(\Delta)$ *then* $|\mu| < |\nu|$.

(*c*) $\| \partial\Delta \| \geq 2n(\Delta)$ *and equality holds if and only if* Δ *is a simple one-layer diagram,* $\langle D_0, \ \ldots \ D_t \rangle$ *in which* $|\partial D_i \cap \partial D_{i+1}| = n(D) - 1$ *for* $i = 0, \ \ldots, \ t-1$.

(*d*) *Let* $\Delta \in \mathcal{R}eg(M')$ *and consider* Δ *as a submap of* M. *Then* Δ *satisfies the (geometrical)* $C(4)\&T(4)$ *condition.*

Remark 2.1. It follows from Theorem 2.1(c) that the results of Proposition 2.1 hold true for CLTTF and extra-large Artin groups. Also, it follows that every CLTTF and every extra-large Artin group has solvable word problem. (This also follows from Ref. 1.)

3. 4-gons

3.1. *4-gons with the V(6) condition*

Definition 3.1. Let $\langle X|\mathcal{R}\rangle$ be the canonical presentation of either a CLTTF or an extra-large Artin group. A 4-gon $M = M(\gamma_1, \alpha, \gamma_2, \beta)$ is a van Kampen diagram with boundary cycle $\gamma_1\alpha\gamma_2^{-1}\beta^{-1}$ made up by four, with respect to their endpoints, shortest paths γ_1, α, γ_2 and β – the sides of M. We occasionally write $M = M(C_1, A, C_2, B)$ in order to emphasize the corresponding labels of the sides.

The following Proposition (with an inessential difference) is due to E. Rips (1979) and nowadays is a folklore in geometric group theory. In principle, it can be deduced from Ref. 5. Since no proof of its statement as we need seems to appear in the literature we include a sketch of its proof.

We need a few definitions: For a path μ in M denote by $o(\mu)$ the initial vertex and by $t(\mu)$ the terminal vertex of $\bar{\mu}$. Denote by $\mathcal{T}(\mu)$ the collection of the terminal non-empty subpaths of μ and by $\mathcal{H}(\mu)$ the collection of the non-empty initial subpaths of μ. For a path μ in M denote by $N_M(\mu)$ (*the neighbourhood of* μ *in* M) the set of all the regions of M whose boundary intersect μ non-trivially. If M is fixed, we shall write $N(\mu)$ for $N_M(\mu)$. Let M be fixed. Since $M \subseteq \mathbb{E}^2$ it makes sense to talk about the *right neighbourhood* $N_r(\mu)$ and the *left neighbourhood* $N_\ell(\mu)$, where $N_r(\mu)$ is the set of all the regions of $N(\mu)$ which are to the right of μ and $N_\ell(\mu)$ is the set of all the regions of $N(\mu)$ which are to the left of μ, when traversing along μ from $o(\mu)$ to $t(\mu)$. (See Fig. 3).

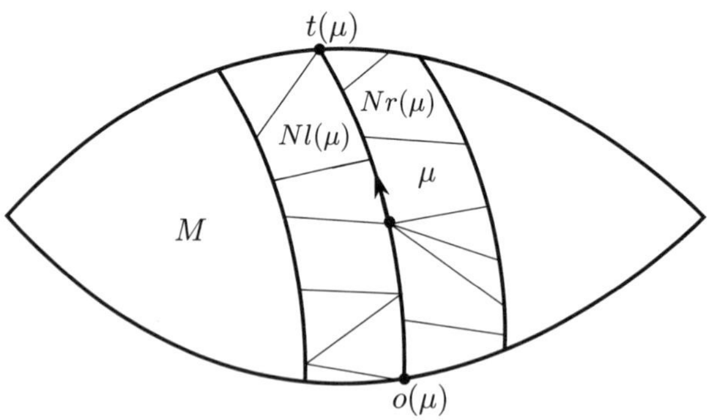

Fig. 3

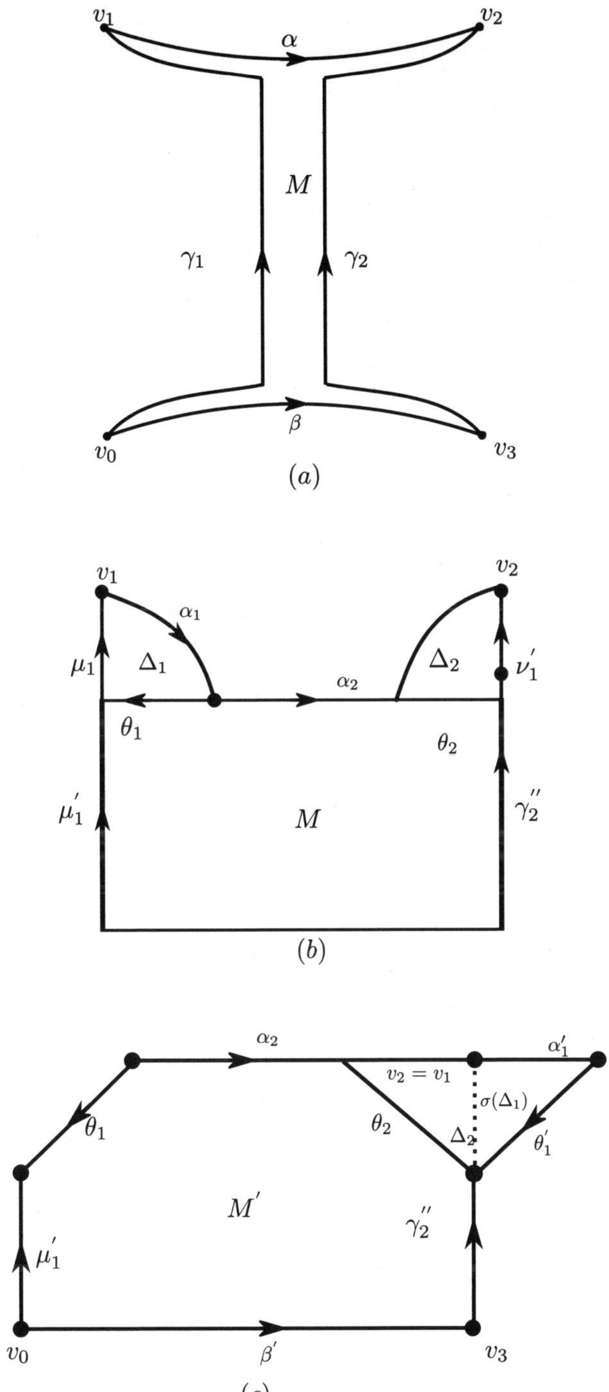

v1
v2
α
M
γ1
γ2
β
v0
v3
(a)
v1
v2
α1
μ1
Δ1
Δ2
ν′1
α2
θ1
θ2
μ′1
M
γ″2
(b)
α2
α′1
v2 = v1
σ(Δ1)
θ1
θ2
Δ2
θ′1
M′
γ″2
μ′1
v0
β′
v3
(c)

Fig. 4

Finally, we remark that following the notation of Ref. 12, Definition 1 (iii), it follows from Ref. 12, Theorem 1 that if $Int(\mathcal{L}_i)$ is not connected than $Int(\mathcal{L}_{i+1})$ are not connected, for $t \geq 1$.

Proposition 3.1. *Let $M(\gamma_1, \alpha, \gamma_2, \beta)$ be a 4-gon. Then*

(a) $N(\xi)$ is a one-layer diagram for $\xi \in \{\gamma_1, \gamma_2, \alpha, \beta\}$;
(b) The following holds $M = N_r(\gamma_1) \cup N_r(\alpha) \cup N_\ell(\gamma_2) = N_r(\gamma_1) \cup N_\ell(\beta) \cup N_\ell(\gamma_2)$.

Proof. Let D_0 be a region in M with maximal number of neighbours. First notice that M may have at most 4 corner regions, one for each vertex v_i, $i = 0, 1, 2, 3$, otherwise one of the corner regions will be contained in one of the sides of M, violating Lemma 2.3. Therefore by Ref. 12, Lemma 2.7 $\mathcal{L}_1 := \mathcal{L}_1(D_0)$ contains at most 4 regions. When constructing the second and further layers $\mathcal{L}_i(D_0)$, we have to take care not to produce more than 4 corner regions in $\mathcal{L}_i(D)$, due to Proposition 2.1 and Remark 2.1. Case by case checking shows that the only possibilities to do so lead to a diagram like the one appearing in Fig. 4(a), where $N(\xi)$ is necessarily a one-layer diagram. □

Corollary 3.1. *Let $D \in \mathcal{R}eg(M)$. Then $\partial D \cap \xi$ is connected, for $\xi \in \{\gamma_1, \gamma_2, \alpha, \beta\}$*

3.2. *The structure of 4-gons with condition V(6).*

Lemma 3.1. *Suppose A, B and C are reduced words in F and A and B are in normal form. See Lemma 2.2. Assume that $A \sim_C B$ in G but in F neither $A \rightharpoonup B$ nor $B \rightharpoonup A$.*

Then there are a subword C^* of C and A^* and B^* in F with $A \rightharpoonup A^*$ and $B \rightharpoonup B^*$, so that $A^* \sim_{C^*} B^*$ in G and in F the word $C^* A^* \sigma(C^*)^{-1} B^{*-1}$ is cyclically reduced.

Proof. Suppose the Lemma is false. Then there is an counter-example (A, B, C) with $|C|$ minimal. We claim that CA is reduced as written. Suppose not. Then $C = C'D$ and $A = D^{-1}A'$ and we do not exclude $A' = 1$ or $C' = 1$. If $C' \neq 1$ then it follows from Lemma 2.2 that $A'\sigma(D)^{-1} \sim_{C'} B$ and $A \rightharpoonup A'\sigma(D)^{-1}$. Since $|C'| < |C|$ the minimality assumption on $|C|$ now implies the conclusion of the Lemma. If $C' = 1$ then $A' \neq 1$ and so $C = D$, and in G we have $1 = CA\sigma(C)^{-1}B^{-1} = A'\sigma(C)^{-1}B^{-1}$, i.e., $A \rightharpoonup_{C^{-1}} B$

in F, contradicting our assumptions. Hence $AC \in F$ is reduced as written. Using the fact that $A \sim_C B$ implies $B \sim_{C^{-1}} A$ and $A^{-1} \sim_{\sigma(C)^{-1}} B^{-1}$ shows that also the expressions CA, $A\sigma(C)^{-1}$, $\sigma(C)^{-1}B^{-1}$ and $B^{-1}C$ are reduced as written in F. As a consequence the word $CA\sigma(C)^{-1}B^{-1}$ is cyclically reduced. □

Proposition 3.2. *Pick $A, B \in F$. Then the images of A and B in G are σ-conjugate if and only if there are σ-cyclic conjugates A^* and B^* and a 4-gon $M^* := M(C^*, A^*, \sigma(C^*), B^*)$ such that $M^{'}$ is a one layer diagram and $||C^*|| = 1$.*

Proof. Using Lemma 3.1 we can assume that A and B are σ-cyclically reduced and are shortest reduced representatives of their images in G. Passing to suitable σ-conjugates of A and B respectively, we can assume that the 4-gon $M = M(C, A, \sigma(C), B)$ (note that neither the word length of A and B did increase nor their respective support) with $|Reg(M)|$ minimal. Let Δ_1 be the derived region of $N_r(\gamma_1)$ which contains v_1 on its boundary and has a non-empty edge μ_1 on γ_1. Similarly, let Δ_2 be the derived region of $N_\ell(\gamma_2)$ which contains v_2 on its boundary and has a non-empty edge ν_1 on γ_2.

First we claim that

$$||\Phi(\mu_1)|| = ||\Phi(\nu_1)|| = 1. \qquad (*)$$

Suppose not. If $\|\Phi(\mu_1)\| > 2$ and $\|\Phi(\nu_1)\| > 2$, then $r(\Delta_2) = \sigma(r(\Delta_1))$. Suppose $|\mu_1| \leq |\nu_1|$ and let μ_2 be the complement of μ_1 on $\partial\Delta_1$. Now, since $r(\Delta_2) = \sigma(r(\Delta_1))$ we can amalgamate $\sigma(\Delta_1)$ with Δ_2 along μ_1 and ν_1' respectively, where ν_1' is the tail of ν_1 with length $|\mu_1|$. Then we get a new 4-gon with sides γ_1', α', γ_2', β' such that each of the following holds: See Figs. 4(b) and (c),

(i) $\boxed{\gamma_1' = \mu_1^{'}\theta_1^{-1}}$ where $\theta_1 = \partial\Delta_1 \cap \partial M_1$, $M_1 := M\backslash\{\Delta_1\}$ and $\gamma_1 = \mu_1'\mu_1$

(ii) $\boxed{\alpha' = \alpha_2\alpha_1'}$ where $\alpha_1 = \partial\Delta_1 \cap \alpha$, $\alpha = \alpha_1\alpha_2$ and , $\Phi(\alpha_1') = \sigma(\Phi(\alpha_1))$.

(iii) $\boxed{\gamma_2' = \gamma_2''\theta_1^{-1}}$, where $\sigma(\Phi(\theta_1)) = \Phi(\theta_2)$, $\gamma_2 = \gamma_2''\nu_1$.

(iv) $\boxed{\beta' = \beta}$.

But then $\Phi(\alpha')$ is a σ-cyclic conjugate of $\Phi(\alpha)$ and has at least one derived region less, because we merged Δ_1 with Δ_2 via σ. This, however, contradicts our supposition on the minimality of $|M'|$. Hence this case cannot happen.

Before considering the remaining cases observe that

If $||\Phi(\mu_1)|| = 1$ *then* Δ_1 *has no neighbours in* $N_r'(\gamma_1)$ *and* $N_r'(\gamma_1) = \Delta_1$

$$(**)$$

and if $||\Phi(\nu_1)|| = 1$ *then* Δ_2 *has no neighbours in* $N_l'(\nu_1)$ *and* $N_l'(\nu_1) = \Delta_2$.

For suppose Δ_1 has a neighbour Δ_3 in $N_r'(\gamma_1)$ and let $\theta = \partial\Delta_1 \cap \partial\Delta_3$. Since $N_r'(\gamma_1)$ is a one-layer derived diagram by Proposition 3.1, Δ_3 is the only neighbour of Δ_1 in M'. Therefore $\mu_1\alpha_1\theta_1$ is a boundary cycle of Δ_1 where $\alpha_1 = \partial\Delta_1 \cap \alpha$ and it is an initial subpath of α. But by assumption $\|\Phi(\mu_1)\| = 1$ and by Theorem 2.1(b) $\|\theta_1\| = 1$. Consequently, $\|\theta_1\mu_1| \leq 2 < n(\Delta_1)$, since by assumption $n(\Delta_1) \geq 3$. Hence, by Appel-Schupp's Lemma 2.4 $|\theta_1\mu_1| < |\alpha_1'|$. This, however, contradicts the minimality of $|\alpha|$. Hence Δ_1 has no neighbour in $N_r'(\gamma_1)$ and $N_r'(\gamma) = \Delta_1$. A similar argument shows that Δ_2 has no neighbour in $N_\ell(\gamma_2)$, hence $N_\ell'(\gamma_2) = \Delta_2$.

Now, if $\|\Phi(\mu_1)\| = 1$ and $\|\Phi(\nu_1\| \geq 2$ then since $N_r(\gamma_1) = \{\Delta_1\}$ by (**) and $\mu_1 = \partial\Delta_1 \cap \gamma_1$ it follows that $\gamma_1 = \mu_1$ and hence $\sigma(\Phi(\mu_1)) = \Phi(\gamma_2)$. Since $\nu_1 \subseteq \gamma_2$ and $\|\Phi(\nu_1)\| \geq 2$, clearly $\|\Phi(\gamma_2)\| \geq 2$, i.e., $\|\sigma(\Phi(\mu_1))\| \geq 2$. But $\|\Phi(\mu_1)\| = 1$, hence also $\|\sigma(\Phi(\mu_1)\| = 1$. Hence (*) holds true and M' is a one-layer diagram. It follows from (**) that $\gamma_1 = \mu_1$ and $\gamma_2 = \nu_2$. Hence by (*) $||\Phi(\gamma_1)|| = ||\Phi(\gamma_2)|| = 1$, hence $||C_1|| = ||\Phi(\gamma_1)|| = 1$. □

Before formulating an important corollary of Proposition 3.2 we recall from Ref. 15, p. 265 a well known method to solve the conjugacy problem in small cancellation groups. We shall need a "σ-variant" of it. Thus, let $\mathcal{P} = \langle \mathcal{Y}|\mathcal{S}\rangle$ be a finite presentation of a group Q and let N be an annular $\mathcal{S}$-diagram which satisfies the condition $C(4)\&T(4)$ with outer boundary cycle ω having label U and inner boundary cycle τ having label V, where U and V are cyclically reduced words. Let $\omega(\mathcal{L}_i)$ be as defined in Ref. 12, Main Theorem (i)(ii) p. 127 with label U_i and let $s = \max\{|R|\,|R \in \mathcal{S}\}$. Then $U = U_k$, $V = V_{-\ell}$ $k, \ell \geq 0$ and are conjugate in Q and moreover we can conjugate U to V step by step as follows. Since $\mathcal{L}_{k-1}$ is a one-layer annular diagram with outer and inner boundary labels U_k and U_{k-1}, respectively, there exists an element h_k in $F(Y)$ of length at most s (which is the label of a boundary path of a region of $\mathcal{L}_{k-1}$) such that $h_kU_kh_k^{-1} = U_{k-1}^*$, where U_{k-1}^* is a cyclic conjugate of U_{k-1}. Also, it follows from Ref. 14, Main Theorem that $|U_{k-i}| \leq s|U_k|$ for $1 \leq i \leq k$. Hence, if we denote by $\mathcal{C}(U)$ the set of all the conjugates of U in Q having length at most $s(|U| + |V|)$

by an element of length at most s then $\mathcal{C}(U)$ contains $\Phi(\omega_i)$, for every i, $0 \leq i \leq k$ where $\omega_i = \omega(\mathcal{L}_i)$. Define $\mathcal{C}(V)$ accordingly. Then U is conjugate to V in G if and only if there exists an annular diagram N as above and this happens if and only if $\mathcal{C}(U) \cap \mathcal{C}(V) \neq \phi$. Since Q has solvable word problem, we can construct $\mathcal{C}(U)$ and $\mathcal{C}(V)$ and decide whether $\mathcal{C}(U) \cap \mathcal{C}(V)$ is empty or not.

In general, if such a procedure is available we say that U *and* V are *step-by-step conjugate with respect to* $\mathcal{P}$.

Taking $Q = \hat{G}$ and $\mathcal{P} = \langle x, t | \mathcal{R}, txt^{-1} = \sigma(x),\ x \in X \rangle$ we say that elements U and V of $\hat{G}$, such that $U = U^{'}t$ and $V = V^{'}t$, with $U^{'},\ V^{'} \in F$ are *step-by-step conjugate with respect to* $\mathcal{P}$ if U_i and V_i as above satisfy $U_i = U_i^{'}t$ and $V_j = V_j^{'}t$, $U_i, V_j \in F$ and $H_i \in F$.

In these terms we have the following corollary.

Corollary 3.2. *Let* $s = max\{|R| \mid R \in \mathcal{R}\}$. *(Thus* $s = max\{n_e \mid$ *e an edge of* $\Gamma\}$.*)*

Suppose A *and* B *are* σ*-conjugate in* G. *Then there are* σ*-cyclically reduced* σ*-cyclic conjugates* A^*, B^* *of* A *and* B *respectively, such that* A^* *and* B^* *are step-by-step* σ*-conjugate with respect to* $\mathcal{P}$.

Proof. Recall from Ref. 14, p.130 Definition 1.2 (c) and (e) the notions of 2-*exceptional* and 2-*corner regions* respectively.

Let $M = M(C^*, A^*, \sigma(C^*), B^*)$ be as in Proposition 3.2. Then $M^{'}$ is a one layer diagrams with $||C^*|| = 1$. Let M_1 be the $\hat{\mathcal{R}}$-diagram with boundary cycle $\gamma_1^{'}\tau_1\gamma_2^{'-1}\tau_2^{-1}$, where $\Phi(\gamma_1^{'}) = C^*$, $\Phi(\gamma_2^{'}) = \sigma(C^*)$, $\Phi(\tau_1) = \Phi(\tau_2) = t$. Then we can merge the diagrams M and M_1 by identifying $\gamma_1^{'}$ with γ_1 and $\gamma_2^{'}$ with γ_2 to got an annular diagram M_0. See (2.2). Considering M_1 as a derived region we get that $M_0^{'}$ is a one-layer derived diagram See Fig. 6. Since $\|\Phi(\gamma_1^{'})\| = \|\Phi(\gamma_2^{'})\| = 1$, by Theorem 2.1(b).

Every inner vertex of γ_1 *and* γ_2 *has valency 3* $\qquad (*)$

Also

Every inner vertex of $\gamma_1^{'}$ *and* $\gamma_2^{'}$ *in* $M_1^{'}$ *has valency 3.* (See Fig. 6) $(**)$

We claim that

M_0 *satisfies the condition C(4) & T(4).* $\qquad (***)$

For $(*)$ and $(**)$ together imply that every inner vertex of γ_1 and γ_2 in M_0 has valency $3+3-2=4$. Now, it follows from Lemma 2.4(d) that every inner vertex of M has valency 4, hence the annular diagram M_0 satisfies the condition T(4). Clearly every inner region in M_1 satisfies the condition C(4) and by parts (c) and (d) of Lemma 2.4 every inner region in M_0 satisfies the condition C(4), as claimed.

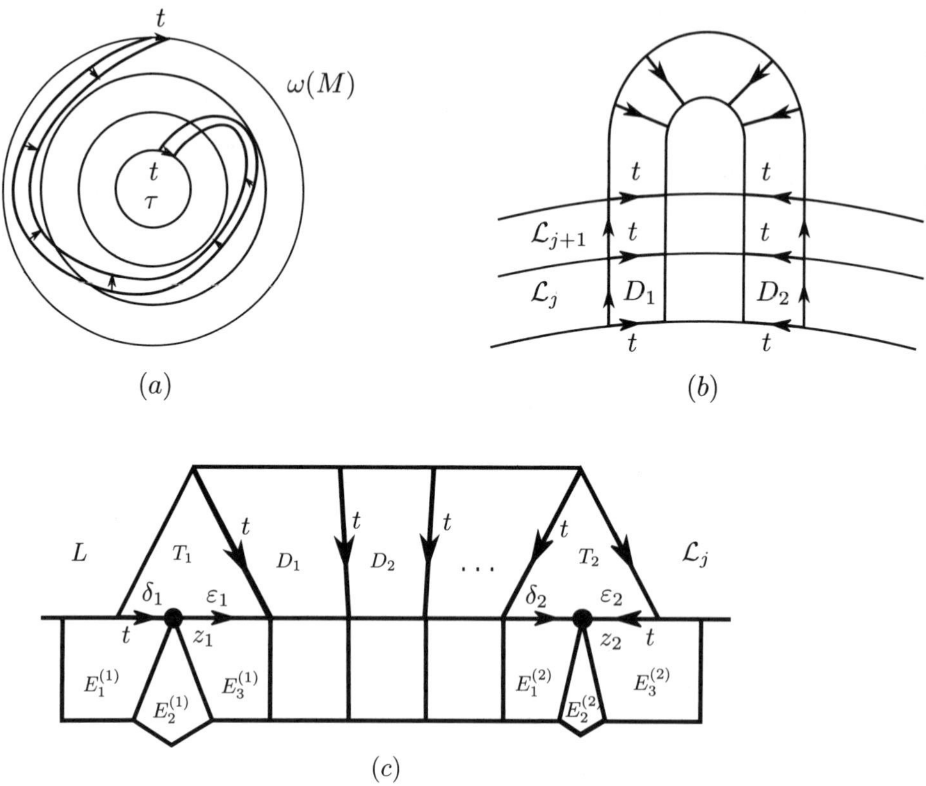

Fig. 5

Next, we claim

$$\textit{every layer } \mathcal{L}_i \textit{ in } M_0 \textit{ contains exactly one } t-\textit{region} \qquad (\square)$$

(i.e., a region with boundary label $tat^{-1}b^{-1}$, for some $a, b \in X \cup X^{-1}$) In principle this need not be the case in general (see Fig. 5(a)).

Suppose that there is a layer which contains at least two t-regions. Then they are part of a t-corridor in M_0, as shown in Fig. 5(b). In particular, there is a layer $\mathcal{L}_j$ which contains a peak of the corridor (see Figs. 5(b) and 5(c)). Without loss of generality we may assume $j \geq 1$. Then $\mathcal{L}_j$ contains a subdiagram $L = \langle T_1, D_1, \ldots D_r, T_2 \rangle$, $r \geq 1$, which is part of the t-corridor, where D_i are $\mathcal{R}$-regions and T_1 and T_2 are t-regions such that the label of $\tau(\mathcal{L}_j) \cap \partial T_i$ contains t, $i = 1, 2$, while $\omega(\mathcal{L}_j) \cap \partial T_i$ contains no t. This last condition forces that $|\tau(\mathcal{L}_j) \cap \partial T_i| = 2$, $i = 1, 2$. Let $\tau(\mathcal{L}_j) \cap \partial T_i = \delta_i z_i \epsilon_i$, $i = 1, 2$, where $|\delta_i| = |\epsilon_i| = 1$, z_i a vertex and $\Phi(\delta_1 \epsilon_1) = t^{\pm 1} a$ for some $a \in X \cup X^{-1}$ and $\Phi(\delta_2 \epsilon_2) = b t^{\pm 1}$ for some $b \in X \cup X^{-1}$. In particular, $d_{\mathcal{L}_j}(z_i) = 2$, $i = 1.2$. See Fig. 5(c). Since by $(***)$ M_0 satisfies the condition $C(4)\&T(4)$, $d_{\mathcal{L}_{j-1}}(z_i) \geq 4$, $i = 1, 2$. By the first part of Proposition 16 on p. 146 of Ref. 14 $d_{\mathcal{L}_{j-1}}(z_i) \leq 4$, hence $d_{\mathcal{L}_{j-1}}(z_i) = 4$. Let $E_1^{(i)}$, $E_2^{(i)}$ and $E_3^{(i)}$ be the regions of $\mathcal{L}_{j-1}$ which contain z_i on their boundary, $i = 1, 2$. Then $E_2^{(1)}$ and $E_2^{(2)}$ are exceptional regions and again due to the $C(4)\&T(4)$ condition and the fact that L is a part of a t-corridor, $\mathcal{L}_{j-1}$ contains no corner regions i.e., between $E_1^{(2)}$ and $E_2^{(2)}$. However, this violates the last part of Proposition 16(c), p. 146 in Ref. 14. Hence every layer contains exactly one t-region, proving $(\square)$.

Now the result follows from $(***)$ and $(\square)$ by the definition of step-by step σ-conjugacy. $\square$

4. Proofs of the Main Theorems

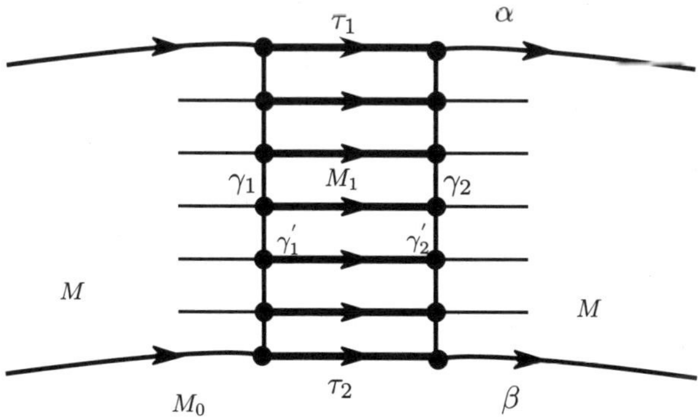

Fig. 6

Proof of Theorem A'. Suppose A and B are σ-conjugate in G. Then by Corollary 3.2 A and B have σ-cyclic conjugates A^* and B^* which are step-by-step σ-conjugate in G. Since G has solvable word problem, we can check whether elements of G represented by words W_1 and W_2 respectively, of length at most $|A|+|B|$ are σ-conjugate in G by an element of length at most s. Therefore we can produce the list $\mathcal{L}(A^*)$ of all the elements which are step-by-step σ-conjugate to A^*. Therefore B^* is σ-conjugate to A^* if and only if $B^* \in \mathcal{L}(A)$. Since we can produce all the cyclic σ-conjugates of A and of B, we can produce the corresponding lists $\mathcal{L}(W)$, where W is a σ-conjugate of A, and check in F whether one of the cyclic conjugates of B is on one of these lists. □

For the proof of Theorem C' we need the following Lemma.

Lemma 4.1. *Let Γ_0 be a subgraph of Γ and let $\tilde{H}$ be the subgroup of F generated by the vertices of Γ_0. Thus $\pi(\tilde{H}) = H$, where π is the natural projection of F on G. Let W be a reduced word in $\tilde{H}$. If W is shortest possible representative of $\pi(W)$ in $\tilde{H}$ then W is a shortest possible representative of $\pi(W)$ in F.*

Proof. Let U be a shortest representative of $\pi(W)$ in F. Then $\pi(UW^{-1}) = 1$ in G, hence there is a van Kampen diagram M with UW^{-1} as a boundary label. Let μ and ν be boundary paths of M with $\Phi(\mu) = U$ and $\Phi(\nu) = W$. Thus, $u\mu v\nu^{-1}$ is a boundary cycle of M and u and v are the endpoints of μ and ν. By Lemma 2.3 μ may contain no corner derived region of M', in the sense that $\partial\Delta \cap \partial M' \not\subseteq \mu$, for every boundary derived region Δ. We claim that ν also cannot contain a corner derived region. Suppose Δ is a corner derived region on ν. Assume first that G is a CLTTF Artin group. Then due to the T(4) condition Δ has at most two neighbours in M', hence it follows from Theorem 2.1 that $\partial\Delta \cap \partial M'$ is the complement of 2 pieces (at most) hence has syllable length at least 4. Hence, certainly $r(\Delta) \in \Gamma_0$ and hence, if we remove Δ from M we get a diagram M_1 with boundary cycle $\mu v \nu_1' v_1 \beta^{-1} v_2 \nu_1'''$, where $\nu = \nu' v_1 \nu'' v_2 \nu'''$ and $v_1 \nu'' v_2 \beta^{-1}$ is a boundary cycle of Δ. But then $\Phi(\beta)$ is a word in $\tilde{H}$, hence $W' := \Phi(\nu' v_1 \beta v_2 \nu''') \in \tilde{H}$ and $\pi(W') = \pi(W)$, and by (Appel-Schupp's) Lemma 2.4 W' is shorter than W, contradicting our assumption concerning W. Hence ν as well contains no corner derived region of M'. If G is an extra-large Artin group then the same arguments apply, this time Δ has at most 3 neighbours in M', while $n(\Delta) \geq 4$. Hence W' is shorter than W. But then for both the CLTTF and the extra-large cases, by Proposition 2.1 M' is a one-layer diagram in which no (boundary) vertex may have valency 4 or more (because then the "middle region" which contains it is a corner region). Hence, the derived regions which have a common edge with U are exactly the same as those which have a common edge with W. Consequently $U \in \tilde{H}$, proving the Lemma. □

Proof of Theorem C'. Choose representatives A and B for a and b, respectively in $\tilde{H}$ with shortest length *in* $\tilde{H}$. Then by Lemma 4.1 we may assume that A and B are shortest possible representations of a and b in F. Let $M_0 = M_1 \cup M$ be a σ-conjugating (annular) t-diagram for A and B. Then by Proposition 3.2 M has one-layer derived diagram in which $M_1 = \langle \Delta_1, \ldots, \Delta_k \rangle$, $k \geq 1$ with $\|\partial\Delta_i \cap \partial\Delta_{i+1}\| = 1$ and $\partial M_1 \cap \partial M = \gamma_1 \cup \gamma_2$ with $\|\Phi(\gamma_1)\| = 1$ and $\|\Phi(\gamma_2)\| = 1$. In particular, $\partial\Delta_i \cap \partial M$ contains an edge. Since A and B belong to $\tilde{H}$, it follows that $r(\Delta) \in \tilde{H}$, for every derived region Δ of M_1. But then there is a σ-conjugating element c of a and b which is the image of $\Phi(\gamma_2) \in H$. □

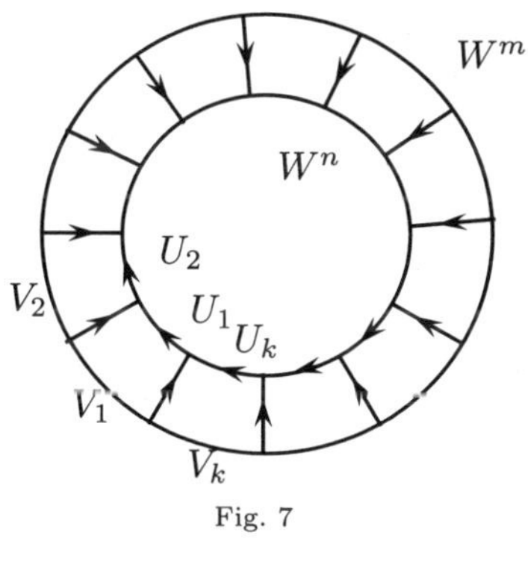

Fig. 7

Proof of Theorem B'. By assumption $|\Gamma| \geq 3$. Therefore, there are x_1, x_2 and x, 3 vertices of Γ such that $\sigma(x)^{-1} \neq x_1$. For $n \geq 1$ define $W_n = (x_1^5 x_2^6 x^7)^n$. Then W_n is σ-cyclically reduced since $\sigma(x)^{-1} \neq x_1$. Also W_n contains no corner regions of any $\mathcal{R}$-diagram which contains W_n as the label of a boundary path because W_n contains no subword of length 4 or more of any relator. Clearly $W_n \neq W_m$ in G, for $n \neq m$, because G is torsion-free. (Its cohomological dimension is at most 2.) Now consider a σ-conjugacy diagram M of W_n to W_m with minimal number of derived regions. Let ω be the outer boundary of M with label W^m and let τ be the inner boundary of M with label W^n. By what we found above, M has one-layer derived diagram, due to Proposition 3.2. Let $M = \langle \Delta_1, \ldots, \Delta_k, \Delta_0 \rangle$, where Δ_i, $1 \leq i \leq k$ are $\mathcal{R}$-derived regions and Δ_0 is the t-diagram M_1 of Fig.5. Since neither ω nor τ may contain corner regions of M', $\partial\Delta_i \cap \partial\Delta_{i+1} \cap \omega$, and $\partial\Delta_i \cap \partial\Delta_{i+1} \cap \tau$ are vertices with valency 3 in M. See Fig. 7. Hence $W_n = U_1 \ldots U_k$ and $W_m = V_1 \ldots V_k$, where each U_i and each V_i is a reduced word in an edge subgroup of G and adjacent U_i (V_i) belong to different edges of Γ. But since $m \neq n$ this is not possible, because then by the Pigeon-hole Principle at least one edge group should contain more than two generators of F, corresponding to the vertices of Γ (in other words, then an edge of Γ should have at least 3 endpoints). Consequently, W_n is not σ-conjugate to W_m as required. $\square$

Acknowledgements. I am very grateful to the referee for useful remarks concerning the presentation of the work. In particular, for the present version of Definition 2.2, Lemma 2.2 and of Lemma 3.1 with its proof.

References

1. K. Appel and P. Schupp, Artin groups and infinite Coxeter groups, *Inventiones Math*, 72, pp. 201-220, 1983.
2. O. Bogopolski, A. Martino, O. Maslakova and E. Ventura, Free-by-cyclic groups have solvable conjugacy problem, *Bulletin of the London Math. Soc.* **38** (5), (2006), 787–794.
3. O. Bogopolski, A. Martino and E. Ventura, The conjugacy problem for some extensions of groups, *Trans. of Amer. Math. Soc.*, 33pps, to appear.
4. J. Crisp, Automorphisms and abstract commensurators of 2-dimensional Artin groups, *Geometry and Topology* **9**, (2005), 1381–1441.
5. T. Delzant, Sous-groupes distinguès et quotients des groupes hyperboliques, *Duke math. J.* 83(3), pp. 661-682, 1996.
6. A.L. Fel'shtyn, The Reidemeister number of any automorphism of a Gromov hyperbolic group is infinite, *J. Math. Sci.*, Vol. 119(1) 117-123, 2004.
7. A.L. Fel'shtyn, Dynamical zeta functions, Nielsen Theory and Reidemeister torsion, *Memoir of Amer. Math. Sco.* Vol. 147, No. 699, 2000.
8. A.L. Fel'shtyn, D. Gonçalves and P. Wong, Twisted conjugacy classes for polyfree groups, arXiv:0802.2937v 2 [math.GR] 21 Feb. 2008.
9. F.R. Gantmakher: Canonical representation of automorphisms of a complex semisimple Lie group, *Math. Sbor.*, 5 pp.101-144, 1939.
10. J. González–Meneses and E. Ventura, Twisted conjugacy in the braid group, in progress.
11. D. Gonçalves and P. Wong, Twisted conjugacy classes in exponential growth groups, *Bull. London Math. Soc.* **35** (2003), 261–268.

12. A. Juhász, Small cancellation theory with unified small cancellation condition, *J. London Math. Soc* 2(40), pp. 57-80, 1989.
13. A. Juhász, Fusion in Artin Groups I, *Journal of the London Math. Society* (2) **44** (1991), 287-300.
14. A. Juhász, Solvability of the conjugacy problem and malnormality of subgroups in certain relative presentations with a weakened small cancellation condition, *Algebra and Geometry in Geneva and Barcelona, "Asymptotic and Probabilistic Methods in Geometric Group Theory"', Geneva, June 2005* and *"Barcelona Group Theory Conference", Barcelona, July 2005* (2007), 121–168.
15. R. C. Lyndon, P. E. Schupp, Combinatorial Group Theory, Springer Verlag, Berlin-Heidelberg-New York, 1977.
16. K. Reidemeister, Automorphismen von Homotopie-Kettenringen, *Math. Ann.* 112, p. 586-593, 1936.
17. V. Shpilrain, An authentication scheme based on twisted conjugacy problems, *LNCS*, Vol. 5037, pp. 366-372, 2008.
18. T.A. Springer, Twisted conjugacy in simply connected groups, *Transformation Groups* Vol. 11, No. 3, p. 539-545, 2006.
19. J. Taback and P. Wong, A note on twisted conjugacy and generalized Baumslag–Solitar groups, arXiv, math.GR|06 06284.

APPLICATIONS OF CLIFFORD'S THEOREM TO FROBENIUS GROUPS OF AUTOMORPHISMS

E. I. KHUKHRO

Institute of Mathematics, Novosibirsk, 630090, Russia

E-mail: khukhro@yahoo.co.uk

We discuss applications of Clifford's theorem in the situation where a finite group G admits a Frobenius group of automorphisms FH with complement H and fixed-point-free kernel F (that is, such that $C_G(F) = 1$). The strong hypothesis makes it easier to apply Clifford's theorem, but the desired results must also have strong conclusions. The aim is proving that various properties of G are close to the corresponding properties of $C_G(H)$. It is shown that the fixed points of H in any FH-invariant section are covered by the fixed points of H in G. This implies a bound for $|G|$ in terms of $|H|$ and $|C_G(H)|$, and a bound for the rank of G in terms of $|H|$ and the rank of $C_G(H)$. It is shown that G is nilpotent if $C_G(H)$ is nilpotent. These results do not need any coprimeness assumptions on the orders of G and FH. Under the additional condition that $(|G|, |H|) = 1$ it is proved that the Fitting height of G is equal to the Fitting height of the fixed-point subgroup $C_G(H)$ and the Fitting series of $C_G(H)$ coincides with its intersections with the Fitting series of G. Further recent results on bounding the exponent and nilpotency class of G are mentioned, which use different Lie ring methods.

Keywords: Clifford's theorem; Frobenius group; automorphism; Fitting height; nilpotent; rank.

Introduction

Clifford's theorem[3] has many applications in group theory. One of the well-known folklore results going back to 1930s is the nilpotency of a finite soluble group G admitting a fixed-point-free automorphism φ of prime order. (Of course, the solubility assumption was eventually removed in Thompson's pioneering work on p-complements[22].) Later Clifford's theory was developed into the modern Hall–Higman–type theorems, originating from the Hall–Higman paper[8], where new methods of analysis of certain minimal situations first appeared. In particular, these methods were applied for bounding the Fitting height (nilpotent length) $h(G)$ of a finite soluble group G in terms of a group of automorphisms $A \leqslant \operatorname{Aut} G$ and its fixed-point subgroup $C_G(A)$. Thompson's theorem[23] gives a bound for $h(G)$ in terms of $h(C_G(A))$ and the number of prime factors whose product is $|A|$ in the case where A is soluble of coprime order. This bound was improved in subsequent papers of various authors, of which we mention a linear bound first obtained by Kurzweil[15] and the best-possible bound by Turull[25]. Dade's theorem[5], in particular, gives a bound for $h(G)$ in the case where A is nilpotent and $C_G(A) = 1$ without the coprimeness condition (but examples by Bell and Hartley[1] show that this is no longer true for any non-nilpotent soluble A).

In the present paper we discuss somewhat 'easier' applications of Clifford's theorem in the situation where a finite group G admits a Frobenius group of automorphisms FH with kernel F and complement H such that $C_G(F) = 1$. Since F is nilpotent being a Frobenius kernel, the condition $C_G(F) = 1$ implies the solubility of G by a theorem of Belyaev and Hartley (Ref. 2, Theorem 0.11) based on the classification of finite simple groups.

The strong hypothesis makes it easier to apply Clifford's theorem, avoiding so-called 'exceptional' Hall–Higman–type situations. But the desired results must also have strong conclusions. The aim is proving that various properties of G are close to the corresponding properties of $C_G(H)$. It is shown that the fixed points of H in any FH-invariant section are covered by the fixed points of H in G (Theorem 1.1). This implies a bound for $|G|$ in terms of $|H|$ and $|C_G(H)|$, and a bound for the rank of G in terms of $|H|$ and the rank of $C_G(H)$ (Theorem 2.1). It is shown that G is nilpotent if $C_G(H)$ is nilpotent (Theorem 3.1). These results do not need any coprimeness assumptions on the orders of G and FH. Under the additional condition $(|G|, |H|) = 1$ it is proved that the Fitting height of G is equal to the Fitting height of $C_G(H)$ and the Fitting series of $C_G(H)$ coincides with its intersections with the Fitting series of G (Theorem 3.2). Thus, much stronger best-possible bounds for the Fitting height of G can be obtained, independent of $|H|$ or $|F|$ — of course, under the stronger hypotheses, all of which are essential, with the possible exception of $(|G|, |H|) = 1$.

We see that, by application of Clifford's theorem, the problem of proving that the properties of G are close to that of $C_G(H)$ is now largely reduced to the case of nilpotent group G. In this area different Lie ring methods are applied to obtain bounds for the nilpotency class and exponent of G under certain additional conditions (some of which are unavoidable); see Refs. 10,13,17,18,20,21. We mention these results in the concluding section.

1. Fixed points and free modules

First we discuss the question of covering the fixed points of a group of automorphisms in an invariant quotient by the fixed points in the group. Let $A \leqslant \operatorname{Aut} G$ for a finite group G and let N be a normal A-invariant subgroup of G. It is well-known that if $(|A|,|N|) = 1$, then $C_{G/N}(A) = C_G(A)N/N$. (The induced group of automorphisms of an invariant section is often denoted by the same letter, which is an abuse of notation, as the action may become non-faithful.) If we do not assume that $(|A|,|N|) = 1$, then the equality $C_{G/N}(A) = C_G(A)N/N$ may no longer be true. But we still have the following well-known result, which is a consequence of the properties of Carter subgroups (see, for example, Ref. 13, Lemma 2.2).

Lemma 1.1. *Let G be a finite group admitting a nilpotent group of automorphisms F such that $C_G(F) = 1$. If N is a normal F-invariant subgroup of G, then $C_{G/N}(F) = 1$.*

We are going to show that, furthermore, when G is a finite group admitting a Frobenius group of automorphisms FH with kernel F and complement H such that $C_G(F) = 1$, the fixed points of H in any FH-invariant section of G are also covered by the fixed points of H in G. This follows from the fact that every such section is a free $\mathbb{F}_pH$-module for a suitable prime p. (Earlier free $\mathbb{F}_pA$-modules were also related to proving, in certain special cases where both G and A are p-groups, that fixed points of $A \leqslant \operatorname{Aut} G$ in G cover fixed points of A in A-invariant sections in the papers by Thompson[24], Evens[6], and the author[9].)

The presence of free $\mathbb{F}_pH$-modules is derived by Clifford's theorem. The following lemma must be known in folklore; similar lemmas are contained in Refs. 11,13. For a group A acting by linear transformations on a vector space V we use the right operator notation va for the image of $v \in V$ under $a \in A$. We also use the centralizer notation for the fixed-point subspace $C_V(A) = \{v \in V \mid va = v \text{ for all } a \in A\}$, and for the kernel $C_A(V) = \{a \in A \mid va = v \text{ for all } v \in V\}$. Recall that for a group A and a field k, a free kA-module of dimension n is a direct sum of n copies of the group algebra kA each of which is regarded as a vector space over k of dimension $|A|$ with a basis $\{v_g \mid g \in A\}$ labelled by elements of A on which A acts in a regular representation: $v_g h = v_{gh}$.

Lemma 1.2. *Suppose that V is a vector space over any field k admitting a finite Frobenius group of linear transformations FH with kernel F and complement H such that $C_V(F) = 0$. Then V is a free kH-module.*

Proof. First we perform reduction to the case of $|F|$ being coprime to the characteristic p of the field k. (No such reduction is needed in the case of characteristic zero.) Recall that F is nilpotent, and let $F = F_p \times F_{p'}$, where F_p is the Sylow p-subgroup of F. Since $C_V(F_{p'})$ is F_p-invariant, we must actually have $C_V(F_{p'}) = 0$. Otherwise the p-group F_p would have nontrivial fixed points on $C_V(F_{p'})$, and clearly $C_{C_V(F_{p'})}(F_p) = C_V(F)$. Thus, the hypothesis of the lemma also holds for V and the Frobenius group $F_{p'}H$, so we can assume that F is a p'-group.

We extend the ground field k to its algebraic closure $\overline{k}$, so that $\overline{V} = V \otimes_k \overline{k}$ is regarded as a $\overline{k}FH$-module. Note that we still have $C_{\overline{V}}(F) = 0$. It suffices to prove that $\overline{V}$ is a free $\overline{k}H$-module. Indeed, by the Deuring–Noether theorem (Ref. 4, Theorem 29.7) two representations over a smaller field are equivalent if they are equivalent over a larger field. Being a free $\overline{k}H$-module, or a free kH-module, means having a basis, of the underlying vector space over the corresponding field, elements of which are permuted by H so that all orbits are regular. In such a basis H is represented by the corresponding permutational matrices, which are defined over k.

Consider an unrefinable series of $\overline{k}FH$-submodules

$$\overline{V} = U_1 > U_2 > \cdots > U_l > U_{l+1} = 0, \tag{1}$$

in which each factor U_i/U_{i+1} is an irreducible $\overline{k}FH$-module. Since an extension of a free $\overline{k}H$-module by a free $\overline{k}H$-module is again a free $\overline{k}H$-module, it is sufficient to prove that each factor U of (1) is a free $\overline{k}H$-module. Since $C_{\overline{V}}(F) = 0$ and the action of F is coprime (semisimple), we also have $C_U(F) = 0$.

By Clifford's theorem, U is the direct sum

$$U = W_1 \oplus \cdots \oplus W_t$$

of the Wedderburn components W_i with respect to F, which are homogeneous $\overline{k}F$-modules transitively permuted by H. None of the W_i is a trivial $\overline{k}F$-module, since $C_U(F) = 0$. Let H_1 be the stabilizer of W_1 in H in its action on the set $\{W_1, \ldots, W_t\}$. If $H_1 = 1$, then the H-orbit of any $\overline{k}$-basis of W_1 gives a basis of U that is the union of several regular H-orbits, so then U is a free $\overline{k}H$-module. But if $H_1 \neq 1$, then H_1 centralizes the centre $Z(F/C_F(W_1))$, which is represented on W_1 by scalar linear transformations. This centre is nontrivial, because $F/C_F(W_1)$ is a nontrivial nilpotent group, as W_1 is a nontrivial $\overline{k}F$-module. Because the action of H_1 on F is coprime, we then obtain nontrivial fixed points of H_1 on F, which is impossible in the Frobenius group FH. □

We now discuss the connections between free modules and fixed points.

Fixed points in free modules have a very clear description as the 'diagonal' elements. Indeed, let M be a free kA-module for a group A and any field k. There is a basis of M as a vector space over k consisting of regular orbits under the action of A. Choose one element in each of these regular A-orbits and let them span a subspace M_1. Then $M = \bigoplus_{a \in A} M_1 a$, where the direct summands $M_a = M_1 a$ are regularly permuted by A, so that $M_a b = M_{ab}$ for all $a, b \in A$. Obviously, the fixed-point subspace is $C_M(A) = \{\sum_{a \in A} ma \mid m \in M_1\}$. In particular,

$$\dim_k M = |A| \dim_k C_M(A). \tag{2}$$

Less obvious is the behaviour of fixed points 'above' free modules.

Lemma 1.3. *Let A be a group of automorphisms of a finite group G, and M an A-invariant elementary abelian p-subgroup of G. If M, regarded as an $\mathbb{F}_p A$-module, is a free $\mathbb{F}_p A$-module, then $C_{G/M}(A) = C_G(A)M/M$.*

Proof. This fact could be derived from the triviality of the first cohomology group $H^1(A, M) = 0$. Indeed, the condition $H^1(A, M) = 0$ is equivalent to the fact that all complements of M in the natural semidirect product MA are conjugate (see, for example, Ref. 19, 11.4.7). If $cM \in C_{G/M}(A)$, then A^c is another complement of M in MA, and if $A^c = A^{am} = A^m$ for some $a \in A$, $m \in M$, then $A^{cm^{-1}} = A$, whence $[A, cm^{-1}] \in A \cap \langle c \rangle M = 1$, so that $cm^{-1} \in C_G(A) \cap cM$.

The existence of a required fixed point can also be shown directly, without references to cohomology theory. First suppose that M is a one-dimensional free $\mathbb{F}_p A$-module, ad let $\{e_g \mid g \in A\}$ be a basis of the $\mathbb{F}_p$-space M regularly permuted by A: $e_g h = e_{gh}$ for $g, h \in A$. Let $cM \in C_{G/M}(A)$. We need to find $x \in M$ such that $(cx)^h = c^h x^h = cx$ for all $h \in A$, that is, $[c, h]x^h = x$, which can be rewritten in the additive notation as

$$x - xh = [c, h] \quad \text{for all } h \in A. \tag{3}$$

Let $[c, a] = \sum_{g \in A} \alpha_g(a) e_g$ for $a \in A$. From $c^{st} = (c^s)^t$ we obtain the formula

$$\alpha_g(st) = \alpha_g(t) + \alpha_{gt^{-1}}(s),$$

which also implies

$$\alpha_g(h^{-1}g) = \alpha_g(g) + \alpha_1(h^{-1}) \tag{4}$$

and

$$0 = \alpha_g(g^{-1}g) = \alpha_g(g) + \alpha_1(g^{-1}),$$

whence

$$\alpha_1(g^{-1}) = -\alpha_g(g). \tag{5}$$

We seek $x = \sum_{g\in A} x_g e_g$ with unknown coefficients x_g. Equations (3) in coordinates give rise to the linear system of $(|A|-1)^2$ equations

$$x_g - x_h = \alpha_g(h^{-1}g); \qquad g,h \in A, \quad g \neq h. \tag{6}$$

We need to show that the system (6) is consistent. For that we need to show that if $\beta(g,h)$ are any coefficients such that the corresponding linear combination of the rows of the coefficient matrix of (6) is trivial, then the right-hand sides of (6) also satisfy this dependence. In other words, if

$$\sum_h \beta(g,h) - \sum_h \beta(h,g) = 0 \qquad \text{for each } g \in A, \tag{7}$$

then $\sum_{g,h} \beta(g,h)\alpha_g(h^{-1}g) = 0$. It is more convenient to rewrite (7) as

$$\sum_h \beta(g,h) = \sum_h \beta(h,g) \qquad \text{for each } g \in A. \tag{8}$$

Using formulae (4), (5), (8), we have

$$\begin{aligned}
\sum_{g,h} \beta(g,h)\alpha_g(h^{-1}g) &= \sum_{g,h} \beta(g,h)(\alpha_g(g) + \alpha_1(h^{-1})) \\
&= \sum_{g,h} \beta(g,h)\alpha_g(g) + \sum_{g,h} \beta(g,h)\alpha_1(h^{-1}) \\
&= \sum_g \alpha_g(g) \sum_h \beta(g,h) + \sum_h \alpha_1(h^{-1}) \sum_g \beta(g,h) \\
&= \sum_g \alpha_g(g) \sum_h \beta(g,h) - \sum_h \alpha_h(h) \sum_g \beta(g,h) \\
&= \sum_g \alpha_g(g) \sum_h \beta(h,g) - \sum_h \alpha_h(h) \sum_g \beta(g,h) = 0,
\end{aligned}$$

as required.

Now suppose that M has a proper one-dimensional free $\mathbb{F}_pA$-submodule M_1; then M/M_1 is also a free $\mathbb{F}_pA$-module of smaller dimension than M. By induction, $C_{G/M}(A) = (C_{G/M_1}(A)M/M_1)/(M/M_1)$, and by the one-dimensional case, this is equal to

$$((C_G(A)M_1/M_1)M/M_1)/(M/M_1) \cong C_G(A)M/M,$$

and the result follows. $\square$

We are now ready to analyse the fixed points of the complement H of a Frobenius group of automorphisms FH. The following theorem was proved in Ref. 11 and Ref. 13.

Theorem 1.1. *Suppose that a finite group G admits a Frobenius group of automorphisms FH with kernel F and complement H. If N is an FH-invariant normal subgroup of G such that $C_N(F) = 1$, then $C_{G/N}(H) = C_G(H)N/N$.*

Proof. Recall that N is soluble by Ref. 2, Theorem 0.11. Consider an unrefinable FH-invariant normal series of G

$$G > N = N_1 > N_2 > \cdots > N_l > N_{l+1} = 1 \tag{9}$$

connecting N with 1; its factors N_i/N_{i+1} are elementary abelian. Each factor of this series can be regarded as an $\mathbb{F}_pFH$-module for a suitable prime p, which is a free $\mathbb{F}_pH$-module by Lemma 1.2. We can now apply Lemma 1.3 in an easy induction on the length of the series (9) to find an element of $C_G(H)$ in any $gN \in C_{G/N}(H)$. For $l > 1$ by induction there is $c_1N_l \in C_{G/N_l}(H) \cap gN/N_l$. By Lemma 1.3 applied with $M = N_l$ there is a required $c \in C_G(H) \cap c_1N_l \subseteq C_G(H) \cap gN$. □

As a corollary, in the case where G is a nilpotent group, the fixed points $C_{L(G)}(H)$ in the associated Lie ring $L(G)$ are described in terms of $C_G(H)$. In particular, this situation arises when GF is also a Frobenius group (so that GFH is a 2-Frobenius group, with not necessarily coprime orders of G and H). The corresponding description of $C_{L(G)}(H)$ was used in the paper by Makarenko and Shumyatsky[18] on 2-Frobenius groups, where Mazurov's problem 17.72(a) from "Kourovka Notebook"[26] was solved in the affirmative.

2. Order and rank

We now establish the connection between the order and rank of G and $C_G(H)$ for a finite group G admitting a Frobenius group of automorphisms FH with fixed-point-free kernel F. (By the rank we mean the minimum number r such that every subgroup can be generated by r elements.)

Here and in subsequent sections we shall need another consequence of the properties of Carter subgroups, which was noted in Ref. 13, Lemma 2.6.

Lemma 2.1. *Suppose that a finite group G admits a Frobenius group of automorphisms FH with kernel F and complement H such that $C_G(F) = 1$.*

Then for each prime p dividing $|G|$ there is a unique FH-invariant Sylow p-subgroup of G.

The following theorem is proved in Ref. 13.

Theorem 2.1. *Suppose that a finite group G admits a Frobenius group of automorphisms FH with kernel F and complement H such that $C_G(F) = 1$. Then*

(a) $|G| = |C_G(H)|^{|H|}$;

(b) *the rank of G is bounded in terms of $|H|$ and the rank of $C_G(H)$.*

Proof. (a) Recall that G is soluble by Ref. 2, Theorem 0.11. Consider an unrefinable FH-invariant normal series of G

$$G = G_1 > G_2 > \cdots > G_l > 1. \tag{10}$$

It is sufficient to prove the required equality for each factor $S = G_i/G_{i+1}$ of the series (10), since $|C_G(H)| = \prod_i |C_{G_i/G_{i+1}}(H)|$ by Theorem 1.1. The factor S can be regarded as an $\mathbb{F}_pFH$-module for a suitable prime p, which is a free $\mathbb{F}_pH$-module by Lemma 1.2, since $C_S(F) = 1$ by Lemma 1.1. This means that $S = \bigoplus_{h\in H} Th$ for some $\mathbb{F}_pH$-submodule T. Therefore, $C_S(H) = \{\sum_{h\in H} th \mid t \in T\}$ and $|C_S(H)| = |T|$, whence $|S| = |T|^{|H|}$. Part (a) is proved.

(b) It is known that the rank of a finite group G is bounded in terms of the maximum rank of its Sylow subgroups[14,16]. Let P be an FH-invariant Sylow p-subgroup of G given by Lemma 2.1. It is known that the rank of a p-group of automorphisms of a finite p-group U is bounded in terms of the rank of U. Let U be a Thompson critical subgroup of P; recall that U is a characteristic subgroup of nilpotency class at most 2 containing its centralizer in P (see, for example, Ref. 7, Theorem 5.3.11). Thus the rank of P is bounded in terms of the rank of U. In turn, since U is nilpotent of class at most 2, the rank of U is bounded in terms of the rank of $S = U/\Phi(U)$. The group S can be regarded as an $\mathbb{F}_pFH$-module, which is a free $\mathbb{F}_pH$-module by Lemma 1.2, since $C_S(F) = 1$ by Lemma 1.1. By (2), the rank of S is equal to $|H|$ times the rank of $C_S(H)$. By Theorem 1.1, $C_S(H)$ is covered by $C_G(H)$; as a result, the rank of S is at most $|H|$ times the rank of $C_G(H)$. □

3. Nilpotency and Fitting height: initial reduction

The following theorem is proved in Ref. 13.

Theorem 3.1. *Suppose that a finite group G admits a Frobenius group of automorphisms FH with kernel F and complement H such that $C_G(F) = 1$. If $C_G(H)$ is nilpotent, then G is nilpotent.*

Under the additional assumption $(|G|, |H|) = 1$ we are able to prove a more general result. Recall that the Fitting series starts with the Fitting subgroup $F_1(G) = F(G)$, the largest normal nilpotent subgroup of a group G, and $F_{i+1}(G)$ is defined as the inverse image of $F(G/F_i(G))$. The smallest number l such that $F_l(G) = G$ is called the Fitting height (or nilpotent length) of a soluble group G. The following theorem is proved in Ref. 12.

Theorem 3.2. *Suppose that a finite group G admits a Frobenius group of automorphisms FH with kernel F and complement H such that $C_G(F) = 1$ and $(|G|, |H|) = 1$. Then*

(a) $F_i(C_G(H)) = F_i(G) \cap C_G(H)$ *for all* i;

(b) *the Fitting height of G is equal to the Fitting height of $C_G(H)$.*

At the moment it is not clear whether the coprimeness condition $(|G|, |H|) = 1$ is essential in Theorem 3.2; note, however, that the order of G is not assumed to be coprime to the order of F.

Recall that in both theorems the group G is soluble by Ref. 2, Theorem 0.11. The first of these theorems would of course follow from the second if not for the additional assumption $(|G|, |H|) = 1$ in the latter. Still, much of the initial reductions in the proofs of both results are the same. Therefore we consider these reductions in this section together, and then finish the proofs in each case separately in the next section. In particular, all the arguments in this section do not use the coprimeness assumption of Theorem 3.2 (so these arguments can be used in the proof of Theorem 3.1), but do preserve this assumption (and so can be used in the proof of Theorem 3.2). In what follows, G is a finite group admitting a Frobenius group of automorphisms FH with kernel F and complement H such that $C_G(F) = 1$.

Since the image of $C_G(H)$ covers $C_{G/N}(H)$ for every FH-invariant normal subgroup N by Theorem 1.1, we have

$$F(C_G(H))N/N \leqslant F(C_{G/N}(H)). \tag{11}$$

It is also obvious that

$$F(C_G(H)) \cap N \leqslant F(C_N(H)). \tag{12}$$

These inclusions will be useful for showing that certain assumptions are inherited by subgroups or quotients of G. By Lemma 1.1 the hypothesis

$C_G(F) = 1$ is also inherited by quotients by F-invariant subgroups and, obviously, by such subgroups.

In both Theorems 3.1 and 3.2 it is sufficient to show that $F(C_G(H)) \leqslant F(G)$. Indeed, in Theorem 3.1, where $C_G(H) = F(C_G(H))$, this inclusion would imply that $C_{G/F(G)}(H) = 1$ due to Theorem 1.1. Then $F(G) = G$, since otherwise H would have nontrivial fixed points in a minimal FH-invariant normal subgroup of $G/F(G)$, which is a free $\mathbb{F}_pH$-module for some prime p by Lemmas 1.1 and 1.2.

For similar reasons, in Theorem 3.2 part (b) follows from part (a). Since $C_{G/F_i(G)}(H) = C_G(H)F_i(G)/F_i(G)$ for every i by Theorem 1.1, part (a) will follow in obvious fashion by induction if we prove that $F(C_G(H)) = F(G) \cap C_G(H)$, which clearly follows from $F(C_G(H)) \leqslant F(G)$.

Arguing by contradiction in both Theorems 3.1 and 3.2, we assume that $F(C_G(H)) \not\leqslant F(G)$ and the semidirect product GFH with this condition has minimal possible order.

Since $F(G) = \bigcap_p O_{p',p}(G)$, there is a prime p such that $F(C_G(H)) \not\leqslant O_{p',p}(G)$. By (11) the image of $F(C_G(H))$ in $G/O_{p'}(G)$ is contained in $F(C_{G/O_{p'}(G)}(H))$. We also have $C_{G/O_{p'}(G)}(F) = 1$ by Lemma 1.1. By the minimality of our counterexample G, we must have $O_{p'} = 1$, so that $F(G) = O_p(G)$.

Since $C_{G/F(G)}(F) = 1$ by Lemma 1.1, by the minimality we have

$$F(C_{G/F(G)}(H)) \leqslant F_2(G)/F(G),$$

and therefore $F(C_G(H)) \leqslant F_2(G)$ by (11). Therefore, by the minimality, we have $G = F_2(G)$ in view of (12). Since $F(G) = O_p(G)$, the quotient $G/F(G) = F_2(G)/F(G)$ is a nilpotent p'-group. Therefore there is a prime $q \neq p$ dividing $|F(C_G(H))|$. Let $Q_0 = O_q(C_G(H))$. Since the quotient $G/O_p(G)$ is nilpotent, $C_G(H)/O_p(C_G(H))$ is also nilpotent. Therefore the Hall q' subgroup of $C_G(H)$ centralizes Q_0. Consequently, $Q_0 \cap Z(C_G(H))) \neq 1$. We choose a nontrivial q-element $c \in Q_0 \cap Z(C_G(H)))$.

Consider $\langle c^{HF}\rangle = \langle c^F\rangle$, the minimal FH-invariant subgroup containing c. The subgroup $N = O_p(G)\langle c^F\rangle$ is also FH-invariant, and together with the (possibly not faithful) group of automorphisms FH is also a counterexample: $c \in F(C_N(H))$, since $c \in F(C_G(H))$, but $c \notin F(N) = O_p(G)$ (the latter equality holds because $F(G) = O_p(G)$ contains its centralizer). Therefore, by the minimality we have $G = O_p(G)\langle c^F\rangle$. Let Q be an FH-invariant Sylow q-subgroup of G, which exists by Lemma 2.1. Since all q-elements of G are contained in $O_p(G)Q$, we have $G = O_p(G)\langle c^F\rangle = O_p(G)Q$. It is easy to see that $C_G(H) = C_{O_p(G)}(H)C_Q(H)$; therefore $C_Q(H)$ is a

Sylow q-subgroup of $C_G(H)$. Hence, $Q_0 = O_q(C_G(H)) \leqslant C_Q(H)$. In particular, $c \in Q$ and we must have the equality

$$Q = \langle c^F \rangle. \tag{13}$$

Since c acts faithfully on $O_p(G)/\Phi(O_p(G))$, in view of (11) and Lemma 1.1 by the minimality we must also have $\Phi(O_p(G)) = 1$, so that $V = O_p(G) = F(G)$ is an elementary abelian p-group, on which the quotient $G/V \cong Q$ acts faithfully. Since c acts non-trivially on the QFH-invariant subgroup $[V, Q]$, by similar minimality argument we must have $V = [V, Q]$, so that $C_V(Q) = 1$.

Note that c centralizes $C_V(H)$ but acts nontrivially on V. Our aim is a contradiction following from these assumptions.

We regard V as an $\mathbb{F}_p QFH$-module. At the same time we reserve the right to regard V as a normal subgroup of the semidirect product $VQFH$. For example, we may use the commutator notation: the subgroup $[V, Q] = \langle [v, g] \mid v \in V,\ g \in Q \rangle$ coincides with the subspace spanned by $\{-v + vg \mid v \in V,\ g \in Q\}$. We also keep using the centralizer notation for fixed points, like $C_V(H) = \{v \in V \mid vh = v \text{ for all } h \in H\}$, and for kernels, like $C_Q(Y) = \{x \in Q \mid yx = y \text{ for all } y \in Y\}$ for a subset $Y \subseteq V$.

We now extend the ground field to a finite field k that is a splitting field for QFH and obtain a $kQFH$-module $\widetilde{V} = V \otimes_{\mathbb{F}_p} k$. Many of the above-mentioned properties of V are inherited by $\widetilde{V}$:

(V1) $\widetilde{V}$ is a faithful kQ-module;
(V2) $C_{\widetilde{V}}(Q) = 0$;
(V3) c acts trivially on $C_{\widetilde{V}}(H)$;
(V4) $C_{\widetilde{V}}(F) = 0$;

Our aim is to show that c centralizes $\widetilde{V}$, which will contradict (V1).

Consider an unrefinable series of $kQFH$-submodules

$$\widetilde{V} = V_1 > V_2 > \cdots > V_n > V_{n+1} = 0. \tag{14}$$

Let W be one of the factors of this series; it is a nontrivial irreducible $kQFH$-module. If c acts trivially on every such W, then c acts trivially on $\widetilde{V}$, as the order of c is coprime to the characteristic p of the field k — this contradicts (V1). Therefore in what follows we assume that c acts nontrivially on W.

The following properties hold for W:

(W1) c acts nontrivially on W;
(W2) $C_W(Q) = 0$;

(W3) c acts trivially on $C_W(H)$;
(W4) $C_W(F) = 0$;
(W5) W is a free kH-module.

Indeed, property (W1) has already been mentioned. Property (W2) follows from (V2) since the order of Q is coprime to the characteristic of the field k. Property (W3) follows from (V3) since $C_{\widetilde{V}}(H)$ covers $C_W(H)$. Property (W4) follows from (V4) by Lemma 1.1 as we can still regard $\widetilde{V}$ as a finite (additive) group on which F acts fixed-point-freely by property (V4). Property (W5) follows from (W4) by Lemma 1.2.

We shall need the following simple remark.

Lemma 3.1. *Let FH be a Frobenius group with kernel F and complement H. In any action of FH with nontrivial action of F the complement H acts faithfully.*

Proof. Indeed, the action kernel K that does not contain F must intersect H trivially: otherwise $K \cap H$ acts trivially on $F/(K \cap F) \neq 1$ and therefore $K \cap H$ has nontrivial fixed points on F, since the action of K on F is coprime. □

The following lemma will be used repeatedly in the proof.

Lemma 3.2. *Suppose that $M = \bigoplus_{h \in H} M_h$ is a free kH-submodule of W, that is, the subspaces M_h form a regular H-orbit: $M_{h_1}h_2 = M_{h_1h_2}$ for $h_1, h_2 \in H$. If the element c leaves invariant each of the M_h, then c acts trivially on M.*

Proof. The fixed points of H in M are the diagonal elements $\sum_{h \in H} mh$ for any m in M_1, where $mh \in M_h$. Since c acts trivially on every such sum by property (W3) and leaves invariant every direct summand M_h, it must act trivially on each mh. Clearly, the elements mh run over all elements in all the summands $M_h = M_1 h$, $h \in H$. □

By applying Clifford's theorem we consider the decomposition

$$W = W_1 \oplus \cdots \oplus W_t$$

of W into the direct sum of the Wedderburn components W_i with respect to Q. We consider the transitive action of FH on the set $\Omega = \{W_1, \ldots, W_t\}$.

Lemma 3.3. *The element c acts trivially on the sum of components in any regular H-orbit in Ω.*

Proof. This follows from Lemma 3.2, because the sum of components in a regular H-orbit in Ω is obviously a free kH-submodule. □

Note that H transitively permutes the F-orbits in Ω. Let $\Omega_1 = W_1^F$ be one of these F-orbits and let H_1 be the stabilizer of Ω_1 in H in the action of H on F-orbits. If $H_1 = 1$, then all the H-orbits in Ω are regular, and then c acts trivially on W by Lemma 3.3. This contradicts our assumption (W1) that c acts nontrivially on W. Thus, we assume that $H_1 \neq 1$.

Lemma 3.4. *The subgroup H_1 has exactly one non-regular orbit in Ω_1 and this orbit is a fixed point.*

Proof. Let $\overline{F}$ be the image of F in its action on Ω_1. If $\overline{F} = 1$, then $\Omega_1 = \{W_1\}$ consists of a single Wedderburn component, which is F-invariant. On this component the quotient Q/K by the corresponding kernel $K = C_Q(W_1)$ acts with its centre $Z(Q/K)$ represented by scalar linear transformations. This centre is nontrivial, since Q/K is nilpotent and nontrivial, as $C_W(Q) = 0$ by property (W2). But then F acts trivially on $Z(Q/K)$, which contradicts Lemma 1.1. Thus, $\overline{F} \neq 1$, and $\overline{F}H_1$ is a Frobenius group with complement H_1. By Lemma 3.1 the subgroup H_1 acts faithfully on Ω_1 and we use the same symbol for it in regard of its action on Ω_1.

Let S be the stabilizer of the point $W_1 \in \Omega_1$ in $\overline{F}H_1$. Since $|\Omega_1| = |\overline{F} : \overline{F} \cap S| = |\overline{F}H_1 : S|$ and the orders $|\overline{F}|$ and $|H_1|$ are coprime, S contains a conjugate of H_1; without loss of generality (changing W_1 and therefore S if necessary) we assume that $H_1 \leqslant S$. Any other stabilizer of a point is equal to S^f for $f \in \overline{F} \setminus S$. We already have a fixed point W_1 for H_1. To prove that all other H_1-orbits are regular, we need to show that $S^f \cap H_1 = 1$, which is the same as $S \cap H_1^{f^{-1}} = 1$. But all the conjugates H_1^x for $x \in \overline{F}$ are distinct and have trivial pairwise intersections, while their union contains all the elements of $\overline{F}H_1$ of orders dividing $|H_1|$; the same is true for the conjugates of H_1 is S. Therefore the only conjugates of H_1 intersecting S are H_1^s for $s \in S$. □

We now consider the H-orbits in Ω. Clearly, the H-orbits of elements of regular H_1-orbits in Ω_1 are regular H-orbits. Thus, by Lemma 3.4 there is exactly one non-regular H-orbit in Ω — the H-orbit of the fixed point W_1 of H_1 in Ω_1. Therefore by Lemma 3.3 the element c acts trivially on all the Wedderburn components W_i that are not contained in the H-orbit of W_1. Since c acts nontrivially on W, it must be nontrivial on the sum over the

H-orbit of W_1. Moreover, since c commutes with H, the element c acts in the same way — and therefore nontrivially — on all the components in the H-orbit of W_1. In particular, c is nontrivial on W_1.

At this point the proofs of Theorems 3.1 and 3.2 diverge.

4. Completion of the proofs

Completion of the proof of Theorem 3.2

Recall that in this theorem on Fitting height we have the condition $(|G|, |H|) = 1$, which is assumed in this subsection; note also that this condition was preserved by all the reductions in the preceding section, so we can use its results. We employ induction on $|H|$ (which can be viewed for definiteness as 'enveloping' induction with respect to the secondary induction on the order of GFH). In the basis of this induction, $|H|$ is a prime, and either $H_1 = 1$, which gives a contradiction as described above, or $H_1 = H$, which is the case dealt with below.

First suppose that $H_1 \neq H$. Then we consider $U = \bigoplus_{f\in F} W_1 f$, the sum of components in Ω_1, which is a $kQFH_1$-module.

Lemma 4.1. *The element c acts trivially on $C_U(H_1)$.*

Proof. Indeed, c acts trivially on the sum over any regular H_1-orbit, because such an H_1-orbit is a part of a regular H-orbit, on the sum over which c acts trivially by Lemma 3.3. By Lemma 3.4 it remains to show that c is trivial on $C_{W_1}(H_1)$. For $x \in C_{W_1}(H_1)$ and some right transversal $\{h_i \mid 1 \leqslant i \leqslant |H : H_1|\}$ of H_1 in H we have $\sum_i xh_i \in C(H)$. Indeed, for any $h \in H$ we have $\sum_i xh_ih = \sum_i xh_{1i}h_{j(i)}$ for some $h_{1i} \in H_1$ and some permutation of the same transversal $\{h_{j(i)} \mid 1 \leqslant i \leqslant |H : H_1|\}$, and the latter sum is equal to $\sum_i xh_{j(i)} = \sum_i xh_i$ as $xh_{1i} = x$ for all i. Since c acts trivially on $\sum_i xh_i \in C(H)$ by property (W3), it must also act trivially on each summand, as they are in different c-invariant components; in particular, $xc = x$. □

In order to use induction on $|H|$, we consider the additive group U on which the group QFH_1 acts as a group of automorphisms. The action of Q and F may not be faithful, but the action of H_1 is faithful by Lemma 3.1, because FH_1 is a Frobenius group and the action of F is nontrivial since $C_U(F) = 0$ by property (W4). Switching to multiplicative notation also for the additive group of U, we now have the semidirect product $G_1 = UQ$ admitting the Frobenius group of automorphisms $(F/C_F(G_1))H_1$ such that

$C_{G_1}(F/C_F(G_1)) = 1$. By Lemma 4.1 we obtain that $c \in F(C_{G_1}(H_1))$, because $C_{G_1}(H_1) = C_U(H_1)C_Q(H_1)$ is a $\{p,q\}$-group, in which $C_U(H_1)$ is a normal p-subgroup centralized by the q-element c. If $H_1 \neq H$, then by induction on $|H|$ we must have $c \in F(G_1)$, which implies that c acts trivially on U. This contradicts the fact that c acts nontrivially on $W_1 \subseteq U$.

Thus, it remains to consider the case where $H_1 = H$, that is, W_1 is H-invariant.

We now focus on the action on W_1, using bars to denote the images of Q and its elements in their action on W_1. We can of course regard H as a group of automorphisms of $\overline{Q}$ (even faithful by Lemma 3.1, though we do not need this fact here). Let $\zeta_2(\overline{Q})$ be the second centre of $\overline{Q}$. We obviously have

$$[[H, \bar{c}], \zeta_2(\overline{Q})] = [1, \zeta_2(\overline{Q})] = 1.$$

We also have

$$[[\bar{c}, \zeta_2(\overline{Q})], H] \leqslant [Z(\overline{Q}), H] = 1,$$

since $Z(\overline{Q})$ is represented on the homogeneous $k\overline{Q}$-module W_1 by scalar linear transformations. Therefore by the Three Subgroup Lemma,

$$[[\zeta_2(\overline{Q}), H], \bar{c}] = 1. \tag{15}$$

By the choice of $c \in Z(C_G(H))$ we also have

$$[C_{\zeta_2(\overline{Q})}(H), \bar{c}] = 1, \tag{16}$$

because $C_Q(H)$ covers $C_{\overline{Q}}(H)$ by Theorem 1.1. Since the action of H on Q is coprime by hypothesis, we have

$$\zeta_2(\overline{Q}) = [\zeta_2(\overline{Q}), H]C_{\zeta_2(\overline{Q})}(H).$$

Therefore equalities (15) and (16) together imply the equality

$$[\bar{c}, \zeta_2(\overline{Q})] = 1. \tag{17}$$

Now let F_1 denote the stabilizer of W_1 in F, so that the stabilizer of W_1 in FH is equal to F_1H. Then for any element $f \in F \setminus F_1$ the component W_1^f is outside the H-orbit of W_1, which is equal to $\{W_1\}$ in the case under consideration. As mentioned above, c acts trivially on all the Wedderburn components outside the H-orbit of W_1. Thus, c acts trivially on W_1^f, which is equivalent to $c^{f^{-1}}$ acting trivially on W_1. In other words, $\overline{c^x} = 1$ in the action on W_1 for any $x \in F \setminus F_1$. (Note that it does not matter that W_1 is not F-invariant: for any $g \in F$ the element c^g belongs to Q, which acts on W_1.) Since $Q = \langle c^F \rangle$ by (13), we obtain that $\overline{Q} = \langle \bar{c}^{F_1} \rangle$. In view of the

F_1-invariance of the section $\overline{Q}$ we can apply conjugation by any $g \in F_1$ to equation (17) to obtain that

$$[\bar{c}^g, \zeta_2(\overline{Q})^g] = [\bar{c}^g, \zeta_2(\overline{Q})] = 1.$$

As a result,

$$[\overline{Q}, \zeta_2(\overline{Q})] = [\langle \bar{c}^{F_1} \rangle, \zeta_2(\overline{Q})] = 1.$$

This means that $\overline{Q}$ is abelian.

The sum of the W_i over all regular H-orbits is obviously a free kH-module. Since the whole W is also a free kH-module by property (W5), the component W_1 must also be a free kH-module, as a complement of the sum over all regular H-orbits. Since $\overline{Q}$ is abelian, c acts on W_1 by a scalar linear transformation. By Lemma 3.2 it follows that, in fact, c must act trivially on W_1, a contradiction with property (W1). Part (a) is proved.

We now derive part (b) of Theorem 3.2. We already know that $F_i(C_G(H)) = F_i(G) \cap C_G(H)$ for all i. In order to obtain that the Fitting height of $C_G(H)$ is equal to the Fitting height of G, it remains to show that if $C_G(H) \leqslant F_n(G)$, then $F_n(G) = G$. Suppose the opposite, $F_n(G) \neq G$. By Lemmas 1.1 and 1.2 any elementary abelian FH-invariant section S of $G/F_n(G)$ is a free $\mathbb{F}_pH$-module (for some prime p), on which H has nontrivial 'diagonal' fixed points. Since $C_G(H)$ covers the fixed points of H in S, this contradicts the inclusion $C_G(H) \leqslant F_n(G)$.

Completion of the proof of Theorem 3.1

In this theorem we do not have any coprimeness restrictions, but instead we have $C_G(H)$ nilpotent. As noted at the beginning of § 3, the inclusion $F(C_G(H)) \leqslant F(G)$ in this case is equivalent to G being nilpotent. By the minimality of the group GFH, since $Z(Q)$ acts faithfully on V, we must have $G = VZ(Q)$, so that Q is abelian.

We now arrive at a contradiction in much the same fashion as above. The sum of the W_i over all regular H-orbits is obviously a free kH-module. Since the whole W is also a free kH-module by property (W5), the sum $Y = \sum_{h\in H} W_1 h$ of components in the H-orbit of W_1 must also be a free kH-module, as a complement of the sum over all regular H-orbits. Since Q is abelian, c acts on W_1 by a scalar linear transformation. Since c commutes with H, we obtain that c acts by the same scalar linear transformation on Y. By Lemma 3.2 it follows that, in fact, c must act trivially on Y, a contradiction with property (W1).

5. Further results on Frobenius groups of automorphisms with fixed-point-free kernel

Again, suppose that a finite group G admits a Frobenius group of automorphisms FH with complement H such that $C_G(F) = 1$. We have seen that, by application of Clifford's theorem, the problem of proving that the properties of G are close to those of $C_G(H)$ is now largely reduced to the case of nilpotent group G. In this area different Lie ring methods are applied to obtain bounds for the nilpotency class and exponent of G under certain additional conditions (some of which are unavoidable). Here we briefly list the recent results in these directions.

Makarenko and Shumyatsky[18] solved in the affirmative Mazurov's problem 17.72(a) from "Kourovka Notebook"[26] on double Frobenius groups: they proved that if in addition GF is also a Frobenius group (so that GFH is a double Frobenius group, with not necessarily coprime orders of G and H), then the nilpotency class of G is bounded in terms of $|H|$ and the nilpotency class of $C_G(H)$. (Earlier the same was proved in Ref. 10 in the case of abelian $C_G(H)$).

Furthermore, this result was extended in Ref. 13 to the case of metacyclic FH (without assuming that GF is also a Frobenius group): if $C_G(H)$ is nilpotent, then the nilpotency class of G is bounded in terms of $|H|$ and the nilpotency class of $C_G(H)$ (here G is necessarily nilpotent by Theorem 3.1).

The Lie ring methods developed in Refs. 10,13,17,18 also yield similar results bounding the nilpotency class of Lie rings and algebras with a metacyclic Frobenius group of automorphisms, with corollaries for connected Lie groups and torsion-free locally nilpotent groups with such groups of automorphisms.

Note that examples show that such nilpotency results are no longer true for non-metacyclic Frobenius groups of automorphisms.

Different Lie ring methods are used in Ref. 13 in the case of metacyclic FH for bounding the exponent of G in terms of $|FH|$ and the exponent of $C_G(H)$. It remains an open problem whether the dependence on $|F|$ can be dropped here, even in the case of double Frobenius group GFH (the latter is Mazurov's problem 17.72(b) in Ref. 26).

It is also unclear if the same results on exponent hold for arbitrary (non-metacyclic) Frobenius group FH, even if the exponent of G were allowed to depend on $|FH|$ (and on the exponent of $C_G(H)$). So far, only the smallest non-metacyclic Frobenius group FH of order 12 has been successfully tackled by Shumyatsky[20].

In the most recent paper Shumyatsky[21] proved that if FH is a metacyclic Frobenius group and $C_G(H)$ satisfies a positive law of degree k, then G satisfies a positive law of degree bounded in terms of k and $|FH|$. This theorem combines the above-mentioned results on exponent and nilpotency class, since a positive law of degree k for a finite group implies that it is an extension of a nilpotent group of k-bounded class by a group of k-bounded exponent, and conversely, an extension of a nilpotent group of class c by a group of exponent n satisfies a positive law of (c, n)-bounded degree.

References

1. S. D. Bell and B. Hartley, A note on fixed-point-free actions of finite groups, *Quart. J. Math. Oxford Ser. (2)* **41** (1990), no. 162, 127–130.
2. V. V. Belyaev and B. Hartley, Centralizers of finite nilpotent subgroups in locally finite groups, *Algebra Logika* **35** (1996), 389–410 (in Russian); English transl., *Algebra Logic* **35** (1996), 217–228.
3. A. H. Clifford, Representations induced in an invariant subgroup, *Ann. Math. (2)* **38** (1937), 533–550.
4. C. W. Curtis and I. Reiner, *Representation theory of finite groups and associative algebras*, Interscience, New York–London, 1962.
5. E. C. Dade, Carter subgroups and Fitting heights of finite solvable groups, *Illinois J. Math.* **13** (1969), 449–514.
6. L. Evens, On a theorem of Thompson on fixed points of groups acting on p-groups, *Math. Z.* **93** (1966) 105–108.
7. D. Gorenstein, *Finite groups*, Harper and Row, New York, 1968.
8. P. Hall and G. Higman, The p-length of a p-soluble group and reduction theorems for Burnside's problem, *Proc. London Math. Soc. (3)*, **6** (1956), 1–42.
9. E. I. Khukhro, Fixed points of p-automorphisms of finite p-groups, *Algebra Logika* **14** (1975), 697–703; English transl., *Algebra Logic* **14** (1976), 417–420.
10. E. I. Khukhro, Graded Lie rings with many commuting components and an application to 2-Frobenius groups, *Bull. London Math. Soc.* **40**, (2008), 907–912.
11. E. I. Khukhro, Fixed points of the complements of Frobenius groups of automorphisms, *Sibirsk. Mat. Zh.*, **51** (2010), 694–699 (in Russian); English transl., *Siberian Math. J.* **51** (2010), 554–558.
12. E. I. Khukhro, Nilpotent length of a finite group admitting a Frobenius group of automorphisms with fixed-point-free kernel, *submitted to Algebra Logika* (2010); English transl., *submitted to Algebra Logic* (2010).
13. E. I. Khukhro, N. Y. Makarenko, and P. Shumyatsky, Frobenius groups of automorphisms and their fixed points, *submitted*, 2010; `arxiv.org/abs/1010.0343`.
14. L. G. Kovács, On finite soluble groups, *Math. Z.* **103** (1968), 37–39.
15. H. Kurzweil, p-Automorphismen von auflösbaren p'-Gruppen, *Math. Z.* **120** (1971) 326–354.

16. P. Longobardi and M. Maj, On the number of generators of a finite group, *Arch. Math. (Basel)* **50** (1988), 110–112.
17. N. Yu. Makarenko, E. I. Khukhro, and P. Shumyatsky, Fixed points of Frobenius groups of automorphisms, *Dokl. Akad. Nauk, to appear*, 2011.
18. N. Y. Makarenko and P. Shumyatsky, Frobenius groups as groups of automorphisms, *Proc. Amer. Math. Soc.* **138** (2010), 3425–3436.
19. D. F. S. Robinson, *A course in the theory of groups*, Springer-Verlag, New York–Heidelberg–Berlin, 1982.
20. P. Shumyatsky, On the exponent of a finite group with an automorphism group of order twelve, *to appear in J. Algebra*, 2011.
21. P. Shumyatsky, Positive laws in fixed points of automorphisms of finite groups, *Preprint*, Univ. of Brasilia, 2010.
22. J. G. Thompson, Finite groups with fixed-point-free automorphisms of prime order, *Proc. Nat. Acad. Sci. USA* **45** (1959), 578–581.
23. J. Thompson, Automorphisms of solvable groups, *J. Algebra* **1** (1964), 259–267.
24. J. G. Thompson, Fixed points of p-groups acting on p-groups, *Math. Z.* **86** (1964), 12–13.
25. A. Turull, Fitting height of groups and of fixed points, *J. Algebra* **86** (1984), 555–566.
26. *Unsolved Problems in Group Theory. The Kourovka Notebook*, no. 17, Institute of Mathematics, Novosibirsk, 2010.

INDUCING π-PARTIAL CHARACTERS WITH A GIVEN VERTEX

M. L. LEWIS

Department of Mathematical Sciences, Kent State University
Kent, Ohio 44242
E-mail: lewis@math.kent.edu

Let G be a solvable group. Let p be a prime and let Q be a p-subgroup of a subgroup V. Suppose $\varphi \in \mathrm{IBr}(G)$. If either $|G|$ is odd or $p = 2$, we prove that the number of Brauer characters of H inducing φ with vertex Q is at most $|\mathrm{N}_G(Q) : \mathrm{N}_V(Q)|$.

Keywords: Brauer characters; Partial characters; Vertices.

1. Introduction

Throughout this note, G is a finite group, and $\mathrm{Irr}(G)$ is the set of irreducible characters of G. Suppose $\chi \in \mathrm{Irr}(G)$ and H is a subgroup of G. It is easy to obtain an upper bound on the number of characters in $\mathrm{Irr}(H)$ that induce χ. Let $\varphi_1, \ldots, \varphi_n \in \mathrm{Irr}(H)$ be the characters so that $\varphi_i^G = \chi$. Evaluating at 1, we obtain $\varphi_i(1) = \chi(1)/|G : H|$ for each i. By Frobenius reciprocity (Lemma 5.2 of Ref. 5), each φ_i is a constituent of χ_H with multiplicity 1. Since there are n such characters occurring as constituents of χ_H, it follows that $n(\chi(1)/|G : H|) \leq \chi(1)$. We deduce that $n \leq |G : H|$, and we have an upper bound. If H is normal in G, this bound is obtained, and it is not particularly difficult to find nonnormal subgroups where this bound is obtained.

We now turn our attention to Brauer characters. Fix a prime p. We will write $\mathrm{IBr}(G)$ for the irreducible p-Brauer characters of G. If $\varphi \in \mathrm{IBr}(G)$, then it is easy to adapt the above proof to show that φ is induced by at most $|G : H|$ Brauer characters of H. However, associated with φ are certain p-subgroups of G called the vertex subgroups. When G is a p-solvable group, a p-subgroup Q of G is defined to be a vertex for φ if there is a subgroup U of G so that φ is induced by a Brauer character of U with p'-degree and Q is a Sylow subgroup of U. It is known that all the vertex subgroups of

φ are conjugate in G. If φ is induced from $\tau \in \mathrm{IBr}(H)$, it is easy to see that a vertex for τ is a vertex for φ. Thus, H contains some vertex Q for φ. Now, different Brauer characters of H that induce φ may have vertex subgroups that are not conjugate in H but are necessarily conjugate in G. Hence, one can ask the following question: Suppose $\varphi \in \mathrm{IBr}(G)$ has vertex Q, and $Q \leq H$, how many characters in $\mathrm{IBr}(H)$ with vertex Q induce φ? When either $|G|$ is odd or G is solvable and $p = 2$, we can obtain an upper bound for this question.

Theorem 1.1. *Let G be a solvable group and p a prime. Assume either $|G|$ is odd or $p = 2$. Let Q be a p-subgroup of H. If $\varphi \in \mathrm{IBr}(G)$, then the number of Brauer characters of H with vertex Q that induce φ is at most $|\mathrm{N}_G(Q) : \mathrm{N}_H(Q)|$.*

At this time, we are not able to determine whether or not this theorem is true if we loosen the hypothesis that either $|G|$ is odd or $p = 2$. In other words, is the conclusion still true if G is a solvable group of even order and p is an odd prime. This result was motivated by our work with J. P. Cossey. If we could prove the conclusion of Theorem 1.1 when p is odd, then we would be able to prove J. P. Cossey's conjecture that the number of lifts of a Brauer character is bounded by the index of a vertex subgroup in the vertex subgroup when p is odd. Our argument can be found in the preprint Ref. 1.

We would like to thank J. P. Cossey and I. M. Isaacs for several helpful discussions while we were preparing this note.

2. Results

We will in the more general setting of irreducible π-partial characters of a π-separable group G. We here briefly mention that if π is a set of primes and G is a π-separable group, one can define (see Ref. 10 for more details) a set of class functions $\mathrm{I}_\pi(G)$ from the set G^o (which consists of the elements of G whose order is divisible by only the primes in π) to $\mathbf{C}$ that plays the role of $\mathrm{IBr}(G)$, and in fact $\mathrm{I}_\pi(G) = \mathrm{IBr}(G)$ if $\pi = \{p'\}$, the complement of the prime p.

We start by considering vertices in Clifford correspondence (see Proposition 3.2 of Ref. 9). Let G be a π-separable group. Let N be a normal subgroup of G. Fix $\varphi \in \mathrm{I}_\pi(G)$. If $\alpha \in \mathrm{I}_\pi(N)$ is a constituent of φ_N, then we write G_α for the stabilizer of α in G, and we write φ_α for the Clifford correspondent of φ with respect to α. In particular, the vertices of the Clifford correspondent form an orbit under the action of the normalizer of a particular vertex.

Lemma 2.1. *Let G be a π-separable group. Let N be a normal subgroup of G. Suppose that $\alpha \in \mathrm{I}_\pi(N)$. Let $\varphi \in \mathrm{I}_\pi(G)$ and $\varphi_\alpha \in \mathrm{I}_\pi(G_\alpha)$ so that $\varphi_\alpha^G = \varphi$. Suppose that Q is a vertex for φ_α. Then Q is a vertex for φ_α^g if and only if there exists $n \in \mathrm{N}_G(Q)$ so that $G_\alpha g = G_\alpha n$.*

Proof. We first suppose that there exists $n \in \mathrm{N}_G(Q)$ so that $G_\alpha g = G_\alpha n$. Thus, $g = tn$ for some $t \in G_\alpha$. We see that $\varphi_\alpha^g = \varphi_\alpha^{tn} = \varphi_\alpha^n$. We see that $Q = Q^n$ is a vertex for $\varphi_\alpha^n = \varphi_\alpha^g$.

Conversely, suppose that Q is a vertex for φ_α^g. Then $Q^{g^{-1}}$ is a vertex for φ_α. Since Q is also a vertex for φ_α, we have $Q^{g^{-1}} = Q^t$ for some $t \in G_\alpha$. It follows that $Q = Q^{tg}$, and so, $tg \in \mathrm{N}_G(Q)$. This implies that $tg = n$ for some $n \in \mathrm{N}_G(Q)$. This implies that $n \in G_\alpha g$, and we conclude that $G_\alpha n = G_\alpha g$. □

We continue to work in the context of the Clifford correspondence. In this case, we can get an exact count of the number of partial characters in N whose Clifford correspondent has vertex Q.

Corollary 2.1. *Let G be a π-separable group. Let N be a normal subgroup of G, let $\varphi \in \mathrm{I}_\pi(G)$ have vertex Q, and suppose that β is an irreducible constituent of φ_N so that φ_β has vertex Q. Then $|\{\alpha \in \mathrm{I}_\pi(N) \mid$* φ_α has vertex $Q\}| = |\mathrm{N}_G(Q) : \mathrm{N}_{G_\beta}(Q)|$.

Proof. By Lemma 2.1, we see that φ_α has Q as a vertex if and only if $\alpha = \beta^g$ where $g \in G$ and $g \in G_\beta n$ for some $n \in \mathrm{N}_G(Q)$. Finally, we observe that $G_\beta n_1 = G_\beta n_2$ if and only if $\mathrm{N}_{G_\beta}(Q)n_1 = \mathrm{N}_{G_\beta}(Q)n_2$ for $n_1, n_2 \in \mathrm{N}_G(Q)$. We have $|\{\alpha \in \mathrm{I}_\pi(N) \mid \varphi_\alpha \text{ has vertex } Q\}| = |\{G_\beta n \mid n \in \mathrm{N}_G(Q)\}| = |\mathrm{N}_G(Q) : \mathrm{N}_{G_\beta}(Q)|$. □

We now look at the conditions of a minimal counterexample. For this we need to review and develop more notation. We make use of the canonical set of π-lifts, $\mathrm{B}_\pi(G)$, that was defined in Ref. 8 by Isaacs. In other words, $\mathrm{B}_\pi(G) \subseteq \mathrm{Irr}(G)$ and the map $\chi \mapsto \chi^o$ is a bijection from $\mathrm{B}_\pi(G)$ to $\mathrm{I}_\pi(G)$. Closely related to this set is the subnormal nucleus which also was defined in Ref. 8. To define the subnormal nucleus, we need the π-special characters defined in Ref. 3. Let G be a π-separable group. A character $\chi \in \mathrm{Irr}(G)$ is π-special if $\chi(1)$ is a π-number and for every subnormal group M of G, the irreducible constituents of χ_M have determinants that have π-order. Many of the basic results of π-special characters can be found in Section 40 of Ref. 2 and Chapter VI of Ref. 13. One result that is proved is that if α is π-special and β is π'-special, then $\alpha\beta$ is necessarily irreducible. We say that χ is **factored** if $\chi = \alpha\beta$ where α is π-special and β is π'-special. We also note that if $\chi \in \mathrm{B}_\pi(G)$ and N is normal in G, then the irreducible constituents of χ_N lie in $\mathrm{B}_\pi(N)$.

If $\chi \in \mathrm{Irr}(G)$, Isaacs constructs the subnormal vertex as follows. Let M be maximal so that M is subnormal in G and the irreducible constituents of χ_M are factored. Let μ be an irreducible constituent of χ_M and let T be the stabilizer of (M, μ) in G. Isaacs proved in Ref. 8 that there is a Clifford theorem for T. In other words, there is a unique character $\tau \in \mathrm{Irr}(T \mid \mu)$ so that $\tau^G = \chi$. He also proved that (M, μ) is unique up to conjugacy, and so, (T, τ) is unique up to conjugacy. If $T = G$, then χ is π-factored and we take (G, χ) to be the subnormal nucleus of χ. If $T < G$, then inductively, the subnormal nucleus for τ is the subnormal nucleus for χ. We write (W, γ) for the subnormal nucleus of χ, and Isaacs showed that $\gamma^G = \chi$, γ is factored, and (W, γ) is unique up to conjugacy. A character $\chi \in \mathrm{Irr}(G)$ is in $\mathrm{B}_\pi(G)$ if and only if the character of its nucleus is π-special.

If Q is a π'-subgroup of G, then we use $\mathrm{I}_\pi(G \mid Q)$ to denote the π-partial characters in $\mathrm{I}_\pi(G)$ that have vertex Q. If $\varphi \in \mathrm{I}_\pi(G)$ and $V \leq G$, then we write $\mathrm{I}_\varphi(V \mid Q) = \{\eta \in \mathrm{I}_\pi(V \mid Q) \mid \eta^G = \varphi\}$. We now find details about properties of a minimal counterexample. We will see that a counterexample cannot occur when either $|G|$ is odd or 2 is not in π. Our goal is find enough information so that we can either find a contradiction or build an example when $|G|$ is even and $2 \in \pi$.

Theorem 2.1. *Let G be a solvable group. Assume $\varphi \in \mathrm{I}_\pi(G)$ has vertex Q, let V be a subgroup of G, and let N be the core of V in G. If G and V are chosen so that $|G| + |G : V|$ is minimal subject to the condition that $|\mathrm{I}_\varphi(V \mid Q)| > |\mathrm{N}_G(Q) : \mathrm{N}_V(Q)|$, then the following are true:*

(1) V is a nonnormal maximal subgroup of G,
(2) $|G : V|$ is a power of 2,
(3) $2 \in \pi$,
(4) $Q \leq V$,
(5) $\varphi_N = a\alpha$ for some $\alpha \in \mathrm{IBr}(N)$,
(6) $\alpha(1)$ is a π-number,
(7) if K is normal in G so that K/N is a chief factor for G, then α is fully ramified with respect to K/N.

Proof. If either $V = G$ or $\mathrm{I}_\varphi(V \mid Q)$ is empty, then we have

$$|\mathrm{I}_\varphi(V \mid Q)| \leq |\mathrm{N}_G(Q) : \mathrm{N}_V(Q)|,$$

contradicting the hypotheses. Thus, $V < G$ and $\mathrm{I}_\varphi(V \mid Q)$ is not empty, and so, $Q \leq V$ and there exist characters in $\mathrm{I}_\pi(V)$ that induce φ and have vertex Q.

We begin by showing that V is a maximal subgroup. Suppose that $V < M < G$ for some subgroup M. Let $\mathrm{I}_\varphi(M \mid Q) = \{\eta_1, \ldots, \eta_m\}$. Using minimality, we have $m = |\mathrm{I}_\varphi(M \mid Q)| \leq |\mathrm{N}_G(Q) : \mathrm{N}_M(Q)|$. Suppose that $\zeta \in \mathrm{I}_\varphi(V \mid Q)$, then $\zeta^M \in \mathrm{I}_\pi(M)$ and ζ^M has Q as a vertex. Since $(\zeta^M)^G = \zeta^G = \varphi$, we see that $\zeta^M \in \mathrm{I}_\varphi(M \mid Q)$. It follows that $\zeta^M = \eta_i$ for some i. We conclude that $|\mathrm{I}_\varphi(V \mid Q)| \leq \sum_{i=1}^{m} |\mathrm{I}_{\eta_i}(V \mid Q)|$. Since this contradicts our hypothesis, we obtain $|\mathrm{I}_{\eta_i}(V \mid Q)| \leq |\mathrm{N}_M(Q) : \mathrm{N}_V(Q)|$. We deduce that

$$|\mathrm{I}_\varphi(V \mid Q)| \leq m|\mathrm{N}_M(Q) : \mathrm{N}_V(Q)| \leq |\mathrm{N}_G(Q) : \mathrm{N}_M(Q)||\mathrm{N}_M(Q) : \mathrm{N}_V(Q)|$$

Since $|\mathrm{N}_G(Q) : \mathrm{N}_M(Q)||\mathrm{N}_M(Q) : \mathrm{N}_V(Q)| = |\mathrm{N}_G(Q) : \mathrm{N}_V(Q)|$, we obtain $|\mathrm{I}_\varphi(V \mid Q)| \leq |\mathrm{N}_G(Q) : \mathrm{N}_V(Q)|$, and as this violates the hypotheses, V is maximal in G.

If V is normal in G, then either φ is induced from V or φ restricts irreducibly to V. If φ is induced from V, then we can apply Corollary 2.1 to see that $|\mathrm{I}_\varphi(V \mid Q)| \leq |\mathrm{N}_G(Q) : \mathrm{N}_V(Q)|$ in violation of the hypotheses. If φ restricts irreducibly, then it cannot be induced from V, and we have seen that this is also a contradiction. We conclude that V is not normal in G.

Suppose $\alpha \in \mathrm{I}_\pi(N)$ is a constituent of φ_N. We use $\varphi_\alpha \in \mathrm{I}_\pi(G_\alpha \mid \alpha)$ to denote the Clifford correspondent for φ with respect to α (see Proposition 3.2 of Ref. 9 again). Write $\{\alpha \in \mathrm{I}_\pi(N) \mid \varphi_\alpha \text{ has vertex } Q\} = \{\alpha_1, \ldots, \alpha_k\}$, and let $\varphi_i = \varphi_{\alpha_i}$ and $G_i = G_{\alpha_i}$. By Corollary 2.1, we know that $k = |\mathrm{N}_G(Q) : \mathrm{N}_{G_i}(Q)|$.

Fix a character $\eta \in \mathrm{I}_\varphi(V \mid Q)$. Enumerate

$$\{\beta \in \mathrm{I}_\pi(N) \mid \eta_\beta \text{ has vertex } Q\} = \{\beta_1, \ldots, \beta_l\},$$

and let $\eta_j = \eta_{\beta_j}$ and $V_j = V_{\beta_j}$. By Corollary 2.1, $l = |\mathrm{N}_V(Q) : \mathrm{N}_{V_i}(Q)|$.

We see that $(\eta_j)^G = ((\eta_j)^V)^G = \eta^G = \varphi$. This implies that $(\eta_j)^{G_{\beta_j}}$ is irreducible and has vertex Q. It follows that $\beta_j = \alpha_{i_j}$ for some i_j. We obtain $G_{\beta_j} = G_{i_j}$ and $(\beta_j)^{G_{i_j}} = \alpha_{i_j}$. Observe that $V_j = G_{i_j} \cap V$, and we denote this subgroup by $V_{i_j}^*$.

Now, we assume that $k > 1$, and we start to count. We see that the character $\eta \in \mathrm{I}_\varphi(G \mid Q)$ is induced by $|\mathrm{N}_V(Q) : \mathrm{N}_{V_{i_j}^*}(Q)|$ partial characters in $\bigcup \mathrm{I}_{\varphi_i}(V_i^* \mid Q)$. Because $G_i < G$, we may use minimality of $|G| + |G : V|$ to deduce $|\mathrm{I}_{\varphi_i}(V_i^* \mid Q)| \leq |\mathrm{N}_{G_i}(Q) : \mathrm{N}_{V_i^*}(Q)|$. We compute

$$|\mathrm{I}_\varphi(V \mid Q)| = \sum_{i=1}^{k} \frac{|\mathrm{I}_{\varphi_i}(V_i^* \mid Q)|}{|\mathrm{N}_V(Q) : \mathrm{N}_{V_i^*}(Q)|} \leq \sum_{i=1}^{k} \frac{|\mathrm{N}_{G_i}(Q) : \mathrm{N}_{V_i^*}(Q)|}{|\mathrm{N}_V(Q) : \mathrm{N}_{V_i^*}(Q)|}.$$

We determine that

$$\frac{1}{|\mathrm{N}_V(Q) : \mathrm{N}_{V_i^*}(Q)|} |\mathrm{N}_{G_i}(Q) : \mathrm{N}_{V_i^*}(Q)| = \frac{|\mathrm{N}_{G_i}(Q)|}{|\mathrm{N}_V(Q)|},$$

for each i. Notice that $|\mathrm{N}_{G_i}(Q)| = |\mathrm{N}_{G_1}(Q)|$ for all i and $k = |\mathrm{N}_G(Q) : \mathrm{N}_{G_1}(Q)|$. This yields

$$|\mathrm{I}_\varphi(V \mid Q)| \leq \sum_{i=1}^{k} \frac{|\mathrm{N}_{G_1}(Q)}{\mathrm{N}_V(Q)}| = \frac{|\mathrm{N}_G(Q) : \mathrm{N}_{G_1}(Q)||\mathrm{N}_{G_1}(Q)|}{|\mathrm{N}_V(Q)|}.$$

We now have $|\mathrm{I}_\varphi(V \mid Q)| \leq |\mathrm{N}_G(Q) : \mathrm{N}_V(Q)|$. This contradicts the hypothesis. We deduce that $k = 1$, and α is invariant in G.

Set $\alpha = \alpha_1$, and let α^* be the character in $\mathrm{B}_\pi(N)$ satisfying $(\alpha^*)^o = \alpha$. Write $(W, \hat{\alpha})$ for the nucleus of α^*, and take T to be the stabilizer of $(W, \hat{\alpha})$ in G. By Lemma 2.3 of Ref. 11, there is a unique character $\hat{\varphi} \in \mathrm{I}(T \mid \hat{\alpha})$ so that $\hat{\varphi}^G = \varphi$ and Q is a vertex for $\hat{\varphi}$. Similarly, if $\eta \in \mathrm{I}_\varphi(V \mid Q)$, then there is a unique character $\hat{\eta} \in \mathrm{I}(T \cap V \mid \hat{\alpha})$ so that $\hat{\eta}^V = \eta$ and Q is a vertex for $\hat{\eta}$. Observe that $\hat{\eta}^T \in \mathrm{I}(T \mid \hat{\alpha})$ and induces φ, so $\hat{\eta}^T = \hat{\varphi}$. It follows that $|\mathrm{I}_\varphi(V \mid Q)| = |\mathrm{I}_{\hat{\varphi}}(T \cap V \mid Q)|$. If $T < G$, then we can use the minimality of $|G| + |G : V|$ to see that $|\mathrm{I}_{\hat{\varphi}}(T \cap V \mid Q)| \leq |\mathrm{N}_T(Q) : \mathrm{N}_{V \cap T}(Q)|$. By the diamond lemma, we have $|\mathrm{N}_T(Q) : \mathrm{N}_{V \cap T}(Q)| = |\mathrm{N}_T(Q) : V \cap \mathrm{N}_T(Q)| \leq |\mathrm{N}_G(Q) : \mathrm{N}_V(Q)|$. This contradicts the hypotheses, and so $T = G$.

We now have that $(W, \hat{\alpha})$ is G-invariant. By the construction of the subnormal, this implies that $W = N$. Since $\alpha^* \in \mathrm{B}_\pi(N)$, the nucleus for α^* has a character that is π-special. Thus, $\hat{\alpha}$ is π-special, and since $W = N$, we see that $\hat{\alpha} = \alpha^*$. In particular, $\hat{\alpha}$ is π-special. We deduce that $\alpha(1)$ is a π-number.

Take K normal in G so that K/N is a chief factor for G. This is the point where we use the fact that G is solvable to see that $G = VK$ and $V \cap K = N$ where K/N is an elementary abelian p-group for some prime p. (This is the only place we use the hypothesis that G is solvable in place of G being π-separable.) Let L/K be a chief factor for G. We know that $(|L : K|, |K : N|) = 1$ and $\mathbf{C}_{(L \cap V)/N}(K/N) = 1$. (See Lemma 5.1 of Ref. 12 for a proof of this.) By Problem 6.12 of Ref. 5, either α^* extends to K or α^* is fully-ramified with respect to K/N.

Suppose first that α^* extends to K. Notice that multiplication by $\mathrm{Irr}(K/L)$ is a transitive action on the irreducible constituents of $(\alpha^*)^K$. Also, $(V \cap K)/L$ acts on compatibly on the irreducible constituents of $(\alpha^*)^K$ and on $\mathrm{Irr}(K/L)$ where the action on $\mathrm{Irr}(K/L)$ is coprime. We can use Glauberman's lemma (Lemma 13.8 of Ref. 5) to see that α^* has a $V \cap L$-invariant extension. The corollary to Glauberman's lemma (Corollary 13.9 of Ref. 5) can be applied to see that α^* has a unique $V \cap L$-invariant extension δ. Since V permutes the $V \cap L$-extensions of α^*, it follows that δ is V-invariant. We now use Corollary 4.2 of Ref. 8 to see that restriction is a bijection from $\mathrm{Irr}(G \mid \beta)$ to $\mathrm{Irr}(V \mid \alpha^*)$.

Let $\eta \in \mathrm{I}_\varphi(V \mid Q)$ so that $\eta^G = \varphi$. We can find $\eta^* \in \mathrm{B}_\pi(V)$ so that $(\eta^*)^o = \eta$. Since $(\eta^{*G})^o = (\eta^{*o})^G = \eta^G = \varphi \in \mathrm{IBr}(G)$, we see that η^G is irreducible. On the other hand, $(\eta^{*o})_N = (\eta_N)^o = b\alpha$ for some integer b. Since the irreducible constituents of ${\eta^*}_N$ lie in $\mathrm{B}_\pi(N)$, we deduce that $\eta^* \in \mathrm{Irr}(V \mid \alpha^*)$. But we saw that this implies that η^* extends to G. Since $V < G$, it is not possible for η^* to both extend to G and induce irreducibly. Therefore, we have a contradiction. We see that α^* (and hence, α) is fully ramified with respect to K/N. Notice that if p is not in π, then Corollary 6.28 of Ref. 5 applies and α^* extends to K. Therefore, $p \in \pi$.

We suppose that p is odd, and we work for a contradiction. Since α^* is fully-ramified with respect to K/N and $|K : N|$ has odd order, main theorem of Ref. 7 implies that no character in $\mathrm{Irr}(V \mid \alpha)$ induces irreducibly to G. (A stronger theorem is proved in Ref. 4.) As in the previous paragraph, this implies that φ is not induced from V which contradicts the assumption that $\mathrm{I}_\varphi(V \mid Q)$ is not empty. (This strongly uses the fact that p is odd. When $p = 2$, it is tempting to try use the correspondence in Ref. 6, but that correspondence does not preclude inducing characters in $\mathrm{Irr}(G \mid \alpha)$ from V. In fact, $\mathrm{GL}_2(3)$ is an example where this occurs.) We conclude that $p = 2$. Since $|G : V| = |K : N|$, we see that $|G : V|$ is a power of 2. This proves the theorem. □

As a corollary, we obtain Theorem 1.1 stated for π-partial characters.

Corollary 2.2. *Let G be a solvable group. Assume either $|G|$ is odd or $2 \notin \pi$. Let Q be a π'-subgroup of G and suppose that $Q \leq V$. If $\varphi \in \mathrm{I}_\pi(G)$, then $|\mathrm{I}_\varphi(V \mid Q)| \leq |\mathrm{N}_G(Q) : \mathrm{N}_V(Q)|$.*

Proof. We suppose the result is not true. Let G be a counterexample with $|G| + |G : V|$ as in Theorem 2.1. By that result, we have that $|G : V|$ is a nontrivial power of 2 which is a contradiction if $|G|$ is odd. We also have

$2 \in \pi$ which is a contradiction to $2 \notin \pi$. This proves the corollary. □

References

1. J. P. Cossey and M. L. Lewis, Counting lifts of Brauer characters, ⟨arXiv:1007.3011v1⟩
2. B. Huppert, "Character Theory of Finite Groups," Walter de Gruyter, Berlin, 1998.
3. D. Gajendragadkar, A characteristic class of characters of finite π-separable groups, *J. Algebra* **59** (1979), 237-259.
4. I. M. Isaacs, Characters of solvable and symplectic groups, *Amer. J. Math.* **95** (1973), 594-635.
5. I. M. Isaacs, "Character Theory of Finite Groups," Academic Press, San Diego, California, 1976.
6. I. M. Isaacs, Character correspondences in solvable groups, *Adv. Math.* **43** (1982), 284-306.
7. I. M. Isaacs, On the character theory of fully ramified sections, *Rocky Mountain J. Math.* **13** (1983), 689-698.
8. I. M. Isaacs, Characters of π-separable groups, *J. Algebra* **86** (1984), 98-128.
9. I. M. Isaacs, Fong characters in π-separable groups, *J. Algebra* **99** (1986), 89-107.
10. I. M. Issacs, Partial characters of π-separable groups. In: Representation Theory of Finite Groups and Finite Dimensional Algebras (Bielefield, 1991) in Progr. Math, vol. 95, Birkhauser, Basel, 1991, 273-287.
11. A. Laradji, On normal subgroups and simple modules with a given vertex in a p-solvable group, *J. Algebra* **308** (2007), 484-492.
12. M. L. Lewis, Characters of maximal subgroups of M-groups, *J. Algebra* **183** (1996), 864-897.
13. O. Manz and T. R. Wolf, "Representation of Solvable Groups," Cambridge University Press, Cambridge, 1993.

GROUPS AND LIE RINGS WITH FROBENIUS GROUPS OF AUTOMORPHISMS

N. YU. MAKARENKO*

Institute of Mathematics, Novosibirsk 630090, Russia and Université de Haute Alsace, Mulhouse 68093 France
E-mail: natalya_makarenko@yahoo.fr

In this survey we discuss recent progress in the study of groups and Lie rings admitting a Frobenius group of automorphisms FH with kernel F and complement H. Our main objective is the case, where a finite group G is acted on by FH in such a way that the fixed-point subgroup of the Frobenius kernel is trivial: $C_G(F) = 1$. Recent results of Khukhro, Makarenko and Shumyatsky show that in this situation various properties of the group G such as the order, rank, nilpotency, nilpotent length, exponent, nilpotency class are close to the corresponding properties of the fixed-point subgroup of the Frobenius complement, $C_G(H)$. Special emphasis is placed on Lie-theoretic methods used for bounding the nilpotency class and exponent of G.

Keywords: Frobenius group; automorphism; finite group; exponent; Lie ring; Lie algebra; connected Lie group; graded; solvable; nilpotent; rank; fixed points; fixed-point-free; locally nilpotent torsion-free group; positive law.

1. Introduction

This survey is devoted to the study of finite groups and Lie rings admitting a Frobenius group of automorphisms. The subject has recently become a new focus of research in the field of investigating the structure of groups and Lie rings via fixed points of their automorphisms. New difficult problems have been raised, some of them have been already solved and new methods have been developed. Our objective here is to discuss this recent progress and to outline various directions for further investigations.

Recall that a *Frobenius group* FH with kernel F and complement H is a semidirect product of a normal subgroup F on which H acts by automor-

*Supported by the Programme of Support of Leading Scientific Schools of the Russian Federation (grant NSh-344.2008.1).

phisms in such a way that $C_F(h) = 1$ for every $h \in H \setminus \{1\}$. We will mainly consider the following situation. Suppose that a finite group G admits a Frobenius group of automorphisms FH with kernel F and complement H such that the fixed-point subgroup of F is trivial: $C_G(F) = 1$. Of course, the fixed-point-free action of the kernel F implies very strong restrictions on the structure of G. For example, G is solvable by the Belyaev–Hartley theorem[3] based on the classification of finite simple groups. If $|F|$ is of prime order, then by Thompson's theorem[58] the group G is nilpotent and by Higman's theorem[16] the nilpotency class of G is bounded in terms of $|F|$. But our aim is to determine what restrictions on the structure of the group are imposed by the action of the Frobenius complement H and by the properties of $C_G(H)$. It turns out that various properties of G are close to the corresponding properties of $C_G(H)$ and "independent" of F. For example, it was shown in Refs. 29,42 that the order of G is bounded in terms of $|H|$ and $|C_G(H)|$, the rank of G is bounded in terms of $|H|$ and the rank of $C_G(H)$, and that G is nilpotent if $C_G(H)$ is nilpotent. Moreover, Khukhro proved in Ref. 26 that the nilpotent length of G is equal to the nilpotent length of $C_G(H)$ if the orders of G and H are coprime.

In the case of metacyclic Frobenius group FH it was shown in Refs. 29,42 that the nilpotency class of G is bounded in terms of $|H|$ and the nilpotency class of $C_G(H)$, and that the exponent of G is bounded in terms of $|FH|$ and the exponent of $C_G(H)$. Recently Shumyatsky[56] has obtained a nice result, which unifies two previous theorems on bounding the nilpotency class and exponent. He has proved that if FH is a metacyclic Frobenius group and $C_G(H)$ satisfies a positive law of degree k, then G satisfies a positive law of degree bounded in terms of k and $|FH|$. The proofs of these results extensively use Lie-theoretic methods. Lie-ring technique based on Lazard's Lie algebra associated with Jennings–Zassenhaus filtration was employed to establish a bound for the exponent of G. The problem of bounding the nilpotency class gives rise to a parallel Lie-theoretic question. Actually, the solution of the corresponding Lie ring problem was crucial in the proof of the group theorem.

In general, the Lie-theoretic methods are an effective tool in studying the problems aiming to find a bound for the nilpotency class. This can be explained by the fact that the associated Lie ring constructed from a group by using its lower series has the same nilpotency class as the group itself. So, if we can find some function bounding the nilpotency class of the associated Lie ring, then the same function will bound the nilpotency class of the group. An advantage of the Lie-theoretic approach is that a Lie ring is a more linear object and so calculations are considerably easier in a Lie ring than in a group. It is worth mentioning that this procedure does not work every time. As an example we cite the following challenging problem:

Problem 1.1. *Let G be a p-group admitting a fixed-point-free automorphism of order n. Is the derived length of G bounded in terms of n?*

In spite of great progress achieved in last decades in the study of (almost) fixed-point-free automorphisms this question is still open for all n except for prime n (Higman's theorem[16]) and $n = 4$ (Kovács[32]). The main problem is that there is no bound in terms of n for the nilpotency class of a Lie ring L admitting a fixed-point-free automorphism of order n if n is not prime. By Kreknin's theorem[34] L is solvable and its derived length is bounded in terms of n, but we cannot derive the same conclusion for the group because, in general, there is no connection between the derived length of the group and that of the associated Lie ring.

Let us now describe the structure of the paper. In Chapter 2 we cite some basic results on groups with fixed-point-free automorphisms, introduce requisite notions and start to determine the connection between the properties of G and $C_G(H)$, stating the facts about the order, rank, nilpotency and nilpotent length of G. Chapter 3 is devoted to double Frobenius groups, which are closely related to the subject matter. We outline the work of Makarenko–Shumyatsky[43] dealing with bounding the nilpotency class of a group G in the particular case where FH acts on G in such a manner that GF is a Frobenius group as well. In Chapter 4 we focus on the problem of bounding the exponent of a group admitting a Frobenius group of automorphisms. Chapter 5 is devoted to Lie rings problems. We present results in the historical order, starting with a result by Khukhro on $\mathbb{Z}/n\mathbb{Z}$-graded Lie rings with many commuting components, and proceeding with more recent result on Lie rings, Lie algebras (as well as some classes of infinite groups) admitting a metacyclic Frobenius group with fixed-point-free Frobenius kernel and nilpotent fixed-point set of Frobenius complement; all these results deal with bounding the nilpotency class. Chapter 6 is devoted to bounding the nilpotency class of a finite group G. Finally, in Chapter 7 we discuss Shumyatsky's theorem[56] on groups with Frobenius group of automorphisms such that the fixed-point subgroup of Frobenius complement satisfies a positive law.

Throughout the paper we will not write down the bounding functions, but this can always be done by working through the proofs.

I am grateful to E. Khukhro and P. Shumyatsky for their comments and the kind permission to include their recent works in this survey.

2. Preliminaries

In this chapter we describe the basic material that we shall need in the subsequent sections.

We use the abbreviation, say, "(m, n)-bounded" for "bounded above in terms of m, n only".

2.1. *Fixed-point-free automorphisms*

We begin with stating classical theorems on groups and Lie rings admitting fixed-point-free automorphisms.

Theorem 2.1 (Thompson, 1959). *If a finite group G admits an automorphism α of prime order such that $C_G(\alpha) = 1$, then G is nilpotent.*

The above result was proved in Ref. 58. The following theorem of Higman[16] gives a bound for the nilpotency class in Thompson's theorem.

Theorem 2.2 (Higman, 1957). *If a nilpotent finite group G admits an automorphism α of prime order p such that $C_G(\alpha) = 1$, then G is nilpotent of p-bounded class.*

The proof of Higman's theorem involves the solution of the corresponding Lie-theoretic problem. The Lie ring analogue of Theorem 2.2 is as follows.

Theorem 2.3 (Higman, 1957). *If a Lie ring L admits an automorphism α of prime order p such that $C_L(\alpha) = 0$, then L is nilpotent of p-bounded class.*

Another proof of Theorems 2.2 and 2.3 including an explicit estimate for the nilpotency class was given by Kreknin and Kostrikin in Refs. 34, 35. Kreknin in Ref. 34 also established the following important fact for automorphisms of not necessarily prime order.

Theorem 2.4 (Kreknin, 1963). *If a Lie ring L admits an automorphism α of finite order n such that $C_L(\alpha) = 0$, then L is solvable of n-bounded derived length.*

Earlier Borel and Mostov[5] proved the solvability of a finite-dimensional Lie algebra with a fixed-point-free automorphism of finite order (without a bound for the derived length).

The following result of Belyaev and Hartley[3] is based on the classification of finite simple groups.

Theorem 2.5 (Belyaev–Hartley, 1996). *Suppose that a finite group G admits a nilpotent group of automorphisms F such that $C_G(F) = 1$. Then G is solvable.*

2.2. *Some definitions*

Let us now introduce some general definitions and notation.

Products in a Lie ring are called commutators. We use the term "span" both for the subspace (in the case of algebras) and for the additive subgroup generated by a given subset. For subsets X, Y we denote by $[X, Y]$ the span of all commutators $[x, y]$, $x \in X$, $y \in Y$; this is an ideal if X, Y are ideals. Terms of the derived series of a Lie ring L are defined as $L^{(0)} = L$; $L^{(k+1)} = [L^{(k)}, L^{(k)}]$. Then L is solvable of derived length at most n if $L^{(n)} = 0$. Terms of the lower central series of L are defined as $\gamma_1(L) = L$; $\gamma_{k+1}(L) = [\gamma_k(L), L]$. Then L is nilpotent of class at most c (sometimes called "nilpotency index") if $\gamma_{c+1}(L) = 0$. We use the standard notation for simple (left-normed) commutators: $[x_1, x_2, \ldots, x_k] = [...[[x_1, x_2], x_3], \ldots, x_k]$ (here the x_i may be elements or subsets).

Let A be an additively written abelian group. A Lie ring L is *A-graded* if

$$L = \bigoplus_{a \in A} L_a \qquad \text{and} \qquad [L_a, L_b] \subseteq L_{a+b}, \quad a, b \in A,$$

where the L_a are subgroups of the additive group of L. Elements of the grading components L_a are called *homogeneous*, and commutators in homogeneous elements *homogeneous commutators*. An additive subgroup H of L is called *homogeneous* if $H = \bigoplus_a (H \cap L_a)$; we then write $H_a = H \cap L_a$.

An element y of a Lie algebra L is called *ad-nilpotent* if there exists a positive integer n such that $(\operatorname{ad} y)^n = 0$, that is, $[x, \underbrace{y, \ldots, y}_{n}] = 0$ for all $x \in L$. If n is the least integer with this property, then we say that y is *ad-nilpotent of index n*.

The induced group of automorphisms of an invariant section is often denoted by the same letter (which is a slight abuse of notation as the action may become non-faithful).

We also mention the following fact, which is helpful in proving nilpotency of a solvable Lie ring L.

Lemma 2.1. *Let K be an ideal of a Lie ring L. If $\gamma_{c+1}(L) \subseteq [K,K]$ and $\gamma_{k+1}(K) = 0$, then $\gamma_{c\binom{k+1}{2}-\binom{k}{2}+1}(L) = 0$.*

This lemma can be regarded as a Lie ring analogue of P. Hall's theorem[12] (with a simpler "linear" proof; see also Ref. 57 for the best possible bound for the nilpotency class of L).

2.3. *Covering the fixed points in a quotient*

We now discuss the problem of covering the fixed points in an invariant quotient by the fixed points in the group. Let G be a finite group, $A \leq \operatorname{Aut} G$ and let N be a normal A-invariant subgroup of G. It is well-known that if $(|A|, |N|) = 1$, then $C_{G/N}(A) = C_G(A)N/N$. If we do not assume that $(|A|, |N|) = 1$, the equality $C_{G/N}(A) = C_G(A)N/N$ may no longer be true. However, sometimes under additional conditions it does hold. For example, Thompson[59], Evens[9], Khukhro[17] proved this equality in some special cases where both G and A are p-groups and certain abelian sections are free $\mathbb{F}_pA$-modules. One more example of a nice behavior of the fixed points is the next elementary fact.

Lemma 2.2. *Let G be a finite group admitting a nilpotent group of automorphisms F such that $C_G(F) = 1$. If N is a normal F-invariant subgroup of G, then $C_{G/N}(F) = 1$.*

Lemma 2.2 is a well-known consequence of the properties of Carter subgroups (see, for example, Lemma 2.2 of Ref. 29).

We now return to the main object of our study, a finite group G that admits a Frobenius group of automorphisms FH with kernel F and complement H such that $C_G(F) = 1$. It turns out that the fixed points of H in any FH-invariant section of G are also covered by the fixed points of H in G.

Theorem 2.6. *Suppose that a finite group G admits a Frobenius group of automorphisms FH with kernel F and complement H. If N is an FH-invariant normal subgroup of G such that $C_N(F) = 1$, then $C_{G/N}(H) = C_G(H)N/N$.*

The above theorem was proved by Khukhro in Ref. 24 under the additional coprimeness assumption $(|N|, |F|) = 1$, the reduction to this case was provided in Ref. 29. The proof relies on the fact that every section of an unrefinable FH-invariant normal series is a free $\mathbb{F}_pH$-module for a suitable prime p. Recall that for a group A and a field k, a *free kA-module of dimension n* is a direct sum of n copies of the group algebra kA, each of which can be regarded as a vector space over k of dimension $|A|$ with a basis $\{v_g \mid g \in A\}$ labelled by elements of A on which A acts in a regular representation: $v_g h = v_{gh}$. The presence of free $\mathbb{F}_p H$-modules in Theorem 2.6 is derived by Clifford's theorem. The similar methods are used to prove the following useful lemma (Lemma 2.4, Ref. 29).

Lemma 2.3. *Suppose that a finite group G admits a Frobenius group of automorphisms FH with kernel F and complement H such that $C_G(F) = 1$. Then $G = \langle C_G(H)^f \mid f \in F\rangle$.*

2.4. *Order, rank, nilpotency, nilpotent length*

In this section we begin to study the connection between the properties of a finite group G admitting a Frobenius group of automorphisms FH with complement H and kernel F such that $C_G(F) = 1$ and the properties of $C_G(H)$. The aim is proving that many features of G, such as the order, rank, nilpotency, nilpotent length are determined by those of $C_G(H)$.

The next result, which is contained in a joint work of Khukhro, Shumyatsky and myself (Theorem 2.7, Ref. 29), establishes the relationship between the order, rank, and nilpotency of G and those of $C_G(H)$.

Theorem 2.7. *Suppose that a finite group G admits a Frobenius group of automorphisms FH with kernel F and complement H such that $C_G(F) = 1$. Then*

(a) $|G| = |C_G(H)|^{|H|}$;

(b) *the rank of G is bounded in terms of $|H|$ and the rank of $C_G(H)$;*

(c) *if $C_G(H)$ is nilpotent, then G is nilpotent.*

(By the rank we mean the minimum number r such that every subgroup can be generated by r elements.)

Note that by the Belyaev–Hartley theorem (Theorem 2.5) the group G is solvable. Part (a) of Theorem 2.7 is a corollary of Theorem 2.6 and the fact that every section of an unrefinable FH-invariant normal series is a free $\mathbb{F}_p H$-module for a suitable prime p. Part (b) reduces to p-groups by a theorem of Kovács[33], which states that the rank of a finite solvable group is bounded in terms of the maximum rank of its Sylow p-subgroups. The further proof involves arguments dealing with a Thompson critical subgroup, and the free $\mathbb{F}_p H$-modules as well. The proof of part (c) requires some more ideas. We will not go into detail and refer the reader to the paper of Khukhro[25] for a much more thorough discussion.

Under the additional assumption $(|G|, |H|) = 1$, a similar result determines the relation between the nilpotent length (Fitting height) of $C_G(H)$ and that of G. Recall that the Fitting series starts with the Fitting subgroup $F_1(G) = F(G)$, the largest normal nilpotent subgroup of a group G, and $F_{i+1}(G)$ is defined as the inverse image of $F(G/F_i(G)$. The smallest number l such that $F_l(G) = G$ is called the nilpotent length of a solvable group G. The following theorem is proved by Khukhro in Ref. 26.

Theorem 2.8. *Suppose that a finite group G admits a Frobenius group of automorphisms FH with kernel F and complement H such that $C_G(F) = 1$ and $(|G|, |H|) = 1$. Then*

(a) $F_i(C_G(H)) = F_i(G) \cap C_G(H)$ *for all* i;

(b) *the nilpotent length of G is equal to the nilpotent length of $C_G(H)$.*

Recall that F is nilpotent being a Frobenius kernel, the condition $C_G(F) = 1$ implies the solvability of G by the Belyaev–Hartley theorem (Theorem 2.5) based on the classification of finite simple groups.

The problem of bounding the nilpotent length of a finite solvable group G in terms of a group of automorphisms $A \leq \operatorname{Aut} G$ and its fixed-point subgroup has a long history. The study was initiated by Thompson in Ref. 60. Thompson's theorem[60] gives a bound for the nilpotent length of a finite solvable group G in terms of $|A|$ and the nilpotent length of $C_G(A)$ in the case, where A is solvable of coprime order. This theorem is based on the techniques of representation theory developed in the Odd Order paper[10] and in the Hall–Higman paper[13]. Among the numerous subsequent papers we mention only those of Berger[4], Hartley and Isaacs[14], Kurzweil[36], Shult[50], Turull[61]. Dade's theorem[7] gives a bound for the nilpotent length of G in the case, where A is nilpotent and $C_G(A) = 1$ without the coprimeness condition (examples by Bell and Hartley[2] show that this is no longer true for any non-nilpotent solvable A).

Theorem 2.8 provides a much stronger best-possible bound for the nilpotent length of G independent of $|H|$ or $|F|$, but under the stronger hypotheses.

Problem 2.1. *Does Theorem 2.8 hold without the assumption* $(|G|, |H|) = 1$*?*

3. Double Frobenius group

Let us now consider a more specific situation. Let FH be a Frobenius group FH with kernel F and complement H. Suppose that the group FH acts on a finite group G in such a manner that GF is also a Frobenius group with kernel G and complement F. Then the product GFH is called a *double Frobenius group* with "upper" complement H. In Ref. 62, Problem 17.72, Mazurov posed the problems:

Problem 3.1.

(a) *Does there exist a bound for the nilpotency class of* G *in terms of the nilpotency class of* $C_G(H)$ *and* $|H|$*?*

(b) *Does there exist a bound for the exponent of* G *in terms of the exponent of* $C_G(H)$ *and* $|H|$*?*

Part (b) is still unsettled. The following result of Makarenko and Shumyatsky[43] provides a positive solution of part (a) of Mazurov's problem.

Theorem 3.1. *Let* GFH *be a double Frobenius group with upper complement* H *of order* q*. Assume that* $C_G(H)$ *is nilpotent of class* c*. Then* G *is nilpotent of* (c, q)*-bounded class.*

Note that by Thompson's theorem[58] (Theorem 2.1) the group G is nilpotent, and by Higman's theorem[16] (Theorem 2.2) the nilpotency class of G is bounded in terms of the least prime divisor of $|F|$ (explicit upper bounds for the nilpotency class were obtained by Kreknin and Kostrikin[34,35]). But Theorem 3.1 asserts that there is a bound depending only on $|H|$ and c.

The case of $c = 1$ and $H = 2$ and 3 of the above theorem is due to Mazurov[45]. In Ref. 22 Khukhro solved the problem for $c = 1$ (and arbitrary $|H|$). His proof is based on the following interesting variation of the Higman–Kreknin–Kostrikin theorem. Let p be a prime number and let L be a $\mathbb{Z}/p\mathbb{Z}$-graded Lie ring with trivial zero component $L_0 = 0$; if for some m each grading component commutes with all but at most m components, then L is nilpotent of nilpotency class bounded in terms of m (see Chapter 5, Theorem 5.1).

The proof of Theorem 3.1 is also based on the corresponding Lie-theoretic result. But at first, we make the following reduction to the case, where F is of prime order. As a Frobenius complement in GF, the group F has no non-cyclic subgroups of order p^2 for any prime p. Hence the Sylow subgroups of F are cyclic or generalized Quaternion. Being the Frobenius kernel of FH, the group F is nilpotent. It follows that the Sylow 2-subgroup of F is trivial, for otherwise its center would contain a non-trivial fixed point for H. Thus, F is cyclic. Let F_1 be a subgroup of F of prime order p. The hypothesis of the theorem is inherited by GF_1H. Therefore we can replace F by F_1 and suppose that F has prime order p.

Assuming F of prime order the reduction to Lie rings is done as follows. We consider the associated Lie ring of the group G (recall that G is nilpotent of bounded class by the Thompson–Higman theorem):

$$L(G) = \bigoplus_{i=1}^{n} \gamma_i/\gamma_{i+1},$$

where n is the nilpotency class of G and the γ_i are the terms of the lower central series of G. The nilpotency class of G coincides with that of $L(G)$. The action of the group FH on G naturally induces an action of FH on $L(G)$. By Lemma 2.2 the Frobenius kernel F acts fixed-point-freely on every quotient γ_i/γ_{i+1}. It follows that $C_{L(G)}(F) = 0$. If c is the nilpotency class of $C_G(H)$, then any group commutator of weight $c+1$ in elements of $C_G(H)$ is trivial; hence any Lie ring commutator of weight $c+1$ in elements of $C_{\gamma_i}(H)\gamma_{i+1}/\gamma_{i+1}$ is also trivial. By Theorem 2.6

$$C_{L(G)}(H) = \bigoplus_{i=1}^{n} C_{\gamma_i/\gamma_{i+1}}(H) = \bigoplus_{i=1}^{n} C_{\gamma_i}(H)\gamma_{i+1}/\gamma_{i+1}.$$

It follows that $C_{L(G)}(H)$ is nilpotent of class at most c. In section 5.2 we will discuss the corresponding Lie-theoretic theorem from which Theorem 3.1 follows: if FH is a Frobenius group with kernel F of prime order and complement H of order q that acts by automorphisms on a Lie ring K in such a way that $C_K(F) = 0$ and $C_K(H)$ is nilpotent of class c, then K is nilpotent of (c, q)-bounded class (see Theorem 5.4).

4. Bounding the exponent of a finite group with a Frobenius group of automorphisms

In this chapter we describe the recent progress in bounding the exponent of a finite group with a Frobenius group of automorphisms. The main question of interest to us is the following.

Problem 4.1. *Let G be a finite group G acted on by a Frobenius group of automorphisms FH with kernel F and complement H such that $C_G(F) = 1$ and $C_G(H)$ has exponent e. Is the exponent of G bounded in terms of e and $|H|$?*

Let us mention that this is a generalization of part (b) of Mazurov's problem (Problem 3.1).

4.1. *Metacyclic Frobenius groups of automorphisms*

We start with a Frobenius group of automorphisms with cyclic kernel. Note that if the kernel F of a Frobenius group FH is cyclic then the complement H is also cyclic. We call such a group FH metacyclic Frobenius. The next theorem was recently proved by Khukhro, Shumyatsky and myself in Refs. 29,42.

Theorem 4.1. *Suppose that a Frobenius group FH with cyclic kernel F and complement H acts on a finite group G in such a manner that $C_G(F) = 1$ and $C_G(H)$ has exponent e. Then the exponent of G is bounded solely in terms of e and $|FH|$.*

Note that the exponent of G depends on $|F|$. It is unclear, even in the case where GFH is a double Frobenius group, if the bound can be made independent of $|F|$.

The first stage of the proof of Theorem 4.1 is a reduction to a finite p-group generated by at most $|FH|$ elements. By the Belyaev–Hartley theorem[3] (Theorem 2.5) G is solvable. By Dade's theorem[7] the nilpotent length of G is bounded in terms of $|F|$. So it suffices to bound the exponents of the factors of the Fittings series of G. Since an arbitrary element x of G is obviously contained in an FH-invariant subgroup of G with at most $|FH|$ generators, we can assume from the outset that G is an $|FH|$-generated finite p-group for some prime p.

The next stage consists of Lie-theoretic constructions. Some ideas were used earlier by Khukhro and Shumyatsky in Ref. 30]. First we define Lazard's Lie algebra $L_p(G)$ associated with the Jennings-Zassenhaus filtration of G. Recall that the *Jennings–Zassenhaus filtration* of the group G is

$$G = D_1 \geq D_2 \geq \cdots ,$$

where

$$D_i = D_i(G) = \prod_{jp^k \geq i} \gamma_j(G)^{p^k}.$$

This series satisfies the inclusions $[D_i, D_j] \leq D_{i+j}$ and $D_i^p \leq D_{pi}$ for all i, j. These properties make it possible to construct a Lie algebra $DL(G)$ over $\mathbb{F}_p$, the field with p elements. Namely, consider the quotients D_i/D_{i+1} as linear spaces over $\mathbb{F}_p$, and let $DL(G)$ be the direct sum of these spaces.

Commutation in G induces a binary operation $[,]$ in $DL(G)$. For homogeneous elements $xD_{i+1} \in D_i/D_{i+1}$ and $yD_{j+1} \in D_j/D_{j+1}$ the operation is defined by

$$[xD_{i+1}, yD_{j+1}] = [x, y]D_{i+j+1} \in D_{i+j}/D_{i+j+1}$$

and extended to arbitrary elements of $DL(G)$ by linearity. It is easy to check that the operation is well-defined and that $DL(G)$ with the operations $+$ and $[,]$ is a Lie algebra over $\mathbb{F}_p$, which is $\mathbb{Z}$-graded with the D_i/D_{i+1} being the grading components.

Let $L_p(G) = \langle D_1/D_2 \rangle$ be the subalgebra of $DL(G)$ generated by D_1/D_2; it is also $\mathbb{Z}$-graded with grading components $L_i = L_p(G) \cap (D_i/D_{i+1})$.

The group FH of automorphisms acts naturally on every factor of the Jennings–Zassenhaus filtration of G. This action induces an action by automorphisms on the Lie algebra $L = L_p(G)$. By Lemma 2.2 F is fixed-point-free on every factor of the Jennings–Zassenhaus filtration. Hence $C_L(F) = 0$. Kreknin's theorem[34] (Theorem 2.4) now tells us that L is solvable of $|F|$-bounded derived length.

Every homogeneous component L_i of the Lie algebra L can be regarded as an FH-invariant subgroup of the corresponding quotient D_i/D_{i+1} of the Jennings–Zassenhaus series. By Lemma 2.3, $L_i = \langle C_{L_i}(H)^f \mid f \in F \rangle$. Moreover, Theorem 2.6 implies that every element of $C_{L_i}(H)^f$ is the image of some element of $C_G(H)^f$, the order of which divides e. By a theorem of Lazard[37] elements of G of order p^k give rise to elements of $L_p(G)$ that are ad-nilpotent of index p^k. It follows that the additive group L_i is generated by elements that are ad-nilpotent of index at most e.

The next lemma is a crucial fact in the proof. It gives a bound for the nilpotency class of a finitely generated solvable Lie algebra "saturated" with ad-nilpotent elements.

Lemma 4.1. *Let $L = \bigoplus L_i$ be a $\mathbb{Z}$-graded Lie algebra over a field such that $L = \langle L_1 \rangle$ and assume that every homogeneous component L_i is spanned by elements that are ad-nilpotent of index at most r. Suppose further that L is solvable of derived length k and that L_1 has finite dimension d. Then L is nilpotent of (d, r, k)-bounded class.*

It follows from Lemma 4.1 that $L = L_p(G)$ is nilpotent of $(e, |FH|)$-bounded class.

The final step is a reduction to powerful groups. Recall that a finite p-group G is *powerful* if $G^p \geq [G, G]$ for $p \neq 2$ (or $G^4 \geq [G, G]$ for $p = 2$). Powerful p-groups were introduced by Lubotzky and Mann in Ref. 40. These groups have many nice properties, so that often a problem becomes much easier once it is reduced to the case of powerful p-groups. The next lemma allows us to perform such a reduction.

Lemma 4.2. *Suppose that X is a d-generator finite p-group such that the Lie algebra $L_p(X)$ is nilpotent of class c. Then X has a powerful characteristic subgroup of (p, c, d)-bounded index.*

This lemma goes back to Lazard[38]; in the present form it can be found, for example, in Ref. 30.

It follows from Lemma 4.2, that G has a powerful characteristic subgroup of $(e, |FH|)$-bounded index. It suffices to bound the exponent of this powerful subgroup, so we can just assume that G is powerful. Powerful p-groups have the property that if a powerful p-group G is generated by elements of order dividing p^k, then the exponent of G also divides p^k (see Lemma 2.2.5, Ref. 8). Combining this fact with Lemma 2.3, which states that our group G is generated by elements of $C_G(H)^f$, $f \in F$, we conclude that G has exponent e.

We close this section with a question. Antonov and Chekanov[1] have constructed an example of a finite 2-group admitting a Frobenius group of automorphisms FH with cyclic kernel F such that $C_G(F) = 1$ and the exponent of G is not equal to the exponent of $C_G(H)$. But there are no known examples demonstrating that under the assumptions of Theorem 4.1 the exponent of G really depends on $|H|$.

Problem 4.2. *Construct examples showing that the exponent of a finite group G satisfying the assumptions of Theorem* 4.1 *increases with the increase of $|H|$.*

4.2. *Non-cyclic Frobenius kernel*

Let us now discuss the problem of bounding the exponent of a group that admits a Frobenius group with non-cyclic kernel. The only known result in this direction is the following theorem of Shumyatsky[55].

Theorem 4.2. *Suppose that a Frobenius group FH of order* 12 *acts coprimely on a finite group G in such a manner that $C_G(F) = 1$ and $C_G(H)$ has exponent e. Then the exponent of G is bounded in terms of e only.*

We will not sketch the proof of Theorem 4.2 here. Let us only note that the solution is more complicated than that of the previous problem dealing with cyclic Frobenius kernel. In particular, unlike Theorem 4.1 the proof of Theorem 4.2 uses Zelmanov's solution of the Restricted Burnside Problem[63,64]. This can be explained by the fact that a Lie algebra admitting a non-cyclic fixed-point-free group of automorphisms is not necessarily solvable (see examples in Ref. 19, p. 149-150) while a Lie algebra admitting a fixed-point-free automorphism of finite order n is solvable of n-bounded derived length by Kreknin's theorem[34] (Theorem 2.4).

Problem 4.3. *In the statement of Theorem 4.2, can we omit the restriction that* $(|G|, 12) = 1$*?*

Finally, we mention a result of Shumyatsky and Romano[48] closely related to the subject matter, although it does not deal with Frobenius groups of automorphisms.

Theorem 4.3. *Let A be isomorphic with either D_8 or S_4. Let V be a normal four-subgroup of A and α an involution in $A \setminus V$. Suppose that A acts on a finite group G in such a manner that $C_G(V) = 1$ and $C_G(\alpha)$ has exponent e. Then*

(a) *if $A \cong S_4$, the exponent of G is bounded in terms of e alone;*
(b) *if $A \cong D_8$, the exponent of G' is bounded in terms of e alone.*

The proof uses Zelmanov's solution of the Restricted Burnside Problem[63,64] and the Khukhro–Shumyatsky theorem[30] that states that the exponent of a finite group acted on by a non-cyclic abelian group A is bounded in terms of $|A|$ and the exponents of $C_G(a)$ for $a \in A \setminus \{1\}$.

5. Lie rings with Frobenius groups of automorphisms

This chapter is devoted to bounding the nilpotency class of Lie rings with a Frobenius group of automorphisms. Let L be a Lie ring (or Lie algebra). Recall that we denote by $C_L(A)$ the fixed-point set for a group of automorphisms A of L. Suppose that a Frobenius group of automorphisms FH with kernel F and complement H acts by automorphisms on L in such a way that $C_L(F) = 0$, and $C_L(H)$ is nilpotent of class c. In view of the previous discussion it is natural to ask whether L is nilpotent, and whether the nilpotency class of L is bounded in terms of c and $|H|$. If F is not cyclic, the following examples show that the answers to both questions are negative.

Example 5.1. The simple 3-dimensional Lie algebra L of characteristic $\neq 2$ with basis e_1, e_2, e_3 and structure constants $[e_1, e_2] = e_3$, $[e_2, e_3] = e_1$,

$[e_3, e_1] = e_2$ admits the Frobenius group of automorphisms FH with non-cyclic F of order 4 and H of order 3 such that $C_L(F) = 0$ and $C_L(H)$ is one-dimensional. Namely, $F = \{1, f_1, f_2, f_3\}$, where $f_i(e_i) = e_i$ and $f_i(e_j) = -e_j$ for $i \neq j$, and $H = \langle h \rangle$ with $h(e_i) = e_{i+1\,(\mathrm{mod}\,3)}$.

One can modify the above example to produce nilpotent Lie rings L of unbounded nilpotency class, and even of unbounded derived length, admitting the same Frobenius group of automorphisms FH of order 12 such that $C_L(F) = 0$ and $C_L(H)$ is abelian ("one-dimensional").

Example 5.2. Let the additive group of L be the direct sum of three copies of $\mathbb{Z}/p^m\mathbb{Z}$ for some prime $p \neq 2$ with generators e_1, e_2, e_3 and let the structure constants of L be $[e_1, e_2] = pe_3$, $[e_2, e_3] = pe_1$, $[e_3, e_1] = pe_2$. The action of FH is "the same": $F = \{1, f_1, f_2, f_3\}$, where $f_i(e_i) = e_i$ and $f_i(e_j) = -e_j$ for $i \neq j$, and $H = \langle h \rangle$ with $h(e_i) = e_{i+1\,(\mathrm{mod}\,3)}$. Then $C_L(F) = 0$ and $C_L(H) = \langle e_1 + e_2 + e_3 \rangle$. It is easy to see that L is nilpotent of class exactly m and that the derived length of L is approximately equal to $\log_2(m+1)$.

In view of the above examples, we can restrict ourself to the case where F is cyclic. (Recall that then the complement H is also cyclic and FH is a metacyclic group.)

5.1. *Graded Lie rings with many commuting components*

We begin the description of the progress in bounding the nilpotency class of Lie rings with a metacyclic Frobenius group with a result of Khukhro[22]. At first sight, it is not directly connected to Frobenius groups of automorphisms, but actually this is the first deep result in the direction and the real starting-point of the whole subject discussed in this paper.

Theorem 5.1. *Let n be a positive integer and let $L = \bigoplus_{i=0}^{n-1} L_i$ be a $(\mathbb{Z}/n\mathbb{Z})$-graded Lie ring, where the L_i are additive subgroups satisfying $[L_i, L_j] \subseteq L_{i+j(\mathrm{mod}\,n)}$. Suppose that the zero-component is trivial, $L_0 = 0$, and for some m every grading component L_k commutes with all but at most m components: $[L_k, L_i] = 0$ unless $i \in N_k \subseteq \{1, \dots, n-1\}$, where $|N_k| \leq m$.*

(a) *Then L is solvable of m-bounded derived length.*

(b) *If, in addition, n is a prime, then L is nilpotent of m-bounded class.*

Let us explain how Theorem 5.1 is related to Frobenius groups of automorphisms. If we consider a Lie ring admitting a single automorphism, say φ, of finite order n, then after the ground ring is extended by a primitive nth root of unity ω, the additive subgroups ("eigenspaces") $L_j = \{a \mid a^\varphi = \omega^j a\}$ behave like the components of a (Z/nZ)-grading: $[L_s, L_t] \leq L_{s+t}$, where $s+t$ is calculated modulo n. For example, Kreknin's theorem (Theorem 2.4) has the following virtually equivalent statement in terms of Lie rings with finite grading: a $(\mathbb{Z}/n\mathbb{Z})$-graded Lie ring L with trivial zero-component $L_0 = 0$ is solvable of n-bounded derived length. Similarly, the Higman–Kreknin–Kostrikin theorem (Theorem 2.3) is virtually equivalent to the fact that for a prime p a $(\mathbb{Z}/p\mathbb{Z})$-graded Lie ring L with trivial zero-component $L_0 = 0$ is nilpotent of p-bounded class. Let now K be a Lie ring admitting a Frobenius group of automorphisms FH with cyclic kernel F of order n and complement H of order q such that $C_K(F) = 0$ and $C_K(H)$ is abelian. By using the idea of eigenspaces and eigenvalues (as above) we perform the reduction to a $(\mathbb{Z}/n\mathbb{Z})$-graded Lie ring L with trivial zero-component $L_0 = 0$. In addition, it can be shown that the assumptions that $C_K(H)$ is abelian and $|H| = q$ yield the condition of "many commuting components" exactly as in Theorem 5.1: for each k there are at most q^2 values of l such that $[L_k, L_l] \neq 0$. So, Theorem 5.1 can be viewed as a technical tool in proving the following result, which is implicitly present in the same paper of Khukhro[22].

Theorem 5.2. *Suppose that a Frobenius group FH with cyclic kernel F of order n and complement H of order q acts by automorphisms on a Lie ring L in such a way that $C_L(F) = 0$ and $C_L(H)$ is abelian.*

(a) *Then L is solvable of q-bounded derived length.*

(b) *If, in addition, n is a prime, then L is nilpotent of q-bounded class.*

Note that, under the assumptions of Theorems 5.1 and 5.2, L is solvable of n-bounded derived length by Kreknin's theorem; and if n is a prime, then L is nilpotent of n-bounded class by the Higman–Kreknin–Kostrikin theorem. But, unlike these n-dependent bounds, the bounds in Theorems 5.1 and 5.2 do not depend on n.

The proof of Theorem 5.1 uses two earlier variations of the Higman-Kreknin-Kostrikin theorem proved by Shalev[49] and Khukhro[21]. Shalev refined Kreknin's theorem by proving that the derived length of a $(\mathbb{Z}/n\mathbb{Z})$-graded Lie ring L with trivial zero-component $L_0 = 0$ is bounded in terms of the number of non-trivial components of the grading. Khukhro proved that if, under the same assumptions, n is a prime, then the nilpotency class of the ring L is bounded in terms of the number of non-trivial components. The proofs in Refs. 21,49 followed the proofs in the Kreknin-Kostrikin papers[34,35], "skipping" the non-existent components.

Part (b) of Theorem 5.2 underlies the proof of Khukhro's theorem on double Frobenius groups GFH with abelian centralizer $C_G(H)$ (see Chapter 3).

Finally, we note that in Ref. 23 Theorem 5.1 was generalized to the case of the finite zero-component of order $|L_0| = m$.

Theorem 5.3. *Let $L = \bigoplus_{i=0}^{n-1} L_i$ be a $(\mathbb{Z}/n\mathbb{Z})$-graded Lie ring (or Lie algebra), where the L_i are additive subgroups (subspaces) satisfying $[L_i, L_j] \subseteq L_{i+j(\mathrm{mod} n)}$. Suppose that the zero-component is finite of order $|L_0| = r$ (respectively, finite-dimensional of dimension $\dim L_0 = r$), and for some m every grading component L_k commutes with all but at most m components: $[L_k, L_i] = 0$ unless $i \in N_k \subseteq \{1, \ldots, n-1\}$, where $|N_k| \leq m$.*

(a) *Then L contains a solvable ideal of m-bounded derived length and of (n,r)-bounded index in the additive group (codimension).*

(b) *If, in addition, n is a prime, then L contains a nilpotent ideal of m-bounded class and of (n,r)-bounded index in the additive group (codimension).*

The proof employs Makarenko's theorem[41] and the Khukhro–Makarenko–Shumyatsky theorem[28] dealing with a $(\mathbb{Z}/n\mathbb{Z})$-graded Lie ring (algebra) with a small number d of nontrivial homogeneous components. Makarenko's theorem asserts that if in such a Lie ring (algebra) L the zero-component is finite of order $|L_0| = r$ (finite-dimensional of dimension $\dim L_0 = r$), then L has a homogeneous solvable ideal of d-bounded derived length and of (r,d)-bounded index in the additive group (codimension). If, in addition, n is a prime, then by the Khukhro–Makarenko–Shumyatsky theorem[28] this ideal can be chosen nilpotent of d-bounded nilpotency class. In turn, these results extend the works of Shalev[49] and Khukhro[21] mentioned above.

We close this section with a question. It would be interesting to obtain a bound for the index (codimension) independent of n in Theorem 5.3.

Problem 5.1. *Under the assumptions of Theorem 5.3 does L contain a solvable ideal of m-bounded derived length and of (r,m)-bounded index in the additive group (respectively, of (r,m)-bounded codimension in the case of a Lie algebra L)? If, in addition, n is a prime, does L contain a nilpotent ideal of m-bounded class and of (r,m)-bounded index in the additive group (respectively, of (r,m)-bounded codimension in the case of a Lie algebra L)?*

5.2. *Metacyclic Frobenius group of automorphisms with kernel of prime order*

In the previous section we have discussed Lie rings (Lie algebras) admitting a metacyclic Frobenius group FH of automorphisms with complement H such that the fixed-point set $C_L(H)$ is abelian. We now pass to the examination of the case, where $C_L(H)$ is nilpotent of arbitrary finite nilpotency class c.

In Ref. 43 Shumyatsky and myself obtained the following result.

Theorem 5.4. *Let FH be a Frobenius group with kernel F of prime order and complement H of order q. Assume that FH acts by automorphisms on a Lie ring L in such a way that $C_L(F) = 0$ and $C_L(H)$ is nilpotent of class c. Then L is nilpotent of (c, q)-bounded class.*

As we have already mentioned in Chapter 3 a group-theoretic consequence of the above theorem gives a positive answer to Mazurov's question, Problem 17.72 a in Ref. 62 (Problem 3.1, part (a)).

Let us briefly outline the proof of Theorem 5.4. We first exploit the idea of eigenvalues and eigenspaces extending the ground ring by a primitive $|F|$-th root of unity in order to reduce the problem to graded Lie rings. Let $\tilde{L} = L \otimes_{\mathbb{Z}} \mathbb{Z}[\omega]$, where ω is a primitive $|F|$-th root of unity. The group FH acts in a natural way on $\tilde{L}$ and the action satisfies the conditions that $C_{\tilde{L}}(F) = 0$ and $C_{\tilde{L}}(H)$ is nilpotent of class c. Let $|F| = p$ and let ϕ be a generator of F. For each $i = 0, \dots, p-1$ we define $K_i = \{x \in \tilde{L} \mid x^{\phi} = \omega^i x\}$ and set $K = K_0 \oplus \cdots \oplus K_{p-1}$ (the sum is direct by Corollary 1.7.3 of Ref. 19). Then $K_0 = 0$ and K is $\mathbb{Z}/p\mathbb{Z}$-graded. Moreover, it is easy to verify that $pL \leq K$. It follows that the nilpotency class of K equals that of L.

The next step consists of deriving the relations implied by the condition of the fixed-point subring $C_L(H)$ being nilpotent of class c. We argue as follows. Since F is cyclic of prime order, H is also cyclic. Let $H = \langle h \rangle$, $\phi^{h^{-1}} = \phi^r$. Then r has multiplicative order q modulo p. The group H permutes the components L_i so that ${L_i}^h = L_{ri}$ for all $i \in \mathbb{Z}/p\mathbb{Z}$. Indeed, if $x_i \in L_i$, then $(x_i^h)^{\phi} = x_i^{h\phi h^{-1}h} = (x_i^{\phi^r})^h = \omega^{ir} x_i^h$.

Given $u_k \in L_k$, we denote $u_k^{h^i}$ by $u_{r^i k}$. The sum over any H-orbit belongs to $C_L(H)$ and therefore

$$u_k + u_{rk} + \cdots + u_{r^{q-1}k} \in C_L(H).$$

Let $x_{a_1}, \dots, x_{a_{c+1}}$ be homogeneous elements in $L_{a_1}, \dots, L_{a_{c+1}}$, respectively.

Consider the elements

$$X_1 = x_{a_1} + x_{ra_1} + \cdots + x_{r^{q-1}a_1},$$

$$\vdots$$

$$X_{c+1} = x_{a_{c+1}} + x_{ra_{c+1}} + \cdots + x_{r^{q-1}a_{c+1}}.$$

Since all of them lie in $C_L(H)$ and since $C_L(H)$ is nilpotent of class c, it follows that

$$[X_1, \ldots, X_{c+1}] = 0.$$

After expanding the brackets, the left-hand side involves the term $[x_{a_1}, \ldots, x_{a_{c+1}}]$. Suppose that the commutator $[x_{a_1}, \ldots, x_{a_{c+1}}]$ is non-zero. Then there must be other non-zero terms in the expanded expression that belong to the same component $L_{a_1+\cdots+a_{c+1}}$. In other words, then

$$a_1 + \cdots + a_{c+1} = r^{\alpha_1}a_1 + \cdots + r^{\alpha_{c+1}}a_{c+1}$$

for some $\alpha_i \in \{0, 1, 2, \ldots, q-1\}$ not all of which are zeros. This means that L satisfies some kind of "weak nilpotency", which we call *the selective c-nilpotency.*

In the particular case $c = 1$ the selective c-nilpotency means that for each k there are at most q^2 values of l such that $[L_k, L_l] \neq 0$. This is exactly the condition of many commuting components in Khukhro's theorem[22] (Theorem 5.1).

For arbitrary c the selective c-nilpotency condition signifies that in a $(\mathbb{Z}/p\mathbb{Z})$-graded Lie ring L

$$[L_{a_1}, L_{a_2}, \ldots, L_{a_{c+1}}] \neq 0$$

only if $a_1+\cdots+a_{c+1} = r^{\alpha_1}a_1+\cdots+r^{\alpha_{c+1}}a_{c+1}$ for some $\alpha_i \in \{0, 1, 2, \ldots, q-1\}$ not all of which are zero. For convenience we introduce a special definition for such sequences of indices. Let $a_1, \ldots, a_k$ be not necessarily distinct non-zero elements of $\mathbb{Z}/p\mathbb{Z}$. We say that the sequence $(a_1, \ldots, a_k)$ is *r-dependent* if

$$a_1 + \cdots + a_k = r^{\alpha_1}a_1 + \cdots + r^{\alpha_k}a_k \tag{1}$$

for some $\alpha_i \in \{0, 1, 2, \ldots, q-1\}$ not all of which are zero. If the sequence $(a_1, \ldots, a_k)$ is not r-dependent, we call it *r-independent.*

In view of the reduction, it suffices to prove the following theorem (Theorem 2.13, Ref. 43), which is crucial for the proof.

Theorem 5.5. *Let q, p, r, c be positive integers such that p is a prime and r is a primitive qth root of* 1 *in* $\mathbb{Z}/p\mathbb{Z}$. *Let* $L = \sum_{i=0}^{p-1} L_i$ *be a* $\mathbb{Z}/p\mathbb{Z}$*-graded Lie ring such that* $L_0 = 0$ *and*

$$[x_{a_1}, x_{a_2}, \dots, x_{a_{c+1}}] = 0 \text{ whenever } (a_1, \dots, a_{c+1}) \text{ is } r\text{-independent,}$$

where $x_{a_1}, \dots, x_{a_{c+1}}$ *are homogeneous elements* $x_{a_i} \in L_{a_i}$. *Then* L *is nilpotent of* (c, q)*-bounded nilpotency class.*

The proof of the above theorem is technical and rather intricate. We use induction on c. First, we show that L is solvable of (c, q)-bounded derived length. This part of the proof eventually reduces to Shalev's theorem[49] on $\mathbb{Z}/n\mathbb{Z}$-graded Lie rings with few non-trivial homogeneous components. Some ideas of Khukhro's proof of Theorem 5.1 are also used. To prove the nilpotency of bounded class we employ induction on the derived length. By the induction hypothesis, $[L, L]$ is nilpotent of bounded class. By the Lie ring analogue of P. Hall's theorem (Lemma 2.1) it suffices to manage the case of metabelian Lie rings. Assume that L is metabelian. In this case $[x, y, z] = [x, z, y]$ for every $x \in [L, L]$ and $y, z \in L$. Using the condition of selective nilpotency we show that if a number n is big enough, in a commutator $[a_1, a_2, \dots, a_n]$ there are sufficiently many elements of the same homogeneous component L_b. Using that $[x, y, z] = [x, z, y]$ we can assume that $a_3 = \cdots = a_w$, where the number w is chosen according to the next lemma.

Lemma 5.1. *Under the assumptions of Theorem* 5.5 *there exists a* (c, q)*-bounded number* w *such that for all* $a, b \in \mathbb{Z}/p\mathbb{Z}$ *we have* $[L_a, \underbrace{L_b, \dots, L_b}_{w}] = 0$.

Thus, the quotient $L/[[L, L], [L, L]]$ is nilpotent of (c, q)-bounded class and consequently L is nilpotent of (c, q)-bounded class. We refer the reader to Ref. 43 for more details.

5.3. *Metacyclic Frobenius group of automorphisms with kernel of non-prime order*

Let now the kernel of the Frobenius group be cyclic of arbitrary non-prime order. Recall that we are considering a Lie ring (Lie algebra) L admitting a metacyclic Frobenius group of automorphisms FH with kernel F such that $C_L(F) = 0$ and $C_L(H)$ is nilpotent of class c. The main question of our interest: is L nilpotent and does there exist a bound on the nilpotency class in terms of $|H|$ and c? The problem has been recently solved in Refs. 29,42 by Khukhro, Shumyatsky and myself. First we cite the theorem on Lie algebras, which looks clearer than a similar result for Lie rings.

Theorem 5.6. *Let FH be a Frobenius group with cyclic kernel F and complement H of order q. Suppose that FH acts by automorphisms on a Lie algebra L in such a way that $C_L(F) = 0$ and $C_L(H)$ is nilpotent of class c. Then L is nilpotent of (c,q)-bounded class.*

Note that the assumption of the kernel F being cyclic is essential: see Examples 5.1 and 5.2 at the beginning of the chapter.

Theorem 5.6 extends the result of Khukhro[22] (Theorem 5.2) and the result of Makarenko–Shumyatsky[43] (Theorem 5.4) discussed above.

By Kreknin's theorem (Theorem 2.4) the condition $C_L(F) = 0$ implies the solvability of L of derived length bounded in terms of $|F|$, but our aim is to obtain nilpotency of class bounded in terms of $|H|$ and the nilpotency class of $C_L(H)$.

Note that we can assume that $n = |F|$ is not divisible by the characteristic of the ground field. Indeed, if the characteristic p is positive, then the Hall p'-subgroup $\langle\chi\rangle$ of F acts fixed-point-freely on L — otherwise the Sylow p-subgroup $\langle\psi\rangle$ of F would have non-trivial fixed points on the ψ-invariant subspace $C_L(\chi)$, and these would be non-trivial fixed points for F. Thus, L admits the Frobenius group of automorphisms $\langle\chi\rangle H$ with $C_L(\chi) = 0$. Replacing F by $\langle\chi\rangle$ we can assume that p does not divide n.

To prove the theorem, we first reduce the problem to $\mathbb{Z}/n\mathbb{Z}$-graded Lie rings satisfying the condition of selective nilpotency in exactly the same way as in the proof of Theorem 5.4. The resulting, in a certain sense combinatorial, fact is as follows.

Theorem 5.7. *Let n, q, r, c be positive integers such that $1 \leq r \leq n-1$ and the image of r in $\mathbb{Z}/d\mathbb{Z}$ is a primitive qth root of 1 for every divisor d of n. Let $L = \sum_{i=0}^{n-1} L_i$ be a $\mathbb{Z}/n\mathbb{Z}$-graded Lie ring such that $L_0 = 0$ and*

$$[x_{a_1}, x_{a_2}, \ldots, x_{a_{c+1}}] = 0 \text{ whenever } (a_1, \ldots, a_{c+1}) \text{ is } r\text{-independent},$$

where $x_{a_1}, \ldots, x_{a_{c+1}}$ are homogeneous elements $x_{a_i} \in L_{a_i}$. Then L is nilpotent of (c,q)-bounded nilpotency class.

Recall, that the sequence $(a_1, \ldots, a_k)$ of not necessarily distinct non-zero elements of $\mathbb{Z}/n\mathbb{Z}$ is r-dependent if it satisfies (1); and if the sequence $(a_1, \ldots, a_k)$ is not r-dependent, it is r-independent.

We now clarify where the condition on the numbers n, q, r comes from. If FH is a Frobenius group with cyclic kernel $F = \langle\varphi\rangle$ of order n and (necessarily cyclic) complement $H = \langle h\rangle$ of order q acting as $\varphi^{h^{-1}} = \varphi^r$ for some $1 \leq r \leq n-1$, then r is a primitive qth root of 1 in $\mathbb{Z}/n\mathbb{Z}$, and, moreover, the image of r in $\mathbb{Z}/d\mathbb{Z}$ is a primitive qth root of 1 for every divisor d of n, since h acts fixed-point-freely on every subgroup of F. Thus, n, q, r satisfy the condition in the hypothesis of Theorem 5.7.

The proof of Theorem 5.7 consists of two stages. First we derive the solvability of L of (c, q)-bounded derived length. This is done by using the same ideas as in the proof of Theorem 5.4 but arguments are more complicated. The second stage is obtaining the nilpotency of (c, q)-bounded class. We use induction on the derived length. If L is abelian, there is nothing to prove. Assume that L is metabelian — this is, in fact, the main part of the proof. When L is metabelian, $[x, y, z] = [x, z, y]$ for every $x \in [L, L]$ and $y, z \in L$. As for the $\mathbb{Z}/p\mathbb{Z}$-graded Lie algebra (Lie ring) with prime p in Theorem 5.4, by using the condition of selective nilpotency, we show that if a number t is big enough, there are sufficiently many elements of the same homogeneous component L_b in a commutator $[a_1, a_2, \ldots, a_t]$. Moving elements we can assume that $a_3 = \cdots = a_m$, where the number m is chosen according to the next lemma.

Lemma 5.2. *There is a (c, q)-bounded number m such that*

$$[L, \underbrace{L_b, \ldots, L_b}_{m}] = 0$$

for every $b \in \mathbb{Z}/n\mathbb{Z}$.

Lemma 5.2 is a key fact in the second stage of the proof. Unlike the analogous lemma for $\mathbb{Z}/p\mathbb{Z}$-graded Lie algebras with prime p (Lemma 5.1) the proof of the above fact is subtle and uses, in particular, induction on $|F|$, Vandermonde-type linear algebra arguments and again the relations implied by the condition of the fixed-point subring $C_L(H)$ being nilpotent of class c but this time in a metabelian Lie algebra. We refer the reader to Ref. 29 for more details.

We now state a similar theorem for Lie rings. The additional conditions on the additive group appear because we no longer have a direct sum of eigenspaces forming a $(\mathbb{Z}/n\mathbb{Z})$-grading.

Theorem 5.8. *Let FH be a Frobenius group with cyclic kernel F of order n and complement H of order q. Suppose that FH acts by automorphisms* *on a Lie ring L in such a way that $C_L(F) = 0$ and $C_L(H)$ is nilpotent of class c. Then*

(a) *the Lie ring L is solvable of (c, q)-derived length;*

(b) *for some functions $u = u(c, q)$ and $v = v(c, q)$ depending only on c and q, the Lie subring $n^u L$ is nilpotent of class v, that is, $\gamma_{v+1}(n^u L) = n^{u(v+1)}\gamma_{v+1}(L) = 0$;*

(c) *moreover, L is nilpotent of (c, q)-bounded class in either of the following cases:*

(i) *L is a Lie algebra over a field;*

(ii) *the additive group of L is periodic (includes the case of L finite);*

(iii) *n is invertible in the ground ring of L;*

(iv) *$nL = L$;*

(v) *n is a prime-power.*

The proof is basically along the same lines as the proof of Theorem 5.6 for algebras; the complications arise because the Lie ring L is not always a direct sum of homogeneous components $L_i = \{x \in L \mid x^\varphi = \omega^i x\}$; only the inclusion $nL \subseteq \sum_{i=0}^{n-1} L_i$ holds.

Theorem 5.8 covers many important cases, including those needed in our group-theoretic applications. Note also that the solvability result does not require the additional conditions: of course, L is solvable of n-bounded derived length by Kreknin's theorem, but we obtain a bound for the derived length in terms of $|H|$ and the nilpotency class of $C_L(H)$.

Problem 5.2. *Does Theorem 5.8 hold without the additional conditions* (i)–(v)*?*

5.4. *Consequences for Lie groups and locally nilpotent torsion-free groups*

By the well-known connection between Lie groups and their Lie algebras, the following theorem is an immediate consequence of Theorem 5.6.

Theorem 5.9. *Suppose that a connected Lie group G (complex or real) admits a finite Frobenius group of automorphisms FH with cyclic kernel F and complement H of order q such that $C_G(F) = 1$ and $C_G(H)$ is nilpotent of class c. Then G is nilpotent of (c, q)-bounded class.*

Another corollary of Theorem 5.6 follows by a similar Lie ring method – the Mal'cev correspondence, which is also based on the Exp and Log functors and the Baker–Campbell–Hausdorff formula.

Theorem 5.10. *Suppose that a locally nilpotent torsion-free group G admits a finite Frobenius group of automorphisms FH with cyclic kernel F and complement H of order q such that $C_G(F) = 1$ and $C_G(H)$ is nilpotent of class c. Then G is nilpotent of (c,q)-bounded class.*

Theorems 5.9 and 5.10 were proved in Ref. 29.

6. Bounding nilpotency class of a finite group

In this chapter we discuss the results related to bounding the nilpotency class of a finite group with a Frobenius group of automorphisms.

6.1. *Fixed-point-free kernel*

Let G be a finite group and let FH be a Frobenius group with kernel F and complement H. Suppose that FH acts by automorphisms on G in such a way that $C_G(F) = 1$ and $C_G(H)$ is nilpotent of class c. By Theorem 2.7 the group G is nilpotent. But does there exist a bound on its nilpotency class in terms of $|H|$ and c?

In general the answer is negative. If F is not cyclic, there are examples showing that there is no bound for the nilpotency class in terms of $|H|$ and c. We modify Example 5.2 at the beginning of Chapter 5 to produce a finite nilpotent p-group P of unbounded derived length admitting a non-metacyclic Frobenius group of automorphisms FH of order 12 such that $C_P(F) = 1$ and $C_P(H)$ is cyclic ("one-dimensional").

Example 6.1. In Example 5.2 we choose $p > m$. Then the Lazard correspondence based on the Baker–Campbell–Hausdorff formula (Ref. 37; also see, for example,§ 10.2 of Ref. 20) transforms L into a finite p-group P of the same (unbounded) derived length as L, which admits the same group of automorphisms FH of order 12 with $C_P(F) = 1$ and $C_P(H)$ cyclic.

If F is cyclic, the problem was recently solved in the affirmative by Khukhro, Shumyatsky and myself in Refs. 29,42.

Theorem 6.1. *Suppose that a finite group G admits a Frobenius group of automorphisms FH with cyclic kernel F and complement H of order q such that $C_G(F) = 1$ and $C_G(H)$ is nilpotent of class c. Then G is nilpotent of (c,q)-bounded class.*

If we "forget" the group H and consider G as a group admitting a fixed-point-free automorphism group F, we can observe that there are two different situations depending on whether the group F is of prime order or not. If $|F|$ is prime, then the nilpotency class of G is bounded in terms of $|F|$ by the Higman–Kreknin–Kostrikin theorem[16,34,35] (Theorem 2.2). The case, where $|F|$ is non-prime is much more complicated. We have already mentioned a long-standing conjecture (see Problem 1.1) that the derived length of a nilpotent finite group admitting a fixed-point-free automorphism must be bounded in terms of $|F|$. In our situation we have a much stronger hypothesis: in addition to the fixed-point-free action of the group F, the group G admits the Frobenius group of automorphisms FH with nilpotent centralizer of the complement H. But the aim is different: bounding the nilpotency class in terms of $|H|$ and the nilpotency class of $C_G(H)$.

The proof is based on a similar result on Lie rings — Theorem 5.8. To reduce the question to a Lie-theoretic problem we consider the associated Lie ring of the group G

$$L(G) = \bigoplus_{i=1}^{m} \gamma_i/\gamma_{i+1},$$

where m is the nilpotency class of G and γ_i are the terms of the lower central series of G. The nilpotency class of G coincides with that of $L(G)$. The action of the group FH on G naturally induces an action of FH on $L(G)$. By the definition of the associated Lie ring,

$$C_{L(G)}(H) = \bigoplus_i C_{\gamma_i/\gamma_{i+1}}(H),$$

which by Theorem 2.6 is equal to

$$\bigoplus_i C_{\gamma_i}(H)\gamma_{i+1}/\gamma_{i+1}.$$

The Lie ring products in $L(G)$ are defined for elements of the γ_i/γ_{i+1} in terms of the images of the group commutators and then extended by linearity. If c is the nilpotency class of $C_G(H)$, then any group commutator of weight $c+1$ in elements of $C_G(H)$ is trivial; hence any Lie ring commutator of weight $c+1$ in elements of $C_{\gamma_i}(H)\gamma_{i+1}/\gamma_{i+1}$ is also trivial. Since these elements generate $C_{L(G)}(H)$, this subring is nilpotent of class at most c. Because F acts fixed-point-freely on every quotient γ_i/γ_{i+1} (Lemma 2.2), it follows that $C_{L(G)}(F) = 0$. We now deduce from Theorem 5.8(c) that $L(G)$ is nilpotent of (c, q)-bounded class. Since the nilpotency class of G is equal to that of $L(G)$, the result follows.

6.2. *Frobenius groups of automorphisms with not necessarily fixed-point free action of the kernel*

Earlier in the paper we always examined the situation, where the kernel F of a Frobenius group FH acts fixed-point-freely on a finite group G. Let us now assume that $C_G(F)$ is non-trivial. As in the previous section, we suppose that $C_G(H)$ is nilpotent of a given class c. What can be said about the structure of G in this situation? If $|F|$ is prime, even without the assumption on the fixed points of the Frobenius complement, by Khukhro's theorem[18] the group G contains a subgroup of index bounded in terms of $|F|$ and $|C_G(F)|$ that is nilpotent of class bounded in terms of $|F|$. But in view of the results stated in the previous sections it makes sense to ask whether the additional conditions, such as an action of a Frobenius group and the nilpotency of $C_G(H)$, yield properties of G "independent" of F. Very little is known concerning this issue.

First we discuss the case of prime $|F|$. In Ref. 23 Khukhro proved the following result.

Theorem 6.2. *Let FH be a Frobenius group with kernel F of prime order p and complement H of order q. Suppose that FH acts coprimely on a finite group G in such a manner that $|C_G(F)| = m$ and $C_G(H)$ is abelian. Then G contains a characteristic subgroup of index bounded in terms of m and p that is nilpotent of class bounded in terms of q only.*

The proof of Theorem 6.2 relies on Theorem 5.3 about $\mathbb{Z}/n\mathbb{Z}$-graded Lie rings with many commuting components.

By the theorems of Fong [11], Hartley–Meixner–Pettet[15,47] and Khukhro[18] the group G contains a subgroup of (p, m)-bounded index that is nilpotent of p-bounded class. Moreover, this subgroup can be chosen characteristic by theorem of Khukhro–Makarenko[27]. But in Theorem 6.2 the bound for the nilpotency class does not depend on $|F|$. It would be interesting to obtain a bound for the index of a nilpotent subgroup independent of $|F|$ as well.

Problem 6.1. *Under the assumptions of Theorem 6.2 does G contain a characteristic subgroup of index bounded in terms of m and q that is nilpotent of class bounded in terms of q?*

Problem 6.2. *Can we omit the assumption $(|G|, |FH|) = 1$ in Theorem 6.2?*

Problem 6.3. *Suppose that a Frobenius group of automorphisms FH with kernel F of prime order and complement H acts on a finite group G in* *such a manner that $C_G(H)$ is nilpotent of class c.*

(a) *Does G contain a characteristic subgroup of index bounded in terms of $|C_G(F)|$ and $|F|$ that is nilpotent of class bounded in terms of $|H|$ and c?*

(b) *Does G contain a characteristic subgroup of index bounded in terms of $|C_G(F)|$, $|H|$ and c that is nilpotent of class bounded in terms of $|H|$ and c?*

To close this section we discuss the progress in bounding the nilpotency class of a group admitting a Frobenius group with non-cyclic and non-fixed-point-free kernel. Khukhro and Shumyatsky[31] studied a group G that admits a Frobenius group of automorphisms FH of coprime order with abelian kernel F of rank at most two and complement H with certain extra conditions.

Theorem 6.3. *Let FH be a Frobenius group with non-cyclic abelian kernel F and with complement H of order q. Suppose that FH acts coprimely on a finite group G in such a manner that $C_G(H)$ is abelian and $[C_G(u), \underbrace{C_G(v), \ldots, C_G(v)}_{k}] = 1]$ for any $u, v \in F \setminus \{1\}$. Then G is nilpotent of (k, q)-bounded class.*

If F is abelian of rank at least three, then the result becomes stronger.

Theorem 6.4. *Let FH be a Frobenius group with abelian kernel F of rank at least three and with complement H of order q. Suppose that FH acts coprimely on a finite group G in such a manner that $C_G(H)$ is abelian and $C_G(f)$ is nilpotent of class c for every $f \in F \setminus \{1\}$. Then G is nilpotent of (c, q)-bounded class.*

Previously, Shumyatsky[51,52] proved that if a finite p'-group G admits an elementary abelian group of automorphisms F of order p^2 (or of order p^3) such that $[C_G(u), \underbrace{C_G(v), \ldots, C_G(v)}_{c}] = 1]$ for any $u, v \in F \setminus \{1\}$ (respectively, $C_G(f)$ is nilpotent of class c for every $f \in F \setminus \{1\}$) then G is nilpotent of (p, c)-bounded class. But in Theorems 6.3 and 6.4 the additional conditions on the fixed points of the Frobenius complement yield the bounds for the nilpotency class independent of $|F|$.

7. Positive laws in fixed points of automorphisms

This last chapter is devoted to a recent, yet unpublished, result of Shumyatsky[56]. We consider finite groups admitting a metacyclic Frobenius group of automorphisms such that the fixed-point subgroup of the Frobenius complement satisfies a positive law.

A positive law in a group can be defined as follows. Let X denote the free group on free generators $x_1, x_2, \ldots$. A positive word in $x_1, x_2, \ldots$ is any non-trivial element of X not involving the inverses of the x_i. A positive (or semigroup) law of a group G is a non-trivial identity of the form $u \equiv v$, where u, v are positive words in $x_1, x_2, \ldots$ holding under every substitution $\{x_1, x_2, \ldots\} \to G$. For example, $x^n = 1$ and $xy = yx$ are positive laws. The maximum of lengths of u and v is called the degree of the law $u \equiv v$.

In 1953 Malcev[44] discovered a positive law of 2 variables and of degree 2^c that holds in any nilpotent group of class c. It follows that an extension of a nilpotent group by a group of finite exponent satisfies a positive law. In general, the converse is not true: Olshansky and Storozhev[46] constructed a 2-generator group satisfying a positive law that is not a periodic extension of a nilpotent group. However, for a large class of groups including all solvable and residually finite groups it is true by the theorem of Burns, Macedońska and Medvedev[6]. They proved that there exist functions $c(n)$ and $e(n)$ of n only, such that any solvable or residually finite group satisfying a positive law of degree n is an extension of a nilpotent group of class at most $c(n)$ by a locally finite group of exponent dividing $e(n)$.

Theorems 4.1 and 6.1 state that if a finite group G admits a metacyclic Frobenius groups of automorphisms FH with kernel F and complement H such that $C_G(F) = 1$, then the exponent of G is bounded in terms of $|FH|$ and the exponent of $C_G(H)$; if, in addition, $C_G(H)$ is nilpotent of class c, then G is nilpotent of $(c, |H|)$-bounded class. The next result of Shumyatsky[56] is a nice extension of these theorems to the case where $C_G(H)$ satisfies a positive law.

Theorem 7.1. *Let FH be a Frobenius group with cyclic kernel F and complement H. Suppose that FH acts on a finite group G in such a way that $C_G(F) = 1$ and $C_G(H)$ satisfies a positive law of degree k. Then G satisfies a positive law of degree bounded in terms of k and $|FH|$.*

The proof of Theorem 7.1 uses various methods, including Lie rings and powerful p-groups. Mainly, it relies on the technique developed in Ref. 29. But several other deep theorems are involved in the proof; these include Zelmanov's Solution of the Restricted Burnside Problem[63,64], the Burns–Macedońska–Medvedev theorem[6] and Kreknin's theorem[34].

Earlier, groups with centralizers of automorphisms satisfying a positive law were investigated in Refs. 39,53,54. These papers deal with elementary abelian groups of automorphisms. In particular, it was shown that if a finite q'-group admits an elementary abelian group A of automorphisms of order at least q^3 such that $C_G(a)$ satisfies a positive law of degree n for any $a \in A \setminus \{1\}$, then G satisfies a positive law of degree bounded by a function of q and n only[53].

As in Theorem 4.1 the estimate in Theorem 7.1 depends on the order of F.

Problem 7.1. *Under assumptions of Theorem* 7.1 *does* G *satisfy a positive law of degree bounded in terms of* k *and* $|H|$ *only?*

In connection with Theorem 7.1 we also rise the following question, which seems to be very difficult.

Problem 7.2. *Let* FH *be a Frobenius group with cyclic kernel* F *and complement* H. *Suppose that* FH *acts on a finite group* G *in such a way that* $C_G(F) = 1$ *and* $C_G(H)$ *is solvable of derived length* d. *Is* G *solvable of* $(d, |H|)$*-bounded derived length?*

Note that G is solvable by the Belyaev–Hartley theorem (Theorem 2.5), since F is nilpotent being a Frobenius kernel.

References

1. V. A. Antonov, S. G. Chekanov, On the Mazurov conjecture, *Siberian Electron. Math. Rep.*, http://semr.math.nsc.ru, **5** (2008), 8–13 (in Russian).
2. S. D. Bell and B. Hartley, A note on fixed-point-free actions of finite groups, *Quart. J. Math. Oxford Ser. (2)*, **41** (1990), 127–130.
3. V. V. Belyaev, B. Hartley, Centralizers of finite nilpotent subgroups in locally finite groups, *Algebra Logika*, **35** (1996), 389–410 (in Russian); English transl., *Algebra Logic*, **35** (1996), 217–228.
4. T. R. Berger, Nilpotent fixed point free automorphism groups of solvable groups, *Math. Z.*, **131** (1973), 305–312.
5. A. Borel, G. D. Mostow, On semi-simple automorphisms of Lie algebras, *Ann. of Math.*, **61** (1955), 389–405.
6. R. G. Burns, O. Macedońska, Yu. Medvedev, Groups satisfying semigroup laws, and nilpotent-by-Burnside varieties, *J. of Algebra*, **195**, 510–525.
7. E. C. Dade, Carter subgroups and Fitting heights of finite solvable groups, *Illinois J. Math.*, **13** (1969), 449–514.
8. J. D. Dixon, M. P. F. du Sautoy, A. Mann, D. Segal, *Analytic pro-p groups* (London Math. Soc. Lecture Note Series **157**), Cambridge Univ. Press., 1991.
9. L. Evens, On a theorem of Thompson on fixed points of groups acting on p-groups, *Math. Z.*, **93** (1966), 105–108.
10. W. Feit and J. G. Thompson, Solvability of groups of odd order, *Pacific J. Math.*, **13** (1963), 775–1029.

11. P. Fong, On orders of finite groups and centralizers of p-elements, *Osaka J. Math.*, **13** (1976), 483-489.
12. P. Hall, Some sufficient conditions for a group to be nilpotent, *Illinois J. Math.*, **2** (1958), 787–801.
13. P. Hall, G. Higman, On the p-length of p-soluble groups and reduction theorems for Burnside's problem, *Proc. London Math. Soc.*, **6** (1956), 1–42.
14. B. Hartley, I. M. Isaacs, On characters and fixed points of coprime operator groups, *J. Algebra*, **131** (1990), 342–358.
15. B. Hartley, T. Meixner, Finite soluble groups containing an element of prime order whose centralizer is small, *Arch. Math. (Basel)*, **36** (1981), 211-213.
16. G. Higman, Groups and rings which have automorphisms without non-trivial fixed elements, *J. London Math. Soc.* (2), **32** (1957), 321–334.
17. E. I. Khukhro, Fixed points of p-automorphisms of finite p-groups, *Algebra Logika*, **14** (1975), 697–703 (in Russian); English transl., *Algebra Logic*, **14** (1976), 417–420.
18. E. I. Khukhro, Groups and Lie rings admitting an almost regular automorphism of prime order, *Mat. Sbornik*, **181** (1990), 1207–1219 (in Russian); English transl., *Math. USSR Sbornik*, **71** (1992), 51–63.
19. E. I. Khukhro, *Nilpotent groups and their automorphisms*, Walter de Gruyter, Berlin, New York, 1993.
20. E. I. Khukhro, *p-Automorphisms of finite p-groups*, LMS Lect. Notes Ser., **246**, Cambridge Univ. Press, 1998.
21. E. I. Khukhro, Finite groups of bounded rank with an almost regular automorphism of prime order, *Sibirsk. Mat. Zh.*, **43** (2002), 1182-1191 (in Russian); English transl., *Siberian Math. J.*, **43** (2002), 955-962.
22. E. I. Khukhro, Graded Lie rings with many commuting components and an application to 2-Frobenius groups, *Bull. London Math. Soc.*, **40** (2008), 907–912.
23. E. I. Khukhro, Lie rings with a finite cyclic grading in which there are many commuting components, *Siberian Electron. Math. Rep.*, http://semr.math.nsc.ru, **6** (2009), 243–250 (in Russian).
24. E. I. Khukhro, Fixed points of the complements of Frobenius groups of automorphisms, *Sibirsk. Mat. Zh.*, **51** (2010), 694-699 (in Russian); English transl., *Siberian Math. J.*, **51** (2010), 554–558.
25. E. I. Khukhro, Applications of Clifford's theorem to Frobenius groups of automorphisms, Proc. of Ischia Group Theory 2010, World Scientific Publishing, Singapore, 2011.
26. E. I. Khukhro, Nilpotent length of a finite group admitting a Frobenius group of automorphisms with fixed-point-free kernel, *submitted to Algebra Logika.*
27. E. I. Khukhro, N. Yu. Makarenko, Large characteristic subgroups satisfying multilinear commutator identities, *J. London Math. Soc.*, **75** (2007), 635-646.
28. E. I. Khukhro, N. Yu. Makarenko, P. Shumyatsky, Nilpotent ideals in graded Lie algebras and almost constant-free derivations, *Commun. Algebra*, **36** (2008), 1869-1882.
29. E. I. Khukhro, N. Yu. Makarenko, P. Shumyatsky, Frobenius groups of automorphisms and their fixed points, preprint, http://arxiv.org/abs/1010.0343.

30. E. I. Khukhro, P. Shumyatsky, Bounding the exponent of a finite group with automorphisms, *J. Algebra*, **212** (1999), 363–374.
31. E. I. Khukhro, P. Shumyatsky, Nilpotency of finite groups with Frobenius groups of automorphisms, *Monatshefte für Mathematik* Online: 12 February 2010, DOI: 10.1007/s00605-010-0198-8
32. L. G. Kovács, Groups with regular automorphisms of order four, *Math. Z.*, **75** (1961), 277–294.
33. L. G. Kovács, On finite soluble groups, *Math. Z.*, **103** (1968), 37–39.
34. V. A. Kreknin, The solubility of Lie algebras with regular automorphisms of finite period, *Math. USSR Doklady*, **4** (1963), 683–685.
35. V. A. Kreknin, A. I. Kostrikin, Lie algebras with regular automorphisms, *Math. USSR Doklady*, **4** (1963), 355–358.
36. H. Kurzweil, p-Automorphismen von auflöosbaren p'-Gruppen, *Math. Z.*, **120** (1971), 326–354.
37. M. Lazard, Sur les groupes nilpotents et les anneaux de Lie, *Ann. Sci. École Norm. Supr.*, **71** (1954), 101–190.
38. M. Lazard, Groupes analytiques p-adiques, *Inst. Hautes Études Sci. Publ. Math.*, **26** (1965) 389–603.
39. A. Lima, P. Shumyatsky, On groups satisfying a positive law in fixed points, *J. Algebra*, **322** (2009), 245–253.
40. A. Lubotzky, A. Mann, Powerful p-groups. I: finite groups, *J. Algebra*, **105** (1987), 484–505.
41. N. Yu. Makarenko, Graded Lie algebras with few nontrivial components, *Sibirsk. Mat. Zh.*, **48** (2007), 116-137 (in Russian); English transl., *Siberian Math. J.*, **48** (2007), 95–111.
42. N. Yu. Makarenko, E. I. Khukhro, and P. Shumyatsky, Fixed points of Frobenius groups of automorphisms, *Doklady Akad. Nauk*, **437** (2011), 1–4.
43. N. Y. Makarenko, P. Shumyatsky, Frobenius groups as groups of automorphisms, *Proc. Amer. Math. Soc.*, **138** (2010), 3425–3436.
44. A. I. Malcev, Nilpotent semigroups, *Uchen. Zap. Ivanovsk. Ped. Inst.*, **4** (1953), 107–111.
45. V. D. Mazurov, Recognition of the finite simple groups $S_4(q)$ by their element orders, *Algebra Logic*, **41** (2002), 93–110.
46. A. Yu. Olshansky, A. Sorozhev, A group variety defined by a semigroup law, *J. Austral. Math. Soc. Ser. A*, **60** (1996), 255–259.
47. M. R. Pettet, Automorphisms and Fitting factors of finite groups, *J. Algebra*, **72** (1981), 404-412.
48. E. Romano, P. Shumyatsky, On the exponent of a finite group with a four-group of automorphisms, Proc. of Ischia Group Theory 2010, World Scientific Publishing, Singapore, 2011.
49. A. Shalev, Automorphisms of finite groups of bounded rank, *Israel J. Math.*, **82** (1993), 395-404.
50. E. E. Shult, On groups admitting fixed point free abelian operator groups, *Illinois J. Math.*, **9** (1965), 701–720.
51. P. Shumyatsky, On locally finite groups and the centralizers of automorphisms, *Boll. Unione Mat. Italiana*, **4** (2001), 731–736.

52. P. Shumyatsky, Finite Groups and the Fixed Points of Coprime Automorphisms, *Proc. Amer. Math. Soc.*, **129** (2001), 3479–3484.
53. P. Shumyatsky, Positive laws in fixed points, *Trans. Amer. Math. Soc.*, **356** (2004), 2081–2091.
54. P. Shumyatsky, Positive laws in derived subgroups of fixed points, *Quarterly Journal of Mathematics*, **60** (2009), 121–132.
55. P. Shumyatsky, On the exponent of a finite group with an automorphism group of order twelve, to appear in *J. Algebra*, 2011.
56. P. Shumyatsky, Positive laws in fixed points of automorphisms of finite groups, to appear in *J. of Pure and Applied Algebra.*
57. A. G. R. Stewart, On the class of certain nilpotent groups, *Proc. Roy. Soc. Ser. A*, **292** (1966), 374–379.
58. J. G. Thompson, Finite groups with fixed-point-free automorphisms of prime order, *Proc. Nat. Acad. Sci. USA*, **45** (1959), 578–581.
59. J. G. Thompson, Fixed points of p-groups acting on p-groups, *Math. Z.*, **86** (1964), 12–13.
60. J. G. Thompson, Automorphisms of solvable groups, *J. Algebra*, **1** (1964), 259–267.
61. A. Turull, Fitting height of groups and of fixed points, *J. Algebra*, **86** (1984), 555-566.
62. *Unsolved Problems in Group Theory. The Kourovka Notebook*, no 17, Institute of mathematics, Novosibirsk, 2010.
63. E. Zelmanov, The solution of the restricted Burnside problem for groups of odd exponent, *Math. USSR Izv.*, **36** (1991), 41–60.
64. E. Zelmanov, The solution of the restricted Burnside problem for 2-groups, *Math. Sb.* **182** (1991), 568–592.

ON INTEGRAL REPRESENTATIONS OF FINITE GROUPS

D. MALININ

Institut des Hautes Études Scientifiques
Le Bois-Marie 35, route de Chartres
91440 Bures-sur-Yvette, France
E-mail: dmalinin@ihes.fr

In this paper we study representations of finite groups stable under Galois operation over arithmetic rings in local and global fields.

We consider a Galois extensions E/F and realization fields of finite subgroups $G \subset GL_n(E)$ stable under the natural operation of the Galois group of E/F; let $E = F(G)$ be a field obtained via adjoining to F all matrix coefficients of all matrices $g \in G$. Though for sufficiently large n and a fixed algebraic number field F every its finite extension E is realizable as $F(G)$ for some group G above, there is only a finite number of possible $F(G)$ if $G \subset GL_n(O_E)$ for the ring O_E of integers of E. We study the possible realization fields for finite extensions of $\mathbf{Q}$, $\mathbf{Q_p}$ and global fields of positive characteristic. In particular, for a finite Galois extension $E/\mathbf{Q}$ and any finite subgroup $G \subset GL_n(O_E)$ which is stable under the natural operation of the Galois group $Gal(E/\mathbf{Q})$ the realization field $\mathbf{Q}(G) = \mathbf{Q}(\zeta_m)$ for an appropriate root ζ_m of 1.

Some related results and conjectures are considered.

Keywords: Algebraic integers; Galois groups; Integral representations; Realization fields.

1. Introduction

Let E/F be a Galois extension of finite degree of global fields, i.e. E, F are finite extensions of the field of rationals $\mathbf{Q}$ or a field of rational functions $R(x)$ with a finite field R.

Let us denote by O_E and O_F the maximal orders of E and F, and let Γ be the Galois group of E/F. Let $E = F(G)$ be a field obtained via adjoining to F all matrix coefficients of all matrices $g \in G$ for some finite subgroup $G \subset GL_n(E)$.

We are interested in 3 basic conditions for the Γ-operation on G and the integrality of G.

A) G is Γ-stable under the natural Galois operation.

B) $G \subset GL_n(O_E)$.

C) A primitive t-root of 1 $\zeta_t \notin E$.

We intend to discuss the following questions:

Question 1. Do the conditions A) and B) imply $G \subset GL_n(FE_{ab})$, where E_{ab} is the maximal abelian subextension of $E/\mathbf{Q}$?

Question 2. Do the conditions A), B) and C) imply $G \subset GL_n(F)$?

Question 3. Is it possible to classify the realization fields $E = F(G)$?

Let us first consider a Galois extension E/F of characteristic 0 and realization fields of finite abelian subgroups $G \subset GL_n(E)$ of a given exponent t. We assume that G is stable under the natural operation of the Galois group of E/F. In Refs. 6–8,10 it is shown that under some reasonable restrictions for n any E can be a realization field of G, while if all coefficients of matrices in G are algebraic integers there are only finitely many fields E of realization having a given degree d for prescribed integers n and t or prescribed n and d.

Below O_E is the maximal order of E and $F(G)$ is an extension of F generated via adjoining to F all matrix coefficients of all matrices $g \in G$, Γ is the Galois group of E over F.

In Ref. 8 we prove the existence of abelian Γ-stable subgroups G such that $F(G) = E$ provided some reasonable restrictions on the fixed normal extension E/F and integers n, t, d hold and study the interplay between the existence of Γ-stable groups G over algebraic number fields and over their rings of integers.

The problems below originate from classification problems of positive definite quadratic lattices and their isometries. There is a number of applications to finite group schemes, arithmetic algebraic geometry and Galois cohomology (see Refs. 2,3,6).

Let K be a totally real algebraic number field with the maximal order O_K, G an algebraic subgroup of the general linear group $GL_n(\mathbf{C})$ defined over the field of rationals $\mathbf{Q}$. Since G can be embedded to $GL_n(\mathbf{C})$, the intersection $G(O_K)$ of $GL_n(O_K)$ and $G(K)$, the subgroup of K-rational points of G, can be considered as the group of O_K-points of an affine group scheme over $\mathbf{Z}$, the ring of rational integers. Assume G to be definite in the following sense: the real Lie group $G(\mathbf{R})$ is compact. The problem which is our starting point is the question:

Does the condition $G(O_K) = G(\mathbf{Z})$ always hold true?

This problem is easily reduced to the following conjecture from the representation theory:

Let $K/\mathbf{Q}$ be a finite Galois extension of the rationals and $G \subset GL_n(O_K)$ be a finite subgroup stable under the natural operation of the Galois group $\Gamma = Gal(K/\mathbf{Q})$. Then there is the following

Conjecture 1. *If K is totally real, then $G \subset GL_n(\mathbf{Z})$.*

There are several reformulations and generalizations of the conjecture.

Definition 1.1. Consider an arbitrary not necessarily totally real finite Galois extension K of the rationals $\mathbf{Q}$ and a free $\mathbf{Z}$-module M of rank n with basis $m_1, \dots, m_n$. The group $GL_n(O_K)$ acts in a natural way on $O_K \otimes M \cong \bigoplus_{i=1}^n O_K m_i$. The finite group $G \subset GL_n(O_K)$ is said to be of A-type, if there exists a decomposition $M = \bigoplus_{i=1}^k M_i$ such that for every $g \in G$ there exists a permutation $\Pi(g)$ of $\{1, 2, \dots, k\}$ and roots of unity $\epsilon_i(g)$ such that $\epsilon_i(g) g M_i = M_{\Pi(g)i}$ for $1 \le i \le k$.

The following conjecture generalizes (and would imply) conjecture 1:

Conjecture 2. *Any finite subgroup of $GL_n(O_K)$ stable under the Galois group $\Gamma = Gal(K/\mathbf{Q})$ is of A-type.*

For totally real fields K conjecture 2 reduces to conjecture 1.

Both conjectures are true (see Ref. 3 for the proof) and have some extra applications.

2. Further results

The following result was obtained in Ref. 8 (see also Ref. 6).

Theorem 2.1 (Finiteness Theorem). *1) For a given number field F and integers n and t, there are only a finite number of normal extensions E/F such that $E = F(G)$ and G is a finite abelian Γ-stable subgroup of $GL_n(O_E)$ of exponent t.*

2) For a given number field F and integers n and d, there is only a finite number of fields E such that $d = [E\colon F]$ and $E = F(G)$ for some finite Γ-stable subgroup G of $GL_n(O_E)$.

In the case of fields of positive characteristic we have

Theorem 2.2 (see Ref. 11, Theorem 2.1). *Let F be a global field of a positive characteristic p, and let E be a splitting field of some irreducible*

polynomial $f(y) \in F[y]$ *whose roots are the conjugates of some element* $t \in E$. *Then* $E = F(G)$ *for any positive integer* n *and an appropriate group* $G \subset GL_n(E)$. *Moreover, if* $t \in E$ *is an element of* O_E, *then* $G \subset GL_n(O_E)$.

Theorem 2.3 (see Ref. 8, Theorem 1). *Let* F *be a field of characteristic 0, let* $d > 1, t > 1$ *and* $n \geq \phi_E(t)d$ *(here* $\phi_E(t)d = [E(\zeta_t) : E]$ *is the generalized Euler function,* ζ_t *is a primitive t-root of 1) be given integers, and let* E *be a given normal extension of* F *having the Galois group* Γ *and degree* d. *Then there is an abelian* Γ*- stable subgroup* $G \subset GL_n(E)$ *of the exponent* t *such that* $E = F(G)$.

In fact, G can be generated by matrices $g^\gamma, \gamma \in \Gamma$ for some $g \in GL_n(E)$.

Theorem 2.4 (see Ref. 8, Proposition 1). *Let* E/F *be a given normal extension of algebraic number fields with the Galois group* Γ, $[E : F] = d$, *and let* $G \subset GL_n(E)$ *be a finite abelian* Γ*-stable subgroup of exponent* t *such that* $E = F(G)$ *and* n *is the minimum possible. Then* $n = d\phi_E(t)$ *and* G *is irreducible under conjugation in* $GL_n(F)$. *Moreover, if* G *has the minimum possible order, then* G *is a group of type* $(t, t, ..., t)$ *and order* t^m *for some positive integer* $m \leq d$.

In the case of quadratic extensions we can give an obvious example.

Example 2.1. Let $d = 2$, $t = 2$. Set $E = \mathbf{Q}(\sqrt{a})$ and $g = \begin{vmatrix} 0 & 1 \\ a^{-1} & 0 \end{vmatrix} \sqrt{a}$ for any $a \in F$ which is not a square in F. Then Γ is a group of order 2 and $G = \{I_2, -I_2, g, -g\}$ is a Γ-stable abelian group of exponent 2.

In the case of unramified extensions the following theorem for integral representations in a similar situation is proven in Ref. 10:

Theorem 2.5. *Let* $d > 1, t > 1$ *be given rational integers, and let* E/F *be an unramified extension of degree* d.
1) If $n \geq \phi_E(t)d$, *there is a finite abelian* Γ*- stable subgroup* $G \subset GL_n(O'_E)$ *of exponent* t *such that* $E = F(G)$ *where* O'_E *is the intersection of valuation rings of all localization rings of* O_E *with respect to primes ramified in* E/F.
2) If $n \geq \phi_E(t)dh$ *and* h *is the exponent of the class group of* F, *there is a finite abelian* Γ*-stable subgroup* $G \subset GL_n(O_E)$ *of exponent* t *such that* $E = F(G)$.
3) If $n \geq \phi_E(t)d$ *and* h *is relatively prime to* n, *then any* G *given in 1) is conjugate in* $GL_n(F)$ *to a subgroup of* $GL_n(O_E)$.
4) If d *is odd, then any* G *given in 1) is conjugate in* $GL_n(F)$ *to a subgroup of* $GL_n(O_E)$.

In all cases above G can be constructed as a group generated by matrices $g^\gamma, \gamma \in \Gamma$ for some $g \in GL_n(E)$.

Some further results for Galois stable groups G with entries in unramified field extensions of characteristic 0 can be found in Refs. 6,10.

The case $F = \mathbf{Q}$, the field of rationals, is specially interesting since there are no unramified extensions of $\mathbf{Q}$. The following theorem was proven in Ref. 3 (see also Ref. 7 for the case of totally real extensions) using the classification of finite flat group schemes over $\mathbf{Z}$ annihilated by a prime p obtained by V. A. Abrashkin and J.- M. Fontaine:

Theorem 2.6. *Let $K/\mathbf{Q}$ be a normal extension with Galois group Γ, and let $G \subset GL_n(O_K)$ be a finite Γ-stable subgroup. Then $G \subset GL_n(O_{K_{ab}})$ where K_{ab} is the maximal abelian over $\mathbf{Q}$ subfield of K.*

A similar result can be expected in the case of local field extensions. Consider a finite Galois extension $K/\mathbf{Q_p}$ of the field $\mathbf{Q_p}$ of rational p-adic numbers for $p \neq 2$ and a free $\mathbf{Z_p}$-module M of rank n with basis $m_1, \ldots, m_n$. The group $GL_n(O_K)$ acts in a natural way on $O_K \otimes M \cong \bigoplus_{i=1}^n O_K m_i$. In this case our definition 1.1 should be modified:

Definition 2.1. Consider a finite Galois extension $K/\mathbf{Q_p}$ for $p \neq 2$ and a free $\mathbf{Z_p}$-module M of rank n with basis $m_1, \ldots, m_n$. The group $GL_n(O_K)$ acts in a natural way on $O_K \otimes M \cong \bigoplus_{i=1}^n O_K m_i$. A finite group $G \subset GL_n(O_K)$ is said to be of A-type, if there exists a decomposition $M = \bigoplus_{i=1}^k M_i$ such that for every $g \in G$ there exists a permutation $\Pi(g)$ of $\{1, 2, \ldots, k\}$ and roots of unity $\epsilon_i(g)$ such that $\epsilon_i(g) g M_i = M_{\Pi(g)i}$ for $1 \leq i \leq k$.

Example 2.2. For a primitive p-root ζ_p of 1 and $\theta = \frac{1}{2}(\zeta_p + \zeta_p^{-1})$ we can consider $K = \mathbf{Q_p}(\theta, \sqrt{1-\theta^2})$ and a Γ-stable subgroup $G \subset GL_n(O_K)$ generated by matrices $g^c, c \in \mathbf{Z}$, where

$$g = \begin{vmatrix} \theta & \sqrt{1-\theta^2} \\ -\sqrt{1-\theta^2} & \theta \end{vmatrix}.$$

Note that $K/\mathbf{Q_p}$ is an abelian tamely ramified extension and G is a cyclic subgroup of $GL_2(O_K)$ of order p. If the odd prime $p \equiv 3(mod4)$, then $\zeta_p \notin K$ since $\zeta_p = \theta + \sqrt{-1} \cdot \theta^{-1}$ and the congruence $x^2 + 1 \equiv 0(modp)$ has no solutions iff $p \equiv 3(mod4)$.

The Main Theorem in Ref. 3 gives a more explicit formulation of the theorem above and states the following:

Theorem 2.7 (Main Theorem). *Let K be a finite Galois extension of* $\mathbf{Q}$ *and G be a finite subgroup of $GL_n(O_K)$ which is stable under the natural operation of the Galois group Γ of the field K. Then G is of A-type and, in particular, $G \subset GL_n(O_{K_{ab}})$ holds.*

Corollary 2.1. *The realization field* $\mathbf{Q}(G) = \mathbf{Q}(\zeta_m)$ *for any G which satisfies the conditions of the Main Theorem and an appropriate root ζ_m of 1.*

The proof of the corollary follows immediately from the Main Theorem and our definition 1.1.

Following the result of the Main Theorem, we can ask 2 questions for the groups G over local fields:

Question 4. *Let K be a finite Galois extension of* $\mathbf{Q_p}$ *and G be a finite subgroup of $GL_n(O_K)$ which is stable under the natural operation of the Galois group Γ of the field K. Is it true that $G \subset GL_n(O_{K_{ab}})$ holds, K_{ab} the maximal abelian subextension of K over* $\mathbf{Q_p}$*?*

It is known (see Refs. 3,6,7) that for global normal field extensions $K/\mathbf{Q}$ the same question can be reduced to the case of elementary abelian Galois stable p-subgroup $G \subset GL_n(O_K)$ of exponent p.

Question 5. *Let K be a finite Galois extension of* $\mathbf{Q_p}$ *with Galois group Γ, and let G be a finite Γ-stable subgroup of $GL_n(O_K)$. Is it possible to classify all fields* $\mathbf{Q_p}(G)$*?*

We can give a positive answer to Question 4 for any elementary abelian Γ-stable p-subgroup $G \subset GL_n(O_K)$. This also shows that for elementary abelian Γ-stable p-groups G above all fields $\mathbf{Q_p}(G)$ are abelian over $\mathbf{Q_p}$.

It follows from example 2.2 that for abelian extensions $K/\mathbf{Q_p}$ of local fields under the conditions of Question 4 G is not always a group of A-type.

Theorem 2.8. *Let $K/\mathbf{Q_p}$ $(p \neq 2)$ be a normal extension of local fields, let Γ be its Galois group, let $G \subset GL_n(O_K)$ be an elementary abelian Γ-stable p-subgroup of exponent p, and let $K = \mathbf{Q_p}(G)$. Then $K/\mathbf{Q_p}$ is an abelian field extension.*

The idea of the proof is to show that $K = \mathbf{Q_p}(G)$ has a special ramification structure over $\mathbf{Q_p}$, in particular, the inertia subgroup of Γ is cyclic for the prime divisor of p. For a certain subfield $E \subset K$ let E/F be a Galois extension of fields with the Galois group $\bar{\Gamma} = \{\sigma_1 = 1, \sigma_2, ..., \sigma_t\}$, let $w_1, w_2, \ldots, w_d$ be a basis of O_E over O_F, and let $\zeta_p \in E$. For the proof of Theorem 2.8 we can use the reduction to the case of group

$G \subset GL_n(E)$, which is irreducible under $GL_n(F)$-conjugation and generated by all $g^\gamma, \gamma \in \bar{\Gamma}$ and some $g \in G$. We can use the following criterion of integrality of G:

Proposition 2.1 (see Ref. 3, Proposition 1 or Ref. 8, Theorem 2.6). *Let $G \subset GL_n(E)$ be irreducible under $GL_n(F)$-conjugation. Then G is conjugate in $GL_n(F)$ to a subgroup of $GL_n(O_E)$ if and only if all determinants*

$$d_k = \det \begin{vmatrix} w_1 & \dots & w_{k-1} & \zeta_{(1)} & w_{k+1} & \dots & w_t \\ w_1^{\sigma_2} & \dots & w_{k-1}^{\sigma_2} & \zeta_{(2)}^{\sigma_2} & w_{k+1}^{\sigma_2} & \dots & w_t^{\sigma_2} \\ \vdots & & & & & & \\ w_1^{\sigma_t} & \dots & w_{k-1}^{\sigma_t} & \zeta_{(t)}^{\sigma_t} & w_{k+1}^{\sigma_t} & \dots & w_t^{\sigma_t} \end{vmatrix}$$

are divisible by $D = det[w_m^{\sigma_k}]_{k,m}$ in the ring O_E for some p-roots $\zeta_{(i)}$ of 1, $i = 1, 2, ..., t$, depending on the generators $g^\gamma, \gamma \in \bar{\Gamma}$ of G.

Let us denote by Γ_0 and Γ_1 the inertia and the first ramification subgroups of Γ respectively, let $G_0 = G^{\Gamma_1}$ be a subgroup of Γ_1-fixed elements of G, let $F = K^{\Gamma_0}$ be the inertia subfield of K, and let $E = F(G_0)$. The proposition above allows us to eliminate the possibility that the ramification index e_0 of $F(G_0)/F$ is not divisible by $p-1$. An extra argument involving a Galois equivariant map $\psi = \psi_m : G' \to GL_{mr}(K)$ for certain $GL_m(F)$-irreducible component G' of G and some $m < n$ (comp. Theorem 2.1) of Ref. 7 and the Main Theorem, case II of the proof in Ref. 3) completes the proof of Theorem 2.8 if e_0 is divisible by $p-1$.

Theorem 2.8 allows us to give a new proof of the Main Theorem stated above and proven in Ref. 3. In the virtue of Theorem 2.8, the proof of the Main Theorem can be reduced to the situation where K is an unramified extension of the maximal abelian subfield of K over $\mathbf{Q}$.

For the proof of the Main Theorem we can first reduce it to the case of elementary abelian group G (see Refs. 3,7), next to apply Theorem 2.8 to prove that the field extension $\mathbf{Q_p}(G)/\mathbf{Q_p}$ is abelian and to use the following theorem proven by Y. Ihara:

Let k be a fixed algebraic number field of finite degree over $\mathbf{Q}$, k^{ab} be the maximal abelian extension of k.

Theorem 2.9 (Y. Ihara, see Ref. 1). *Let L be a finite Galois extension of k. Then, Lk^{ab} is unramified over k^{ab} if and only if, for any prime divisor of L its decomposition group in L/k is commutative.*

We can use Theorem 2.9 for the case $k = \mathbf{Q}$. The proof of Theorem 2.9 is given in Ref. 1, Proposition 1.

Finally we use the special ramification properties of the field $K = \mathbf{Q_p}(G)$ in Theorem 2.8 to prove the Main Theorem above using Theorem 2.8 and Theorem 2.9.

3. The case of relative extensions of number fields

It is known that if E/F has unramified subextensions E_1/F, $E_1 \subset E$, then there exist examples of Galois stable finite groups $G \subset GL_n(O_E)$ (see Ref. 10 for an explicit construction). This is completely different from the situation where $F = \mathbf{Q}$ and there are no unramified extensions of the ground field $\mathbf{Q}$. We can consider the role of the group of units of the ring of integers O_E for the existence of finite $Gal(E/F)$-stable subgroups $G \subset GL_n(O_E)$.

It is also difficult to transfer the idea of reduction to the case of abelian Galois stable groups G of composite order.

Example 3.1. For $p \neq 2$ consider the following matrix:

Let

$$g_2 = \begin{vmatrix} 0 & \sqrt[p]{u} \\ (\sqrt[p]{u})^{-1} & 0 \end{vmatrix}$$

and $g := diag(g_2, I_{p-2}) \in GL_p(O_K)$. Then $g^\gamma, \gamma \in \Gamma$ and $\zeta_p I_p$ generate a finite Γ-stable nonabelian subgroup of $GL_p(O_K)$ of order divisible by 2 and p.

Example 3.2. Let

$$g = \begin{vmatrix} \sqrt{3+\sqrt{2}} & -\sqrt{2+\sqrt{2}} \\ \sqrt{2+\sqrt{2}} & -\sqrt{3+\sqrt{2}} \end{vmatrix},$$

let $E = F(\sqrt{3+\sqrt{2}})$, $F = \mathbf{Q}(\sqrt{3+\sqrt{2}} \cdot \sqrt{2+\sqrt{2}})$. Then E/F is ramified at 2, the ramification is wild, and $G = \{g, -g, I_2, -I_2\} \subset GL_2(O_E)$ is a Γ-stable subgroup of order 2 and exponent 2.

The difficulties to extend the result of the Main Theorem to the case of relative extensions over a ground field F ramified over $\mathbf{Q}$ can be illustrated using the following construction:

Example 3.3. If there exist an intermediate extension $L = F(\sqrt[p]{u}) \subset E$

for some unit $u \in O_E$, we can put

$$g = \begin{vmatrix} 0 & \sqrt[p]{u} & 0 \dots & 0 \\ 0 & 0 & \sqrt[p]{u} \dots & 0 \\ \vdots & \ddots & \ddots & \\ 0 & \dots & 0 & \sqrt[p]{u} \\ \sqrt[p]{u}^{1-p} & \dots & 0 & 0 \end{vmatrix}$$

Then $g^\gamma, \gamma \in \Gamma$ and $\zeta_p I_p$ generate a finite Γ-stable subgroup of $GL_p(O_E)$.

Hence for relative extensions $E/F, L \neq F$ and some units u it may happen that neither $F(\zeta_p, \sqrt[p]{u}) \subset FE_{ab}$, nor $F(\sqrt[p]{u})/F$ is unramified, when $L = F(\zeta_p)$.

However, some progress is still possible to give a positive answer for relative extensions of number fields that satisfy the following

Assumption. *Consider relative extensions K/F which are of the form $K = TF$. Here we assume: T is a finite Galois over $\mathbf{Q}$ and unramified outside the rational primes $p_1, p_2, ...p_k$, and $F/\mathbf{Q}$ is a number field unramified in $p_1, p_2, ..., p_k$. So we suppose that $(d(F/\mathbf{Q}), p_i) = 1$ for all indices i and the discriminant $d(F/\mathbf{Q})$ of $F/\mathbf{Q}$. We consider finite subgroups G of $GL_n(O_K)$ that are stable under the natural operation of the Galois group $Gal(K/F)$.*

It is possible to reduce our considerations to the case of the only one prime $p_1 = p$.

Theorem 3.1. *Let F be a number field of discriminant $d(F)$ not divisible by an odd prime p and let T be a finite Galois extension of $\mathbf{Q}$ of discriminant coprime to $d(F)$. Set $K = TF$. If G is a finite $Gal(K/F)$-stable p-subgroup of $GL_n(O_K)$ then $G \subset GL_n(FT_{ab})$ where T_{ab} is the maximal abelian subextension of $T/\mathbf{Q}$ (or equivalently, the commutator subgroup of $Gal(K/F)$ acts trivially on G).*

The proof of Theorem 3.1 is given in Ref. 4.

Remark 3.1. Under the assumptions of Theorem 3.1 for K and F there do not exist unramified intermediate extensions between K and F.

4. Rarity of Γ-stable representations.

Let $K = \mathbf{Q}$ or $K = \mathbf{Q}(\sqrt{d})$ and d is a negative rational integer. We consider the set $O(N) = \{\alpha \in O_K \big| \, |N_{K/\mathbf{Q}}(\alpha)| \leqslant N\}$ where $N_{K/\mathbf{Q}}$ is the norm map. The proof of the following theorem (see Ref. 8, Theorem 2.5) is based on the result by S. D. Cohen (see Theorem 1 in Ref. 5) combined with some asymptotic estimates for the number of integral polynomials having bounded coefficients with respect to the norm and reducible over $K = (\sqrt{b})$ (b is contained in a finite set of elements from O_K):

Theorem 4.1. *Let $v(N)$ denote the total number of polynomials of degree m with coefficients in $O(N)$, and let $\psi(N)$ denote the number of those polynomials whose splitting fields do not contain any fields $K(G) \neq K$ for $G \subset GL_n(O_E), E \supset K$ and fixed n. Then*

$$\lim_{N \to \infty} \frac{\psi(N)}{v(N)} = 1.$$

So "almost all" fields are not realizable via adjoining all matrix coefficients of all matrices $g \in G$ for Γ-stable groups G to K. Note that the proof of Theorem 9 gives an effective bound for $v(N) - \psi(N)$; we can estimate the error term in Theorem 9, for instance, if $K = \mathbf{Q}$, we have $v(N) - \psi(N) = o(N^{m+0.5}(lnN)^2)$ (see Ref. 11, Theorem 2.3).

References

1. M. Asada, On unramified Galois extensions over maximum abelian extensions of algebraic number fields, Mathematische Annalen, vol. 270, nr. 4 (1985), p. 477–487.
2. H.-J. Bartels, Zur Galois Kohomologie definiter arithmetischer Gruppen, J. reine angew. Math., vol. 298 (1978), p. 89–97.
3. H.-J. Bartels, D. Malinin, Finite Galois stable subgroups of GL_n. In: Noncommutative Algebra and Geometry, Edited by C. de Concini, F. van Oystaeyen, N. Vavilov and A. Yakovlev, Lecture Notes In Pure And Applied Mathematics, vol. 243 (2006), p. 1–22.
4. H.-J. Bartels, D. Malinin, Finite Galois stable subgroups of GL_n in some relative extensions of number fields, Journal of Algebra and Its Applications, vol. 8, nr. 4 (2009), p. 493–503.
5. S. D. Cohen, The distribution of the Galois groups of integral polynomials, Illinois J. Math. vol. 23 (1979), p. 135–152.
6. D. Malinin, Finite arithmetic groups, a monograph, Minsk, 2009.
7. D. Malinin, Galois stability for integral representations of finite groups. Algebra i analiz, vol. 12 (2000), p. 106–145.
8. D. Malinin. Galois stability, integrality and realization fields for representations of finite Abelian groups, Algebras and representation theory, vol. 6, nr. 2 (2003), p. 215–237.
9. D. Malinin, Integral representations of finite groups with Galois operation, Dokl. Russ. Akad. Nauk, vol. 349 (1996), p. 303–305.
10. D. Malinin, On the existence of finite Galois stable groups over integers in unramified extensions of number fields. Publ. Mathem. Debrecen, vol. 60, nr. 1-2 (2002), p. 179–191.

11. D. Malinin, Some integral representations of finite groups and their arithmetic applications, Algebraic Geometry and Its Applications, World Scientific (2008), p. 467–480.

ON p-GROUPS OF SMALL POWERFUL CLASS

A. MANN

Einstein Institute of Mathematics, Hebrew University, Jerusalem
E-mail: mann@math.huji.ac.il

The *powerful class* of a finite p-group G measures the departure of G from being powerful, in the same way that the nilpotence class measures the departure from being abelian. We prove that if G has powerful class at most $p-1$, then products of pth powers are pth powers and G^p is powerful. The analogous result for elements of a given order does not hold, but a weaker result can be proved.

Keywords: p-groups; Powerful groups; Power structure.

1. Introduction

In this note all groups are finite p-groups, for some prime p. A standard reference for these groups is Ref. 1. If G is a p-group, we let G^{p^n} be the subgroup which is generated by all p^nth powers, let $\Omega_n(G)$ be the subgroup generated by all elements of order at most p^n, and let $exp(G)$ be the exponent of G. In Ref. 3 a p-group is termed *power abelian*, if it satisfies the following three properties:

(1) For all n, products of p^nth powers are again p^nth powers. Thus G^{p^n} is the set of all p^nth powers.

(2) $Exp(\Omega_n(G)) \leq p^n$. Thus $\Omega_n(G)$ is the set of all elements of order at most p^n.

(3) $|G^{p^n}| = |G : \Omega_n(G)|$.

The systematic study of these properties, which are obvious for abelian groups, starts with Ref. 4, where P.Hall defined *regular* p-groups, and showed that they are power abelian. Groups of exponent p are trivially regular. Two other classes of p-groups were shown in Refs 4,5 to be regular: groups of nilpotence class less than p, and groups in which $|G : G^p| < p^p$. The class of regular p-groups is closed under taking subgroups or factor groups, but not necessarily under direct products. In Refs 7,8 we discussed groups in which all sections (factor groups of subgroups) are power abelian, or satisfy one of (1) - (3). Next, a group is *powerful*, if $G' \leq G^p$ (for $p = 2$, if

$G' \leq G^4$). Here G' is the commutator subgroup, and $\gamma_i(G)$ is the ith term of the lower central series, while the terms of the upper central series are denoted by $Z_i(G)$, with $Z_1(G) = Z(G)$ being the centre. In Ref. 6 it was shown that powerful groups satisfy (1), and the more difficult proofs that they also satisfy (2) and (3), and hence are power abelian, were given by L.Wilson (Ref. 9). A considerably simpler proof was provided by G.A.Fernàndez-Alcober (Ref. 2). J.González-Sánchez and A.Jaikin-Zapirain, generalizing the notion of powerful groups, say that G is *potent*, if either G is a powerful 2-group, or $p > 2$ and $\gamma_{p-1}(G) \leq G^p$. For an odd p, they show that if $N \lhd G$, then N is power abelian, and if moreover $N \leq G^p$, then N is powerful (Ref. 3).

In this note we introduce a different generalization of powerful groups. Recall that a normal subgroup $N \lhd G$ is *powerfully embedded* in G, if $[N, G] \leq N^p$ (for $p = 2$, if $[N, G] \leq G^4$). Thus G is powerful iff it is powerfully embedded in itself. In any case, G contains a unique maximal powerfully embedded subgroup, which is denoted by $\eta(G)$. We define the *upper η-series* of G by $\eta_0(G) = 1$, and $\eta_i(G)/\eta_{i-1}(G) = \eta(G/\eta_{i-1}(G))$. The least c such that $\eta_c(G) = G$ is termed the *powerful class* of G, denoted by $pwc(G)$ (the nilpotence class is denoted by $cl(G)$). An ascending normal series $\{N_i\}$ ($i = 0, ..., k$), with $N_0 = 1$, is an *η-series*, if N_{i+1}/N_i is powerfully embedded in G/N_i, for each $0 \leq i < k$. Given $N \lhd G$, the least length k of an η-series with $N_k = N$ is termed the *powerful height*, or just *height*, of N, denoted by $pwh(N)$. If $N = G$, then $pwh(G) = pwc(G)$. We say that N has *small height*, if $pwh(N) < p$, and G has *small powerful class*, if $pwc(G) < p$. Like the properties of powerfulness or potency, having small powerful class is inherited by factor groups and direct products, but not necessarily by subgroups. Note that if $exp(N) = p$, then N is powerfully embedded iff it is central. Thus for groups of exponent p, the powerful class coincides with the nilpotence class. Therefore a group of powerful class c satisfies $\gamma_{c+1}(G) \leq G^p$. In particular, if $c \leq p - 2$, then G is potent. Any group of exponent p and nilpotence class $p - 1$ has small powerful class but it is not potent.

For $p = 2$, having small powerful class is the same as being powerful, so we are interested mostly in odd primes.

Theorem 1.1. *Let G have small powerful class. Then for each n, the product of p^nth powers in G is again a p^nth power, and G^p is powerful.*

But groups of small powerful class need not be power abelian. Let H be the wreath product of two groups of order p. Then $|H| = p^{p+1}$, $cl(H) = p$, and $H^p = Z(H)$ has order p. H can be generated by two elements, say x, y, of order p, but contains elements of order p^2. Let w be one of these. Then $z := w^p$ generates H^p. Let G be the central product of H and a cyclic group $C = \langle c \rangle$ of order p^2, with $c^p = z$. Then G is a group of exponent p^2 which can be generated by the three elements x, y, cw^{-1}, of order p. Let $Z = Z_2(H)$, a group of order p^2, then $N := Z_2(G) = CZ$ is powerfully embedded in G, and $cl(G/N) = p-2$, therefore $pwc(G) = p-1$, and G is a group of small powerful class which does not satisfy (2).

(This example shows also that the powerful class of a subgroup can exceed that of the ambient group. Specifically, H above has powerful class p).

The following weaker result holds.

Theorem 1.2. *A p-group G of small powerful class which can be generated by two elements of order at most p^e has exponent at most p^e.*

2. Proofs

Lemma 2.1. *If N is powerfully embedded in G and H is a powerful subgroup of G, then HN is powerful.*

Proof. $[HN, HN] = [H, H][H, N][N, N] \leq H^p N^p \leq (HN)^p$. □

Lemma 2.2. *If G is not powerful, then $G/\eta(G)$ is not cyclic.*

Proof. Otherwise, $G = \eta(G)\langle x \rangle$, for some $x \in G$, and G is powerful, by Lemma 2.1. □

Lemma 2.3. *Let $cl(G) \leq p$, $N \lhd G$. Then $exp([N, G]) \leq p$ if and only if $N^p \leq Z(G)$.*

Proof. Let $x \in N, y \in G$. We have $[N, G] \leq G' \leq Z_{p-1}(G)$, therefore $cl(\langle x, [N, G] \rangle) \leq p-1$, and that subgroup contains both x and x^y. Since groups of class less than p are regular, $[y, x] = (x^{-1})^y \cdot x$ has order $\leq p$ iff $x^p = (x^p)^y$ (see Ref. 1, Theorems 7.1 and 7.2), i.e. commutators involving elements of N have order $\leq p$ iff the pth powers of elements of N are central. □

Lemma 2.4. *Let $Z \leq Z(G)$, and let R/Z be powerfully embedded in G/Z. Then $[R, G]$ is powerfully embedded in G. If $exp(Z) = p$, then R^p is powerfully embedded in G.*

This is a slight extension of Theorem 2.1(a),(b) in Ref. 6, which is the case $R = G$.

Proof. We start with $[R, G]$. We may assume that $[R, G]^p = 1$, and want to derive $[R, G, G] = 1$. We may also assume that $[R, G, G, G] = 1$, i.e. $[R, G] \leq Z_2(G)$. Let $x \in G$, $y \in R$. Then $cl(\langle x, x^y \rangle) = cl(\langle x, [x, y] \rangle) \leq cl(\langle x, [R, G] \rangle) \leq 2$. Then $[x, y] = (y^{-1})^x \cdot y$ having order p implies $y^p = (y^p)^x$, i.e. $y^p \in Z(G)$ and $R^p \leq Z(G)$. Since R/Z is powerfully embedded, we obtain $[R, G] \leq R^p Z \leq Z(G)$.

Now assume that $Z^p = 1$. To show that R^p is powerfully embedded, we may assume that $(R^p)^p = 1$. Since $[R, G]$ and R/Z are powerfully embedded, we have $[R, G, G] \leq [R, G]^p \leq (R^p Z)^p = 1$. Thus $[R, G]$ is central, and for x, y as above we have $(y^p)^x = (y^x)^p = (y[y, x])^p = y^p$, i.e. $R^p \leq Z(G)$. □

Lemma 2.5. *If $N \lhd G$, then $pwh([N, G]) \leq pwh(N)$ and $pwh(N^p) \leq pwh(N)$.*

Proof. If $\{N_i\}$ $(i = 0, ..., k)$ is an η-series for $N = N_k$, then Lemma 2.4 shows that $[N_i, G]$ and N_i^p are η-series for $[N, G]$ and N^p, respectively. □

Proof of Theorem 1.1, first claim. We first consider the case $n = 1$. Let $N = \eta_{c-1}(G)$ be the penultimate term in the upper η-series of G, and let $H/N = (G/N)^p$. By Lemma 2.2, G/N is not cyclic, and therefore G/H is also not cyclic. Moreover, H/N is powerful, even powerfully embedded in G/N, and the upper η-series of G with H replacing G, is an η-series for H, therefore $pwc(H) < p$, and by induction, each element of H^p is a pth power in H.

Assume first that $H^p \neq 1$, and let $x, y \in G$. By induction, $x^p y^p = z^p w$, for some $w \in H^p$, i.e. $w = u^p$, with $u \in H$. Write $K = \langle z, H \rangle$. Then K/N is powerful, therefore, as for H, $pwc(K) < p$, and $K \neq G$, hence by induction $x^p y^p = z^p u^p$ is a pth power.

We thus assume that $exp(H) = p$. Since a subgroup of exponent p is powerfully embedded only if it is central, the η-series of G, up to N, is actually a central series, and since H/N is also of exponent p and powerfully embedded in G/N, with G/H being abelian, we obtain that $cl(G) \leq p$. Moreover, $G' \leq H$, and so $exp(G') = p$, and the pth powers are central in G, i.e. $G^p \leq Z(G) \leq \eta(G) \leq N$. Thus G/N is abelian, and $cl(G) \leq p - 1$. Now our claim follows from well known properties of groups of small class.

For $n > 1$, the previous part implies that the p^nth powers are exactly the p^{n-1}th powers in G^p. By Lemma 2.5, $pwc(G^p) < p$, and we apply induction.

Corollary 2.1. *Let G have small powerful class. Then for any m, n, $(G^{p^n})^{p^m} = G^{p^{m+n}}$.*

We derive more properties of groups of small powerful class.

Proposition 2.1. *Let $N \lhd G$ have small powerful height. Then $[N^p, G] = [N, G]^p \geq [N, G^p]$*

Proof. It suffices to show that if either $[N^p, G]$ or $[N, G]^p$ is trivial, so are the other two subgroups. At first assume that $[N^p, G] = 1$. That means that N^p is central in G, and N/N^p being of small height in G/N^p means that $N/N^p \leq Z_{p-1}(G/N^p)$ and $N \leq Z_p(G)$. If $x \in G, y \in N$, then $cl(\langle N, x\rangle) \leq p$, and $y^p = (y^p)^x$ implies $[y, x]^p = 1$, by Lemma 2.3. Moreover, $[N, G] \leq Z_{p-1}(G)$, and the generators of $[N, G]$ have order p, and thus $exp([N, G]) = p$ (we are assuming that $[N, G] \neq 1$). A similar argument shows that $[N, G^p] = 1$.

Now assume that $[N, G]^p = 1$. Since $[N, G]$ has small height, we obtain $[N, G] \leq Z_{p-1}(G)$ and $N \leq Z_p(G)$, and continue as before. □

Corollary 2.2. *Let N be as above. Then $[N^{p^k}, G] = [N, G]^{p^k} \geq [N, G^{p^k}]$.*

Proof. Writing $N^{p^{k-1}}$ for N in the previous proposition, and applying induction, we have $[N^{p^k}, G] = [N^{p^{k-1}}, G]^p = ([N, G]^{p^{k-1}})^p = [N, G]^{p^k}$. The remaining inclusion is derived in the same way. □

Corollary 2.3. *Let N be as before, and assume also that $N \leq G^p$. Then N is powerfully embedded in G^p.*

Proof. $[N, G^p] \leq [N, G]^p \leq N^p$. □

Proposition 2.2. *If G has small powerful class, then the inclusion $[G^{p^r}, G^{p^s}] \leq [G, G]^{p^{r+s}}$ holds, for any r, s. In particular, the subgroups G^{p^r}, for $r \geq 1$, are powerful.*

Proof. If either $r = 0$ or $s = 0$, this follows from Corollary 2.2, with an equality rather than inclusion. For positive exponents, we may assume that $r \leq s$. Then $H = G^{p^{r-1}}$ also has small powerful class. Take $N = G^{p^s}$, and apply induction and Proposition 2.1 to get $[G^{p^r}, G^{p^s}] = [H^p, N] \leq [H, N]^p = [G^{p^{r-1}}, G^{p^s}]^p \leq ([G, G]^{p^{r+s-1}})^p = [G, G]^{p^{r+s}}$. □

This proposition completes the proof of Theorem 1.1. Theorem 1.2 follows from the following more general result.

Proposition 2.3. *Let $N \lhd G$ have small height, and suppose that N is the normal closure of two elements of order at most p^e. Then $exp(N) \leq p^e$.*

Proof. First take $e = 1$. Let Z be a normal subgroup of order p. Considering G/Z and applying induction, we may assume that $Z = N^p$, and Z is the only minimal normal subgroup of G. Since N/Z has small height and exponent p, we have $N/Z \leq Z_{p-1}(G/Z)$, therefore $N \leq Z_p(G)$ and $[N,G] \leq Z_{p-1}(G)$. It follows that if $x \in N$, then $K := \langle x, [N,G]\rangle \lhd G$ and $cl(K) \leq p-1$. If $y \in G$, then $x, x^y \in K$, and since $N^p = Z \leq Z(G)$, we have $x^p = (x^y)^p$ and $[x,y]^p = 1$. Thus $exp([N,G]) = p$ (here we also assume that $[N,G] \neq 1$). For any normal subgroup $L \lhd G$, write $\Phi_G(L) = [L,G]L^p$. Then some elements generate L as a normal subgroup iff they generate it, in the usual sense, *modulo* $\Phi_G(L)$. Since N^p is the unique minimal normal subgroup of G, we have $\Phi_G(N) = [N,G]$. Let N_1 be the first term in an η-series for N. Then $pwh(N/N_1) \leq p-2$, and since $N^p = Z \leq N_1$, it follows that $cl(N/N_1) = pwh(N/N_1) \leq p-2$. If $N_1 \leq Z(G)$ then $cl(N) \leq p-1$, and N has exponent p. We thus assume that $N_1 \not\leq Z(G)$, however N_1/Z is powerfully embedded in G/Z and of exponent p, implying $N_1/Z \leq Z(G/Z)$ and $N_1 \leq Z_2(G)$. On the other hand $N_1^p \geq [N_1, G] \neq 1$, and N_1 contains an element x of order p^2. Thus $x \notin [N,G]$, and there exists $y \in N$ such that N is the normal closure of x and y. Now $x \in Z_2(G) \leq Z_{p-1}(G)$, therefore $x, y \in L := \langle y, Z_{p-1}(G) \cap N\rangle \lhd G$, implying $N = L$ and $cl(N) \leq p-1$, and finally $exp(N) = p$.

Now let $e > 1$. Let x, y be normal generators of N of order $\leq p^e$, and let R be the normal closure of x^p and y^p. These are elements of order $\leq p^{e-1}$ in the powerful group G^p, therefore $exp(R) \leq p^{e-1}$, while $exp(G/R) = p$, by the previous case, and thus $exp(G) \leq p^e$ (and $R = G^p$). □

References

1. Y.Berkovich, *Groups of prime power order 1*, Walter de Greuyter, Berlin 2008.
2. G.A.Fernández-Alcober, Omega subgroups of powerful p-groups, Israel J. Math. 162 (2007), 75-79.
3. J.González-Sánchez and A.Jaikin-Zapirain, On the structure of normal subgroups of potent p-groups, J.Alg. 276(2004), 193-209.
4. P.Hall, A contribution to the theory of groups of prime power order, Proc. London Math. Soc. (2) 36 (1933), 29-95.
5. P.Hall, On a theorem of Frobenius, Proc. London Math. Soc. (2) 40 (1936), 468-501.

6. A.Lubotzky and A.Mann, Powerful p-groups. I. Finite groups, J.Alg. 105 (1987), 484-505.
7. A.Mann, The power structure of p-groups I, J.Alg. 42 (1976), 121-135.
8. A.Mann, On the power structure of some P-groups, Proc. 2nd Int. Group Th. Conf. Bressanone/Brixen 1989, in Supp. Rend. Cir. Mat. Palermo (2) 23 (1990), 227-235.
9. L.Wilson, On the power structure of powerful p-groups, J. Group Th. 5 (2002), 129-144.

LIFTING $(2, k)$-GENERATORS OF LINEAR GROUPS

A. MARÓTI

MTA Alfréd Rényi Institute of Mathematics
Reáltanoda utca 13-15, H-1053, Budapest, Hungary
e-mail: maroti@renyi.hu

C. TAMBURINI BELLANI

Dipartimento di Matematica e Fisica
Università Cattolica
Via Musei 41, Brescia, Italy
e-mail: c.tamburini@dmf.unicatt.it

Dedicated to Karl Gruenberg

Let $\ell = kh$, where k, h are orders of arbitrary elements of $\mathrm{SL}_2(q)$ subject to $k \geq 3$, $h \geq 3$ and $(k, h) = 1$. For q even allow also $k = 4$ or $h = 2$. We describe $(2, \ell)$-generating pairs of $\mathrm{PSL}_n(q)$, for all $n > 5$ and $q > 2$.

Keywords: (2,ℓ)-generating pairs; Finite simple groups.

1. Introduction

A $(2, \ell)$-generating pair of a group G consists of two elements, of respective orders 2 and ℓ, which generate G. Clearly $\ell \geq 3$, unless G is abelian or dihedral. The authors of Ref. 2 study the problem of finding uniform $(2, k)$-generating pairs for the finite classical groups $\mathrm{PSL}_4(q)$, $\mathrm{PSp}_4(q)$ and $\mathrm{PSU}_4(q^2)$, with $k \geq 3$ the order of some element of $\mathrm{SL}_2(q)$, including $k = 4$ when q is even. In Theorem 3.1 of this paper we lift their $(2, k)$-generating pairs of $\mathrm{PSL}_4(q)$ to $(2, \ell)$-generating pairs of $\mathrm{PSL}_n(q)$, for all $n \geq 5$. Here $\ell = kh$, where k, h are orders of arbitrary elements of $\mathrm{SL}_2(q)$ subject to $k \geq 3$, $h \geq 3$ and $(k, h) = 1$. For q even we allow also $k = 4$ or $h = 2$. Most likely the same construction, with $\ell = k$ and σ in (2) of order h dividing k, produces $(2, k)$-generating pairs of $\mathrm{PSL}_n(q)$, $n \geq 5$. This would be the best possible generalization. But the proof becomes much more intricate.

Let $\mathbb{F}_q$ be the Galois field of order $q = p^a$, where p is a prime, and

$\mathbb{F}_q^*$ be the set of its non-zero elements. For k as above, except in the case $(k,q) = (4, 2^a)$, let $\gamma = \begin{pmatrix} 0 & -1 \\ 1 & s \end{pmatrix}$ be a rational canonical form of $\mathrm{SL}_2(q)$ having order k, and consider the matrices:

$$x = \begin{pmatrix} 0 & 0 & 1 & 0 \\ 0 & 0 & 0 & 1 \\ d & 0 & 0 & 0 \\ 0 & d & 0 & 0 \end{pmatrix}, \quad d = \pm 1, \quad y = \begin{pmatrix} 1 & 0 & 0 & 0 \\ 0 & 1 & 0 & r \\ 0 & 0 & 0 & -1 \\ 0 & 0 & 1 & s \end{pmatrix}, \quad r \in \mathbb{F}_q^*. \tag{1}$$

Clearly $x^2 = dI$. Moreover y has the same order k of γ, except when $s = 0$ and $q = 2^a$, in which case y has order 4. When necessary, we identify x, y with their projective images, of respective orders 2 and k.

Lemma 1.1. *If $q > 3$, for fixed $s \in F_q$ and $d = \pm 1$, there exists $r \in \mathbb{F}_q^*$ such that $\mathbb{F}_q = \mathbb{F}_p\left[s, r^2\right]$ and $r \neq \pm\sqrt{d}\,(s-2)$.*

The easy proof can also be deduced from Lemma 5.3 of Ref. 2, where the following result is proved (Theorem 11.1):

Theorem 1.1. *Assume that x, y are defined as in (1) with $s \in \mathbb{F}_q$, $r \in \mathbb{F}_q^*$ such that $\mathbb{F}_q = \mathbb{F}_p\left[s, r^2\right]$ and $r \neq \pm\sqrt{d}\,(s-2)$. Then*

$$\langle x, y \rangle = \mathrm{SL}_4(q).$$

In particular the groups $\mathrm{SL}_4(q)$, $q > 3$, and $\mathrm{PSL}_4(q)$, $q > 2$, are $(2,k)$-generated for all $k \geq 3$ which correspond to the order of some element of $\mathrm{SL}_2(q)$, including $k = 4$ when q is even.

For the reader's convenience, we note that the assumptions on k are equivalent to the following: $k \geq 3$, k divides $q-1$ or k divides $q+1$ or $k \in \{p, 2p\}$.

2. Definition of the $(2,\ell)$-generating pairs

Let $\ell = kh$, where k, h are orders of arbitrary elements of $\mathrm{SL}_2(q)$ subject to the conditions $k \geq 3$, $h \geq 3$ and $(k,h) = 1$. For q even allow also $k = 4$ or $h = 2$. For all $n \geq 5$, we lift any $(2,k)$-generating pair (x,y) of $\mathrm{PSL}_4(q)$ to a $(2,\ell)$-generating pair (X,Y) of $\mathrm{PSL}_n(q)$ via the following blocks:

$$\sigma := \begin{pmatrix} 0 & -1 \\ 1 & t \end{pmatrix}, \quad \pi_\lambda := \begin{pmatrix} 0 & 1 \\ \lambda & 0 \end{pmatrix}, \quad \lambda = \pm 1. \tag{2}$$

Here $t \in \mathbb{F}_q$ is such that σ has order h.

Denoting by $e_1, \ldots, e_4$ the canonical basis of $\mathbb{F}_q^4$, let y_3 be the restriction of y in (1) to $\langle e_2, e_3, e_4 \rangle$, namely:

$$y_3 := \begin{pmatrix} 1 & 0 & r \\ 0 & 0 & -1 \\ 0 & 1 & s \end{pmatrix}, \quad r \in \mathbb{F}_q^*. \tag{3}$$

For $n = 2m + 3 \geq 5$, in (2) take $\lambda = 1$ and define

$$X := \begin{pmatrix} \pm 1 & & & & \\ & \pi_1 & & & \\ & & \cdots & & \\ & & & \pi_1 & \\ & & & & x \end{pmatrix}, \qquad Y := \begin{pmatrix} \sigma & & & \\ & \cdots & & \\ & & \sigma & \\ & & & y_3 \end{pmatrix} \tag{4}$$

where x is as in (1) with $d = 1$, π_1 and σ are as in (2), and the sign $\pm$ is chosen so that $\det X = 1$. In (4) the number of blocks π_1 is $m - 1$ and the number of blocks σ is m.

For $n = 2m + 4 \geq 6$, in (2) take $\lambda = 1$ if $n \equiv 0 \pmod 4$, $\lambda = -1$ if $n \equiv 2 \pmod 4$, and define:

$$X := \begin{pmatrix} \pi_\lambda & & & \\ & \cdots & & \\ & & \pi_\lambda & \\ & & & x \end{pmatrix}, \qquad Y := \begin{pmatrix} 1 & & & & \\ & \sigma & & & \\ & & \cdots & & \\ & & & \sigma & \\ & & & & y_3 \end{pmatrix} \tag{5}$$

where x is as in (1) with $d = \lambda$, σ and π_λ are as in (2). In (5) the number of blocks π_λ and σ is m.

Note that X^2 is scalar and Y has order $\ell - kh$.

3. The result

For each m such that $0 < m < n$, we consider the subgroup $S_m(q)$ of $\mathrm{SL}_n(q)$ defined as follows:

$$S_m(q) := \begin{pmatrix} I_{n-m} & \\ & \mathrm{SL}_m(q) \end{pmatrix}.$$

For the reader's convenience, we give a direct proof of a fact which is well known, namely:

Lemma 3.1. *Let σ and π_{-1} be defined as in (2). For all $n \geq 4$, set*

$$\Sigma := \begin{pmatrix} \sigma & \\ & I_{n-2} \end{pmatrix}, \quad \Pi := \begin{pmatrix} \pi_{-1} & \\ & I_{n-2} \end{pmatrix}.$$

Then $\mathrm{SL}_n(q) = \langle S_{n-1}(q), \Sigma\rangle = \langle S_{n-1}(q), \Pi\rangle$.

Proof. Consider the elementary transvection $\tau_1 := I + E_{2,3}$ and let $g \in \{\Sigma, \Pi\}$. Then $\tau_1^g = I + E_{1,3}$. Using the transitivity of $\mathrm{SL}_{n-1}(q)$ on the non-zero vectors of $\mathbb{F}_q^{n-1}$, it is easy to see that the conjugates of $I + E_{1,3}$ under $S_{n-1}(q)$ include all root subgroups $I + \mathbb{F}_q E_{1,j}$, $2 \le j \le n$.

In a similar way, consider the elementary transvection $\tau_2 := I + E_{3,2}$. Then $\tau_2^g = I + E_{3,1} \pmod{S_{n-1}(q)}$. As above, the conjugates of $I + E_{3,1}$ under $S_{n-1}(q)$ include all root subgroups $I + \mathbb{F}_q E_{j,1}$, $2 \le j \le n$. Since $\mathrm{SL}_n(q)$ is generated by the elementary root sugroups $I + \mathbb{F}_q E_{i,j}$, $i \ne j$, (see, e.g. Ref. 1) our claim follows. □

Theorem 3.1. *Assume* $n \ge 5$. *Define* X, Y *respectively as in* (4) *or* (5) *according to* n *odd or even. If* $r \in \mathbb{F}_q^*$ *is such that* $\mathbb{F}_q = \mathbb{F}_p\left[s, r^2\right]$ *and* $r \ne \pm\sqrt{d}\,(s-2)$, *then:*

$$\langle X, Y\rangle = \mathrm{SL}_n(q).$$

In particular the group $\mathrm{SL}_n(q)$, $q > 3$ *and* $n \not\equiv 2 \pmod 4$ *if* q *is odd, is* $(2,\ell)$*-generated. The group* $\mathrm{PSL}_n(q)$, $q > 2$, *is* $(2,\ell)$*-generated.*

Proof. The subspace $U = \langle e_1, \ldots, e_{n-3}\rangle$, generated by the first $n-3$ vectors of the canonical basis, is Y-invariant. So we define:

$$Y_{n-3} := \begin{pmatrix} Y_{|U} & \\ & I_3 \end{pmatrix}, \qquad Y_3 := \begin{pmatrix} I_{n-3} & \\ & y_3 \end{pmatrix}.$$

By the assumption that k and h are coprime, we have:

$$\langle Y\rangle = \langle Y_{n-3}\rangle \times \langle Y_3\rangle\,.$$

It follows that $\langle X, Y\rangle$ contains the subgroup $H := \langle X, Y_3\rangle \le C_2 \times \mathrm{SL}_4(q)$. Using Theorem 1.1 and the fact that the group $\mathrm{SL}_4(q)$ is perfect, we get that $H' = S_4(q) \le \langle X, Y\rangle$. By induction we may assume that

$$S_{n-1}(q) \le \langle X, Y\rangle\,.$$

Noting that $\Sigma^{-1}Y \in S_{n-2}(q)$ if n is odd, and that either $\Pi X \in S_{n-1}(q)$ or $\Pi^{-1}X \in S_{n-1}(q)$ if n is even, we deduce $\Sigma \in \langle X, Y\rangle$ if n is odd, $\Pi \in \langle X, Y\rangle$ if n is even. Our claim follows from Lemma 3.1. □

References

1. R. Carter, Simple groups of Lie type, Wiley and Sons, London, 1972.
2. M. Pellegrini, C. Tamburini Bellani and M.A. Vsemirnov, Uniform $(2,k)$-generation of 4-dimensional classical groups (submitted).

FIXED POINT SUBGROUPS AND CHARACTER TABLES

G. NAVARRO

Departament d'Àlgebra, Universitat de València,
46100 Burjassot, SPAIN
E-mail: gabriel.navarro@uv.es

If a q-group A acts coprimely on G, we would like to understand how the group theoretical properties of the fixed point subgroup $C = \mathbf{C}_G(A)$ are reflected in the set of irreducible characters of G, and vice-versa.

Keywords: Coprime Action; Character Tables.

1. Introduction

Let q be a prime. Suppose that a finite q-group A acts as automorphisms on a finite group G of order not divisible by q. We would like to understand how much the set $\mathrm{Irr}(G)$ of irreducible complex characters of G *knows* about the structure of the fixed point subgroup $C = \mathbf{C}_G(A)$ and vice-versa. For instance, we have already shown that $\mathrm{Irr}(G)$ detects the set of prime divisors of C (Ref. 3) or if C is cyclic (Ref. 4).

We write $\mathrm{Irr}_A(G)$ for the set of A-invariant characters in $\mathrm{Irr}(G)$.

Theorem A. *Suppose that a q-group A acts coprimely on G. Let $C = \mathbf{C}_G(A)$. Then C is abelian if and only if for every $\chi \in \mathrm{Irr}_A(G)$ we have that 1_G is the only A-invariant irreducible constituent of $\chi\bar{\chi}$ with multiplicity not divisible by q.*

In order to detect p-nilpotency of the fixed point subgroup C, we need to go deeper and consider Brauer characters. Recall that if $\chi \in \mathrm{Irr}(G)$, then the restriction χ^0 of χ to the p-regular elements of G is a Brauer character of G.

Theorem B. *Suppose that a q-group A acts coprimely on a p-solvable group G. Let $C = \mathbf{C}_G(A)$. Then C has a normal p-complement if and only if for every $\chi \in \mathrm{Irr}_A(G)$ the Brauer character χ^0 has a unique A-invariant* *irreducible constituent* $\varphi \in \mathrm{IBr}(G)$ *(depending on* χ*) with multiplicity not divisible by* q.

We give an easy consequence of these results.

Corollary C. *Suppose that the* q*-groups* A_i *act coprimely on* K_i*, for* $i = 1, 2$*, and assume that the semidirect products* K_1A_1 *and* K_2A_2 *have the same character table.*

(a) Then $\mathbf{C}_{K_1}(A_1)$ *is abelian if and only if* $\mathbf{C}_{K_2}(A_2)$ *is abelian.*

(b) Assume that K_i *is* p*-solvable for* $i = 1, 2$*. Then* $\mathbf{C}_{K_1}(A_1)$ *has a normal* p*-complement if and only if* $\mathbf{C}_{K_2}(A_2)$ *has a normal* p*-complement.*

We have been unable to remove the solvability hypothesis in the previous results, or to extend them to the general case where $\gcd(|A|, |G|) = 1$.

2. Proofs

We follow the notation of Ref. 2 for ordinary characters and the notation of Ref. 5 for modular characters.

We start with a nearly trivial result.

Lemma 2.1. *Suppose that* G *is a* p*-solvable finite group, and suppose that* $T \subseteq G$ *has* p'*-index. Then the Brauer character* $(1_T)^G$ *contains the principal Brauer character* 1_G *with multiplicity 1.*

Proof. Let H be a p-complement of G, so that $G = TH$. The multiplicity of $\varphi \in \mathrm{IBr}(G)$ in a Brauer character Δ of G is the ordinary inner product $[\Delta_H, \delta]$, where $\delta \in \mathrm{Irr}(H)$ is Fong for φ. (This follows from Theorem 10.17 of Ref. 5.) Now 1_H is a Fong character for the principal Brauer character of G, (by Theorem 10.18 of Ref. 5, for instance.) Hence the multiplicity we want is

$$[(1^G)_H, 1_H] = [(1_{T\cap H})^H, 1_H] = [1_{T\cap H}, 1_{T\cap H}] = 1\,,$$

which proves the lemma. □

The following is inspired by a Theorem of Pahlings[6]. We do not know if the p-solvability condition is really necessary.

Theorem 2.2. *Let* G *be a* p*-solvable finite group, and let* q *be a prime not dividing* $|G|$*. Then* G *has a normal* p*-complement if and only if for every* $\chi \in \mathrm{Irr}(G)$ *there exists some* $\varphi \in \mathrm{IBr}(G)$ *such that*

$$\chi^0 = e\varphi + q\Delta\,,$$

where Δ is some Brauer character of G or zero.

Proof. Suppose first that G has a normal p-complement K, and let $\chi \in \mathrm{Irr}(G)$. Now, let $\theta \in \mathrm{Irr}(K)$ be under χ and let T be the stabilizer of θ in G. Since $(|G : K|, |K|) = 1$, then there exists $\eta = \hat{\theta} \in \mathrm{Irr}(T)$ extending θ (Corollary 6.28 of Ref. 2). Now, $\theta \in \mathrm{IBr}(K) = \mathrm{Irr}(K)$ and since $\theta = (\eta_K)^0 = (\eta^0)_K$ is irreducible, we have that $\eta^0 \in \mathrm{IBr}(T)$. By the Clifford correspondence for Brauer characters (Theorem 8.9 of Ref. 5), we have that $\varphi = (\eta^0)^G \in \mathrm{IBr}(G)$. Now, $\chi = (\beta\eta)^G$, where $\beta \in \mathrm{Irr}(T/K)$ by Gallagher's theorem (Corollary 6.17 of Ref. 2). Since T/K is a p-group, $\beta^0 = \beta(1)1_T$ and

$$\chi^0 = \beta(1)\varphi ,$$

as desired.

Suppose now that for every $\chi \in \mathrm{Irr}(G)$ there exists some $\varphi \in \mathrm{IBr}(G)$ such that $\chi^0 = e\varphi + q\Delta$ where Δ is some Brauer character of G or zero. We prove that G has a normal p-complement by induction on $|G|$. Since the hypothesis is clearly inherited by factor groups, then we may assume that $\mathbf{O}_{p'}(G) = 1$. Also, if $N = \mathbf{O}_p(G)$, then G/N has a normal p-complement K/N. Let $P \in \mathrm{Syl}_p(G)$, and let $\lambda \in \mathrm{Irr}(P)$ be linear. Let $\nu = \lambda_N$ and let T be the stabilizer of ν in G, so that $P \subseteq T \subseteq G$. Now let $\hat{\nu} \in \mathrm{Irr}(T \cap K)$ be the canonical extension (using again Corollary 6.28 of Ref. 2). By uniqueness, we have that $\hat{\nu}$ is P-invariant, and therefore restriction defines a bijection from $\mathrm{Irr}(T|\hat{\nu}) \to \mathrm{Irr}(P|\nu)$ (Corollary 4.2 of Ref. 1). In particular, we see that λ extends to T. Since this extension, say ϵ, is linear and $(\epsilon_{p'})_P = 1$, then we may assume that $\epsilon = \epsilon_p$ has p-power order. In particular, $\epsilon^0 = 1_T$. Suppose that $T < G$, and let $\chi = \epsilon^G \in \mathrm{Irr}(G)$ (by the Clifford correspondence). By hypothesis, we have that

$$(1_{T^0})^G = \chi^0 = e\varphi + q\Delta$$

for some $\varphi \in \mathrm{IBr}(G)$. By Lemma 2.1, we conclude that $\varphi = 1$. Now, suppose that $T < G$. Then $T \cap K < K$ (using that $G = TK$) and therefore there exists

$$g \in K - \bigcup_{k \in K} (T \cap K)^k .$$

Hence $x = g_{p'}$ lies in no G-conjugate of T. By the induction formula, we conclude that

$$(1_{T^0})^G(x) = 0 .$$

Hence, $0 = e + q\Delta(x)$, and since $\Delta(x)$ is an algebraic integer, we conclude that q divides e. Therefore q divides $\chi(1)$, which contradicts the fact that $|G|$ is not divisible by q. We conclude that $T = G$ and therefore that λ extends to G. Hence, all linear characters of P extend to G. Then, by using Tate's theorem, it is well-known that G has a normal p-complement. (To have a proof of this fact, use that every $\lambda \in \mathrm{Irr}(P)$ has a p-power order extension, and then use Theorem 6.31 of Ref. 2, with $N = P$.) □

(We would like to point out now that it is not difficult to show that a π-version of Pahling's theorem holds true for π-separable groups by using Isaacs π-partial characters (see Ref. 1 for a definition of those): A finite π-separable group G has a normal Hall π-subgroup if and only for every $\chi \in \mathrm{Irr}(G)$, the restriction χ^0 of χ to the π-elements of G is a multiple of some irreducible Isaacs π-partial character $\varphi \in \mathrm{I}_\pi(G)$ that depends of χ.)

In order to prove Theorem B of the introduction, we shall need to use the Glauberman correspondence. As it is well-known, the Glauberman correspondence is a bijection

$$^* : \mathrm{Irr}_A(G) \to \mathrm{Irr}(C)$$

between the set $\mathrm{Irr}_A(G)$ of A-invariant irreducible complex characters of G and the irreducible characters of $C = \mathbf{C}_G(A)$. If A is a q-group and $\theta \in \mathrm{Irr}_A(G)$, then θ^* is the unique irreducible constituent of θ_C that appears with multiplicity e not divisible by q. (See Chapter 13 of Ref. 2.) In fact

$$e \equiv \pm 1 \bmod q\,.$$

But we shall also need its version for Brauer characters. This result is due to K. Uno: if A acts coprimely on a p-solvable group, then there is a bijection

$$^* : \mathrm{IBr}_A(G) \to \mathrm{IBr}(C)$$

between the set $\mathrm{IBr}_A(G)$ of A-invariant irreducible Brauer characters of G and the irreducible Brauer characters of $C = \mathbf{C}_G(A)$. If A is a q-group and $\theta \in \mathrm{IBr}_A(G)$, then θ^* is the unique irreducible constituent of θ_C that appears with multiplicity not divisible by q. (See Ref. 7.)

Now, we are ready to prove Theorem B which we restate for the reader's convenience.

Theorem 2.3. *Suppose that a q-group A acts coprimely on a p-solvable group G. Let $C = \mathbf{C}_G(A)$. Then C has a normal p-complement if and only* *if for every $\chi \in \mathrm{Irr}_A(G)$ the Brauer character χ^0 has a unique A-invariant irreducible constituent with multiplicity not divisible by q.*

Proof. Suppose first that C has a normal p-complement. Let $\chi \in \mathrm{Irr}_A(G)$. We can write

$$\chi^0 = e_1\mu_1 + \cdots + e_s\mu_s + q\Delta + \Xi\,,$$

where Δ and Ξ are Brauer characters of G (or zero), $\mu_i \in \mathrm{IBr}_A(G)$ are all distinct, no irreducible constituent of Ξ is A-invariant, and q does not divide e_i for every i. (With certain abuse of language, we allow $s = 0$.) We wish to prove that $s = 1$.

Since the irreducible constituents of Ξ (if any) are permuted by A, it easily follows that $\Xi_C = q\Xi^*$, where Ξ^* is a Brauer character of C (or zero). By the Uno correspondence, we can also write

$$(\mu_i)_C = e_i^*(\mu_i)^* + q\Delta_i\,,$$

where q does not divide e_i^* and Δ_i is a Brauer character of C or zero. Hence, we can write

$$(\chi^0)_C = e_1^*\mu_1^* + \cdots + e_s^*\mu_s^* + q\Lambda\,,$$

where Λ is a Brauer character of C or zero. Now, by the Glauberman correspondence, we can write

$$\chi_C = d\chi^* + q\rho\,,$$

where q does not divide d, and ρ is an ordinary character (or zero). Since C has a normal p complement, it follows by the first paragraph of the proof of Theorem 2.2, that $(\chi^*)^0 = v\varphi$, where v is not divisible by q and $\varphi \in \mathrm{IBr}(C)$. Since $(\chi^0)_C = (\chi_C)^0$, we have that

$$e_1^*\mu_1^* + \cdots + e_s^*\mu_s^* + q\Lambda = dv\varphi + q\rho^0\,.$$

Since $(\mu_i)^* \neq (\mu_j)^*$ if $i \neq j$, we conclude that $s = 1$ by the linear independence of the irreducible Brauer characters.

To prove the converse, we use Theorem 2.2. Let $\delta \in \mathrm{Irr}(C)$. Then $\delta = \chi^*$ for some $\chi \in \mathrm{Irr}_A(G)$. By hypothesis, we can write

$$\chi^0 = e\varphi + q\rho + \Xi\,,$$

where e is not divisible by q, $\varphi \in \mathrm{IBr}_A(G)$ and no irreducible constituent of Ξ is A-invariant. Hence, we can write

$$(\chi_C)^0 = e^*\varphi^* + q\Psi\,,$$

where e^* is not divisible by q, and Ψ is a Brauer character of C (or zero). Since $\chi_C = \pm\chi^* + q\Delta$, then we have that

$$\pm(\chi^*)^0 + q\Delta^0 = e^*\varphi^* + q\Psi\,.$$

Hence

$$(\chi^*)^0 = \pm e^*\varphi^* + q\Phi\,,$$

where Φ is a difference of Brauer characters. Since $(\chi^*)^0$ is a non-negative integer linear combination of $\mathrm{IBr}(C)$, which are linearly independent, and q does not divide $\varphi^*(1)$ (because $\varphi^*(1)$ divides $|C|$, using the Fong-Swan Theorem 10.1 of Ref. 5), we can write

$$\delta^0 = (\chi^*)^0 = d\varphi^* + q\tau$$

for some Brauer character τ of C (or zero). Now, we can apply Theorem 2.2 to finish the proof. □

Theorem 2.4. *Suppose that a q-group A acts coprimely on G, and let $\chi \in \mathrm{Irr}_A(G)$. Then χ^* is linear if and only if $\chi\bar{\chi} = 1 + q\Delta + \Xi$, where Δ and Ξ are characters of G (or zero) and no irreducible constituent of Ξ is A-invariant.*

Proof. By the Glauberman correspondence, write

$$\chi_C = e\chi^* + q\rho\,,$$

where q does not divide e and ρ is a character of C (or zero).

Suppose first that $\chi\bar{\chi} = 1 + q\Delta + \Xi$, where no irreducible constituent of Ξ is A-invariant. Now, we easily deduce that

$$\chi_C\bar{\chi}_C = 1 + q\Xi^* = e^2\chi^*\bar{\chi}^* + q\Psi\,,$$

where Ξ^* and Ψ are characters of C (or zero). If χ^* is not linear, by Burnside's theorem, there exists $c \in C$ such that $\chi^*(c) = 0$. We then deduce that q divides 1, a contradiction.

Suppose now that χ^* is linear. Then we have that $\chi_C\bar{\chi}_C = e^2 1_C + q\tau$, where τ is some character of C or zero. Now write

$$\chi\bar{\chi} = 1_G + a_1\delta_1 + \cdots + a_s\delta_s + q\Delta + \Xi\,,$$

where $\delta_i \in \mathrm{Irr}_A(G)$ are different and q does not divide a_i for all i. By restricting to C and using the Glauberman correspondence, we easily conclude that $s = 0$. □

Now, the proof of Theorem A easily follows from Theorem 2.4. The proof of Corollary C follows from Theorems A and B.

References

1. I. M. Isaacs, Characters of π-separable groups, *J. Algebra* **86**, (1984) 98-128.
2. I. M. Isaacs, *Character Theory of Finite Groups*, Dover, New York, 1994.
3. I. M. Isaacs, G. Navarro, Character tables and Sylow normalizers, *Arch. Math.* **78** (2002), no. 6, 430–434.
4. A. Moretó, G. Navarro, Field equivalent finite groups. *Math. Z.* **248** (2004), no. 4, 795–803.
5. G. Navarro, *Characters and Blocks of Finite Groups*, Cambridge University Press, 1998.
6. H. Pahlings, Normal p-complements and irreducible characters. *Math. Z.* **154** (1977) 243–246.
7. K. Uno, Character correspondences in p-solvable groups, *Osaka J. Math.* **20** (1983), 713-725.

PERMUTABILITY AND SERIALITY IN LOCALLY FINITE GROUPS

D. J. S. ROBINSON

Department of Mathematics
University of Illinois at Urbana-Champaign
Urbana, IL 61801, USA
E-mail: dsrobins@illinois.edu

The role of Sylow-permutability in locally finite groups and the connection with serial subgroups are explored.

Keywords: Sylow permutable subgroup; locally finite group.

1. Permutability and Sylow Permutability

A subgroup H is said to be *permutable* in a group G if $HK = KH$ for all subgroups K of G: we denote this by

$$H \text{ per } G.$$

More generally, H is called π*-permutable*, where π is a set of primes, if $HP = PH$ for all Sylow p-subgroups P of G with $p \in \pi$. If π is the set of all primes, then H is *Sylow-permutable* (or *S-permutable*) in G, in symbols

$$H \text{ S-per } G.$$

If H is $\pi(H)$-permutable in G, then we will say that H is *weakly Sylow-permutable* (or *wS-permutable*) in G, and write

$$H \text{ wS-per } G.$$

Permutability and Sylow permutability in finite groups were first studied by Ore (1939) and Kegel (1962) respectively. The following results are basic.

Theorem 1.1 (Ore[10]). *If H per G where G is a finite group, then H is subnormal in G, (H sn G).*

This result was generalized to infinite groups by Stonehewer[18].

Theorem 1.2. *If H per G for an arbitrary group G, then H is ascendant in G, (H asc G).*

Thus permutability and ascendance are seen to be associated properties.

Ore's theorem was extended to S-permutability by Kegel[8].

Theorem 1.3. *If H wS-per G where G is a finite group, then H is subnormal in G.*

We will indicate a proof of Kegel's theorem; first here are some simple results that are invaluable in dealing with S-permutable subgroups.

Lemma 1.1. *Let H be a wS-permutable (S-permutable) subgroup of a group G.*

(*i*) *If $H \leq K \leq G$, then H is wS-permutable (S-permutable) in K.*
(*ii*) *If G is locally finite, $N \lhd G$ and G/N has countable Sylow subgroups, then HN/N is wS-permutable (S-permutable) in G/N.*

Proof. (i) Let $P \in \mathrm{Syl}_p(K)$ and let $P \leq P_1 \in \mathrm{Syl}_p(G)$; thus $P = K \cap P_1$. If $HP_1 = P_1H$, then intersecting both sides with K yields $HP = PH$.
(ii) It is well known that every Sylow p-subgroup of G/N has the form PN/N where $P \in \mathrm{Syl}_p(G)$. The result follows at once. □

Proof of Theorem 1.3. Let $\pi = \pi(H)$. It is enough to show that if H is a maximal proper π-permutable subgroup of G, then $H \lhd G$, so assume that this is false. We claim that there is a $p \in \pi$ and a $P \in \mathrm{Syl}_p(G)$ such that $H^P \neq H$. For suppose this is not the case, so that all such P normalize H. Then clearly H^G normalizes H and thus $H \lhd H^G \lhd G$, so Hsn G. Since H^G is π-permutable in G, maximality of H shows that $H \lhd G$. Therefore there is a $p \in \pi$ and an element x in a Sylow p-subgroup P of G such that $H \neq H^x$. Then $H \neq J =< H, H^x >$ and J is π-permutable in G, so $J = G$ by maximality of H. Therefore $G = HP$, from which it follows that $\pi(G) = \pi$.

Next let $q \in \pi$ be different from p and let $Q \in \mathrm{Syl}_q(G)$. Then $H \leq HQ \leq HP = G$, so $|HQ : H|$ divides $|HP : H|$. Since $p \neq q$, we deduce that $Q \leq H$. Thus the subgroup L generated by all the p'-elements of G is contained in H. Now $L \lhd G$ and G/L is a finite p-group, so H sn G, which is a contradiction. □

Infinite groups

If H S-per G and G is infinite, what can be said about H? In general the answer is nothing: for example, if G is torsion-free or a p-group, there is no condition on H. It is therefore reasonable to restrict attention to locally finite groups. The question arises as to what the corresponding result to Theorem 1.3 for locally finite groups should be.

Recall that a subgroup H of a group G is called *serial* if there is a series of general order type between H and G: denote this by

$$H \text{ ser } G.$$

The following important criterion for seriality in locally finite groups is due to Hartley[7].

Theorem 1.4. *Let G be a locally finite group and let H be a subgroup such that $H \cap F$ is subnormal in F for every finite subgroup F. Then H is serial in G.*

This is proved by using the theory of local systems and a standard procedure due to Mal'cev. In each finite subgroup F of G, the information that $H \cap F$ is subnormal in F can be encoded by means of a function. By passing to the Mal'cev limit of these functions one obtains a function defined on G which decodes into a series from H to G: for details see Ref. 12, 8.2. Notice the consequence: *in a locally finite p-group G every subgroup H is serial.* Now trivially every subgroup of G is S-permutable. This suggests that we should reformulate the question raised above as follows.

Problem

If H is an wS-permutable subgroup of a locally finite group G, is it true that H ser G?

This question is still open and it appears to be difficult. One obstacle to progress is the following Sylow pathology: if G is locally finite and $N \lhd G$, a Sylow p-subgroup of G/N need not have the form PN/N with P a Sylow p-subgroup of G. This phenomenon makes it difficult to pass to quotients.

However, the problem has a positive solution in many special situations, as we will shortly detail. In any event the evidence is that S-permutability and seriality should be considered as corresponding properties, just like permutability and ascendance. The first result is quite simple.

Theorem A. *If H is a finite wS-permutable subgroup of a locally finite group G, then H is serial in G.*

Proof. Let F be a finite subgroup of G and put $J = < H, F >$, which is finite. Now H wS-per G implies that H wS-per J by Lemma 1.1, so H sn J by Kegel's theorem. Hence $H \cap F$ sn F, and it follows by Hartley's theorem that H ser G. □

The next result is due to Ballester-Bolinches, Kurdachenko, Otal and Pedraza[4] in the case of S-permutable subgroups.

Theorem 1.5. *If H is a wS-permutable subgroup of a hyperfinite group G, then H is ascendant in G.*

Proof. Let $\{G_\alpha \mid \alpha \leq \beta\}$ be an ascending normal series in G with finite factors. We will argue that H asc HG_α by transfinite induction on α. Assume that α is the smallest ordinal such that H is not ascendant in HG_α. Suppose first that α is a limit ordinal. Then if $\beta < \alpha$, we have H asc $HG_{\beta+1}$ and hence HG_β asc $HG_{\beta+1}$. Therefore H is ascendant in HG_α, a contradiction which shows that α cannot be a limit ordinal.

Next H asc $HG_{\alpha-1}$. Also H wS-per HG_α. Let K be the normal core of $HG_{\alpha-1}$ in HG_α. Then HG_α/K is finite since $G_\alpha/G_{\alpha-1}$ is finite. Furthermore, HK/K wS-per HG_α/K by Lemma 1.1, which by Kegel's theorem implies that HK sn HG_α. Therefore $HG_{\alpha-1}$ snHG_α and H is ascendant in HG_α, a contradiction which completes the proof. □

Notice that we cannot conclude in Theorem A that H is ascendant in G since there are examples of non-trivial locally finite p-groups which have no non-trivial ascendant cyclic subgroups, (see Ref. 12, 6.27).

The following are some more recent results on the seriality problem.

Theorem B. *Let G be a locally finite group satisfying min-p for all primes p. If H is wS-permutable in G, then H is ascendant in G.*

Proof. Let $\pi = \pi(H)$. Consider first the case where G has finite Sylow subgroups. The key step is to show that if H is any a proper π-permutable subgroup, there is an π-permutable subgroup K properly containing H such that $H \triangleleft K$. To see this assume that H is not normal in G and argue as in the proof of Theorem 1.3 that there is a $p \in \pi$ and $P \in \mathrm{Syl}_p(G)$ such that $H^P \neq H$. Since P is finite, H sn HP. Form the series of successive normal closures of H in $J = HP$,

$$H = H_k \triangleleft H_{k-1} \triangleleft \cdots \triangleleft H_1 \triangleleft H_0 = J.$$

Here $k \geq 2$. Put $K = H_{k-1}$; then $K = H^{H_{k-2}}$ is π-permutable in G and $H \triangleleft K$. Applying this result repeatedly, we find that H asc G.

Now consider the general case, where each Sylow subgroup of G satisfies min, i.e., it is a Černikov group. There is a unique maximal divisible abelian subgroup D such that G/D has finite Sylow subgroups, (see Ref. 5, 2.5.14,

3.5.15). By Lemma 1.1 we see that HD/D is wS-permutable in G/D and therefore HD asc G.

For a fixed prime p set

$$D(i) = \{x \mid x \in D, x^{p^i} = 1\}.$$

Thus $D(i) \lhd G$ is finite and H sn $HD(i)$ by the normal core argument. Hence $HD(i-1)$ sn $HD(i)$ for $i = 1, 2. \dots,$ so that H asc HD_p for all p.

Finally, if $p_1, p_2, \dots$ is the sequence of primes, we have

$$H \text{ asc } HD(p_1) \text{ asc } HD(p_1)D(p_2) \text{ asc} \dots,$$

so H asc HD and H asc G. □

Finally, we mention without proof a recent result which generalizes Theorem 1.5.

Theorem C. *Let G be a locally finite group which is hyper-(abelian or finite) and let H be wS-permutable in G. Then H is serial in G*

2. Transitivity of Permutability

A group G is called a *PT-group* if permutability is transitive in G, i.e.,

$$H \text{ per } K \text{ and } K \text{ per } G \Rightarrow H \text{ per } G$$

Also G is called a *PST-group* if S-permutability is transitive, i.e.,

$$H \text{ S-per } K \text{ and } K \text{ S-per } G \Rightarrow H \text{ S-per } G$$

The following are simple consequences of the theorems of Ore and Kegel.

Theorem 2.1.

(*i*) *A finite group is a PT-group if and only if the subnormal subgroups and the permutable subgroups are the same.*
(*ii*) *A finite group is a PST-group if and only if the subnormal subgroups and the S-permutable subgroups are the same.*

Finite PT-groups and PST-groups have been studied intensively, especially in the soluble case. There are two principal structure theorems.

Theorem 2.2 (Agrawal[1]). *Let G be a finite group. Then G is a soluble PST-group if and only if there is an abelian normal subgroup L of odd order such that:*

(*i*) *G/L is nilpotent;*
(*ii*) $\pi(L) \cap \pi(G/L) = \emptyset$;

(*iii*) *elements of G induce power automorphisms in L.*

Theorem 2.3 (Zacher[19]). *Let G be a finite group. Then G is a soluble PT-group if and only if there is an abelian normal subgroup L of odd order such that:*

(*i*) *G/L is a nilpotent group which is modular, (i.e., its Sylow subgroups are Iwasawa groups);*
(*ii*) $\pi(L) \cap \pi(G/L) = \emptyset$;
(*iii*) *elements of G induce power automorphisms in L.*

Recall that an *Iwasawa group* is a p-group of the form $< x > A$ where A is an abelian group of finite exponent p^e, $a^x = a^{1+p^m}$, $a \in A$, with $e > m \geq 1$, and $m \geq 2$ if $p = 2$. We call these *theorems of Gaschütz type*, after W. Gaschütz's classification of finite soluble T-groups (Ref. 6). So the difference between finite soluble PT-groups and PST-groups is that the former have modular Sylow subgroups.

We mention that there are also structure theorems for finite insoluble T-, PT-, PST-groups, which depend on the truth of the Schreier Conjecture and hence on the classification of finite simple groups: for details see Ref. 13. (Apparently groups in which weak Sylow-permutability is transitive have not been studied).

3. Locally finite PT- and PST-groups

If we wish to construct a theory of locally finite PST- and PT-groups along the lines of the finite theory, there are significant obstacles, namely the lack of a good Sylow theory, eg., failure of conjugacy of Sylow subgroups, and the falsity of the Schreier Conjecture for infinite simple groups. However, some progress has been made in special cases.

(a) Periodic soluble T- and PT-groups are studied in Refs. 9,11 respectively; however, periodic soluble PST-groups have not yet received attention.

(b) Hyperfinite radical PST-groups: for these groups some necessary conditions are known (Ref. 4).

Here we present two recent results about locally finite groups whose finite subgroups are PT-groups or PST -groups: details of the proofs will appear elsewhere (Ref. 15).

Theorem D. *Let G be a locally finite group. Then the following conditions are equivalent.*

(*i*) *Finite subgroups of G are PST-groups.*
(*ii*) *There is an abelian normal subgroup L containing no involutions such that G/L is locally nilpotent, $\pi(L) \cap \pi(G/L = \emptyset$ and elements of G induce power automorphisms in L.*
(*iii*) *In each section of G the S-permutable subgroups and the serial subgroups are the same.*
(*iv*) *Each section of G is a PST-group.*

Theorem E. *Let G be a locally finite group. Then the following conditions are equivalent.*

(*i*) *Finite subgroups of G are PT-groups.*
(*ii*) *There is an abelian normal subgroup L containing no involutions such that G/L is a modular locally nilpotent group, $\pi(L) \cap \pi(G/L) = \emptyset$ and elements of G induce power automorphisms in L.*
(*iii*) *In each section of G the permutable subgroups and the ascendant subgroups are the same.*
(*iv*) *Each section of G is a PT-group.*

Notice that the conditions (ii) in these theorems are of Gaschütz type. Observe once again the conjunction of the properties S-permutable and serial in Theorem D.

Corollary. *A locally finite minimal non-PST-group or PT-group is finite.*

This fills a gap in the proofs in Ref. 14, where the finite minimal non-PST and PT-groups are classified: this was pointed out in Ref. 3. (The finite minimal non-PT-groups have also been classified by Scoppola[17]).

The main step in the proof of Theorem D is (ii) $\Rightarrow$ (iii), for which the following fact is critical.

Lemma 3.1. *Let $G = AB$ be a periodic group where A, B are abelian p-groups. Then G is a locally finite p-group.*

This is a consequence of a theorem of Sysak and Černikov (see Ref. 2, 3.2.7), and it can also be deduced from the following result in Ref. 16:

Theorem 3.1. *Let $G = AB$ where A, B are abelian groups. Then each chief factor of G is centralized by either A or B.*

References

1. R. K. Agrawal, Finite groups whose subnormal subgroups permute with all Sylow subgroups, Proc. Amer. Math. Soc. **47** (1975), 77-83.

2. B. Amberg, S. Franciosi and F. de Giovanni, Products of Groups, Oxford, 1992.
3. A. Ballester-Bolinches, L.A. Kurdachenko, J. Otal and T. Pedraza, Infinite groups with many permutable subgroups, Rev. Mat. Iberoamericana **24** (2008), 745-64.
4. A. Ballester-Bolinches, L.A. Kurdachenko, J. Otal and T. Pedraza, Infinite groups with Sylow permutable subgroups, preprint.
5. M. R. Dixon, Sylow Theory, Formations and Fitting Classes in Locally Finite Groups, World Scientific, Singapore, 1994.
6. W. Gaschütz, Gruppen, in denen das Normalteilersein transitiv ist, J. reine angew. Math. **198** (1957), 87-92.
7. B. Hartley, Serial subgroups of locally finite groups, Proc. Cambridge Philos. Soc. **71** (1972), 199-201.
8. O. H. Kegel, Sylow-Gruppen und Subnormalteiler endlicher Gruppen, Math. Z. **78** (1962), 205-221.
9. F. Menegazzo, Gruppi nei quali la relazione di quasi-normalità è transitiva I, II, Rend. Sem. Mat. Univ. Padova **40** (1968), 347-361, *ibid* **42** (1970), 389-399.
10. O. Ore, Contributions to the theory of groups of finite order, Duke Math. J. **5** (1939), 431-460.
11. D. J. S. Robinson, Groups in which normality is a transitive relation, Proc. Cambridge Philos. Soc. **60** (1964), 21-38.
12. D. J. S. Robinson, Finiteness Conditions and Generalized Soluble Groups, Springer-Verlag, Berlin 1972.
13. D. J. S. Robinson, The structure of finite groups in which permutability is a transitive relation, J. Austral. Math. Soc. **70** (2001), 143-159.
14. D. J. S. Robinson, Minimality and Sylow-permutability in locally finite groups, Ukr. Math. J. **54** (2002), 1038-1049.
15. D. J. S. Robinson, Sylow permutability in locally finite groups, Ricerche di Matematica **59** (2010), 313-318.
16. D. J. S. Robinson and S.E. Stonehewer, Triple factorizations by abelian groups, Arch. Math. (Basel) **60** (1993), 223-232.
17. C. M. Scoppola, Su una classe di gruppi finiti, Atti Accad. Naz. Lincei Rend. Cl. Fis. Mat. Natur. (8) **62** (1977), 579-583.
18. S. E. Stonehewer, Permutable subgroups of infinite groups, Math. Z. **125** (1972), 1-16.
19. G. Zacher, I gruppi in cui i sottogruppi di composizione coincidano con i sottogruppi quasinormali, Atti Accad. Naz. Lincei. Cl. Sc. Fis. Mat. Natur. **37** (1964), 150-154.

ON THE EXPONENT OF A FINITE GROUP WITH A FOUR-GROUP OF AUTOMORPHISMS

E. ROMANO

Dipartimento di Matematica e Informatica, Università di Salerno
Via Ponte don Melillo, 84084 Fisciano (SA) Italy
E-mail: eromano@unisa.it

P. SHUMYATSKY*

Department of Mathematics, Universidade de Brasilia
70.919 Brasilia - DF Brazil
E-mail: pavel@mat.unb.br

Let A be a group isomorphic with either S_4 or D_8. Let V be a normal four-subgroup of A and α an involution in $A \setminus V$. Suppose that A acts on a finite group G in such a manner that $C_G(V) = 1$ and $C_G(\alpha)$ has exponent e. We show that if $A \cong S_4$ then the exponent of G is e-bounded and if $A \cong D_8$ then the exponent of G' is e-bounded. This work was motivated by recent results on the exponent of a finite group admitting an action by a Frobenius group of automorphisms.

Keywords: Automorphisms; Lie algebras; Finite groups.

1. Introduction

Let G be a group admitting an action of a group A. We denote by $C_G(A)$ the set $C_G(A) = \{x \in G; x^a = x \text{ for any } a \in A\}$, the centralizer of A in G (the fixed-point group). It is known that in many cases the properties of G are determined by those of $C_G(A)$. Special attention was recently given to the situation where a Frobenius group acts by automorphisms on another group. Recall that a Frobenius group FH with kernel F and complement H can be characterized as a finite group that is a semidirect product of a normal subgroup F by H such that $C_F(h) = 1$ for every $h \in H \setminus \{1\}$. By Thompson's theorem[19] the kernel F is nilpotent, and by Higman's theorem[2] the nilpotency class of F is bounded in terms of the least prime divisor of

*The author was supported by CNPq-Brazil.

$|H|$ (explicit upper bounds for the nilpotency class are due to Kreknin and Kostrikin[7,8]). Suppose that the Frobenius group FH acts on a finite group G in such a way that $C_G(F) = 1$. It was shown in Ref. 4 that the order and rank of G are bounded in terms of $|H|$ and the order and rank of $C_G(H)$, respectively. Further, it was shown that if F is cyclic, then the nilpotency class of G is bounded in terms of $|H|$ and the nilpotency class of $C_G(H)$. In the case when GF is also a Frobenius group with kernel G and complement F (so that GFH is a double Frobenius group) the latter result was obtained earlier in Ref. 12. This solved in the affirmative Mazurov's problem 17.72(a) in Kourovka Notebook[5].

The other problem of Mazurov about double Frobenius groups – Problem 17.72(b) in Kourovka Notebook – is whether in a double Frobenius group GFH the exponent of G is bounded in terms of $|H|$ and the exponent of $C_G(H)$ only. That problem seems to be very hard and so far no viable approach to it has been found. We will quote just one result from Ref. 4 that indirectly addresses the problem:

Theorem 1.1. *Suppose that a Frobenius group FH with cyclic kernel F and complement H acts on a finite group G in such a manner that $C_G(F) = 1$ and $C_G(H)$ has exponent e. Then the exponent of G is bounded solely in terms of e and $|FH|$.*

The proof of the above theorem uses Lazard's Lie algebra associated with the Jennings–Zassenhaus filtration and its connection with powerful p-groups.

It is natural to ask if the theorem remains valid without assuming that F is cyclic. The case of the smallest Frobenius group whose kernel is non-cyclic was treated in Ref. 18. The group in question is of course the non-abelian group of order 12 also known as the Alternating group A_4 of degree 4. The following theorem is the main result of Ref. 18.

Theorem 1.2. *Suppose that the Frobenius group FH of order 12 acts coprimely on a finite group G in such a manner that $C_G(F) = 1$ and $C_G(H)$ has exponent e. Then the exponent of G is bounded in terms of e only.*

Recall that an action of a finite group A on a finite group G is coprime if $(|G|, |A|) = 1$. It is amazing that tools required for the treatment of the situation where FH has order 12 are more sophisticated than those employed in the proof of Theorem 1.1. In particular, the proof of Theorem 1.2 uses in the very essential way the solution of the Restricted Burnside Problem[20,21] while the proof of Theorem 1.1 is based on more simple techniques. Some explanation of this phenomenon can be found in the study of automorphisms of Lie algebras. By the Kreknin Theorem[7] a Lie algebra that admits a fixed-point-free automorphism of finite order n is soluble with n-bounded derived length. On the other hand, a Lie algebra that admits a non-cyclic fixed-point-free group of automorphisms can be unsoluble or soluble with arbitrarily large derived length (see examples in Ref. 3, p. 149–150). It is the necessity to work with Lie algebras of unbounded derived length that accounts for the complexity of the proof of Theorem 1.2.

We do not know whether the assumption that $(|G|, 12) = 1$ is necessary in Theorem 1.2. It seems likely that the theorem remains valid without that hypothesis. Our attempts to deal with the problem led us to study in more detail some questions about the exponent of a finite group admitting a fixed-point-free four-group of automorphisms. In the first result that we would like to mention we consider groups acted on by S_4, the symmetric group on 4 symbols. In what follows we denote by V the maximal normal 2-subgroup of S_4. Of course V is the non-cyclic group of order 4.

Theorem 1.3. *Let A be isomorphic with S_4 and let α be an involution in $A \setminus V$. Suppose that A acts on a finite group G in such a manner that $C_G(V) = 1$ and $C_G(\alpha)$ has exponent e. Then the exponent of G is bounded in terms of e only.*

Thus, we were not able to prove Theorem 1.2 without the coprimeness assumption but at least the related result on the action of S_4 does not require coprimeness at all. It is interesting and somewhat unusual that Theorem 1.3 involves only a hypothesis on the exponent of $C_G(\alpha)$ rather than the centralizer of the subgroup of order three as in Theorem 1.2. It is well-known that a Sylow 2-subgroup of S_4 is isomorphic with D_8, the dihedral group of order 8. When studying the action of D_8 on G satisfying the conditions similar to the ones in Theorem 1.3 we discovered a new phenomenon – such an action actually has strong impact on the exponent of G', the derived group of G. Write D_8 as a product $V\langle\alpha\rangle$, where V is a four-group and α an involution.

Theorem 1.4. *Let A be isomorphic with D_8 and suppose that A acts on a finite group G in such a manner that $C_G(V) = 1$ and $C_G(\alpha)$ has exponent e. Then the exponent of G' is bounded in terms of e only.*

The proofs of both Theorem 1.3 and Theorem 1.4 use the solution of the Restricted Burnside Problem. Another important tool used in the proof of the above results is the theorem that the exponent of a finite group acted on by a non-cyclic abelian group A is bounded in terms of $|A|$ and the exponents of $C_G(a)$, where $a \in A\setminus\{1\}$[6]. As a part of the proof of Theorem 1.4 we obtained the following result that seems to be of independent interest.

Theorem 1.5. *Let G be a finite group acted on by the four-group V in such a manner that $C_G(V) = 1$. Suppose that the centralizers $C_G(v_1)$ and $C_G(v_2)$ of two involutions $v_1, v_2 \in V$ have exponent e. Then the exponent of G' is bounded in terms of e only.*

In view of the above results a number of questions about the exponent of a finite group with automorphisms can be asked. In particular, it would be interesting to see if similar results hold in the situation where V is an elementary abelian p-group for an odd prime p. This could be a subject of some future research project.

2. Some Lie ring machinery

The results mentioned in the introduction rely heavily on the use of so called Lie methods. In this section we will describe some key Lie-theoretic tools required for the proofs. Throughout the paper we use the expression "(m,n)-bounded" for "bounded above in terms of m, n only".

Let G be a group and p a prime. We set

$$D_i = D_i(G) = \prod_{jp^k \geq i} \gamma_j(G)^{p^k}.$$

The subgroups D_i form the Jennings–Zassenhaus filtration

$$G = D_1 \geq D_2 \geq \cdots$$

of the group G. The series satisfies the inclusions $[D_i, D_j] \leq D_{i+j}$ and $D_i^p \leq D_{pi}$ for all i, j. These properties make it possible to construct a Lie algebra $DL(G)$ over $\mathbb{F}_p$, the field with p elements. Namely, consider the quotients $K_i = D_i/D_{i+1}$ as linear spaces over $\mathbb{F}_p$, and let $DL(G)$ be the direct sum of these spaces. Commutation in G induces a binary operation $[,]$ in $DL(G)$. For elements $xD_{i+1} \in K_i$ and $yD_{j+1} \in K_j$ the operation is defined by

$$[xD_{i+1}, yD_{j+1}] = [x,y]D_{i+j+1} \in K_{i+j}$$

and extended to arbitrary elements of $DL(G)$ by linearity. It is easy to check that the operation is well-defined and that $DL(G)$ with the operations $+$ and $[,]$ is a Lie algebra over $\mathbb{F}_p$.

For any $x \in D_i \setminus D_{i+1}$ let $\bar{x}$ denote the element xD_{i+1} of $DL(G)$.

Lemma 2.1 (Lazard[9]). *For any $x \in G$ we have $(\mathrm{ad}\,\bar{x})^p = \mathrm{ad}\,\overline{x^p}$. Consequently, if x is of finite order p^t, then $\bar{x}$ is ad-nilpotent of index at most p^t.*

Recall that an element a of a Lie algebra L is called ad-nilpotent if there exists a positive integer n such that $[x, \underbrace{a, \ldots, a}_{n}] = 0$ for all $x \in L$. If n is the least positive integer with the above property then we say that a is ad-nilpotent of index n. If $X \subseteq L$ is a subset of L, we denote by $\langle X \rangle$ the subalgebra generated by X. Let $L_p(G) = \langle K_1 \rangle$ be the subalgebra of $DL(G)$ generated by K_1. The following lemma goes back to Lazard[10]; in the present form it can be found, for example, in Ref. 6.

Lemma 2.2. *Suppose that X is a d-generator finite p-group such that the Lie algebra $L_p(X)$ is nilpotent of class c. Then X has a powerful characteristic subgroup of (p, c, d)-bounded index.*

Recall that powerful p-groups were introduced by Lubotzky and Mann in Ref. 11: a finite p-group G is *powerful* if $G^p \geq [G, G]$ for $p \neq 2$ (or $G^4 \geq [G, G]$ for $p = 2$). These groups have many nice properties, so that often a problem becomes much easier once it is reduced to the case of powerful p-groups. The above lemma is quite useful as it allows us to perform such a reduction. Precisely, we will require the property that if $G = \langle g_1, \ldots, g_d \rangle$ is a powerful p-group and $e = p^k$, then $G^e = \langle {g_1}^e, \ldots, {g_d}^e \rangle$.

Using this it is not really difficult to prove Theorems 1.3, 1.4 and 1.5 in the particular case where G is a powerful p-group. Thus, it follows from Lemma 2.2 that we need a criterion of nilpotency for $L_p(G)$ where G is a p-group satisfying the relevant hypothesis. We will now proceed to describe our main result on nilpotency of Lie algebras with automorphisms. Though the hypotheses of Proposition 2.1 below may look bizarre, the proposition is sufficient (and perhaps even necessary) for possible group-theoretic applications. In what follows the term "Lie algebra" means a Lie algebra over some commutative ring with unity in which 2 is invertible. If $X \subseteq L$ is a subset of L, by a commutator of weight 1 in elements of X we mean just any element of X. We define inductively commutators in X of weight $w \geq 2$ as elements of the form $[x, y]$, where x and y are commutators in X of weight w_1 and w_2 respectively such that $w_1 + w_2 = w$.

We will now work with a Lie algebra L such that $L = \bigoplus_{1 \leq i \leq 3} L_i$, where L_i are abelian subalgebras with the property that

$$[L_1, L_2] \leq L_3, \quad [L_2, L_3] \leq L_1, \quad [L_3, L_1] \leq L_2. \tag{1}$$

The next lemma is taken from Ref. 15.

Lemma 2.3. *(Ref. 15, Proposition 1.1) Let $a \in L_1 \cup L_2 \cup L_3$ and suppose that a is ad-nilpotent of index m. Then the ideal that a generates in L is nilpotent of class at most $2m-1$.*

We assume additionally that L admits an automorphism α of order two such that

$$L_1{}^{\alpha} = L_2 \quad \text{and} \quad L_3{}^{\alpha} = L_3. \tag{2}$$

We denote by L_α the fixed-point subalgebra for α. Given a set $X \subseteq L$ such that $L = \langle X \rangle$, an element of L is said to be homogeneous (of weight w) with respect to the generating set X if it can be written as a homogeneous Lie polynomial (of degree w) in elements of X. Further, we assume the following hypothesis.

Hypothesis 2.1. The algebra L is generated by two elements $x_1 \in L_1$ and $x_2 = x_1{}^{\alpha} \in L_2$ for which there exist positive integers m and n such that every homogeneous element contained in L_α is ad-nilpotent in L of index at most m and every pair of homogeneous elements contained in L_α generates a subalgebra that is nilpotent of class at most n.

Our main result on nilpotency of Lie algebras is as follows.

Proposition 2.1. *Let $L = \bigoplus L_i$ be a Lie algebra satisfying (1) and admitting an automorphism α of order two as in (2). Assume Hypothesis 2.1. Then L is nilpotent of (m,n)-bounded class.*

The proof of the above proposition is somewhat technical. We intend to present it in full detail in a separate paper. It is somewhat similar to the proofs of Proposition 2.11 and Theorem 2.12 from Ref. 18. In the next section we show how the proposition can be used in the proof of Theorems 1.3, 1.4 and 1.5.

3. Some proofs

In this section we will show how Lie-theoretic results from Section 2 can be used in the study of groups with automorphisms. First we recall some facts about finite groups admitting a fixed-point-free four-group of automorphisms. Without further references we use the well-known fact that if a finite group A acts coprimely on a finite group G and N is a normal A-invariant subgroup of G, then $C_{G/N}(A) = C_G(A)N/N$. Let G be a finite group and $V = \{1, v_1, v_2, v_3\}$ the non-cyclic group of order 4 acting fixed-point-freely on G. Then G has odd order. Put $G_i = C_G(v_i)$, for $i \in \{1, 2, 3\}$. The proofs of the next few lemmas can be found in Ref. 14.

Lemma 3.1. *Each G_i is abelian and if $i \neq j$ then v_j acts on G_i by the rule $x^{v_j} = x^{-1}$ for each $x \in G_i$; $i, j \in \{1, 2, 3\}$. It follows that every subgroup generated by a subset of $G_1 \cup G_2 \cup G_3$ is V-invariant.*

Lemma 3.2. *Let x be an element of G such that $x^{v_i} = x^{-1}$ for some i. Suppose that $x \in S$, where S is some V-invariant subgroup of G. Then there exists a unique pair of elements $y \in G_j \cap S$ and $t \in G_k \cap S$ such that $x = yty$ and $\{i, j, k\} = \{1, 2, 3\}$.*

Lemma 3.3. $G = G_1G_2G_3$.

Lemma 3.4. $G' = \langle G_1, G_2\rangle \cap \langle G_2, G_3\rangle \cap \langle G_3, G_1\rangle$. *In particular, the subgroups $\langle G_i, G_j\rangle$ are normal and contain G'.*

For any $x \in G_i$ and $y \in G_j$ with $i \neq j$ it is clear that v_i sends y^x to the inverse. Therefore Lemma 3.2 guarantees that there exists a unique pair $(s, t) \in G_k \times G_j$ such that $y^x = sts$ and $\{i, j, k\} = \{1, 2, 3\}$. Thus we can define $x * y = s$. According to the same lemma, if y^x is an element of a V-invariant subgroup S, then $x * y \in S$. The next lemma is taken from Ref. 16.

Lemma 3.5. *Let $x \in G_i, y \in G_j$ and $H = \langle x, y\rangle$. Then $\langle (x * y)^H\rangle = H'$.*

Let us define

$$R_1 = \langle a * b \mid a \in G_2, b \in G_3\rangle,$$

$$R_2 = \langle a * b \mid a \in G_1, b \in G_3\rangle,$$

$$R_3 = \langle a * b \mid a \in G_1, b \in G_2\rangle.$$

We will require the following result.

Lemma 3.6. *Suppose G is a p-group. Then $G_i \cap G' = R_i$ for $i = 1, 2, 3$.*

Proof. Let G be a counterexample of minimal order. Set $K = G'$ and $K_i = K \cap G_i$ for $i = 1, 2, 3$. If N is any normal V-invariant subgroup of G, by induction we have $K_iN = R_iN$. Let $Z = Z(G)$. If Z contains a nontrivial element of the form $a * b$ for some $a \in G_i, b \in G_j$, we put $N = \langle a * b\rangle$. It is clear that N is a normal V-invariant subgroup contained in $R_1 \cup R_2 \cup R_3$

and in view of the above assumption we get a contradiction. Thus, assume that Z contains no nontrivial elements of the form $a * b$.

Let $T = Z_2(G)$ be the second term of the upper central series of G and set $T_i = T \cap G_i$ for $i = 1, 2, 3$. Choose arbitrarily $a \in G_i$ and $b \in T_j$. Then, because of Lemma 3.5, it follows that $a * b \in Z(G)$. Therefore we conclude that a and b commute. This happens for every choice of $a \in G_i$ and $b \in T_j$. Hence $T = Z(G)$ and G is abelian, a contradiction. □

Another useful fact is provided by the next lemma.

Lemma 3.7. *Suppose that N is a normal subgroup of G such that $N \leq G_3$. Then $N \leq Z(G)$.*

Proof. Choose $b \in N$ and $a \in G_1$ (or $a \in G_2$). Since $b^a \in G_3$, it follows that $a * b = 1$ and so by Lemma 3.5 $[a, b] = 1$. Thus an arbitrary element of N commutes with an arbitrary element of $G_1 \cup G_2$. Using that G_3 is abelian and $N \leq G_3$ we conclude that $N \leq Z(G)$. □

Sketch of the proof of Theorem 1.5. Suppose that the centralizers G_1 and G_2 have exponent e. We wish to prove that the exponent of G' is bounded in terms of e only. In view of the main result of Ref. 6 it is sufficient to prove that $G_3 \cap G'$ has bounded exponent.

Suppose first that G is a p-group. Taking into account Lemma 3.6 we notice that the fact that $G_3 \cap G'$ has bounded exponent could be deduced by showing that $a * b$ has bounded order for every $a \in G_1$ and $b \in G_2$. Thus, choose $a \in G_1$ and $b \in G_2$. Without loss of generality we can assume that $G = \langle a, b \rangle$. Let $L = L_p(G)$. Combining Lemma 2.1 and Lemma 2.3, we deduce that L is nilpotent of e-bounded class. Therefore by Lemma 2.2 G has a powerful characteristic subgroup H of e-bounded index. Since H is generated by the centralizers $H \cap G_i$, we can choose generators $g_1, \ldots, g_d$ of H such that $g_1, \ldots, g_d \in G_1 \cup G_2 \cup G_3$. Since ${G_1}^e = {G_2}^e = 1$ and $H^e = \langle {g_1}^e, \ldots, {g_d}^e \rangle$, we conclude that $H^e \leq G_3$. Thus, Lemma 3.7 tells us that $H^e \leq Z(G)$. We see that $G/Z(G)$ has bounded order and using the Schur Theorem (Ref. 13, p. 102) we conclude that G' has e-bounded order. In particular, it follows that the order of $a * b$ is e-bounded.

Thus, the theorem is valid for p-groups. In general, a finite group G admitting a fixed-point-free four-group of automorphisms need not be nilpotent but it is known that G' is nilpotent (Ref. 1, 10.5.3). (Stronger results on nilpotency of G' can be found in Refs. 16,17.) We therefore deduce that G'' has e-bounded exponent. Passing to the quotient G/G'', we can assume that G is metabelian. Because it is sufficient to bound the exponent of $G_3 \cap G'$ we also assume that $G = \langle G_1, G_2 \rangle$. Let $N = G'$. Since N is generated by the centralizers $N \cap G_i$ and ${G_1}^e = {G_2}^e = 1$, we conclude that $N^e \leq G_3$. Thus, Lemma 3.7 tells us that $N^e \leq Z(G)$. Hence, G is a metabelian group such that $G/Z(G)$ has bounded exponent. It follows that G' has bounded exponent, as well. □

Sketch of the proof of Theorem 1.4. Recall that now the group $A = V\langle\alpha\rangle$ isomorphic with D_8 acts on a finite group G in such a manner that $C_G(V) = 1$ and $C_G(\alpha)$ has exponent e. We wish to show that the exponent of G' is bounded in terms of e only. We use the notation as in the previous proof and assume that ${G_1}^\alpha = G_2$ and G_3 is α-invariant.

Let H be any A-invariant subgroup of G and N the minimal A-invariant subgroup of H containing all conjugates of $C_H(\alpha)$. Then α is fixed-point-free on the quotient H/N. It follows that A induces an abelian group of automorphisms of H/N. In particular v_3 must act trivially on H/N. The conclusion is that $H = NC_H(v_3)$ and a similar decomposition holds for all A-invariant subgroups of G and all A-invariant quotients.

In view of Theorem 1.5 the exponent of G' is bounded by some number that depends only on the exponents of G_1 and G_2. Since ${G_1}^\alpha = G_2$, it is sufficient to show that an arbitrary element x_1 of G_1 has e-bounded order. Put ${x_1}^\alpha = x_2$. The subgroup $\langle x_1, x_2 \rangle$ is A-invariant so without loss of generality we can assume that $G = \langle x_1, x_2 \rangle$. Since v_3 has no fixed-points in G/G', it follows that G/G' is generated by elements of order dividing e. Since the rank of G/G' is at most two, we conclude that the order of G/G' is e-bounded.

Suppose first that G is p-group and let $L = L_p(G)$. The group A naturally acts on L and one can show that L satisfies all the hypothesis of Proposition 2.1. For example, the fact that every homogeneous element contained in L_α is ad-nilpotent in L of index at most e follows from Lazard's Lemma 2.1 while the existence of an e-bounded number n such that every pair of homogeneous elements in L_α generates a subalgebra that is nilpotent of class at most n follows from Zelmanov's solution of the restricted Burnside problem[20,21]. Thus, Proposition 2.1 tells us that L is nilpotent with e-bounded class. According to Lemma 2.2 G has a powerful characteristic subgroup H of e-bounded index. Since $H = NC_H(v_3)$, we can choose generators $g_1, \ldots, g_d$ of H each of which belongs either to a conjugate of $C_H(\alpha)$ or to $H \cap G_3$. Taking into account that $C_H(\alpha)$ has exponent e we derive that $H^e \leq G_3$. It follows now from Lemma 3.7 that $G/Z(G)$ has e-bounded order and using the Schur Theorem we conclude that G' has

e-bounded order. We know that the order of G/G' is e-bounded, as well. So G has e-bounded order. This proves the theorem in the case where G is a p-group.

The rest of the proof is similar to that of Theorem 1.5. Since G' is nilpotent, we deduce from the previous paragraph that G'' has e-bounded exponent. Passing to the quotient G/G'', we can assume that G is metabelian. Let $M = G'$. Since M is generated by $M \cap G_3$ and elements of order e, we conclude that $M^e \leq G_3$. Thus, Lemma 3.7 tells us that $M^e \leq Z(G)$. Therefore G' has e-bounded order. Combining this with the fact that the order of G/G' is likewise e-bounded, the theorem follows. □

Proof of Theorem 1.3. Recall that now the group A is isomorphic with S_4 and A acts on a finite group G in such a manner that $C_G(V) = 1$ and $C_G(\alpha)$ has exponent e. Here V is the maximal normal 2-subgroup of A and α an involution such that $\alpha \notin V$. The group $V\langle\alpha\rangle$ is isomorphic with D_8 and so we can use all the information that we obtained in the proof of Theorem 1.4. It follows that the exponent of G_1 is e-bounded. Of course in the group S_4 all involutions contained in V are conjugate so we conclude that G_1, G_2 and G_3 have the same exponent. Now the main result of Ref. 6 tells us that the exponent of G is e-bounded, as required. □

References

1. D. Gorenstein, *Finite groups*, Harper and Row, New York, 1968.
2. G. Higman, Groups and rings which have automorphisms without non-trivial fixed elements, J. London Math. Soc. (2) **32** (1957), 321–334.
3. E. I. Khukhro, "Nilpotent Groups and their Automorphisms", de Gruyter–Verlag, Berlin, 1993.
4. E. I. Khukhro, N. Yu. Makarenko and P. Shumyatsky, Frobenius groups of automorphisms and their fixed points, preprint, http://arxiv.org/abs/1010.0343.
5. E. I. Khukhro and V. D. Mazurov (Eds.), *Unsolved Problems in Group Theory. The Kourovka Notebook*, no. 17, Institute of Mathematics, Novosibirsk, 2010.
6. E. I. Khukhro and P. V. Shumyatsky, Bounding the exponent of a finite group with automorphisms, *J. Algebra* **212** (1999), 363–374.
7. V. A. Kreknin, The solubility of Lie algebras with regular automorphisms of finite period, Math. USSR Doklady, **4** (1963), 683–685.
8. V. A. Kreknin and A. I. Kostrikin, Lie algebras with regular automorphisms, Math. USSR Doklady, **4**, 355–358.
9. M. Lazard, Sur les groupes nilpotents et les anneaux de Lie, *Ann. Sci. École Norm. Supr.* **71** (1954), 101–190.

10. M. Lazard, Groupes analytiques p-adiques, *Inst. Hautes tudes Sci. Publ. Math.* **26** (1965) 389–603.
11. A. Lubotzky and A. Mann, Powerful p-groups. I: finite groups, *J. Algebra*, **105** (1987), 484–505.
12. N. Y. Makarenko and P. Shumyatsky, Frobenius groups as groups of automorphisms, Proc. Amer. Math. Soc., **138** (2010), 3425–3436.
13. D. J. S. Robinson, Finiteness Conditions and Generalized Soluble Groups, Part 1, Springer-Verlag, Berlin, 1972.
14. P. Shumyatsky, *Groups with regular elementary 2-groups of automorphisms*, Algebra and Logic **27**, (1988), 447-457.
15. P. Shumyatsky, Nilpotency of Some Lie Algebras Associated with p-Groups, Canad. J. Math. **51** (1999), 658-672.
16. P. Shumyatsky, C. Sica, On groups admitting a fixed-point-free four-group of automorphisms, J. Group Theory, **12** (2009), 401-406.
17. P. Shumyatsky, C. Sica, On groups admitting a fixed-point-free elementary 2-group of automorphisms, to appear in Communications in Algebra.
18. P. Shumyatsky, On the exponent of a finite group with an automorphism group of order twelve, J. Algebra, (2010), doi: 10.1016/j.jalgebra.2010.09.038.
19. J. G. Thompson, Finite groups with fixed-point-free automorphisms of prime order, Proc. Nat. Acad. Sci. USA, **45** (1959), 578–581.
20. E. Zelmanov, The solution of the restricted Burnside problem for groups of odd exponent, Math. USSR Izv. **36** (1991), 41-60.
21. E. Zelmanov, The solution of the restricted Burnside problem for 2-groups, Math. Sb. **182** (1991), 568-592.

EXAMPLES OF MARKOV CHAINS ON SPACES WITH MULTIPLICITIES

F. SCARABOTTI

Dipartimento SBAI, Università La Sapienza,
Via A. Scarpa 8, Rome, 00161, Italy
E-mail: scarabot@dmmm.uniroma1.it

F. TOLLI

Dipartimento di Matematica, Università Roma TRE,
Largo S. Leonardo Murialdo 1, 00146 Rome, Italy
E-mail: tolli@mat.uniroma3.it

We show how to compute the spectra of several random walks that are invariant for the action of a group G in the case that the action is either not transitive or not multiplicity-free. This extends the classical analysis of stochastic processes developed by Diaconis *et. al.* in a Gelfand pair setting. In this expository note, we do not present the general theory, but we show the main ideas through examples, avoiding the technical results.

Keywords: Ehrenfest diffusion model; Bernoulli–Laplace diffusion model; Lamplighter random walks; Markov spectrum; Wreath product; Permutation representation.

1. Introduction: random walks on multiplicity-free homogeneous spaces

In the 80's Persi Diaconis and his collaborators provided innovative techniques to analyze several stochastic processes, including two famous classical models of gas diffusion.

The Bernoulli–Laplace diffusion process consists of $2n$ balls placed in two urns, n balls in each urn. At each step, two balls are randomly chosen, one from each urn, and then switched. This process can be represented by a Markov chain whose state space is given by the n-subsets of $\{1, 2, \ldots, 2n\}$. Indeed, the configuration of the system is uniquely determined by the content of one of the two urns, say the first.

In the Ehrenfest model, we have m balls distributed in two urns: at each step a ball is randomly chosen and moved to the other urn. The state space X is given by set of all k-subsets of $\{1, 2, \ldots, m\}$. There is natural identification of the space X with the set of the vertices of an m-hypercube and, in fact, the Ehrenfest diffusion model is equivalent to the simple random walk on such a hypercube.

From the classical theory of Markov chains we know that, for both of these processes, the distribution of probability $\nu^{(k)}$ at time k converges to the uniform distribution π, in total variation norm, as k tends to infinity.

From an applicative point of view, this result is not completely satisfactory: for instance, in the Bernoulli–Laplace model the main question is to establish what is the sufficient and necessary number of switches needed to mix the content of both urns.

In 1987 Diaconis and Shahshahani[1] proved the following estimates for the Bernoulli–Laplace model:

Theorem 1.1. *There exists a universal positive constant a such that if* $k = \frac{n}{4}(\log(2n) + c)$, *with* $c \geq 0$, *then*

$$\|\nu^{(k)} - \pi\|_{TV} \leq ae^{-c/2}.$$

Moreover if n is sufficiently large and $k = \frac{1}{4}n[\log(2n) - c]$ *with* $0 < c < \log(2n)$, *then*

$$\|\nu^{(k)} - \pi\|_{TV} \geq 1 - 32e^{-c}.$$

From these estimates we deduce the existence of a critical value $k = \frac{1}{4}n\log(2n)$ such that if we make additional cn switches then the balls are well mixed, while if we make less than $\frac{1}{4}n\log(2n) - cn$ switches the balls are not mixed. In other words, the total variation distance remains close to the initial value for a while, then it drops rapidly in correspondence of the critical value and finally it has an exponential tail. This phenomenon (see Ref. 2 for a formal definition) is now called *cutoff*.

In his seminal book[3], Diaconis showed that the cutoff phenomenon holds for the Ehrenfest model and does not hold for the simple random walk on the discrete circle. In this case, if the number of vertices n is odd, n^2 steps are sufficient to reach the uniformity, but the transition from order to chaos is gradual.

In these examples, a group G acts transitively on the state space X and the associated Markov operator is invariant for this action. Thus, there exists a probability measure μ on G such that the distribution at time k can be identified with the kth convolution power of μ.

In the abelian setting, the Fourier analysis has been used to prove several results about te asymptotic behaviour of convolution powers of a probability measure. A complete nonabelian Fourier analysis has been developed for compact groups, but even for finite groups the expression of the corresponding Fourier transform is complicated, since it depends on the knowledge of the representations of the group.

In the above examples the computation of the Fourier transform is easier since the group G and the space X form a so-called *Gelfand pair.* To define it, first observe that the group G acts on $L(X) = \{f : X \to \mathbb{C}\}$ by setting $[gf](x) = f(g^{-1}x)$ with $g \in G$, $f \in L(X)$ and $x \in X$: this is the *permutation representation.* We say that (G, X) is a Gelfand pair if the permutation representation $L(X)$ is *multiplicity-free*, i.e. it decomposes into a direct sum of irreducible pairwise inequivalent subrepresentations V_i. When (G, X) is a Gelfand pair the spectral analysis of invariant operators on $L(X)$ follows easily from the decomposition of the permutation representation. Indeed, an invariant operator can be written as a linear combination of the orthogonal projections onto the subspaces V_i's.

An entire book devoted to Markov chains and Gelfand pairs is Ref. 2.

2. A family of random walks on the hypercube

In this section, we consider the case when the Markov chain is invariant for the action of a group G that does not act transitively on the state space X, so that X decomposes into a finite number of G-orbits Ω_i (particular cases were considered in Ref. 4). In Ref. 5 we gave a complete analysis of the G-invariant operators on $L(X)$ in the case the action of G on each Ω_i gives rise to a Gelfand pair: differently from the multiplicity-free case, the knowledge of the irreducible representations of the group does not give the spectral analysis of an invariant operator, but it simply allows to reduce the determination of the spectrum to a series of eigenvalue problems of smaller dimensions.

In Ref. 5 this analysis is applied to two classes of posets, but in this note we limit ourselves to random walks on the m-hypercube X. If we identify its vertices with the the set $\{A : A \subset \{1, 2, \ldots, m\}\}$, the natural inclusion relation between subsets defines a partial order on such a hypercube. Moreover, the vertices that correspond to subsets of equal cardinality constitute the level Ω_i. The symmetric group S_m acts naturally on X and the orbits of its action are exactly the Ω_i's. This stratification allows us to decompose a generic function $f \in L(X)$ as $f = \sum_{i=0}^{m} f_i$ where $f_i \in L(\Omega_i)$.

We can introduce two fundamental operators, called *Radon transforms*,

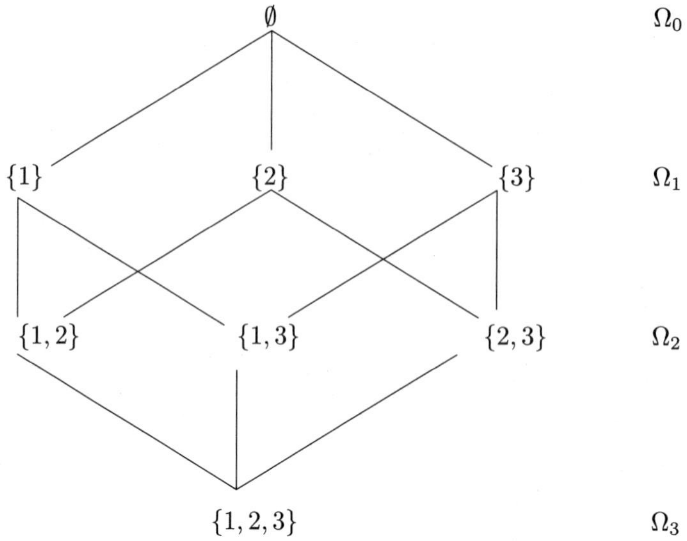

Fig. 1. The 3-hypercube.

on $L(X) = \oplus_{i=0}^{n} L(\Omega_i)$:

- $[df_{h+1}](x) = \sum_{z\in\Omega_{h+1}:z\sim x} f_{h+1}(z)$, $x \in \Omega_h$
- $[d^* f_{h-1}](x) = \sum_{y\in\Omega_{h-1}:y\sim x} f_{h-1}(y)$, $x \in \Omega_h$,

where $y \sim x$ means that x is connected to y and $df_0 = d^* f_n = 0$. Note that d^* is the adjoint of d and we have the fundamental relation $dd^* f_h = d^* df_h + (m-2h)f_h$ for all $f_h \in L(\Omega_h)$.

We consider random walks on an "enlarged" hypercube, i.e. we add a loop to each vertex and we connect by an edge two vertices that are in the same level and have (graph-)distance two in the original hypercube. For instance, for the 3-hypercube in Fig. 1 all pairs of vertices that belong to the same level are connected.

More precisely, if we are at a vertex x of the level Ω_h we move to an adjacent distinct vertex w of the same level with probability b'_h, to a vertex z of the level Ω_{h+1} with probability c_h and to a vertex y of the level Ω_{h-1} with probability a_h. Finally we remain in the same vertex x with probability b''_h. As the parameters a_h, b'_h, b''_h, c_h depend only on the levels of the hypercube, these random walks are invariant for the action of the symmetric group S_m. We should observe that the Ehrenfest model, that

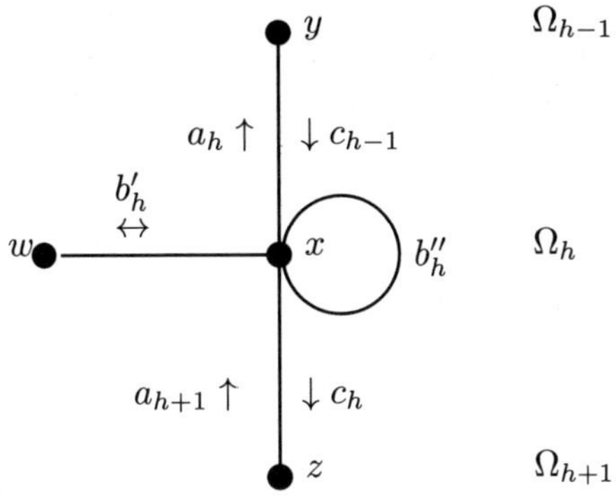

Fig. 2. The random walk on the hypercube.

is described by $a_h = \frac{h}{m+1}, b''_h = \frac{1}{m+1}, c_h = \frac{m-h}{m+1}$ and $b'_h = 0$, is invariant for a larger group: the wreath product of S_2 with S_m. This group acts transitively on the hypercube and indeed it gives rise to a Gelfand pair.

The Markov operator has the following expression in terms of the Radon transforms,

$$(Mf)_h = a_h d^* f_{h-1} + b'_h \Delta_h f_h + b''_h f_h + c_h d f_{h+1},$$

for $h = 0, 1, 2, \ldots, m$ and where $\Delta_h f_h = \Delta_h f_h = (d^* d - hI)$ is the Laplacian.

The Radon transforms come into play also in the decomposition of the permutation representation of S_m on each space $L(\Omega_i)$. Indeed, if we set $W_0 = L(\Omega_0)$ and we define W_k as the kernel of the operator d restricted to $L(\Omega_k)$, $k = 0, 1, \ldots, [m/2]$, we have the following decompositions (with m even, say equal to $2n$):

$$
\begin{array}{lcccccccccc}
L(\Omega_0) & = & W_0 & & & & & & & & \\
L(\Omega_1) & = & d^*(W_0) & \oplus & W_1 & & & & & & \\
L(\Omega_2) & = & (d^*)^2(W_0) & \oplus & d^*W_1 & \oplus & W_2 & & & & \\
\vdots & = & \vdots & \vdots & \vdots & \vdots & \vdots & & \cdots & \vdots & \\
L(\Omega_n) & = & (d^*)^{(n)}(W_0) & \oplus & (d^*)^{(n-1)}W_1 & \oplus & (d^*)^{(n-2)}W_2 & \oplus & \cdots & \oplus W_n & \\
\vdots & = & \vdots & \vdots & \vdots & \vdots & \vdots & & \cdots & \vdots & \\
L(\Omega_{2n-2}) & = & (d^*)^{(2n-2)}(W_0) & \oplus & (d^*)^{(2n-3)}W_1 & \oplus & (d^*)^{(2n-4)}W_2 & & & & \\
L(\Omega_{2n-1}) & = & (d^*)^{(2n-1)}(W_0) & \oplus & (d^*)^{(2n-2)}W_1 & & & & & & \\
L(\Omega_{2n}) & = & (d^*)^{2n}W_0 & & & & & & & & \\
 & & U_0 & & U_1 & & U_2 & & \cdots & U_n &
\end{array}
$$

In the above table the subspaces in the same row are irreducible inequivalent subrepresentation of S_m; this corresponds to the fact that $(G, L(\Omega_i))$ is a Gelfand pair. Moreover, the subspaces that belong to the same column are all equivalent: their direct sum U_k constitutes an *isotypic component* of $L(X)$. A G-invariant operator preserves each isotypic component, but in general does not act trivially on it. In particular the restriction of the Markov operator M to each isotypic component gives rise to an eigenvalue problem for three-diagonal matrices that sometimes leads to equations of hypergeometric type.

To make a concrete example, consider a real parameter $0 < \epsilon < 1$ and set

$$a_h = \frac{h}{m^2}, b'_h = \epsilon\frac{2}{m^2}, b''_h = (1-\epsilon)\frac{2h(m-h)}{m^2}, c_h = \frac{m-h}{m^2}.$$

We call the corresponding Markov chain *second order Ehrenfest model*, since it has the following interpretation in terms of mixing of urns.

We have m balls distributed in two urns. At each step, a random ordered pair of balls is chosen: if both balls belong to the same urn the first ball is moved to the other urn (this is an Ehrenfest move); if the balls are not in the same urn they are either switched with probability ϵ (this is a Bernoulli-Laplace move) or they are left in their original position with probability $(1-\epsilon)$. We allow the possibility that the same ball is chosen twice. In this case the hypergeometric equation is solved by a family of Hahn polynomial and this leads to an explicit spectral decomposition of the Markov operator M_ϵ of this random walk:

Theorem 2.1 (Scarabotti and Tolli, 2007). *The eigenspaces of* M_ϵ

are:

$$V_t^k = \left\{ \sum_{h=k}^{m-k} E_t(m, m-2k, m-k, h) \frac{(d^*)^{h-k}}{(h-k)!} f : \qquad f \in W_k \right\}$$

with corresponding eigenvalues $\lambda_t^k = \frac{1}{m^2}[(m-2t-2\epsilon k)(m-k+1)+t(t+1)-m]$, $0 \le k \le [m/2]$ *and* $t = 0, 1, \ldots, m-2k$, *where*

$$E_t(a,b,c,x) = \sum_{j=\max\{0,x-c+t\}}^{\min\{t,x\}} (-1)^j \binom{t}{j} (b-t+1)_j (-x)_j (a-t+1)_{t-j} (x-c)_{t-j}.$$

For this model we have also proved the presence of the cutoff phenomenon with critical value $\frac{1}{2(2\epsilon+1)} m \log m$. Note that for $\epsilon = \frac{1}{2}$ we get the same critical value for the Bernoulli–Laplace diffusion model: this is not a case since for such a choice of the parameter ϵ the second order Ehrenfest model has the same eigenvalues of the Bernoulli–Laplace model.

3. Markov chains on wreath products

In this section, we consider Markov chains that are invariant for a group of symmetries that acts transitively on the state space, but the decomposition of the permutation representation is not multiplicity-free. As we have already observed in the previous section, a G-invariant operator T acts trivially on an isotypic component if and only if the operator is in the center of the commutant. An explicit example is given by the Bernoulli-Laplace with m urns and nm balls analyzed in Ref. 6: at each step a pair of urns is chosen, a ball is selected from each selected urn and then the balls are switched. In this example the homogenous space X is given by the m-tuples of n disjoint subsets of $\{1, 2, \ldots, nm\}$. The group S_{nm} acts transitively on X, but the permutation representation $L(X)$ is not multiplicity-free. Nevertheless the operator is in the center of the commutant and each isotypic component is a global eigenspace for the Markov operator and therefore its spectrum can be deduced from the representation theory of the symmetric groups.

The general case has been investigated in Refs. 7 and 8: in the first paper we show how to obtain an explicit decomposition of the isotypic componenent in the case of exponentiation or crested product of permutation representations.

The strategy for the decomposition for the exponentiation action was suggested by our analysis of the lamplighter random walk in Ref. 9. Let X be a locally finite graph and suppose that on each vertex there is a lamp that can be on or off. There is lamplighter that performs the simple random walk on X and every time he moves from the vertex x to the vertex y he randomizes the state of the lamps at x and y. This process can be represented by a simple random walk on a new graph $\mathcal{L}(X)$ given by the cartesian product $X \times \{0,1\}^X = \{(x,\omega) : x \in X, \omega : X \to \{0,1\}\}$ (in the following we will identify $\{0,1\}$ with the group S_2). The first coordinate denotes the position of the lamplighter, while the function ω gives the configuration of the lamps: $\omega(x) = 0$ (resp. $\omega(x) = 1$) means that the lamp at x is off (resp. on). We say that two vertices (x,ω) e (y,σ) are connected if the vertices x and y are connected and ω and σ coincide on $X \setminus \{x,y\}$.

This process was introduced by Kaimanovich and Vershik in 1983 and since then it has been deeply studied: Grigorchuk and Żuk[10], Bartholdi and Woess[11], have determined the spectrum in the case the base graph is the discrete line $\mathbb{Z}$, while Peres and Revelle[12] and Häggström, Jonasson[13] have studied several examples when the base graph is finite, but only from a probabilistic point of view.

This random walk is invariant for the action of the wreath product $S_2 \wr Aut(X) = S_2^X \rtimes Aut(X)$ where $Aut(X)$ is the group of automorphisms of the graph X and S_2^X is the group of symmetries of the lamps. This action is transitive but does not produce a Gelfand pair. To overcome this difficulty, we consider only the invariance for the group S_2^X and we look for eigenfunctions of the Markov operator $\mathcal{M}_X$ of the form $\chi_\theta \otimes f$, where, for $\theta \in S_2^X$, χ_θ is the character of S_2^X defined by $\chi_\theta(\omega) = (-1)^{\sum_{x\in X} \theta(x)\omega(x)}$ and f is a generic function on X. These functions span $L(\mathcal{L}(X))$ since $L(\mathcal{L}(X)) = L(X \times S_2^X) = L(X) \otimes L(S_2^X)$.

For $\theta \in S_2^X$, we consider the operator

$$(M_\theta f)(x) = \begin{cases} \frac{1}{\deg(x)} \sum_{\substack{y\in X_\theta: \\ y\sim x}} f(y) & \text{if } \theta(x) = 0 \\ 0 & \text{if } \theta(x) = 1, \end{cases}$$

where $X_\theta = \{x \in X : \theta(x) = 0\}$.

If X is regular, then M_θ is a multiple of the adjancency operator for the graph X_θ, obtained from the graph X by erasing the vertices on which the corresponding lamp was on. Simple computations (see Ref. 9) show that

$$\mathcal{M}_X(\chi_\theta \otimes f) = \chi_\theta \otimes M_\theta f.$$

This way the spectral analysis of the Markov operator is reduced to the spectral analysis of the operators M_θ. When X is either the complete graph or the discrete circle or the segment, the spectral analysis of all operators M_θ can be fully achieved and therefore we obtained an explicit description of the spectrum of the lamplighter random walk (see Ref. 9 for detailed results). In particular, for the lamplighter random walk associated with the complete graph on n vertices we proved that the cutoff phenomenon holds with critical value $k = \frac{1}{2}n\log n$. What is relevant is that we use no results from the representation theory of wreath products of groups, but we apply our techniques to compute the spectra of random walks to construct concrete realizations of the irreducible representations of $S_2 \wr Aut(X)$ that appear in the decomposition of the permutation representation $L(\mathcal{L}(X))$.

For instance, in the case of the complete graph on n vertices, so that $Aut(X) = S_n$, we have a very explicit decomposition of the permutation representation. Define

$$\begin{aligned}
W^1_{k;0,0} &= \langle \chi_\theta \otimes f : \quad |X_\theta| = k,\ f|_{X_\theta} \in S_1^{(k)}(X_\theta) \text{ and } f|_{X\setminus X_\theta} \equiv 0\rangle,\\
W^0_{k;0,0} &= \langle \chi_\theta \otimes f : \quad |X_\theta| = k,\ f|_{X_\theta} \equiv 0 \text{ and } f|_{X\setminus X_\theta} \in S_1^{(n-k)}(X\setminus X_\theta)\rangle,\\
W^1_{k;1,0} &= \langle \chi_\theta \otimes f : \quad |X_\theta| = k,\ f|_{X_\theta} \in S_1^{k-1,1}(X_\theta) \text{ and } f|_{X\setminus X_\theta} \equiv 0\rangle,\\
W^0_{k;0,1} &= \langle \chi_\theta \otimes f : \quad |X_\theta| = k,\ f|_{X_\theta} \equiv 0 \text{ and } f|_{X\setminus X_\theta} \in S_1^{n-k-1,1}(X\setminus X_\theta)\rangle.
\end{aligned}$$

where, for a finite set A of cardinality k, $S_1^{(k)}(A)$ are the constant functions, while $S_1^{k-1,1}(A)$ is made up of the functions on A satisfying $\sum_{a\in A} f(a) = 0$. Then

Theorem 3.1 (Scarabotti and Tolli, 2008).
The decomposition of $L(S_2^X \times X)$ into irreducible $S_2 \wr S_n$-representations is:

$$L(S_2^X \times X) = \left(W^0_{0;0,0} \oplus W^0_{0;0,1}\right) \oplus \left[\bigoplus_{k=1}^{n-1} \left(W^0_{k;0,0} \oplus W^1_{k;0,0} \oplus W^0_{k;0,1} \oplus W^1_{k;1,0}\right)\right] \oplus \left(W^1_{n;0,0} \oplus W^1_{n;1,0}\right).$$

Remark 3.1. The general theory is presented in Refs. 5,7 and 8. We hope that this short note, that is a natural sequel of Ref. 14, could stimulate the interest of the reader towards these quite technical papers.

Acknowledgements

The authors thank the referee for his (her) useful comments and corrections.

References

1. P. Diaconis and M. Shahshahani, Time to reach stationarity in the Bernoulli–Laplace diffusion model, *SIAM J. Math. Anal.* **18** (1987), pp. 208–218.

2. T. Ceccherini-Silberstein, F. Scarabotti and F. Tolli, *Harmonic analysis on finite groups: Representation theory, Gelfand pairs and Markov chains*, Cambridge studies in advanced mathematics **108** (Cambridge University Press, 2008).
3. P. Diaconis, *Group Representations in Probability and Statistics* (IMS Hayward, CA. 1988).
4. S. Boyd, P. Diaconis, P. Parrillo and L. Xiao, Symmetry analysis of reversible Markov chains, *Internet Math.* **2** (2005), no. 1, pp. 31–71.
5. F. Scarabotti and F. Tolli, Spectral analysis of Markov chains with spherical symmetries. *Adv. in Appl. Math.* **38** (2007), no. 4, pp. 445–481.
6. F. Scarabotti, Time to reach stationarity in the Bernoulli-Laplace diffusion model with many urns, *Adv. in Appl. Math.* **18** (1997), no. 3, pp. 351–371.
7. F. Scarabotti and F. Tolli, Harmonic analysis on a finite homogeneous space, *Proc. Lond. Math. Soc.*(3) **100** (2010), no. 2, pp. 348–376.
8. F. Scarabotti and F. Tolli, Harmonic analysis on a finite homogeneous space II: the Gelfand-Tsetlin decomposition, *Forum Math.* **22** (2010), no. 5, pp. 879–911.
9. F. Scarabotti and F. Tolli, Harmonic analysis of finite lamplighter random walks, *J. Dyn. Control Syst.* **14** (2008), no. 2, pp. 251–282.
10. R. I. Grigorchuk and A. Żuk, The lamplighter group as a group generated by a 2-state automaton, and its spectrum, *Geom. Dedicata* **87** (2001), pp. 209–244.
11. L. Bartholdi and W. Woess, Spectral computations on lamplighter groups and Diestel–Leader graphs, *J. Fourier Anal. Appl.* **11** (2005), no. 2, pp. 175–202.
12. Y. Peres and D. Revelle, Mixing times for random walks on finite lamplighter groups, *Electron. J. Probab.* **9** (2004), no. 26, pp. 825–845.
13. O. Häggström and J. Jonasson, Rates of convergence for lamplighter processes, *Stochastic Process. Appl.* **67** (1997), no. 2, pp. 227–249.
14. T. Ceccherini-Silberstein, D. D'angeli, A. Donno, F. Scarabotti and F. Tolli, Finite Gelfand pairs: examples and applications, in *Ischia group theory 2008*, eds. M. Bianchi, P. Longobardi, M. Maj and C. M. Scoppola (World Scientific 2009) pp. 7–41.

ON THE ORDER AND THE ELEMENT ORDERS OF FINITE GROUPS: RESULTS AND PROBLEMS *

Wujie SHI

Department of Mathematics and Statistics,
Chongqing University of Arts and Sciences,
Chongqing 402160, P.R. China
and
School of Mathematics, Suzhou University,
Suzhou 215006, P.R. China
E-mail: wjshi@suda.edu.cn

By comparing the influence of the order $|G|$ of finite groups G to the structure of G, the influence of the set $\pi_e(G)$ of all orders of elements in G to groups G is researched. In this article, the relative results are surveyed, and some unsolved problems are posed.

Keywords: Finite group; Order of a group; Set of element orders; Conjugacy class.

1. Introduction

The concepts of the order of a group and its element orders are the most fundamental in group theory. They play an important role in the quantitative structure of groups. For example, Sylow theorems, Lagrange theorem, $p^a q^b$ theorem and odd order theorem are well-known in the theory of finite groups.

In 1902 W. Burnside put forward the celebrated problem[1]: Let G be a finitely generated group. If for every $g \in G$, the order $|g|$ of g is finite, whether G is necessarily finite? It stressed the role of the orders of elements in the structure of groups.

Notation[2]. Let G be a group. We use $\pi_e(G)$ to denote the set of all orders of elements in G, which is also called the spectrum of G.

*The project is supported by the Natural Science Foundation of China (Grant No. 10871032) and the SRFDP of China (Grant No. 20060285002).

In 1937, B.H. Neumann discussed groups G with $\pi_e(G) = \{1, 2, 3\}$ (see Ref. 3). 20 years later, G. Higman researched the finite groups G all of whose elements are of prime power order[4], that is, $\pi_e(G) = \{p^n, q^m, \cdots\}$, where p, q, $\cdots$ are primes. For convenience, we call such groups finite EPPO-groups. The subject of this article is to study the influence on the group structure from the arithmetical conditions of the order of a group, its element orders or its conjugacy classes etc..

2. Preliminaries

In this section we collect the fundamental definitions, lemmas and theorems which we need later on.

Definition 2.1. A group G is called an EPPO-groups if $\pi_e(G) = \{p^n, q^m, \cdots\}$, where p, q, $\cdots$ are primes and $n, m, \cdots$ are non-negative integers.

Definition 2.2. A group G is called an EPO-groups if $\pi_e(G) = \{1, p, q, \cdots\}$, where p, q, $\cdots$ are primes.

In 1981, the finite EPO-groups and EPPO-groups are researched in the Master's Thesis of author. We got an interesting result[5]: we may characterize A_5 only using the element orders. That is, let G be a finite group. Then $G \cong A_5$ if and only if $\pi_e(G) = \{1, 2, 3, 5\}$.
Generally we got the following theorem.

Theorem 2.1. *Let G be a finite EPO-group. Then*
(1) G is nilpotent if and only if G is a p-group of exponent p.
(2) G is solvable and nonnilpotent if and only if G is a Frobenius group with kernel $P \in Syl_p(G)$, with P a p-group of exponent p and complement $Q \in Syl_q(G)$, with $|Q| = q$. Moreover, if $|G| = p^n q$, then G has a chief series $G = C_0 > P = C_1 > C_2 > \cdots > C_k > C_{k+1} = 1$ such that for every $1 \leq i \leq k$ one has $C_i/C_{i+1} \leq Z(P/C_{i+1})$, Q acts irreducibly on C_i/C_{i+1}, and $|C_i/C_{i+1}| = p^b$, where b is the exponent of p (mod q).
(3) G is nonsolvable if and only if $G \cong A_5$.

Let G be a finite EPPO-group. If G is solvable, then $|\pi(G)| \leq 2$ (see Ref. 4). Moreover G has a special chief series. If $|G| = p^\alpha q^\beta$, P_1 is the maximal normal p-subgroup of G and $|P_1| = p^\gamma$, then the Sylow subgroups of G/P_1 are cyclic or generalized quaternion, $p^{\alpha-\gamma} \mid q - 1$ and $\gamma = kb$ where b is the exponent of p (mod q). See Ref. 6, Theorem 2.4, for the details.
If G is simple, then $G \cong L_2(q)$, q = 5, 7, 8, 9 , 17; $L_3(4)$, $Sz(8)$ or $Sz(32)$

(see Ref. 7, Theorem 16).
If G is nonsolvable, then G is simple, $G \cong M_{10}$, or G has an elementary Abelian 2-subgroup P, P is normal in G and $G/P \cong L_2(q)$, $q = 5, 8$; $Sz(8)$ or $Sz(32)$ (see Ref. 8, Theorem 3; Ref. 6, Theorem 3.1).
In Ref. 6, the authors proved that the following proposition:

Proposition 2.1. *If G is a finite EPPO-group, $H < G$ and $(|H|, d) = 1$, then $|H|$ divide the number of elements of order d in G.*

Question 2.1. Let G be a finite group. If for every proper subgroup H of G and every element order d coprime to $|H|$, we have $|H|$ divide the number of elements of order d in G, is G an EPPO-group?

Question 2.2. What are the infinite (periodic) EPO-groups and the infinite (periodic) EPPO-groups?

For the locally finite EPPO-groups we may refer to Ref. 9, and the periodic groups G with $\pi_e(G) \subseteq \{1, 2, 3, 4, 5\}$ can be found in Ref. 10.

3. On the orders of finite groups

The order $|G|$ of a group G is an integral property of groups. For the special number $|G|$ we can determine the group G, for example, $|G| = p$, p is a prime. That is, there is only one group of order p which is an Abelian simple group. Let G be a finite group. Denote $|G| = n$ and $f(n)$ the number of distinct groups G with $|G| = n$. How many different groups with the given order exist? The question was first posed by Cayley in 1854. The first study of the values of $f(n)$ seems to be in the work of G. A. Miller in the 1920's. O. Hölder determines all groups of squarefree order and gives an explicit formula for their number. Many specialists of group theory researched also it, as J.G. Thompson, G. Higman, C.C. Sims and others. We may refer to Refs. 11,12 and the references listed in Refs. 11,12.
If assume G is simple, then the number of distinct simple groups G with $|G| = n$, $f_{simple}(n) = 0$, 1 or 2 (see Theorem 2.7 below).
$f(n) = 1$ if and only if $(n, \varphi(n)) = 1$, where $\varphi(n)$ is Euler function of n. Also the solutions for $f(n) = 2$, 3 and 4 is found[13].
Discuss the inverse function f^{-1}, that is, for given k, which n satisfies $f(n) = k$?

Conjecture 3.1 (Ref. 13, Question; Ref. 12, Conjecture 2). *For any $k \in Z^+$, there exist an integer n, such that $f(n) = k$.*

From the order of G we have obtained some famous theorems above-mentioned, and from the orders of minimal simple groups[14] we may get the following sufficient condition: If $(|G|, 15) = 1$, then G is solvable. Moreover, 2 and 15 are only the two smallest numbers such that all groups G whose order is prime to 2 or 15 are always solvable[15].

As early as 1940s R. Brauer and H.F. Tuan characterized some projective special linear groups using the conditions of the group order $|G|$ and the simplicity of G. Then many mathematicians characterized many simple groups from various point of view. Using the number $|\pi(G)|$, the number of prime factors of $|G|$, M. Herzog obtained the following result[16].

Theorem 3.1. *Let G be a finite simple group. If $|\pi(G)| = 3$, then G is isomorphic to one of the following groups: A_5, $L_2(7)$, $L_2(8)$, A_6, $L_2(17)$, $L_3(3)$, $U_3(3)$ or $U_4(2)$.*

D. Gorenstein called above eight simple groups as simple K_3-groups and pointed out "that classification is obtained only as a corollary of the complete classification of all finite simple groups"[14]. We determined all simple K_4-groups using the classification theorem[17].

Theorem 3.2. *Let G be a simple K_4-group. Then G is isomorphic to one of the following groups: A_n, $n = 7, 8, 9, 10$; M_{11}, M_{12}, J_2; $L_2(q)$, $q = 16, 25, 49, 81$; $L_3(q)$, $q = 4, 5, 7, 8, 17$; $L_4(3)$; $O_5(q)$, $q = 4, 5, 7, 9$; $O_7(2)$, $O_8{}^+(2)$, $G_2(3)$; $U_3(q)$, $q = 4, 5, 7, 8, 9$; $U_4(3)$; $U_5(2)$; ${}^3D_4(2)$; ${}^2F_4(2)'$; $Sz(8)$, $Sz(32)$; and*

$L_2(r)$, r being prime and satisfying the following equation:

$$r^2 - 1 = 2^a 3^b u^c. \tag{1}$$

where $a \geq 1$, $b \geq 1$, $c \geq 1$, u prime, $u > 3$;

$L_2(2^m)$ and satisfying the following equations:

$$\begin{cases} 2^m - 1 = u \\ 2^m + 1 = 3t^b \end{cases} \tag{2}$$

where $m \geq 1$, u, t primes, $t > 3$, $b \geq 1$.

$L_2(3^m)$ and satisfying the following equations:

$$\begin{cases} 3^m + 1 = 4t \\ 3^m - 1 = 2u^c \end{cases} \tag{3}$$

or

$$\begin{cases} 3^m + 1 = 4t^b \\ 3^m - 1 = 2u \end{cases} \tag{4}$$

where $m \geq 1$, u, t *odd primes*, $c \geq 1$, $b \geq 1$.

Remark 3.1. In Ref. 18, the authors investigated these Diophantine systems and proved that the equations (2), (3) and (4) have no other solution except $m = 5$, $u = 11$, $c = 2$ in (3) when the exponents b and c are greater than 1.

Question 3.1. The number of simple K_4-groups is determined by the number of solution of equations (1)-(4). But it is unknown whether the number of solution is finite or infinite. In other words, is the number of simple K_4-groups finite or infinite?

The author believes the number of simple K_4-groups is infinite. For the simple K_5, K_6-groups, the related results are listed in Ref. 19.

H.F. Tuan had pointed that it is very difficult to determine all finite simple groups whose order does not contain a prime to the first power. For the simple K_4-groups although we do not know its number, we have the following conclusion from Theorem 2.2.

Corollary 3.1. *Let* G *be a simple* K_4*-group. Then the order of* G *contains a prime to the first power except* $G \cong S_4(7)$.

Problem 3.1 (H.F. Tuan). *Determine all finite simple groups whose order does not contain a prime to the first power.*

For the orders of finite simple groups Z.M. Chen proved the following theorem[20].

Theorem 3.3. *The order of any finite simple group* G *is not the* k*th power,* $k \geq 3$. *The order of* G *is square if and only if* $G \cong S_4(p)$, *where* p *is a prime satisfying the following condition:*

$$p = 1 + 2C_{2n+1}^{2} + 2^2 C_{2n+1}^{4} + \cdots + 2^n C_{2n+1}^{2n}, n \geq 1, \tag{5}$$

where C_{2n+1}^{r} *is the binomial coefficient. If* $p < 10^{15}$, *then the above equation has only solutions* $p = 7, 41, 239, 9369319, 63018038201$ *(see Ref. 21).*

Problem 3.2. *How many primes satisfy* (5)*?*

In 1955 E. Artin studied the orders of the known types of finite simple groups. He pointed out that it is interesting to study simple groups of order $|G|$ which contain a Sylow subgroup of order greater than $|G|^{1/3}$ (see Ref. 22). The simplest question of this type is the following one which E.

Artin asked R. Brauer in a letter: "What are the simple groups G whose order is divisible by a prime $p > |G|^{1/3}$?" R. Brauer and W.F. Reynolds solved the simplest case and proved the following theorem[23].

Theorem 3.4. *Let G be a finite non-Abelian simple group whose order $|G|$ is divisible by a prime $p > |G|^{1/3}$. Then G is isomorphic either to $L_2(p)$ where $p > 3$ is a prime or to $L_2(p-1)$ where $p > 3$ is a Fermat prime.*

Using the classification theorem of all finite simple groups the author solved E. Artin's problem and proved the following two theorems[24]:

Theorem 3.5. *Let G be a finite non-Abelian simple group. If $p^k \parallel |G|$, i.e., $p^k \mid |G|$ and $p^{k+1} \nmid |G|$, where p is an odd prime, and $|G| < p^{3k}$, then G is isomorphic to one of the following groups:*
(1) A simple group of Lie type of characteristic p;
(2) A_5 ($k = 1$, $p = 5$), A_6 ($k = 2$, $p = 3$), A_9 ($k = 4$, $p = 3$); or
(3) $L_2(p-1)$ ($k = 1$, p is a Fermat prime), $L_2(8)$ ($k = 2$, $p = 3$), $U_5(2)$ ($k = 5$, $p = 3$).

Theorem 3.6. *Let G be a finite non-Abelian simple group. If $2^k \parallel |G|$ and $|G| < 2^{3k}$, then G is isomorphic to one of the following groups:*
(1) A simple group of Lie type of characteristic 2 or ${}^2F_4(2)'$;
(2) $L_2(r)$ where r is a Fermat prime or a Mersenne prime; or
(3) A_6 ($k = 3$), $U_3(3)$ ($k = 5$), A_8, A_9, M_{12}, A_{10}, M_{22}, J_2 ($k = 7$), HS ($k = 9$), M_{24} ($k = 10$), Suz ($k = 13$), Ru ($k = 14$), Fi_{22} ($k = 17$), Co_2 ($k = 18$), Co_1 ($k = 21$), B ($k = 41$).

As a corollary of the above theorem we have (also see Ref. 25):

Corollary 3.2. *For any prime p with $p \geq 5$, the smallest nonabelian simple groups of order divisible by p is $L_2(p)$. For $p = 2$ and 3, the smallest nonabelian simple groups is A_5.*

Making use of the above theorems one can prove that almost all finite simple groups can be characterized by the conditions of the simplicity of G and $|G|$ (see Refs. 26,27).

Theorem 3.7. *Let G be a simple group and $|G| = |M|$, where M is a known finite simple group. Then one of the following conclusions holds.*
(1) If $|M| = |A_8| = |L_3(4)|$, then $G \cong A_8$ or $L_3(4)$;
(2) If $|M| = |B_n(q)| = |C_n(q)|$ where $n \geq 3$ and q is odd , then $G \cong B_n(q)$ or $C_n(q)$;or
(3) If $|M|$ is not the above case (1) or (2), then $G \cong M$.

By inspecting the orders of all the known finite simple groups we proved the following statement[27].

Theorem 3.8. *Let G be a finite non-Abelian simple group. Then G contains a proper subgroup H such that $|G| < |H|^2$.*

Theorem 3.9. *Let G be a finite group and $|G| \neq p$, p is prime. Then G contains a proper subgroup H such that $|G| \leq |H|^2$.*

4. On the set of element orders

Now we investigate the set $\pi_e(G)$, obviously, $\pi_e(G)$ is a subset of the set Z^+ of positive integers, and it is a most difficult problem which subsets of Z^+ can become the sets of element orders of groups.
First we give the following function h defined on subsets of positive integers.

Definition 4.1. For a set Γ of positive integers let $h(\Gamma)$ be the number of isomorphism classes of finite group G such that $\pi_e(G) = \Gamma$.

For a given group G, we have $h(\pi_e(G)) \geq 1$. And a group G is called recognizable if $h(\pi_e(G)) = 1$. From the classification of finite EPPO-groups we proved the following theorem:

Theorem 4.1. *Let G be a finite simple EPPO-group. Then $h(\pi_e(G)) = 1$ except $G \cong A_6$. Moreover $h(\pi_e(A_6)) = \infty$.*
In other words, we proved the following results: Let G be a finite group. Then
$G \cong A_5$ if and only if $\pi_e(G) = \{1,2,3,5\}$ (Ref. 2),
$G \cong L_2(7)$ if and only if $\pi_e(G) = \{1,2,3,4,7\}$ (Ref. 28),
$G \cong L_2(8)$ if and only if $\pi_e(G) = \{1,2,3,7,9\}$ (Ref. 29),
$G \cong L_2(17)$ if and only if $\pi_e(G) = \{1,2,3,4,8,9,17\}$ (Ref. 8),
$G \cong L_3(4)$ if and only if $\pi_e(G) = \{1,2,3,4,5,7\}$ (Ref. 30),
$G \cong Sz(8)$ if and only if $\pi_e(G) = \{1,2,4,5,7,13\}$ (Ref. 31),
$G \cong Sz(32)$ if and only if $\pi_e(G) = \{1,2,4,5,25,31,41\}$ (Ref. 31).
Moreover, if $\pi_e(G) = \pi_e(A_6) = \{1,2,3,4,5\}$, then $G \cong A_6$ or $G \cong [N]Q$, where $Q \cong A_5$ and N is an elementary abelian 2-group and a direct sum of natural $SL(2,4)$-modules[32].

Furthermore, we have the following conclusions[2,5,28–70]

Theorem 4.2. *Let G be one of the following groups:*
(1) *A_n, $n = 5, 16, 22, p$, $p+1$, $p+2$, where $p \geq 7$ is a prime.*

(2) S_n, $n = 7, 9, 11, 12, 13, 14$.
(3) *A sporadic simple groups except* J_2.
(4) *Simple groups of Lie type* $L_2(q)$, $q \neq 9$, $L_3(2^m)$, $m \geq 1$, $L_3(q)$, *where* $3 < q \equiv 2$ *(mod 5) and* $(6, (q-1)/2) = 1$, $U_3(2^m)$, $m \geq 2$, $L_4(2^m)$, $m \geq 1$, $U_4(2^m)$, $m \geq 2$, $Sz(2^{2m+1})$, $m \geq 1$, $R(3^{2m+1})$, $m \geq 1$ *and* ${}^2F_4(2^{2m+1})$, $m \geq 1$, $S_4(3^{2m+1})$, $m \geq 0$, $B_p(3)$, $p > 3$ *is a prime,* $C_p(3)$, p *is an odd prime,* $D_p(5)$, p *is an odd prime,* $D_n(q)$, $q = 2, 3$ *or* 5 *for some* n, $G_2(3^m)$, ${}^2F_4(2)'$, $L_3(7)$, $L_4(3)$, $L_n(2)$, $n \geq 3$, $L_5(3)$, $U_3(9)$, $U_3(11)$, $U_4(3)$, $U_6(2)$, $G_2(3)$, $G_2(4)$, $S_6(3)$, $O_8^-(2)$, $O_{10}^-(2)$, $F_4(2)$, ${}^3D_4(2)$ *and* ${}^2E_6(2)$.
(5) *Nonsolvable groups* M_{10}, $L_3(4).\langle\beta\rangle$, *where* β *is a unitary automorphism of* $L_3(4)$, $Aut(L_5(3))$ *and* $PGL_2(p^m)$, $m > 1$, $p^m \neq 9$.
Then $h(\pi_e(G)) = 1$.

Remark 4.1. In case (3) of Theorem 3.2, for Janko's second simple group J_2 we have $h(\pi_e(J_2)) = h(\pi_e(S_8)) = \infty$.

Question 4.1. For the sufficient large n, does A_n is recognizable?

In the proof of Theorem 3.2 we often reduce the existence of some orders of elements to the action of a group on another group. However, only some sufficient conditions of the existence of some orders are established.

In Refs. 71,72, the authors researched the case which the prime graphs (The prime graph of G is a graph whose vertex set is the set $\pi(G)$ of prime divisors of $|G|$ in which two vertices $p, q \in \pi(G)$ are joined by an edge if and only if G contains an element of order pq.) of finite groups are connection and found a new approach which makes possible to study the case of a finite simple group with the connected prime graph (see Ref. 73). Moreover, some finite simple groups with connected prime graph are characterized by the element orders (for example, $L_n(2)$, $n > 2$; $B_p(3)$, etc. See Ref. 64 and the references listed in Ref. 64).

If we divide the set $\pi_e(G)$ into 1, the set $\pi_e'(G)$ consisting of primes and the set $\pi_e''(G)$ consisting of composite numbers, then we may give a criterion for the simplicity of finite groups using the number among these sets[74].

Theorem 4.3. *Let* G *be a finite group. Then* $|\pi_e'(G)| \leq |\pi_e''(G)| + 3$, *and if the equality holds, then* G *is simple. Moreover, these simple groups are all recognizable.*

Problem 4.1. *For infinite periodic groups, does the analogous inequality hold?*

A group G is called C_{pp} group if the centralizers of its elements of order p in G are p-groups. The characteristic property of some simple groups by their element orders depends on the results of simple C_{22} groups and simple C_{33} groups.
A classification of simple C_{pp} groups, p is prime and $p = 2^\alpha 3^\beta + 1$, $\alpha \geq 0$, $\beta \geq 0$ or $p = 2^\alpha 5^\beta + 1$, $\alpha \geq 0$, $\beta \geq 0$ was given in Refs. 75,76.

Problem 4.2. *For every prime p, classify all finite simple C_{pp}-groups.*

5. The influence of $\pi_e(G)$ and $|G|$ on the structure of G

Now we consider the influence of $\pi_e(G)$ or $|G|$ (or both) to G.

A. Existence

For all $n \in Z^+$, there always exists G, such that $|G| = n$.
But for any $\Gamma \subset Z^+$, there does not always exist groups G, such that $\pi_e(G) = \Gamma$.
For example, if $1 \notin \Gamma$, then G does not exist. If $\Gamma = \{1, 2, 3, 5\}$, we have $G \cong A_5$. If $\Gamma = \{1, 2, 3, \cdots, k\}$, $k \geq 9$ and G is finite, G does not exist (see Ref. 32). Since $\pi_e(G)$ is closed by divisibility, it clearly is partially ordered as well.

Problem 5.1. *Which closed subset of Z^+ under divisibility can become $\pi_e(G)$ of a group C?*

I think it is a very difficult problem.

B. Quantity

For $n \in Z^+$, let $f(n)$ be the number of isomorphism classes of G such that $|G| = n$. We have discussed this function f in Section 3, and give the h function in Section 4. For a given group G, we have $h(\pi_e(G)) \geq 1$.
Now we give a new classification for groups using the h function.

Definition 5.1. A group G is called recognizable if $h(\pi_e(G)) = 1$, that is, G is uniquely determined by its spectrum. A group G is called irrecognizable if $h(\pi_e(G)) = \infty$. A group G is called almost(k)-recognizable if $h(\pi_e(G)) = k$ $(1 < k < \infty)$.

Recognizable groups

For finite recognizable groups we have obtained many results, which can be summed up as follows.

Theorem 5.1. *The groups listed in Theorem 4.2 are recognizable.*

Refs. 10,77 extended the characterization of the above some groups from finite groups to periodic groups.

For some infinite groups we may also characterize them by their element orders.

Theorem 5.2 (see Ref. 78). *The locally finite simple group G is linear if and only if there is a natural number n such that $n \notin \pi_e(G)$.*

Theorem 5.3 (see Ref. 77). *Let G be a group(finite or infinite) and $\pi_e(G) = \{1, 2, 3, 5\}$. Then $G \cong A_5$.*

Almost recognizable groups

In the case of k-recognized groups, we have found infinite pairs of 2-recognizable groups as follows.

Theorem 5.4. *The following almost simple groups are 2-recognized groups:*
(1) $L_3(q)$, $L_3(q)\langle\theta\rangle$ where θ is a graph automorphism of $L_3(q)$ of order 2, $q = 5, 29, 41$ or $q \equiv 2 \pmod 5$ and $(6, (q-1)/2) = 2$ (Refs. 38,79);
(2) $L_3(9)$, $L_3(9).2_1$ (Ref. 80);
(3) $S_6(2)$, $O_8{}^+(2)$ (Refs. 39,81);
(4) $O_7(3)$, $O_8{}^+(3)$ (Ref. 81);
(5) $L_6(3)$, $L_6(3)\langle\theta\rangle$ where θ is a graph automorphism of $L_6(3)$ of order 2 (Refs. 82,83);
(6) $U_4(5)$, $U_4(5)\langle\gamma\rangle$ where γ is a graph automorphism of $U_4(5)$ of order 2 (Ref. 82).

For any $r > 0$, $h(\pi_e(L_3(7^n))) = r + 1$, with $n = 3^r$ and these $r + 1$ groups are $L_3(7^n)\langle\rho\rangle$, where ρ is a field automorphism of $L_3(7^n)$ (see Ref. 84). This is an example for any k, there are exist the k-recognizable groups.

Considering the above all k-recognizable groups we may easily find that one of each pair is the simple section of another group. So we have the following problem.

Problem 5.2. *Does there exists such two finite k-recognizable groups which are section-free (that is, one of each pair is not the simple section of another group)?*

Irrecognizable groups

Z.M.Chen has proved the following lemma:

Lemma 5.1. *Let N be a minimal normal subgroup of finite group G and N be an elementary abelian p-group. Then G is a irrecognizable group.*

Corollary 5.1. *The finite solvable groups are irrecognizable.*

Theorem 5.5 (see Ref. 85). *Let G be an infinite group. If G has a normal elementary abelian subgroup, then G is irrecognizable. In particular, any solvable infinite group is irrecognizable.*
The list of known almost simple irrecognizable groups includes S_5, A_6, S_6, S_8, A_{10}, $U_3(3)$, $U_4(2)$, $U_5(2)$, $U_3(5)$, $U_3(7)$, J_2, $L_3(4) : 2_2$ and infinitely many groups $PGL_p(r^s)$ (if p, r are odd primes, $r-1$ is divisible by p but not by p^2 and s is a natural number non-divisible by p).

In Ref. 49 the authors proved the following result.

Theorem 5.6. *For every integer $m \geq 2$, there exists an extension E of an elementary 2-group by $L_2(2^{2m})$ such that $\pi_e(E) = \pi_e(S_4(2^m))$.*

This theorem gives the answer: there exist infinite many simple irrecognizable groups.

6. The quantitative characterization of all finite simple groups

Now we give the pure quantitative characterization for all finite simple groups. We replace the condition of the simplicity of G with a quantitative set $\pi_e(G)$ for these cases occurred in Section 2 (Theorem 2.7). That is, we characterize the finite simple groups using only the sets $|G|$ and $\pi_e(G)$. In 1987, the author wrote a letter to Professor John G. Thompson and posed the following conjecture:

Conjecture 6.1. *Let G be a group and M a finite simple group. Then $G \cong M$ if and only if (a) $\pi_e(G) = \pi_e(M)$, and (b) $|G| = |M|$.*

In the reply, Thompson pointed that, "Good luck with your conjecture about simple groups. I hope you continue to work on it", "This would certainly be a nice theorem."[86]. The author thinks it gives an integral characterization to all finite simple groups and establishes the connection between the pure quantities and the finite simple groups. Up to 2003, the author and his students proved the following theorem[87–94]:

Theorem 6.1. *Let G be a group and M one of the following simple groups:*
(1) A cyclic simple group Z_p;
(2) An alternating group A_n, $n \geq 5$;
(3) A sporadic simple group;
(4) A Lie type group except B_n, C_n, D_n (n even and $q \neq 2$); or
(5) A simple group with order $< 10^8$.
Then $G \cong M$ if and only if (a) $\pi_e(G) = \pi_e(M)$, and (b) $|G| = |M|$.

The simple groups $B_n(q)$ and $C_n(q)$ for q odd are of equal order. However, as has been proved in Refs. 95,96, they have different spectrum. Hence these groups are not conterexamples to Conjecture 5.1.

In 2009, A. V. Vasilev, M. A. Grechkoseeva, and V. D. Mazurov proved that Conjecture 5.1 is correct for B_n, C_n and D_n (n even) (see Ref. 97). They analyzed the problem of previous proofs, researched the finite groups isospectral to simple symplectic and orthogonal groups[98] and established a number of assertions pertinent to recognition by spectrum for these simple groups. Thus we proved Conjecture 5.1 finally, and all finite simple groups can determined by their "two orders".

Besides the finite simple groups we may characterize some nonsolvable groups G using only the conditions of $\pi_e(G)$ and $|G|$. For example, in Ref. 99, the author proved the following theorem:

Theorem 6.2. *Let G be a group. Then $G \cong S_n$, $n \geq 3$ if and only if (a) $\pi_e(G) = \pi_e(S_n)$, and (b) $|G| = |S_n|$.*

Generally we can not characterize the solvable groups, even p-groups, using only the conditions of $\pi_e(G)$ and $|G|$. From this view we may say the simple groups are simple.

Problem 6.1. *Let G_1, $G_2, \cdots$, G_k be finite groups. If $\pi_e(G_1)=\pi_e(G_2)=\cdots=\pi_e(G_k)$, whether or not the order of simple group G_i(if exsits) is the smallest?*

7. Thompson's two problems and related topics

For each finite group G and each integer $d \geq 1$, let $G(d) = \{x \in G; x^d = 1\}$.

Definition 7.1. G_1 and G_2 are of the same order type if and only if $|G_1(d)| = |G_2(d)|$, $d = 1, 2, \cdots$.

A problem related to Conjecture 6.1 is the following open problem put forward by J.G. Thompson[86] in 1987.

Problem 7.1. *Suppose G_1 and G_2 are groups of the same order type. Suppose also that G_1 is solvable. Is it true that G_2 is also necessarily solvable?*

In Thompson's private letter he pointed out that "The problem arose initially in the study of algebraic number fields, and is of considerable interest". In generally, the same order type groups are not isomorphic. So it is necessary to study recognizable groups.
J.G. Thompson has given an example of the same order type groups which are not isomorphic: $G_1 = 24 : A_7$ and $G_2 = L_3(4) : 2_2$, both are maximal subgroups of M_{23}.
In fact, the groups of the same order type are such groups whose "order equations" are the same. That is, according to the orders of elements as the equivalent relation, the group G is partitioned into the same order classes. The "order equation" is $|G| = n_1 + n_2 + \cdots + n_k$, where n_i denotes the number of elements of order i in G. Obviously the same order class is an union of several conjugacy classes and the partition of "order equation" is rougher than the "class equation".
If G_1 and G_2 are finite groups of the same order type, then $\pi_e(G_1) = \pi_e(G_2)$ and $|G_1| = |G_2|$. Thus we have the following corollary since Conjecture 6.1 is correct:

Corollary 7.1. *Let G_1 and G_2 be finite groups of the same order type. If G_1 is solvable, then G_1 is not simple.*

If G_1 and G_2 are finite groups of the same order type, then $\pi_e(G_1) = \pi_e(G_2)$ and $\tau_e(G_1) = \tau_e(G_2)$, where $\tau_e(G)$ denote the set of numbers of the same order elements in G. We have discussed the influence of $\pi_e(G)$ to the structure of G in this article, what about the action of $\tau_e(G)$ to G? In Ref. 100 the following lemma is proved.

Lemma 7.1. *Let G be a group containing more than two elements. If the maximal number s of elements of the same order in G is finite, then G is finite and $|G| \leq s(s^2 - 1)$.*

Using the above lemma the authors of Ref. 100 proved the following result: Let G be a group. Then $G \cong A_n$ if and only if $\tau_e(G) = \tau_e(A_n)$, where $n = 4$, 5, 6.
Given a finite group G, how can recognize the solvability of G? The famous

Odd Order Theorem proved that all finite groups of odd order are solvable. For the group of even order, how recognize them? If the answer of the above Thompson's problem is positive, thus we may recognize the solvability for all finite groups G using the order equation of G.

If G_1 and G_2 are of the same order type, and assume that G_1 is nilpotent, we clearly have that G_2 is also nilpotent. In Ref. 101, the author proved that if G_1 possess Sylow tower (or G_1 is supersolvable), then G_2 also possess Sylow tower. Lastly, we proved that let G_1 be a solvable group whose prime graph $\Gamma(G_1)$ (see Ref. 73) is not connected. If G_1 and G_2 are the groups of same order type, then G_2 is also solvable[102].

Another conjecture of Thompson which also aims at characterizing all finite simple groups by a number of theoretic quantity is the following[103]:

If G is a finite group, set N(G)=$\{n \in N;\ G$ has a conjugacy class C with $|C| = n\}$.

Conjecture 7.1. *If G and M are finite groups and $N(G) = N(M)$, and if in addition, M is a non-Abelian simple group while the center of G is 1, then G and M are isomorphic.*

On Conjecture 6.1, G.Y. Chen has dealt with the situation that M is a finite simple group having a non-connected prime graph. He has proved the following two lemmas:

Lemma 7.2. *Let G be a finite group with $Z(G) = 1$ and M a non-connected prime graph non-Abelian simple group such that $N(G) = N(M)$. Then $|G| = |M|$.*

Lemma 7.3. *Let G and M be finite groups having same orders. If $N(G) = N(M)$, then the number of components of prime graph of G is equal to that of M. And the prime graph components of G are equal to those of M, where the prime graph components are viewed as the sets of vertices.*

By using above lemmas Guiyun Chen has proved that Thompson's conjecture holds for all non-abelian simple groups with non-connected prime graphs. Hence Thompson's conjecture holds for all non-abelian simple groups except $A_n(q)$, $B_n(q)$, $C_n(q)$, $D_n(q)$, ${}^2A_n(q)$, ${}^2D_n(q)$, $E_7(q)$ and the alternating groups A_n. Some of them published in several papers: for sporadic simple groups, see Ref. 104; for finite simple groups with at least three prime graph components, see Ref. 105; for $G_2(q)$, see Ref. 106.

Definition 7.2. Let G be a finite group and $\pi_1, \pi_2, \cdots, \pi_t$ the connected components of prime graph $\Gamma(G)$ (see Ref. 73). Then $|G|$ can be expressed as a product of $m_1, m_2, \cdots, m_t$, where m_i are positive integers with $\pi(m_i) = \pi_i$. These m_i's are called the order components of G.

Theorem 7.1 (see Refs. 107–112). *Let G be a finite group and M one of the following simple groups: A sporadic group, $G_2(q)$, $q \equiv 0 (mod\ 3)$, $E_8(q)$, a Suzuki-Ree group, $PSL_2(q)$, A_p, where p and $p-2$ are primes. Then $G \cong M$ if and only if G and M have the same order components.*

Recently, M. Khademi gave a summary of results on characterization of finite simple groups using their order components[113]. The conditions of same"two order", that is"$|G| = |M|$, and $\pi_e(G) = \pi_e(M)$" mean "G and M have the same order components". Also, the conditions of G and M are of "same order type" mean G and M have the same "two order".
There exist a series of generalization for the characterization of finite simple groups using "two orders". Since the set of element orders is the set of orders of cyclic subgroups, we may generalize the orders of "cyclic" subgroups to the orders of "abelian", "maximal abelian", "nilpotent" or "solvable" subgroups, etc.(for example, we may refer to Refs. 114–116. Furthermore, related the characterization of "two orders", we have "OC", "OD", "OG" characterizations (see Refs. 113,117–120).

Lastly we pose the following question:

Question 7.1. What properties does the set X=$\{m$; there exist (finite) groups G with $\pi_e(G) = m\}$ have?

Acknowledgments

The author would like to thank the referee for his/her comments and help.

References

1. W. Burnside, On a unsettled question in the theory of discontinuous groups, Quart. J. Pure Appl. Math., 33 (1902), 230-238.
2. W.J. Shi, A characteristic property of A_5, J. Southwest-China Normal University, 3 (1986), 11-14. (Chinese)
3. B.H. Neumann, Groups whose elements have bounded orders, J. London Math. Soc., 12 (1937), 195-198.
4. G. Higman, Finite groups in which every element has prime power order, J. London Math. Soc., 32 (1957), 335-342.
5. W.J. Shi and W.Z. Yang, A new characterization of A_5 and the finite groups in which every non-identity element has prime order, J. Southwest-China Teachers College, 1 (1984), 36-40. (Chinese)

6. W.J. Shi and W.Z. Yang, The finite groups all of whose elements are of prime power order, J. Yunnan Educational College, 1 (1986), 2-10. (Chinese)
7. M. Suzuki, On a class of doubly transitive groups, Ann. Math., 75 (1962), 105-145.
8. R. Brandl, Finite groups all of whose elements are of prime power order, Boll. U. M. I. (5) 18-A (1981), 491-493.
9. A.I. Starostin, On groups whose centralizers have partitions, Izv, Akad. Nauk SSSR Ser. Mat. 29 (1965), 605-614.
10. V.D. Mazurov and W.J. Shi, On periodic groups with prescribed orders of elements, Science in China (Ser. A), 52:2(2009), 311-317.
11. L. Pyber, Enumerating finite groups of given order, Ann. of Math. 137(1993), 203-220.
12. R.K. Dennis, The number of groups of order n, to appear.
13. G.M. Wei, Enumeration Formulae for the number of finite groups and their application, Southeast Asian Bull. Math., 22(1998), 93-102.
14. D. Gorenstein, Finite Simple Groups, Plenum Press, New York and London, 1982.
15. J.H. He and W. Pu, On the number n which make any group with order prime to n solvable, J. of Southwest-China Normal Univ., 24:6 (1999), 612-614. (Chinese)
16. M. Herzog, On finite simple groups of order divisible by three prime only, J. Algebra, 10 (1968), 383-388.
17. W.J. Shi, On simple K_4-groups, Chinese Science Bulletin, 36 (1991), 1281-1283. (Chinese)
18. Y. Bugeaud, Z. Cao and M. Mignotte, On simple K_4-groups, J. Algebra 241 (2001), 658-668.
19. A. Jafarzadeh and A. Iranmanesh, On simple K_n-groups for $n = 5, 6$, Groups St Andrews 2005, Vol.2, London Math. Soc. Lecture Notes Ser. 340, 517-526.
20. Z.M. Chen, On the orders of the finite simple groups, Acta Math. Sinica, 30(1987), 605-613. (Chinese)
21. Q. Sun, An unsolved diophantine equation which arises in the theory of finite groups, J. of Math. Res. And Exposition, 6 (1986), 20. (Chinese)
22. E. Artin, The orders of the classical simple groups, Comm. Pure Appl. Math, 8(1955), 355-365.
23. R. Brauer and W. F. Reynolds, On a problem of E. Artin, Ann. Math., 68 (1958), 713-720.
24. W.J. Shi, On a problem of E. Artin, Acta. Math. Sinica, 35:2 (1992), 262-265. (Chinese)
25. G. Mendoza, On some minimality conditions involving elements of prime order in a group G, Ph. D. Thesis, State Univ. of New York at Binghamton, 2007.
26. W. Kimmerle, R. Lyons, R. Sandling and D.N. Teague, Composition factors from the groups ring and Artin's theorem on orders of simple groups, Proc. London Math. Soc. (3) 60 (1990), 89-122.
27. W.J. Shi, On the orders of the finite simple groups, Chinese Science Bulletin, 38:4 (1993), 296-298. (Chinese)

28. W.J. Shi, A characteristic property of $PSL_2(7)$, J. Austral. Math. Soc. (Ser. A), 36 (1984), 354-356.
29. W.J. Shi, A characteristic property of J_1 and $PSL_2(2^n)$, Adv. in Math., 16 (1987), 397-401. (Chinese)
30. W.J. Shi, A characterization of some projective special linear groups, J. of Math (PRC), 5 (1985), 191-200.
31. W.J. Shi, A characterization of Suzuki's simple groups, Proc. Amer. Math. Soc., 114:3 (1992), 589-591.
32. R. Brandl and W. J. Shi, Finite groups whose element orders are consecutive integers, J. Algebra, 143 (1991), 388-400.
33. J.B. An and W.J. Shi, The characterization of finite simple group with no elements order six by their element orders, Comm. in Algebra, 28:7(2000).
34. R. Brandl and W. J. Shi, The characterization of $PSL(2,q)$ by its element orders, J. Algebra, 163(1994), 109-114.
35. R.Brandl and W. J. Shi, A characterization of finite simple groups with abelian sylow 2-subgroups, Ricerche di Matematica, 42 (1993), 193-198.
36. M.R. Darafsheh and A.R. Moghaddamfar, A characterization of some finite groups by their elements orders, Algebra Colloq., 7:4(2000),467-476.
37. H.L. Li and W.J. Shi, A characteristic property of some sporadic simple groups, Chin. Ann. Math. 14A:2 (1993), 144-151. (Chinese)
38. V.D. Mazurov, The set of orders of elements in a finite group, Algebra and Logic, 33:1(1994), 49-55.
39. V.D. Mazurov, Characterization of finite groups by sets of orders of their elements, Algebra and Logic, 36:1(1997), 23-32.
40. V.D. Mazurov, Recognition of finite simple groups $S_4(q)$ by their element orders, Algebra and Logic, 41:2(2002), 93-110.
41. V.D. Mazurov, Characterizations of groups by arithmetic properties, Pro ceedings of the International Conference on Algebra, Algebra Colloq., 11 (2004), 129-140.
42. C.E. Praeger and W.J. Shi, A characterization of some alternating and symmetric groups, Comm. in Algebra, 22:5(1994), 1507-1530.
43. W.J. Shi, The characterization of the sporadic simple groups by their elements orders, Algebra Colloquium, 1:2 (1994), 315-326.
44. W.J. Shi and H.L. Li, A characteristic property of M_{12} and $PSU(6,2)$, Acta Math. Sinica, 32 (1989), 758-764. (Chinese)
45. M.R. Darafsheh, A.R. Moghaddamfar and A.R. Zokayi, A recognition of simple groups $PSL(3,q)$ by their element orders, Acta Mathematica Scientia, 21:1(2004), 45-51.
46. H.W. Deng and W.J. Shi, The characterization of Ree groups ${}^2F_4(q)$ by their element orders, J. Algebra, 217:1(1999), 180-187.
47. A.S. Kondrat'ev, V.D. Mazurov. Recognition of alternating groups of prime degree from their element orders., J. Siberian Math., 41:2(2000), 294-302.
48. S. Lipschutz and Wujie Shi, Finite groups whose element orders do not exceed twenty, Progress in Natural Science 10 (2000), 11-21.
49. V.D. Mazurov, M.C. Xu and H.P. Cao, Recognition of finite simple groups $L_3(2^m)$ and $U_3(2^m)$ by their element orders, Algebra and Logic, 39:5(2000),

324-334.
50. W.J. Shi, A characteristic property of A_8, Acta Math. Sin., New Ser., 3 (1987), 92-96.
51. W.J. Shi, A characteristic property of Mathieu groups, Chinese J. of Contemporary Mathematics, 9 (1988), 317-326.
52. W.J. Shi, A characterization of the Higman-Sims group, Houston J. of Math., 16 (1990), 597-602.
53. W.J. Shi, A characterization of the Conway simple group Co_2, J. of Math. (PRC), 9 (1989), 171-172. (Chinese)
54. W.J. Shi, A class of special finite groups, Northeast Math. J., 9:1 (1993), 1-4.
55. W.J. Shi, A Characterization of $U_3(2^n)$ by their element orders, J. of Southwest-China Normal University, 25:4(2000), 353-360.
56. V.D. Mazurov and W.J. Shi, A Note to the Characterization of Sporadic Simple Groups, Algebra Colloq., 5:3(1998), 285-288.
57. W.J. Shi and S. Lipschutz, A new classification of finite simple groups, The Mathematical Foundation of Informatics, Proceeding of the Conference (MFI '99), 2005, 173-182.
58. A.V. Vasil'ev, Recognizability of groups $G_2(3^n)$ by their element orders, Algebra and Logic, 41:1(2002), 74-80.
59. A.V. Zavarnitsin, Recognition of alternating groups of degrees $r+1$ and $r+2$ for prime r and the group of degree 16 by their element order sets. Algebra and Logic, 39:6(2000), 370-377.
60. F. Zhou, A characteristic property of some unitary groups, J. of Shanghai Jaiotong Univ., 28 (1994), 106-109. (Chinese)
61. A.V. Zavarnitsin and V.D. Mazurov, Element orders in coverings of symmetric and alternating groups, Algebra and Logic, 38:3(1999), 159-170.
62. V.D. Mazurov and W.J. Shi, Groups whose elements have given orders, in "Proc. of Groups, St. Andrews 1997 in Bath, II", LMS. Lecture Notes Ser. 261, Cambridge University Press, 1999. 532-537.
63. W.J. Shi, Arithmetical property of finite groups, in "Groups, St. Andrews 2005 Vol. 2", LMS. Lecture Notes Ser. 340, Cambridge University Press, 2007. 646-653.
64. M.A. Grechkoseeva, W.J. Shi and A.V. Vasil'ev, Recognition by spectrum for finite simple groups of Lie type, Front. Math. China, 3:2(2008), 275-285.
65. H.Y. He, W.J. Shi, Recognition of alternating group A_{22} by the set of its element orders, J. Southwest-China Normal University, 31 (2009), 121-123.
66. H.Y. He, W.J. Shi, Recognition of some finite simple groups of type $D_n(q)$ by spectrum, International Journal of Algebra and Computation, 19:5(2009), 681-698.
67. H.Y. He, W.J. Shi, A note on the adjacency criterion for the prime graph and the characterization of $C_p(3)$, to be published in Algebra Colloq.
68. R.L. Shen, W.J. Shi, M.R. Zinov'eva, Recognition of simple groups $B_p(3)$ by the set of element orders, Siberian Mathematical Journal, 51(2010), 244-254.
69. R.L. Shen, W.J. Shi, Characterization of simple groups $D_p(5)$ by the set of element orders, to appear.
70. V.D. Mazurov and G.Y. Chen, Recognizability of finite simple groups $L_4(2^m)$

and $U_4(2^m)$ by spectrum, Algebra and Logic, 47(2008), 49-55.
71. A.V. Vasil'ev, On connection between the structure of finite groups and properties of its prime graph, Sib. Math. J., 46:3(2005), 396-404.
72. A.V. Vasil'ev and E.P. Vdovin, An adjacency criterion for the prime graph of a finite simple group, Algebra and Logic, 44:6(2005), 381-406.
73. J.S. Williams, Prime graph components of finite groups, J. Algebra, 69(1981), 487-513.
74. H.W. Deng and W.J. Shi, A simplicity criterion for finite groups, J. Algebra, 191:1(1997), 371-381 .
75. Z.M. Chen and W.J. Shi, On simple C_{pp} groups, J. Southwest-China Normal University, 18:3 (1993), 249-256. (Chinese)
76. M.H. Li, All the C_{pp} simple groups on certain conditions, Master Thesis of Southwest Univ. Chongqing, July, 2008. (Chinese)
77. A.K. Zhurtov and V.D. Mazurov, A recognition of simple groups $L_2(2^m)$ in the class of all groups, Siberian Math. J., 40(1999), 62-64.
78. O.H. Kegel and W.J. Shi, On the orders of elements of a locally finite simple group, C.R. Math. Rep. Acad. Sci. Canada, 13 (1991), 253-254.
79. A.R. Moghaddamfar and M.R. Darafsheh, A family of finite simple groups which are 2-recognizable by their elements order, Comm. in Algebra, 32(2004), 4507-4513.
80. N. Chigira and W.J. Shi, More on the set of element orders of finite groups, Northeast. Math. J. 12(1996), 257-260.
81. W.J. Shi, C.Y. Tang, A characterization of some orthogonal groups, Progress in Natural Sci. 7(1997), 155-162.
82. A.V. Vasil'ev, On recognition of all finite simple groups with orders having prime divisor at most 13, Siberian Math. J. 46(2005).
83. M.R. Darafsheh and Y. Farjami, On groups with the same set of order elements, International Mathematical Forum, 1(2006), 1325-1334.
84. A.V. Zavarnitsin, Recognition of the simple groups $L_3(q)$ by element orders, J. Group Theory, 7(2004), 81-97.
85. H.W. Deng and W.J. Shi, A class of irrecognizable infinite groups, J. Southwest China Normal University, 25:4(2000), 361-363.
86. J.G. Thompson, Private Communication, (1987).
87. W.J. Shi, A new characterization of the sporadic simple groups, Group Theory, in "Proc. of the 1987 Singapore Conf". Walter de Gruyter. Berlin, New York, 1989, 351-340.
88. W.J. Shi, A new characterization of some simple groups of Lie type, Contemporary Mathematics, 82 (1989), 171-180.
89. W.J. Shi and J.X. Bi, A characteristic property for each finite projective special linear groups, Lecture Notes in Math., 1456 (1990), 171-180.
90. W.J. Shi and J.X. Bi, A characterization of Suzuki-Ree groups, Science in China (Ser. A), 34 (1991), 14-19.
91. W.J. Shi and J.X. Bi, A new characterization of the alternating groups, SEA Bull. Math., 16:1 (1992), 81-90.
92. W.J. Shi, The pure quantitative characterization of finite simple groups (I), Progress in Natural Sci., 4 (1994), 316-326.

93. H.P. Cao and W.J. Shi, Pure quantitative characterization of finite projective special unitary groups, Science in China (Ser. A), 45:6(2002), 761-772.
94. M.C. Xu and W.J. Shi, Pure quantitative characterization of finite simple groups ${}^2D_n(q)$ and $D_l(q)$ (l odd), Algebra Colloq., 10:3(2003), 427-443.
95. W.J. Shi, Pure quantitative characterization of finite simple groups, Front. Math. China, 2(2007), 123-125.
96. M.A. Grechkoseeva, On difference between the spectra of the simple groups $B_n(q)$ and $C_n(q)$, Siberian Mathematical Journal, 48(2007), 73-75.
97. A.V. Vasil'ev, M.A. Grechkoseeva, and V.D. Mazurov, Characterization of the finite simple groups by spectrum and order, Algebra and Logic, 48(2009), 385-409.
98. A.V. Vasil'ev, M.A. Grechkoseeva, and V.D. Mazurov, On finite groups isospectral to simple symplectic and orthogonal groups, Siberian Mathematical Journal, 50(2009), 965-981.
99. J.X. Bi, A characterization of symmetric groups, Acta. Math. Sinica, 33 (1990), 70-77. (Chinese)
100. R.L. Shen, C.G. Shao, Q.H. Jiang, W.J. Shi and V.D. Mazurov, A new characterization of A_5, Monatsh Math 160 (2010), 337-341.
101. Xu M., The same order typical groups, Natural Science J. Hannan University, 4(2)(1996), 103-105. (Chinese)
102. R.L. Shen and W.J. Shi, On Thompson problem, Scientia Sinica Mathematica, 40(2010), 533-538. (Chinese)
103. J.G. Thompson, Private Communication, (1988).
104. G.Y. Chen, On Thompson's conjecture for sporadic groups, in "Proc. China Assoc. Sci. and Tech., First Academic Annual Meeting of Youths" , Chinese Sci. and Tech. Press, Beijing, 1992, 1-6. (Chinese)
105. G.Y. Chen, On Thompson's conjecture, J. Algebra, 185(1996), 184-193.
106. G.Y. Chen, Further reflections on Thompson's conjecture, J. Algebra, 218(1999), 276-285.
107. G.Y. Chen, A new characterization of sporadic simple groups, Algebra Colloq., 3:1(1996), 49-58.
108. G.Y. Chen, A new characterization of $G_2(q)$, $q \equiv 0 (\mathrm{mod}\ 3)$, J. Southwest China Normal Univ., Special Issue(1996), 47-51.
109. G.Y. Chen, A new characterization of $E_8(q)$, J. Southwest China Normal Univ., 21:3(1996), 215-217. (Chinese)
110. G.Y. Chen, A new characterization of Suzuki-Ree groups, Sci. in China (Ser. A), 40(1997), 807-812.
111. G.Y. Chen, A new characterization of $PSL_2(q)$, Southeast Asian Bull. of Math. 22(1998), 257-263.
112. A. Iranmanesh and S.H. Alavi, A new characterization of A_p where p and $p-2$ are primes, Korean J. Comput. Appl. Math., 8:3(2001), 889-897.
113. M. Khademi, Characterizability of finite simple groups by their order components: A summary of results, International Journal of Algebra, 4(2010), 413-420.
114. W.J. Shi, Finite groups whose proper subgroup orders are consecutive integers, J. Math. Res. and Expo., 14(1994), 165-166.

115. G. Y. Chen, Characterization of alternating groups by the set of orders of maximal subgroups, Siberian Mathematical Journal, 47(2006), 594-596.
116. Z.J. Han, G.Y. Chen and X.Y. Guo, A characterization theorem of sporadic simple groups, Siberian Mathematical Journal, 49(2008): 1138-1146.
117. L.C. Zhang and W.J. Shi, OD-characterization of simple K_4-groups, Algebra Colloquium, 16:2(2009), 275-282.
118. J.T. Shi, W.J. Shi and C. Zhang, The type of conjugacy classes of maximal subgroups and characterization of finite groups, Communications in Algebra, 38(2010), 143-153.
119. L.C. Zhang and W.J. Shi, OD-Characterization of all simple groups whose orders are less than 108, Front. Math. China, 3(2008), 461-474.
120. Q.L. Zhang and W.J. Shi, Quasirecognition of prime graphs of simple groups $G_2(q)$ and ${}^2B_2(q)$, to be published in the Journal of Algebra and its Applications.

ON LOCAL FINITENESS OF VERBAL SUBGROUPS IN RESIDUALLY FINITE GROUPS

P. SHUMYATSKY*

Department of Mathematics, University of Brasilia,
Brasilia, DF, 70910-900, Brazil
E-mail: pavel@mat.unb.br

We describe some results related to the following problem. Let n be a positive integer and w a word. Assume that G is a residually finite group such that any w-value in G has order dividing n. Does it follow that the verbal subgroup $w(G)$ is locally finite? In particular, we focus on the theorem that if w is a multilinear commutator and G a residually finite group in which every product of 896 w-values has order dividing n, then $w(G)$ is locally finite.

Keywords: Groups; Commutators; Restricted Burnside Problem.

1. A generalization of the Restricted Burnside problem

One immediate corollary of the solution of the Restricted Burnside Problem is that every residually finite group of finite exponent is locally finite. The solution of the problem for nilpotent groups was obtained by Zelmanov in Ref. 29,30 while the reduction to the nilpotent case was performed by Hall and Higman using the classification of finite simple groups[13].

The Lie-theoretic methods used in the solution happened to be very effective in treatment of other problems in group theory. In Ref. 20 we used the methods to prove the following theorem.

Theorem 1.1. *Let n be a prime-power and G a residually finite group satisfying the identity $[x,y]^n \equiv 1$. Then G' is locally finite.*

Note that in general a periodic residually finite group need not be locally finite. The corresponding examples have been constructed in Ref. 2,8,10,11, 27. Moreover, if the assumption that G is residually finite is dropped from the hypothesis of Theorem 1.1, the derived group need not even be periodic.

*The author was supported by CNPq and FAPDF.

Deryabina and Kozhevnikov showed that for sufficiently big odd integers n there exist groups G in which all commutators have order dividing n such that G' has elements of infinite order[5]. Independently, this was also proved by Adian[1]. Later Ivanov constructed similar examples for arbitrary sufficiently big integers n (Ref. 15).

In view of Theorem 1.1 we raised in Ref. 21 the next problem.

Problem 1.1. *Let n be a positive integer and w a word. Assume that G is a residually finite group such that any w-value in G has order dividing n. Does it follow that the verbal subgroup $w(G)$ is locally finite?*

We also would like to mention the following related problem that was suggested in Ref. 22.

Problem 1.2. *Let $n \geq 1$ and w a group-word. Consider the class of all groups G satisfying the identity $w^n \equiv 1$ and having $w(G)$ locally finite. Is that a variety?*

It is easy to see that whenever Problem 1.2 has a positive answer so does Problem 1.1. A more complicated question is whether a positive answer to Problem 1.1 yields a positive answer to Problem 1.2. This is unknown. So far in all cases where we were able to give a positive answer to Problem 1.1 a positive answer to Problem 1.2 has also been obtained.

If w is a word in variables $x_1, \ldots, x_t$ we think of it primarily as a function of t variables defined on any given group G. The corresponding verbal subgroup $w(G)$ is the subgroup of G generated by the values of w. The word w is commutator if the sum of the exponents of any variable involved in w is zero. According to the solution of the Restricted Burnside Problem the answer to Problem 1.1 is positive if $w(x) = x$. In fact it is easy to see that the answer is positive whenever w is any non-commutator word. Indeed, suppose $w(x_1, \ldots, x_t)$ is such a word and that the sum of the exponents of x_i is $r \neq 0$. Now, given a residually finite group G, substitute the unit for all the variables except x_i and an arbitrary element $g \in G$ for x_i. We see that g^r is a w-value for all $g \in G$. Hence G satisfies the identity $x^{nr} = 1$ and therefore is locally finite by the result of Zelmanov.

Hence, Problem 1.1 is essentially about commutator words. In Ref. 21 the problem was solved positively in the case where n is a prime-power and w a multilinear commutator (outer commutator word). A word w is a multilinear commutator if it can be written as a multilinear Lie monomial. Particular examples of multilinear commutators are the derived words, defined by the equations:

$$\delta_0(x) = x,$$

$$\delta_k(x_1, \ldots, x_{2^k}) = [\delta_{k-1}(x_1, \ldots, x_{2^{k-1}}), \delta_{k-1}(x_{2^{k-1}+1} \ldots, x_{2^k})],$$

and the lower central words:

$$\gamma_1(x) = x,$$

$$\gamma_{k+1}(x_1, \ldots, x_{k+1}) = [\gamma_k(x_1, \ldots, x_k), x_{k+1}].$$

In the case where n is not a prime-power Problem 1.1 seems to be very hard. We mention a theorem obtained in Ref. 23.

Theorem 1.2. *Let n be a positive integer that is not divisible by p^2q^2 for any distinct primes p and q. Let G be a residually finite group satisfying the identity $([x_1, x_2][x_3, x_4])^n \equiv 1$. Then G' is locally finite.*

Some further progress has been made in Ref. 24 where the following theorem was proved.

Theorem 1.3. *For any positive integer n there exists t depending only on n such that if w is a multilinear commutator and G is a residually finite group in which every product of t values of w has order dividing n, then $w(G)$ is locally finite.*

More recently it was shown that actually the number t in the above theorem can be taken a constant independent of n. The next theorem was proved in Ref. 26.

Theorem 1.4. *If G is a residually finite group in which every product of 68 commutators has order dividing n, then G' is locally finite.*

In the present article we will outline the proof of a similar result for arbitrary multilinear commutators.

Theorem 1.5. *Let n be a positive integer and w a multilinear commutator. Let G be a residually finite group in which every product of 896 w-values has order dividing n. Then $w(G)$ is locally finite.*

The constant 896 in the theorem comes from the famous results of Nikolov and Segal on commutator width of finite groups. In the course of proving Theorem 1.5 we need to consider subgroups of a finite soluble group that can be generated by 4 w-values. In the paper of Segal[19] it was shown that every element in the derived group of a finite soluble d-generated group is a product of at most $72d^2+46d$ commutators. A better bound can be obtained working through the proofs given in Ref. 17. It follows that every element of the derived group of a finite soluble d-generated group is a product of at most

$$\min\{d(6d^2+3d+4), 8d(3d+2)\}$$

commutators. In the case that $d=4$ this is 448. We apply this result in the situation where each commutator is a product of 2 w-values. So the constant 896 comes about.

Thus, the theorem of Nikolov and Segal plays an important role in the proof of Theorem 1.5. The proof also relies on the classification of finite simple groups as well as on the Lie-theoretic techniques that Zelmanov created in his solution of the Restricted Burnside Problem. We also use recent result, essentially due to Flavell, Guest and Guralnick, that an element a of a finite group G belongs to $F_h(G)$ if and only if every 4 conjugates of a generate a soluble subgroup of Fitting height at most h[7]. It is the necessity to use this result that accounts for the difference between the constants 896 in Theorem 1.5 and 68 in the case of simple commutators $[x,y]$ in Theorem 1.4. When dealing with simple commutators $[x,y]$ it is enough to bound Fitting height of some 2-generated subgroups while, as was already mentioned, the case of general multilinear commutators requires considering 4-generated subgroups.

Perhaps the most interesting commutator words that are not multilinear commutators are the Engel words $[y,{}_kx]$. These are defined inductively by

$$[y,{}_0x]=y \text{ and } [y,{}_kx]=[[y,{}_{k-1}x],x], \text{ for } k=1,2,\dots.$$

An element x is called a (left) k-Engel element if $[g,{}_kx]=1$ for all $g\in G$. A group G is called k-Engel if all elements of G are k-Engel. There is an old problem whether any k-Engel group is locally nilpotent. This is unknown. Following the solution of the Restricted Burnside Problem Wilson showed that the answer is positive for residually finite groups[28]. Later Burns and Medvedev proved the following result[4].

There exist numbers $c(k)$ and $e(k)$ depending only on k such that if G is a (residually) finite k-Engel group, then G is an extension of a group of exponent dividing $e(k)$ by a nilpotent group of class at most $c(k)$.

This was used in the proof of the following theorem[25].

Theorem 1.6. *Let n and k be positive integers. There exists s depending only on n and k such that in every residually finite group G satisfying the* *identity* $([x_1, {}_k y_1] \dots [x_s, {}_k y_s])^n \equiv 1$ *the verbal subgroup corresponding to the* k*th Engel word is locally finite.*

Unlike the situation with multilinear commutators, in the case of Engel words we were not able to show that s can be chosen independent of k and n.

To see what is the main difficulty in dealing with the Engel words one has to look at Lie theoretic tools used in treatment of the Restricted Burnside Problem. A theorem of Zelmanov says that if L is a Lie algebra generated by finitely many elements in which all products are ad-nilpotent, and if L satisfies a polynomial identity, then L is nilpotent. To be able to use that theorem we need a situation where the group G has generators $a_1, a_2, \dots, a_m$ such that all commutators in $a_1, a_2, \dots, a_m$ have finite (bounded) order. Some words w have the property that whenever x, y are w-values so is $[x, y]$, that is, sometimes the set of w-values is commutator-closed. This happens, for example, in the case where w is γ_k or δ_k. Also, sometimes, the problem can be reduced to the case that the set of w-values is commutator-closed. However the Engel words do not have the commutator-closed property. That is why they are difficult to handle.

In the next two sections we describe some parts of the proof of Theorem 1.5. Details will be available in a separate paper.

2. Bounding the order of a finite group

We call an element a of a group G a δ_k-commutator if it is a value of the word δ_k in G. Theorem 1.5 can be deduced from the following result on finite groups.

Theorem 2.1. *Let* k, m, n *be positive integers and* G *a finite group in which every product of 896* δ_k*-commutators is of order dividing* n. *Assume that* G *can be generated by* m *elements* $g_1, g_2, \dots, g_m$ *such that each* g_i *and all commutators of the forms* $[g, x]$ *and* $[g, x, y]$, *where* $g \in \{g_1, g_2, \dots, g_m\}$, $x, y \in G$, *have orders dividing* n. *Then the order of* G *is bounded by a function depending only on* k, m, n.

Of course, bounding the order of a finite group is the quintessence of the activities around the RBP. It is interesting that the general plan of the proof of the above theorem is the same as in the solution of the RBP.

If G is nilpotent, Zelmanov's techniques work just fine. In this case Theorem 2.1 can be proved using Zelmanov's approach[21].

If G is soluble, Theorem 2.1 can be reduced to the nilpotent case by bounding the Fitting height of G. In what follows we denote by $F_k(G)$ the kth term of the Fitting series of a group G and by $h(G)$ the Fitting height of G. The following important tool is essentially due to Flavell, Guest and Guralnick.

Proposition 2.1. *An element a of a finite group G belongs to $F_k(G)$ if and only if every 4 conjugates of a generate a soluble subgroup of Fitting height at most k.*

The proposition follows from certain results that can be found in the papers Refs. 3,7. The fact of crucial importance is that if every four conjugates of an element a in a group G generate a soluble subgroup, then a belongs to the soluble radical of G. This was established independently in Ref. 7 and also in Gordeev, Grunewald, Kunyavskii and Plotkin[9]. The proof uses the classification of finite simple groups. A well-known corollary of the Hall-Higman theory[13] says that the Fitting height of a finite soluble group of exponent n is bounded by a number depending only on n. We will denote the number by $h(n)$. We have

Lemma 2.1. *Let $k, n \geq 1$ and G a finite soluble group in which every product of 896 δ_k-commutators has order dividing n. Then $h(G) \leq h(n) + k + 1$.*

Proof. Let a_1 be a δ_k-commutator in G and a_2, a_3, a_4 some conjugates of a. Put $H = \langle a_1, a_2, a_3, a_4 \rangle$ and $h = h(n)$. We know from Ref. 17 that every element of H' is a product of 448 commutators of the form $[x, a_i]$ for suitable $x \in G$. Each commutator of the form $[x, a_i]$ is a product of 2 δ_k-commutators so every element of H' is a product of 896 δ_k-commutators. Hence, H' is of exponent n and so the Fitting height of H' is at most h. It follows that the Fitting height of H is at most $h + 1$. We now deduce from Proposition 2.1 that every δ_k-commutator of G is contained in $F_{h+1}(G)$. Therefore $G^{(k)} \leq F_{h+1}(G)$ and the lemma follows. □

The reduction of Theorem 2.1 from the general case to the case where G is soluble is somewhat tricky.

In what follows $X_k(G)$ denotes the set of all δ_k-commutators obtained using elements of the group G. Note that if P is a subgroup of G, in general $X_k(G) \cap P \neq X_k(P)$.

Let G be a finite group and k a positive integer. We will associate with G a triple of numerical parameters $n_k(G) = (\lambda, \mu, \nu)$ where the parameters

λ, μ, ν are defined as follows.

If G is of odd order, we set $\lambda = \mu = \nu = 0$. Suppose that G is of even order and choose a Sylow 2-subgroup P in G. If the derived length $dl(P)$ of P is at most $k+1$, we define $\lambda = dl(P) - 1$. Put $\mu = 2$ if $X_\lambda(P)$ contains elements of order greater than two and $\mu = 1$ otherwise. We let $\nu = \mu$ if $X_\lambda(P) \not\subseteq Z(P)$ and $\nu = 0$ if $X_\lambda(P) \subseteq Z(P)$.

If the derived length of P is at least $k+2$, we define $\lambda = k$. Then μ will denote the number with the property that 2^μ is the maximum of orders of elements in $X_k(P)$. Finally, we let 2^ν be the maximum of orders of commutators $[a, b]$, where $b \in P$ and a is an involution in a cyclic subgroup generated by some element from $X_k(P)$.

The set of all possible triples $n_k(G)$ is naturally endowed with the lexicographical order. We call a group G monolithic if it has a unique minimal normal subgroup which is non-abelian simple. Usefulness of $n_k(G)$ can be explained by the following lemmas.

Lemma 2.2. *Let $k \geq 1$ and G be a group of even order such that G has no nontrivial normal soluble subgroups. Then G possesses a normal subgroup L such that L is residually monolithic and $n_k(G/L) < n_k(G)$.*

Lemma 2.3. *There exist $\{k, n\}$-bounded numbers λ_0, μ_0, ν_0 with the property that if G is a group in which every δ_k-commutator is of order dividing n, then $n_k(G) \leq (\lambda_0, \mu_0, \nu_0)$.*

Thus, Theorem 2.1 can be proved by induction on $n_k(G)$. By Lemma 2.3 there are only boundedly many possibilities for $n_k(G)$. If $n_k(G) = (0, 0, 0)$, then G is of odd order and so is soluble[6]. Lemma 2.1 shows that the Fitting height of G is bounded and this reduces the proof to the nilpotent case. If $n_k(G) \neq (0, 0, 0)$, we use Lemma 2.2. Monolithic groups can be handled using the classification of finite simple groups while for groups R such that $n_k(R) < n_k(G)$ Theorem 2.1 is valid by the induction hypothesis.

This was a brief description of the proof of Theorem 2.1. In the next section we show how Theorem 1.5 can be deduced from Theorem 2.1.

3. Proof of Theorem 1.5

It is convenient first to prove Theorem 1.5 in the case where w is a δ_k-commutator.

Theorem 3.1. *Let n be a positive integer and G a residually finite group in which every product of 896 δ_k-commutators has order dividing n. Then $G^{(k)}$ is locally finite.*

Proof. Let T be any finite subset of $G^{(k)}$. Clearly one can find finitely many δ_k-commutators $h_1, h_2, \ldots, h_m \in G$ such that T is contained in $H = \langle h_1, h_2, \ldots, h_m \rangle$. It is sufficient to prove that the subgroup H is finite. The order of each h_i divides n. Moreover, if $h \in \{h_1, h_2, \ldots, h_m\}$ and $x, y \in H$, then the commutator $[h, x]$ is a product of two δ_k-commutators and the commutator $[h, x, y]$ is a product of four δ_k-commutators. So the order of each of the commutators divides n. If Q is any finite quotient of G, by Proposition 2.1 the order of the image of H in Q is finite and $\{k, m, n\}$-bounded, so it follows that this order actually does not depend on Q. Since G is residually finite, we conclude that H is finite, as required. □

Theorem 1.5 easily follows from Theorem 3.1. We quote here a couple of lemmas from Ref. 21.

Lemma 3.1. *Let G be a group and w a multilinear commutator of weight k. Then every δ_k-commutator in G is a w-value.*

Lemma 3.2. *Let w be a multilinear commutator, G a soluble group in which all w-values have finite order. Then the verbal subgroup $w(G)$ is locally finite.*

Theorem 1.5 is now straightforward.

Proof. Indeed, suppose that G satisfies the hypothesis of Theorem 1.5. By Lemma 3.1 there exists $k \geq 1$ such that any δ_k-commutator is a w-value. Hence any product of 896 δ_k-commutators in G has order dividing n. Theorem 3.1 now tells us that $G^{(k)}$ is locally finite. It is immediate from Lemma 3.2 that $w(G)/G^{(k)}$ is likewise locally finite, as required. □

References

1. S. I. Adian, On groups with periodic commutators, Doklady Mathematics, **62** (2000), 174–176.
2. S. V. Aleshin, Finite automata and the Burnside problem for periodic groups, Math. Notes **11** (1972), 199–203.
3. A. Al-Roqi and P. Flavell, On the Fitting height of a soluble group that is generated by a conjugacy class of 3-elements, Bull. London Math. Soc., **39** (2007), 973-981.
4. R. G. Burns and Y. Medvedev, A note on Engel groups and local nilpotence, *J. Austral. Math. Soc.* (Series A), **64** (1998), 92–100.
5. G. S. Deryabina, P. A. Kozhevnikov, The derived subgroup of a group with commutators of bounded order can be non-periodic, Comm. Algebra, **27(9)** (1999), 4525–4530.

6. W. Feit, J. Thompson, Solvability of groups of odd order, Pacific J. Math. **13** (1963), 773–1029.
7. P. Flavell, S. Guest, R. Guralnick, Characterizations of the solvable radical, Proc. Amer. Math. Soc., **138** (2010), 1161–1170.
8. E. S. Golod, On nil-algebras and residually finite groups, Izvestia Akad. Nauk SSSR, Ser Mat., **28** (1964), 273–276.
9. N. Gordeev, F. Grunewald, B. Kunyavskii and E. Plotkin, A description of Baer–Suzuki type of solvable radical of a finite group, Journal of Pure and Applied Algebra, **213** (2009), 250–258.
10. R. I. Grigorchuk, On the Burnside problem for periodic groups, Funct. Anal. Appl., **14** (1980), 53–54
11. N. Gupta and S. Sidki, On the Burnside problem for periodic groups, Math. Z., **182** (1983), 385–386.
12. M. Hall, "The Theory of Groups", Macmillan, New York, 1959.
13. P. Hall, G. Higman, The p-length of a p-soluble group and reduction theorems for Burnside's problem, Proc. London. Math. Soc. (3), **6** (1956), 1–42.
14. B. Huppert, Endliche Gruppen I, Springer-Verlag, Berlin, 1967.
15. S. Ivanov, On groups with periodic products of commutators, in "Topological and asymptotic aspects of group theory", 143-147, Contemp. Math., **394**, Amer. Math. Soc., Providence, RI, 2006.
16. G. A. Jones, Varieties and simple groups, J. Austral. Math. Soc., **17** (1974), 163–173.
17. Nikolay Nikolov and Dan Segal, On finitely generated profinite groups, I: strong completeness and uniform bounds, Annals of Mathematics, **165** (2007), 171-238.
18. D.J.S. Robinson, "Finiteness Conditions and Generalized Soluble Groups", Part 1, Springer Verlag, Berlin-New York, 1972.
19. Dan Segal, Closed subgroups of profinite groups, Proc. London Math. Soc. (3) **81** (2000), 29–54.
20. P. Shumyatsky, Groups with commutators of bounded order, Proc. Amer. Math. Soc. **127** (1999), 2583–2586.
21. P. Shumyatsky, Verbal subgruops in residually finite groups, The Quarterly Journal of Mathematics, **51** (2000), 523–528; doi:10.1093/qjmath/51.4.523.
22. P. Shumyatsky, On varieties arising from the solution of the Restricted Burnside Problem, J. Pure Appl. Alg. **171** (2002), 67–74.
23. P. Shumyatsky, Commutators in residually finite groups, Monatsh. Math. **137** (2002), 157–165.
24. P. Shumyatsky and J. C. Silva, The Restricted Burnside Problem for Multilinear Commutators, Math. Proc. Camb. Phil. Soc. **146** (2009), 603–613.
25. P. Shumyatsky and J. C. Silva, Engel Words and the Restricted Burnside Problem, Monatsh. Math., **159** (2010), 397–405.
26. P. Shumyatsky, Commutators in residually finite groups, Israel Journal of Mathematics, to appear.
27. V.I. Sushchansky, Periodic p-elements of permutations and the general Burnside problem, Dokl. Akad. Nauk SSSR, **247** (1979), 447–461.
28. J. S. Wilson, Two-generator conditions for residually finite groups, *Bull. Lon-*

don Math. Soc. **23** (1991), 239–248.

29. E. Zelmanov, The solution of the restricted Burnside problem for groups of odd exponent, Math. USSR Izv. **36** (1991), 41-60.
30. E. Zelmanov, The solution of the restricted Burnside problem for 2-groups, Math. Sb. **182** (1991), 568-592.

THE ADJOINT GROUP OF RADICAL RINGS AND RELATED QUESTIONS

Ya. P. SYSAK

Institute of Mathematics National Academy of Sciences of Ukraine, Kiev, 01601, Ukraine
E-mail: sysak@imath.kiev.ua

The aim of this article is to survey some results and problems concerning the adjoint group of radical rings and related topics. A main attention is focused on the structure of this group, on the relationship between the adjoint and Lie structures in such rings and on the structure of radical rings whose adjoint group is finitely generated. Furthemore, some generalizations of radical rings to so-called braces and radical modules and their natural connection with triply factorized groups are considered. The article also contains proofs of some crucial results.

Keywords: Radical ring; Adjoint group; Associated Lie ring; Engel condition; Nil ring; Brace; Radical module; Triply factorized group.

1. Introduction

Let $R = (R, +, \cdot)$ be an associative ring, not necessarily with an identity element. Then the set of all elements of R forms a monoid with neutral element $0 \in R$ under the "circle" operation $r \circ s = r + s + rs$ for all $r, s \in R$ which is called the *adjoint monoid* of R. The group of all invertible elements of this monoid is said to be the *adjoint group* of R and usually denoted by R°. The ring R is called *radical* (in sense of Jacobson) if $R = R^\circ$ which means that R coincides with its Jacobson radical.

If R is a ring with an identity 1, then the set of all non-zero elements of R forms the *multiplicative monoid* of R whose group of all invertible elements is called the *multiplicative group* of R and denoted by R^*. In this case the mapping $r \mapsto 1 + r$ with $r \in R^\circ$ determines a group isomorphism from R° onto R^*. In fact, $R^* = 1 + R^\circ$. Recall that R is a *division ring* if $R^* = R \setminus \{0\}$. It is well-known that every ring R can be embedded in a ring R_1 with 1, for instance, if we put $R_1 = R \oplus \mathbb{Z}$ and extend the multiplication in R_1 by distributivity.

The best known examples of radical rings are the nilpotent rings. A ring R is *nilpotent* if there exists a positive integer n such that every product of n elements from R is 0. The least n with this property is usually called the *index of nilpotency* of R. For instance, rings of strictly upper triangular $(n \times n)$ matrices over a ring are well-known examples of nilpotent rings of index at most n. If every finitely generated subring of R is nilpotent, then R is *locally nilpotent.* It is obvious that the adjoint group of every (locally) nilpotent ring is (locally) nilpotent.

A ring R is *nil,* if every element r of R is nilpotent, i.e. there exists a positive integer $n = n(r)$ with $r^n = 0$. Every locally nilpotent ring is nil. In 1964, for any prime p, Golod[25] constructed a finitely generated nil algebra over the Galois field $GF(p)$ which is not nilpotent and whose adjoint group is a residually finite p-group containing finitely generated infinite periodic subgroups. A considerable simplification of this construction was found by Olshanskii[43] only thirty years later.

All considered rings are associative but usually without identities. If R is a radical ring, then R_1 denotes the ring with 1 containing R as an ideal such that $R_1 = R \oplus \mathbb{Z}/n\mathbb{Z}$ where n is the least positive integer for which $nR = 0$ if it exists or $n = 0$ otherwise. Furthermore, if R is an algebra over a field F, then $R_1 = R \oplus F$.

2. The structure of the adjoint group of radical rings

We present first a list of elementary assertions concerning radical rings and their adjoint groups. They are either trivial or well-known whose proofs can for instance be found in Jacobson's book[29].

Proposition 2.1. *Let R be a radical ring. Then the following statements hold:*

1) *a subring of R is radical if and only if its adjoint monoid is a subgroup of R°;*
2) *each one-sided ideal and the centralizer $C_R(X)$ of any subset X of R is a radical subring of R;*
3) *if I is an one-sided ideal of R and I° is a normal subgroup of R°, then I is an ideal of R;*
4) *if I is an ideal of R, then the factor ring R/I is radical, I° is a normal subgroup of R° and the adjoint group $(R/I)^\circ$ is isomorphic to the factor group R°/I°;*
5) *$C_{R^\circ}(X) = C_R(X)$ for each subset X of R; in particular, every maximal abelian subgroup and the center $Z(R^\circ)$ of R° coincide with a*

maximal commutative subring and the center $Z(R)$ of R, respectively;

6) *if M is an R_1-module and m is a non-zero element of M, then $m \notin mR$; in particular, every simple R_1-module is annihilated by R;*
7) *if $M = MR$, then M cannot be finitely generated as an R_1-module;*
8) *if R is an algebra over a field F and $r \in R$, then the element r is either nilpotent or transcendental over F.*

For each prime p, the ring $p\mathbb{Q}_p$ of all rational numbers whose numerators are divisible by p and denominators are coprime with p is radical, but its non-zero finitely generated subrings are not. Thus subrings of a radical ring R need not be radical and it is easy to see that R is nil if and only if every subring of R is radical. If X is a subset of R such that no proper radical subring of R contains X, then we say that R is *radical on the set X* or R is *generated by X as a radical ring*. In this case R need not be generated by X as a ring excepting the case when R is nil. For instance, the only radical subring of $p\mathbb{Q}_p$ containing p is $p\mathbb{Q}_p$ itself, so that p generates $p\mathbb{Z}$ as a ring and $p\mathbb{Q}_p$ as a radical ring.

As an immediate consequence of statement 7), we have the following assertion.

Corollary 2.1. *If R is a non-zero radical ring on finitely many generators, then $R^2 \neq R$. In particular, if $R^2 \neq 0$, then R cannot be a simple ring.*

A first example of a simple radical ring R with $R^2 = R$ was found by Sasiada in 1961 and published in Ref. 47. Afterwards Cohn[19] obtained the following result which we reformulate as follows.

Theorem 2.1. *Every radical subalgebra of a division ring can be embedded in a simple radical algebra R (again contained in a division ring) whose adjoint group R° has only two conjugacy classes of elements.*

It should be noted that the original theorem in Ref. 19 concerns radical subrings of a division ring and in such a form it is not correct. Indeed, the adjoint group of the radical subring $2\mathbb{Q}_2$ of the field $\mathbb{Q}$ of rationals contains -2 as an element of order 2 because $(-2) \circ (-2) = -2 - 2 + 4 = 0$ and so this group cannot be embedded in a group with two conjugacy classes of elements.

Note also that a long-standing question whether there exists a non-trivial simple nil ring was positively solved by Smoktunowicz[53] well after.

It is well-known that every group (ring, algebra, etc.) can be embedded in a simple group (simple ring, simple algebra, etc.). As a contrast to this, Valitskas[59] answering a question of Cohn[19] proved the following.

Theorem 2.2. *For any field F there exists a radical F-algebra R which cannot be embedded into a simple radical ring.*

Clearly the adjoint group R° in the conclusion of Theorem 2.1 is simple and torsion-free. On the other hand, taking into account statements 7) and 8) of Proposition 2.1, it was shown in Ref. 2 that the adjoint group of any radical ring is an SN-group, i.e. it has a series with abelian factors and so is generalized soluble.

Theorem 2.3. *The adjoint group R° of a radical ring R is an SN-group in which every finite subgroup is nilpotent.*

In particular, from here and Theorem 2.1 it follows that there exist SN-groups in which every two non-identity elements are conjugate. Recall that first examples of groups with two conjugacy classes of elements were constructed by Higman, Neumann and Neumann.[28] However it is not clear whether their construction can result to SN-groups.

Using Zelmanov's theorem on the Restricted Burnside Problem[63,64], the following assertion is deduced from Theorem 2.3.

Corollary 2.2. *Let G be a subgroup of R° and suppose that one of the following conditions holds:*

1) *G is locally finite,*
2) *G has finite exponent,*
3) *G is an n-Engel group for some $n \geq 1$,*
4) *G is locally artinian.*

Then the group G is locally nilpotent.

Let $\mathfrak{U}$ be the class of all groups each of which can be embedded in the adjoint group of a radical ring. The following properties of this class follow immediately from Proposition 2.1 and well-known properties of finitely generated nilpotent groups.

Proposition 2.2. *The following statements hold:*

1) *the class $\mathfrak{U}$ is closed with respect to forming subgroups and cartesian products;*

2) *if G is a group whose finitely generated subgroups belong to $\mathfrak{U}$, then G is also contained in $\mathfrak{U}$;*
3) *all locally nilpotent and residually nilpotent groups belong to $\mathfrak{U}$;*

Proof. Clearly statement 1) follows from statement 1) of Proposition 2.1. If Σ is a local system of the group G and $\Pi_{S\in\Sigma}S$ denotes the cartesian product of groups from Σ, then G is isomorphic to a subgroup of the ultraproduct $(\Pi_{S\in\Sigma})S/\mathcal{F}$ for a suitable filter $\mathcal{F}$ on the set Σ by assertion 1.L.7 of the book[31] of Kegel and Wehrfritz. Therefore statement 2) follows from statement 1) and 4) of Proposition 2.1. In order to prove 3), note first that every finitely generated nilpotent group is a subcartesian product of finite p-groups for a set of primes p. As every finite p-group is embedded in the adjoint group of the nilpotent algebra of all strictly upper triangular $n \times n$ matrices over the Galois field $GF(p)$ for a suitable positive integer n, statement 3) follows from 1) and 2). □

Note that polycyclic groups which are embedded in the adjoint group of a radical ring and so belong to the class $\mathfrak{U}$ need not be nilpotent or residually nilpotent. This can be shown by the following example of Neroslavskii[42].

Example. Let A be the algebra of formal power series without constant terms in two non-commuting variables x and y over a Galois field $GF(p)$ with $p > 2$. Then the ideal I of A generated by the elements $(1+x)^{-1}y(1+x) - y^2 - 2y$ and y^p does not contain y and the factor algebra $R = A/I$ is radical with generators $a = x+I$ and $b = y+I$. It is easy to see that $a^n \neq 0$ for each $n \geq 1$, $b^p = 0$ and $(1+a)^{-1}(1+b)(1+a) = (1+b)^2$. Therefore the subgroup G of R° generated by the elements a and b is metacyclic, but not residually nilpotent.

In a connection with this example the following question is natural:

Question 2.1. Is every polycyclic group whose finite subgroups are nilpotent embedded in the adjoint group of a radical ring?

As Golod's examples show, periodic groups contained in the class $\mathfrak{U}$ need not be locally nilpotent. Nevertheless the following holds (see Ref. 56).

Theorem 2.4. *If G is a periodic group contained in $\mathfrak{U}$ and Z is the center of G, then the factor group G/Z is the direct product of its primary components. In particular, every two elements of coprime orders of G are permutable.*

It seems that the question *whether every periodic group in $\mathfrak{U}$ is a direct product of its primary components* is open. It can be reduced to the following more general problem which is of independent interest.

Question 2.2. Does every central extension G of a cyclic p-group by a periodic p'-group split if G is an SN-group?

It should also be noted that if G is not an SN-group, then an example of Adjan (Ref. 1, p. 276) shows that there exists such a non-split extension.

In respect to normal subgroups of the adjoint group of a radical ring, the following result of Catino, Miccoli and the author[15] can be regarded as an analogue of Scott's result[48] concerning normal subgroups of the multiplicative group of a division ring.

Theorem 2.5. *If a radical ring R has no non-zero nilpotent ideal and Z is the center of R, then the factor group R°/Z° has neither abelian nor finite non trivial normal subgroups.*

As an immediate consequence, we have

Corollary 2.3. *Let G be a soluble-by-finite normal subgroup of R°. Then the derived subgroup G' is locally nilpotent.*

It is trivial and noted by statement 5) of Proposition 2.1 that the centers of a radical ring R and its adjoint group R° coincide. Using a subtle argument, Du[21] proved that the same holds for the n-th terms of the upper central series of R and R°, respectively, with any positive integer n. Recall that for each ordinal number α the α-th term $Z_\alpha = Z_\alpha(R)$ of the upper central series of the ring R is defined inductively as follows:

$$Z_0 = 0, \qquad Z_\alpha = \{x \in R \mid rx - xr \in Z_{\alpha-1} \text{ for each } r \in R\}$$

if α is not a limit ordinal and $Z_\alpha = \cup_{\beta<\alpha} Z_\beta$ otherwise.

In a similar definition of $Y_\alpha = Z_\alpha(R^\circ)$ we have $Y_\alpha/Y_{\alpha-1} = Z(R^\circ/Y_{\alpha-1})$. Furthermore, regarded R as an additive group on which R° acts by conjugation, we put also $C_0 = 0$, $C_\alpha = C_{R/C_{\alpha-1}}(R^\circ)$ if α is not a limit ordinal and $C_\alpha = \cup_{\beta<\alpha} C_\beta$ otherwise. Then C_α is the α-th term of the upper R°-stable series of R defined by Catino[14]. Based on the above-mentioned result of Du, he proved that $C_n = Z_n$ for each $n \geq 1$. Combining both results, we have.

Theorem 2.6. *If R is a radical ring, then $C_n = Y_n = Z_n$ for each positive integer n.*

It should be noted that the equality $Y_n = Z_n$ has been conjectured and proved for $n = 2$ by Laue[36]. Afterwards Du[22] proved also that if the radical ring R is 2-torsion-free, i.e. $2r = 0$ implies $r = 0$ for each $r \in R$, then $C_\alpha = Y_\alpha = Z_\alpha$ for each ordinal number α. It is unknown at present *whether the condition that R is 2-torsion-free can be removed.*

3. Relations between the adjoint and Lie structures in radical rings

It is well-known that every ring R can be viewed as a Lie ring under the Lie multiplication $[r, s] = rs - sr$ for all $r, s \in R$ which is called the *associated Lie ring* of R and denoted by $R^{(-)}$. The ideals of this Lie ring are called Lie-ideals of R. Thus an additive subgroup L of R is a Lie-ideal if $[L, R] \subseteq L$ where $[L, R]$ is the additive subgroup of R generated by all Lie-commutators $[x, r]$ with $x \in L$ and $r \in R$. Evidently every ideal of R is a Lie-ideal but not conversely.

If $r_1, r_2, \ldots$ are elements of R, the Lie-commutators $[r_1, \ldots, r_{n+1}]$ are defined inductively by $[r_1, \ldots, r_{n+1}] = [[r_1, \ldots, r_n], r_{n+1}]$ for all $n \geq 2$. The ring R is called *Lie-nilpotent* if $R^{(-)}$ is nilpotent, i.e. there exists a positive integer n such that $[r_1, \ldots, r_{n+1}] = 0$ for all $r_1, \ldots r_{n+1}$ of R. The least n with this property is the *class of Lie-nilpotency* of R. We will also say that R is *locally Lie-nilpotent* if every finitely generated subring of R is Lie-nilpotent.

The question of a relationship between the group structure of R° and the Lie structure of R for a radical ring R was initially raised by Jennings[30]. He proved that *R is Lie-nilpotent if and only if the adjoint group R° is nilpotent* and conjectured that the classes of nilpotency of both structures coincide. As it has been shown above, this conjecture was confirmed by Du[21] even in more strong form.

A natural generalization of the concept of nilpotency of groups is that of nilpotency of semigroups in the sense of Mal'cev[38] or Neumann and Taylor[41] whose precise definition can be given as follows.

Let X be a countable set of non-commuting indeterminates and let the sequence $W_1(x, y)$, $W_2(x, y, z_1)$,...,$W_n(x, y, z_1, \ldots, z_{n-1})$, ... of words in the indeterminates x, y, $z_1, \ldots$, $z_n, \ldots$ of X be defined by the rule

$W_1(x, y) = xy$ and

$$W_{n+1}(x, y, z_1, \ldots, z_n) = W_n(x, y, z_1, \ldots, z_{n-1}) z_n W_n(y, x, z_1, \ldots, z_{n-1})$$

for every $n \geq 1$.

A *semigroup* A is said to be *nilpotent* if there exists a positive integer n such that

$$W_n(a, b, c_1, \dots, c_{n-1}) = W_n(b, a, c_1, \dots, c_{n-1})$$

for any elements $a, b, c_1, \dots, c_{n-1}$ of A. The least n with this property is called the *class of nilpotency* of the semigroup A.

Answering a question posed by Krasil'nikov[34] and independently by Riley and Tasić[45], Amberg and the author[7] proved a semigroup version of Jenning's conjecture for any associative ring.

Theorem 3.1. *If R is a ring and n is a positive integer, then the adjoint monoid of R is nilpotent of class n if and only if R is Lie-nilpotent of class n.*

Two years later Riley[44] using a characterization of minimal non-Engel varieties gave another proof of this theorem. At the same time a locally nilpotent analogue of Theorem 3.1 was proved in[9].

Theorem 3.2. *The adjoint monoid of a ring R is locally nilpotent if and only if R is locally Lie-nilpotent.*

It is obvious that if R is locally Lie-nilpotent, then the Lie ring $R^{(-)}$ is locally nilpotent. In connection with Theorems 3.1 and 3.2 the following question arises:

Question 3.1. Is an associative ring R locally Lie-nilpotent if its associated Lie ring $R^{(-)}$ is locally nilpotent?

Recall that a ring R is *Engel* if $[r, s, \dots, s] = 0$ for each pair of elements r and s in R, and *n-Engel* if s appears exactly n times. Engel and n-Engel groups are defined by the analogous group commutator identity. An essential contribution in studying Engel groups was made by Gruenberg[26] who proved in particular that every soluble Engel group is locally nilpotent. It was for a long time known that every finite Engel group is nilpotent. Whether or not the n-Engel condition in groups can be defined by semigroup identities is an open problem of Shirshov (see Problem 2.82 in[58]). Shirshov gave a positive solution to his question in the case $n \leq 3$(see [18]). More recently, this was extended to include $n = 4$ by Longobardi and Maj[37] and Traustason[57]. In this connection a monoid A was called by Riley[44] *positively n-Engel* if

$$W_{n+1}(a, b, 1, c, \dots, c^{n-1}) = W_{n+1}(b, a, 1, c, \dots, c^{n-1})$$

for any elements a, b, c of A and the following two results were established.

Theorem 3.3. *Let R be an arbitrary ring. Then the following statements are equivalent:*

1) the ring R is n-Engel for some $n \geq 1$;
2) the adjoint monoid of R is positively m-Engel for some $m \geq 1$;
3) the multiplicative semigroup of R is l-Engel for some $l \geq 1$.

Note that in this and the next theorem the numbers n, m and l are in fact dependent on one another.

Theorem 3.4. *Let R be a radical ring. Then the following statements are equivalent:*

1) the adjoint group R° is n-Engel for some $n \geq 1$;
2) the adjoint group R° is positively m-Engel for some $m \geq 1$;
3) the ring R is l-Engel for some $l \geq 1$.

In particular, the equivalence of statements 1) and 2) gives a positive solution of Shirshovs problem for the class of all groups arising as the adjoint group of a radical ring. It should also be noted that the equivalence of statements 1) and 3) of this theorem was already proved in[6].

A dependence between the n-Engel structure of rings and that of their adjoint group for more wider classes than the radical rings was studied in[11]. Recall that rings which are artinian modulo their Jacobson radical are called *semilocal.*

Theorem 3.5. *Let R be a semilocal ring whose adjoint group R° is n-Engel for some integer $n \geq 1$. Then R° is a locally nilpotent group. Moreover, R is m-Engel for some integer $m \geq 1$ depending only on n, provided that R is generated by R°.*

It should be noted that the condition that R is generated by R° cannot be omitted. A counterexample is given by the ring of all upper triangular 3×3-matrices over the Galois field $GF(2)$.

For any additive subgroups V and W of R, let $[V, W]$ be the additive subgroup of R generated by all Lie-commutators $[v, w]$ with $v \in V$ and $w \in W$. The derived chain of $R^{(-)}$ is defined inductively as

$$\delta_0(R) = R \qquad \text{and} \qquad \delta_{n+1}(R) = [\delta_n(R), \delta_n(R)]$$

for each integer $n \geq 0$. The ring R is called *Lie-soluble of length at most* m if $\delta_m(R) = 0$. Lie-soluble rings of length at most 2 are called *Lie-metabelian.*

Zalesskii and Smirnov[62] and independently Sharma and Srivastava[49] proved that every Lie-soluble ring R has a nilpotent ideal I whose factor ring R/I is center-by-metabelian as a Lie ring. However the adjoint group R° of R need not be soluble in this case.

Indeed, if R is the ring of all 2×2-matrices over a commutative radical domain S of characteristic 2, say $S = \{\frac{xf(x)}{1+xf(x)} \mid f(x) \in \mathbb{F}_2[x]\}$, then $R^{(-)}$ is center-by-metabelian and R° contains free subgroup generated by the matrices $\begin{pmatrix} 1 & t \\ 0 & 1 \end{pmatrix}$ and $\begin{pmatrix} 1 & 0 \\ t & 1 \end{pmatrix}$. On the other hand, Krasil'nikov[33] and independently Sharma and Srivastava[50] proved that the adjoint group of every Lie-metabelian ring is metabelian. Furthermore, Krasil'nikov[33] proved that every nil ring with metabelian adjoint group is Lie-metabelian. As in the nilpotent case, the question arises whether every radical ring with soluble adjoint group is Lie-soluble. A complete answer to this question is given by the following theorem of Amberg and the author[8].

Theorem 3.6. *Let R be a radical ring. Then the adjoint group R° is soluble if and only if the following statements hold:*

(1) R is Lie-soluble, and
(2) there exists a chain

$$0 = I_0 \subseteq I_1 \subseteq \cdots \subseteq I_m = R$$

of ideals of R such that every factor I_i/I_{i-1} is generated by commutative ideals of R/I_{i-1} for $1 \leq i \leq m$.

Moreover, if R° is soluble of length n for some $n \geq 1$ and L the Levitzki radical of R, then there exist positive integers k, l and m depending only on n such that:

(3) R satisfies the identity $[x, y]^k = 0$ for all $x, y \in R$;
(4) the factor ring R/L is commutative and L satisfies the identity $[x, y, \ldots, y] = 0$ with y repeated l times, i.e. L is an l-Engel ring;
(5) the derived subgroup of R° is an m-Engel group.

Furthermore, answering a question of Krasilnikov, Sharma and Srivastava[33,50], Amberg and the author[10] proved the following.

Theorem 3.7. *The adjoint group R° of a radical ring R is metabelian if and only if R is Lie-metabelian.*

Finally, the ring R is said to be *Lie-supersoluble* if R has an ascending series of Lie-ideals whose factors are cyclic as additive groups. The next assertion is proved by Catino, Miccoli and the author[16].

Theorem 3.8. *Let R be an associative ring. Then the following statements hold:*

(1) if R is radical, then the adjoint group R° is locally supersoluble if and only if R° is locally nilpotent and so R is locally Lie-nilpotent;
(2) if R is semilocal and its adjoint group R° is locally supersoluble, then R is locally Lie-supersoluble and contains a locally Lie-nilpotent ideal I of finite index such that the factor ring R/I is a direct sum of ideals each of which is isomorphic either to the Galois field $GF(p)$ of a prime order p or to the matrix ring $M_2(GF(2))$.

Note the adjoint group of a (locally) Lie-supersoluble ring need not be (locally) supersoluble. For instance, the ring of all (2×2)-matrices of the form $\begin{pmatrix} a & b \\ 0 & a^2 \end{pmatrix}$ over the Galois field $GF(4)$ is Lie-supersoluble but its multiplicative and so adjoint group is isomorphic to the alternating group of degree 4 and so is not supersoluble.

4. Finiteness condition on the adjoint group and nil rings

It is well-known and actually follows from the Hilbert Nullstellensatz that every homomorphic image of the polynomial ring $\mathbb{Z}[x_1, \ldots, x_n]$ in n commuting variables $x_1, \ldots, x_n$ has the property that its Jacobson radical is nil. Since finitely generated commutative nil rings are nilpotent, this results to the following assertion.

Lemma 4.1. *If a commutative radical ring R is finitely generated as a ring, then R is nilpotent. In particular, this holds if the adjoint group of R is finitely generated.*

Answering an old question of Amitsur[12], Smoktunovicz and Puczylowski[54] have shown that finitely generated non-commutative radical rings need not be nil.

It is easy to see that both additive and adjoint groups of any finitely generated nilpotent ring are also finitely generated. An extension of this assertion was obtained by Watters[60] and independently by Kruse[35].

Theorem 4.1. *The adjoint group of a radical ring R is finitely generated nilpotent if and only if R is a nilpotent ring with a finitely generated additive group.*

A further step was done by Amberg and the author[8]. Recall that an abelian group G is said to be of *finite torsion-free rank* if it has a finitely generated torsion-free subgroup A such that the factor group G/A is periodic.

Theorem 4.2. *If the adjoint group R° is soluble and its commutator factor group has finite torsion-free rank, then the ring R and so the group R° is locally nilpotent. In particular, every radical ring whose adjoint group is finitely generated and soluble must be nilpotent.*

In this connection a natural question arises:

Question 4.1. Is every radical ring R with finitely generated adjoint group nilpotent?

Although it is similar to the long-standing question about the finitely generated multiplicative group of a division ring, it seems that its positive solution is more available. At least for the group R° with two generators this question can be answered affirmatively. In fact, the same answer can be obtained in the general case, provided that there exists a bound for dimensions of finite nilpotent algebras depending on the number of generators of their adjoint group. For brevity, we will say a *p-algebra* instead of an algebra over a Galois field $GF(p)$ and mark out two related questions.

Question 4.2. Is the dimension of every nilpotent p-algebra whose adjoint group is generated by d elements bounded by a number $f(d,p)$ depending only on d and p?

Recall that a group G has *finite Prüfer rank* $r = r(G)$ if every finitely generated subgroup of G can be generated by r elements and r is a minimal number with this property.

Question 4.3. Does there exist a bound for Prüfer rank of the adjoint group of every finite nilpotent p-algebra which depends only on p and the minimal number of generators of this group?

It is easy to show that Question 4.2 admits a positive answer if and only if so does Question 4.3.

Indeed, let A be a finite nilpotent p-algebra whose adjoint group A° is generated by d elements. Clearly Prüfer rank of A as an additive group coincides with $\dim A$. If Question 4.2 is positively answered, then $\dim A \leq f(d,p)$ and so $r(A^\circ) \leq 3r(A) = 3\dim(A) \leq 3f(d,p)$ by a result of Dickenschied[20]. Conversely, it was proved by Zaitsev[61] that for every soluble group G of the form $G = BC$ with subgroups B and C the Prüfer rank $r(G)$ is bounded by a polynomial function of $r(B)$ and $r(C)$. Since A is an A°-module under the action $a^b = a + ab$ for all $a, b \in A$ and in the semidirect product $G = A \rtimes A^\circ$ there exists a subgroup B and C isomorphic to A° such that $G = A \rtimes B = A \rtimes C = BC$ (see Ref. 55 for details), the Prüfer rank $r(A)$ is bounded by a polynomial function $g(r(A^\circ))$. Thus if Question 4.3 is positively answered, then there exists a bound $h(d,p)$ such that $r(A^\circ) \leq h(d,p)$ and hence we can put $f(d,p) = g(h(d,p))$, as desired.

In particular, if the algebra A is commutative, then $r(A^\circ) \leq d$ and so the number $f(d,p)$ in Question 4.2 always exists.

Proposition 4.1. *A positive solution of Question 4.1 follows from that of Question 4.2.*

Proof. Indeed, assume that a radical ring R in question is non-nilpotent. Then R is finitely generated and so there exists an ideal P of R which is maximal in respect to the condition that the factor ring R/P is non-nilpotent. Thus, using Corollary 2.1, the question is reduced to the case when R is residually nilpotent and every proper factor ring of R is nilpotent. Since the additive group of each factor ring R/R^n with $n \geq 1$ is finitely generated and $\bigcap_{n=1}^{\infty} R^n = 0$, there exist a prime p such that $R \neq pR$.

Let d be the number of generators of the adjoint group R°. If $pR = 0$, then R is an p-algebra and hence $\dim R/R^n \leq f(d,p)$ for all $n \geq 1$ by the hypothesis. But then $R^n = 0$ for some n, contrary to the assumption. Therefore the factor ring R/pR is finite and thus there exists a finitely generated additive subgroup X of R such that $R = X + pR$. This implies $R = X + p^n R$ for every $n \geq 1$ and so each factor ring $R/p^n R$ satisfies a polynomial identity whose degree depends only on the number of generators of X. Put $I = \bigcap_{n=1}^{\infty} p^n R$. If $I = 0$, then R satisfies the same polynomial identity and so is nilpotent by Ref. 13, contrary to the assumption. Thus the factor ring R/I is nilpotent by the choice of R and so $R^m \subseteq I$ for some $m > 1$. Therefore I is finitely generated as a radical ring and hence $I \neq I^2$ by Corollary 2.1. As the factor ring R/I^2 is also nilpotent and so its additive group is finitely generated, it follows that the intersection $\bigcap_{n=1}^{\infty} p^n(R/I^2) = I/I^2$ is of finite order, say k. Then $kI \subseteq I^2$ and $I = Y + kI$ for some finitely generated additive subgroup Y of I from which it follows that $k^nI \subseteq I^{n+1}$ and $I = B + k^nI$ for each $n \geq 1$. Since $\bigcap_{n=1}^{\infty} k^nI \subseteq \bigcap_{n=1}^{\infty} I^{n+1} = 0$, this means that as above I satisfies a polynomial identity. But then R does so and thus R must be nilpotent, a final contradiction. □

Note that the bound $f(d,p)$ for dimensions of nilpotent p-algebras with adjoint group having d generators exists if $d \leq 2$ and its exact values are known. It was long ago known that $f(1,2) = 2$ and $f(1,p) = 1$ if $p \neq 2$ (see for instance Ref. 24). Furthermore, a view years ago Amberg and Kazarin[4] proved that $f(2,2) = 5$ and $f(2,p) = 3$ for $p \neq 2$. Using these values and the above proof, we have the following result.

Theorem 4.3. *If R is a radical ring whose adjoint group R° is generated by two elements, then R and so R° is nilpotent.*

The case of commutative nilpotent p-algebras is of special interest because in this case the function $f(d,p)$ is defined for all $d \geq 1$. Clearly in such a p-algebra A the set $A^{(p)} = \{a^p \mid a \in A\}$ is a subalgebra. A conjecture of Eggert[23] says that

if A is finite-dimensional, then $\dim A \geq p \dim A^{(p)}$.

In the cases when $\dim A^{(p)} \leq 4$ it was affirmed by several authors whose results are summed up in[3]. Some other partial cases for which this conjecture is valid were considered in Refs. 39,40. An attempt to prove Eggert's conjecture was made by L. Hammoudi in Ref. 27 but his proof has an unrepairable error which was afterwards admitted by himself. One of most general results in this direction was recently presented by M. Korbelar in Ref. 32. He proved that the conjecture holds if the subalgebra $A^{(p)}$ is generated by two elements.

Now, taking d as the minimal number of generators of the adjoint group A°, we obtain $\dim A = d + \dim A^{(p)}$ because $(A^\circ)^p = A^{(p)}$. Therefore from $d + \dim A^{(p)} \geq p \dim A^{(p)}$ it follows that $\dim A \leq \frac{pd}{p-1}$ and conversely. Thus in terms of the introduced function $f(d,p)$ the Eggert conjecture can also be reformulated as follows.

Question 4.4. Is it true that $f(d,p) \leq \frac{pd}{p-1}$ for each commutative nilpotent p-algebra whose adjoint group has d generators?

As it has been shown above, the function $f(d,p)$ is bounded by a polynomial function $g(d)$ which does not depend on p. In fact, in the commutative case $g(d)$ can even be chosen as a quadratic function. However it unknown at present *whether there exists such a linear function.*

It is intuitively clear that there exists a connection between the adjoint group of nil rings and Engel conditions. For lack of suitable results we can only restrict ourselves to some remarks concerning an influence of (unbounded) Engel conditions on the structure of nil rings and their adjoint group. First of all, the following simple assertion should be mentioned.

Proposition 4.2. *Every nil ring is Engel.*

Proof. Indeed, if x and y are elements of a ring R and $[x, {}_m y]$ denotes the commutator $[x, y, \dots, y]$ in which y appears exactly m times, then

$$[x, {}_m y] = \sum_{i=0}^{m} (-1)^{i+1} \binom{m}{i} y^i x y^{l-i}$$

for each $m \geq 1$. Thus if R is nil, then $y^{2m} = 0$ for some $m \geq 1$ and so $[x, {}_m y] = 0$, as desired. □

In this connection the following question is of interest.

Question 4.5. Is the adjoint group of every nil ring Engel?

Note that the adjoint group of Golod's nil algebra mentioned in the Introduction is Engel but not locally nilpotent.

If R is a nil algebra over an uncountable field, then it was proved by Amitsur[12] that the polynomial algebra $R[t]$ in one commutative variable t is nil and so Engel by Proposition 4.2. On the other hand, for every countable field F there exists a nil F-algebra R such that the polynomial algebra $R[t]$ is not nil by a known result of Smoktunowicz[52]. Thus it is natural to ask whether the polynomial ring $R[t]$ in one commutative variable t over a nil R is also Engel.

Theorem 4.4. *Let R be a nil ring whose additive group is periodic. Then the polynomial ring $R[t]$ is Engel if and only if $R[t]$ is nil.*

Proof. Clearly it suffices to restrict ourselves to the case when the ring R is finitely generated and the ring $R[t]$ is Engel.

Assume first that R is an algebra over a field of prime characteristic p. Then $[x, {}_{p^n} y] = [x, y^{p^n}]$ for each $x, y \in R[t]$ and every $n \geq 1$. Since $R[t]$ is Engel, for each pair of elements x, y there exists a positive integer

$n = n(x,y)$ such that $[x, y^{p^n}] = 0$ and so $y^{n(x,y)} \in C_{R[t]}(x)$ for each $y \in R[t]$. If the elements $x_1, \dots, x_m$ generate $R[t]$ and $k(y) = n(x_1, y) \cdots n(x_m, y)$, then $Z = \cap_{i=1}^m C_{R[t]}(x_i)$ is the center of $R[t]$ and $y^{k(y)} \in Z$ for every $y \in R[t]$. Furthermore, if $f = a_0 + a_1 t + \cdots + a_n t^n \in Z$ for some $a_0, \dots, a_n$ of R, then for each $a \in R$ it follows that $0 = [f, r] = [a_0, a] + [a_1, a]t + \cdots + [a_n, a]t^n$ and thus $[a_i, a] = 0$ for all $1 \leq i \leq n$. Therefore $a_i \in Z(R)$ and hence $f \in Z(R)[t]$. As commutative nil rings are locally nilpotent, so is the polynomial ring $Z(R)[t]$ and this implies that $R[t]$ is nil.

Now let the additive group of R be a p-group for a prime p and assume that $f = a_0 + a_1 t + \cdots + a_n t^n$ is a non-nilpotent polynomial of $R[t]$. Evidently without loss of generality we can suppose that R is generated by the coefficients $a_0, a_1, \dots, a_n$ of f. Then there exists a positive integer k such that $p^k a_i = 0$ for all $1 \leq i \leq n$ and so $p^k R = 0$. Since the factor ring R/pR is an algebra over the Galois field $GF(p)$, the polynomial ring $(R/pR)[t]$ is nil by proved above and so $f^l \in pR$ for an integer $l \geq 1$. Therefore $f^l = px$ for some $x \in R$ and hence $f^{kl} = p^k x^k \in p^k R = 0$, contrary to the assumption. Thus $R[t]$ is nil because R is the direct sum of its primary components. □

We do not now at present, *whether the condition of periodicity of the additive group in this theorem can be omitted.*

5. Braces, radical modules and triply factorized groups

Let P be a right ideal of a radical ring R, so that $M = R/P$ is a right R-module. Then $A = R^\circ$ operates on M via the rule $m^a = m + ma$ for a in A and m in M. If $G = M \rtimes A$ is the semidirect product of M by A and $B = \{am \mid m = a + P,\ a \in A\}$, then B is a subgroup of G and

$$G = M \rtimes A = M \rtimes B = AB$$

with the intersection $A \cap B$ isomorphic to P°. This was a reason for the author[55] to say that a module M over a group A is *radical* if the semidirect product $G = M \rtimes A$ has the above triple factorization.

Recall that if M is an A-module, then a mapping $d : A \to M$ is called a *derivation* (or 1-cocycle) if $d(ab) = d(a)^b + d(b)$ for all $a, b \in A$. We say that M is of exponent n if $nM = 0$ for a positive integer n which is minimal with this property or of exponent 0 otherwise. Note that if M is of exponent $n \geq 0$ and $K = \mathbb{Z}/n\mathbb{Z}$, then every A-module can be regarded as a module over the group ring KA.

The next theorem proved in Ref. 55 describes radical modules from two standpoints, that of the ring theory and that of the group theory, and motives their study.

Theorem 5.1. *Let A be a group, M an A-module of exponent $n \geq 0$ and $\Delta_K(A)$ the augmentation ideal of the group ring KA with $K = \mathbb{Z}/n\mathbb{Z}$. Then the following statements are equivalent:*

(1) the module M is radical;
(2) there exists a surjective derivation $d : A \to M$;
(3) the module M as a right KA-module is isomorphic to the module $\Delta_K(A)/\mathfrak{a}$ for a right ideal $\mathfrak{a}$ of KA such that

$$\Delta_K(A) = A - 1 + \mathfrak{a};$$

(4) in the semidirect product $G = M \rtimes A$ there exists a subgroup B such that $G = M \rtimes A = M \rtimes B = AB$

Moreover, the right ideal $\mathfrak{a}$ and the subgroup B can be chosen such that $\ker d = A \cap (1 + \mathfrak{a}) = A \cap B$.

Of course, if $\mathfrak{a}$ is an ideal of KA, for instance when A is abelian, then the module $\Delta_K(A)/\mathfrak{a}$ is in fact a radical factor ring of the ring $\Delta_K(A)$. In particular, if in addition $A \cap (1 + \mathfrak{a}) = 1$, then A and M are isomorphic to the adjoint and additive group of the radical ring $\Delta_K(A)/\mathfrak{a}$, respectively. However in general the right ideal $\mathfrak{a}$ need not be an ideal. For example, if $K = \mathbb{Z}/2\mathbb{Z}$ and $A = \langle a, b \mid a^4 = b^2 = (ab)^2 = 1\rangle$ is the dihedral group of oder 8, then the right ideal $\mathfrak{p} = (1 + a + a^2 + ab)KA$ is not any ideal in KA and it is easy to verify that $\Delta_K(A) = A - 1 + \mathfrak{p}$ and $A \cap (1 + \mathfrak{a}) = 1$.

A detailed exposition of results obtained by an application of this technique of radical modules for studying factorized groups can be found in Ref. 5.

The following notion which was introduced by Rump in Ref. 46 shows that the case when $A \cap (1 + \mathfrak{a}) = 1$ and so there exists a bijective derivation $d : A \to M$ is of special interest.

Definition 5.1. An additive abelian group V with a multiplication $V \times V \to V$ is called a (right) *brace* if for all $u, v, w \in A$ the following statements hold:

(1) $(u + v)w = uw + vw$;
(2) $u(vw + v + w) = (uv)w + uv + uw$;
(3) the map $v \mapsto vu + v$ is bijective.

Note that soon after this a similar definition named as a *radical circle algebra* was independently introduced by Catino and Rizzo[17].

Clearly every radical ring R is a brace under the addition and multiplication in R. As in R, the set of all elements of any brace V forms a group with neutral element 0 under the *adjoint multiplication* $u \circ v = u + v + uv$ which will also be called the *adjoint group* of V and denoted by V°. On the other hand, every brace whose multiplication is two-sided distributive is in fact a radical ring.

Example. Let A be a group and $\mathfrak{a}$ a right ideal of the group ring $\mathbb{Z}A$ such that $A \cap (1 + \mathfrak{a}) = 1$ and $\Delta_{\mathbb{Z}}(A) = A - 1 + \mathfrak{a}$. Then the factor group $V = \Delta_{\mathbb{Z}}(A)/\mathfrak{a}$ is an A-module and the mapping $d : A \to V$ given by $d(a) = a - 1 + \mathfrak{a}$ for $a \in A$ determines a bijective derivation from A onto V. It is easily verified that V becomes a brace under the multiplication $uv = ud^{-1}(v) - u$ for all $u, v \in V$ and its adjoint group V° is isomorphic to A. In this case we will say that the brace V is *determined by the right ideal* $\mathfrak{a}$ *of the group ring* $\mathbb{Z}A$.

The following immediate consequence from Theorem 5.1 shows in particular that each brace V can be characterized as a brace determined by a right ideal of the integer group ring $\mathbb{Z}V^{\circ}$ and so it is in fact a special case of our concept of radical modules.

Corollary 5.1. *Let A be a group and V an A-module. Then the following statements are equivalent:*

(1) V is a brace whose adjoint group is isomorphic to A;
(2) there exists a bijective derivation $d : A \to V$ such that $u \cdot v = ud^{-1}(v) - u$ for all $u, v \in V$;
(3) the integer group ring $\mathbb{Z}A$ contains a right ideal $\mathfrak{a}$ such that V is the brace determined by $\mathfrak{a}$;
(4) in the semidirect product $G = V \rtimes A$ there exists a subgroup B such that $G = V \rtimes A = V \rtimes B = AB$ and $A \cap B = 1$.

The concepts of *subbraces*, *right ideals* and a *brace homomorphisms* for braces are defined by the same way as for rings. In particular, if V and W are braces and $\lambda : V \to W$ is a brace homomorphism, then the kernel $\ker \lambda$ and the image $\lambda(V)$ are subbraces of V and W, respectively. Furthermore, it is also easily seen that the adjoint group of $\ker \lambda$ is a normal subgroup of V°. A subbrace I of V is an *ideal* in V if $I = \ker \lambda$ for some brace homomorphism λ. The following proposition whose proof is a routine describes these entities in the language of group rings.

Proposition 5.1. *Let* $V = \Delta_{\mathbb{Z}}(A)/\mathfrak{a}$ *be a brace determined by a right ideal* $\mathfrak{a}$ *of the integer group ring* $\mathbb{Z}A$*, so that* $\Delta_{\mathbb{Z}}(A) = A-1+\mathfrak{a}$ *and* $A \cap (1+\mathfrak{a}) = 1$*. If* I *is an additive subgroup of* V *and thus* $I = \mathfrak{b}/\mathfrak{a}$ *for some subgroup* $\mathfrak{b}$ *of* $\Delta_{\mathbb{Z}}(A)$*, then* $\mathfrak{b} = X-1+\mathfrak{a}$ *for a subset* X *of* A *and the following statements hold:*

(1) I *is a subbrace of* V *if and only if the set* X *is a subgroup of* A*;*
(2) I *is a right ideal of* V *if and only if* $\mathfrak{b}$ *is a right ideal of* $\mathbb{Z}A$*;*
(3) I *is an ideal of* V *if and only if* I *is a right ideal of* V *and* X *is a normal subgroup of* A*.*

As it follows from Corollary 5.1, each so called *involutive Yang-Baxter group* (IYB group) G introduced in Ref. 18 is nothing more than a complement in a triple factorized semidirect product $M \rtimes G = M \rtimes H = GH$ with an abelian normal subgroup M because G is isomorphic to the adjoint group of a brace. It should be noted that several properties of IYB groups proved in Ref. 18 follow immediately from this fact and properties of factorized groups. A different definition of G can also be given in the language of group rings, namely

G *is an IYB group if and only if there exists an one-sided ideal* $\mathfrak{g}$ *of the integer group ring* $\mathbb{Z}G$ *such that* $\Delta_{\mathbb{Z}}(G) = G-1+\mathfrak{g}$ *and* $G \cap (1+\mathfrak{g}) = 1$*.*

Of course, if G is finite, then we can replace the ring $\mathbb{Z}$ by $\mathbb{Z}/n\mathbb{Z}$ for a suitable integer n which is a divisor of the order of G.

References

1. S. I. Adjan, The Burnside problem and identities in groups, Springer-Verlag, Berlin (1978).
2. B. Amberg, O. Dickenschied and Ya. P. Sysak, Subgroups of the adjoint group of a radical ring, Can. J. Math., 50 (1998), 3 - 15.
3. B. Amberg, L. Kazarin, Nilpotent p-algebras and factorized p-groups, Groups St. Andrews 2005. Vol. 1, 130 - 147, London Math. Soc. Lecture Note Ser., 339, Cambridge Univ. Press, Cambridge, 2007.
4. B. Amberg, L. Kazarin, On the central series of the adjoint group of a nilpotent p-algebra, Publ.Math. Debrecen, 63 (2003), 473 - 482.
5. B. Amberg, Ya. P. Sysak, Radical rings and products of groups, in: Groups St. Andrews 1997 in Bath, I, 1 - 19, London Math. Soc. Lecture Note Ser., 260, Cambridge Univ. Press, Cambridge, 1999.
6. B. Amberg, Ya.P. Sysak, Radical rings with Engel conditions, J. Algebra, 231 (2000), 364 - 373.
7. B. Amberg, Ya. P. Sysak, Associative rings whose adjoint semigroup is locally nilpotent, Arch. Math. 76 (2001), 426 - 435.

8. B. Amberg, Ya.P. Sysak, Radical rings with soluble adjoint groups, J. Algebra, 247 (2002), 692 - 702.
9. B. Amberg, Ya.P. Sysak, On associative rings with locally nilpotent adjoint semigroup, Comm. Algebra 31 (2003), 123 - 132.
10. B. Amberg, Ya.P. Sysak, Associative rings with metabelian adjoint group, J. Algebra, 277 (2004), 456 - 473.
11. B. Amberg, Ya. P. Sysak, Semilocal rings with n-Engel multiplicative group, Arch. Math. 83 (2004), 416 - 421.
12. S. A. Amitsur, Radicals of polynomials rings, Canad. J. Math. 8 (I956), 355 - 361.
13. A. Braun, The nilpotency of the radical in finitely generated PI-rings, J. Algebra 89 (1984), 375 - 396.
14. F. Catino, On the centers of a radical ring, Arch. Math. 60 (1993), 330 - 333.
15. F. Catino, M. M. Miccoli and Ya. P. Sysak, On the adjoint group of semiprime rings, Comm. Algebra, 35 (2007), 265 - 270.
16. F. Catino, M. M. Miccoli and Ya. P. Sysak, Semilocal rings whose adjoint group is locally supersoluble, Arch. Math. 95 (2010), 213 - 224
17. F. Catino, R. Rizzo, Regular subgroups of the affine group and radical circle algebras, Bull. Aust. Math. Soc. 79 (2009), 103 - 107.
18. F. Cedó, E. Jespers, Á. del Río, Involutive Jang-Baxter groups, Trans. Amer. Math. Soc. 362(2010), 2541 - 2558.
19. P. M. Cohn, The embedding of radical rings in simple radical rings, Bull. London Math. Soc., 3 (1971), 185 - 188 (Corr., ibid 4 (1972), 54; ibid 5 (1973), 322).
20. O. Dickenschied, On the adjoint group of some radical rings, Glasgow Math. J. 39 (1997), 35 - 41.
21. X. Du, The centers of a radical ring, Can. Math. Bull. 35 (1992), 174 - 179.
22. X. Du, The upper central chain of a radical ring, Northeast. Math. J. 17 (2001), 130 - 132.
23. N. H. Eggert, Quasi-regular groups of finite commutative nilpotent algebras, Pacific J. Math. 36 (1971), 631 - 634.
24. I. Fischer, K. Eldridge, Artinian rings with cyclic quasiregular group, Duke Math. J. 36 (1969), 43 - 47.
25. E. S. Golod, On nil-algebras and residually finite groups, Izv. Akad. Nauk SSSR Ser. Mat., 28 (1964), 273 - 276
26. K. W. Gruenberg, The Engel elements of a soluble group, Illinois J. Math. 3 (1959), 151 - 168.
27. L. Hammoudi, Eggert's conjecture on the dimensions of nilpotent algebras, Pacific J. Math. 202 (2002), 93 - 97 (Corr., ibid. 220 (2005), 197).
28. G. Higman, B. H. Neumann and H. Neumann, Embedding theorems for groups, J. London Math. Soc. 24 (1949), 247 - 254.
29. Jacobson N., Structure of rings, AMS Coll. Publ. 37, Providence R.I., 1964.
30. S.A. Jennings, Radical rings with nilpotent associated groups, Trans. Roy. Soc. Can. (ser. III) 49 (1955), 31 - 38.
31. O. H. Kegel, B. A. F. Wehrfritz, Locally finite groups, North-Holland, Amsterdam, 1973.

32. M. Korbelár, 2-Generated nilpotent algebras and Eggert's conjecture, J. Algebra 324 (2010), 1558 - 1576.
33. A. N. Krasil'nikov, On the group of units of a ring whose associated Lie ring is metabelian, Russian Math. Surveys 47 (1992), 214 - 215.
34. A. N. Krasil'nikov, On the semigroup and Lie nilpotency of associative algebras, Mat. Zametki 62 (1997), 510 - 519.
35. R. L. Kruse, A note on the adjoint group of a finitely generated radical ring, J. London Math. Soc. (2), 1 (1969), 743 - 744.
36. H. Laue, On the associated Lie ring and the adjoint group of a radical ring, Canad. Math. Bull. 27 (1984), 215 - 222.
37. P. Longobardi, M. Maj, Semigroup identities and Engel groups, in: Groups, St. Andrews, 1997 in Bath, II, 527 - 531, London Math. Soc. Lecture Note Ser., 261, Cambridge Univ. Press, 1999.
38. A. I. Mal'cev, Nilpotent semigroups, Uch. Zap. Ivanov. Gos. Ped. Inst., Ser. Fiz.-Mat. Nauki, 4 (1953), 107 - 111.
39. K.R. McLean, Eggert's conjecture on nilpotent algebras, Comm. Algebra 32 (2004), 997 - 1006.
40. K.R. McLean, Graded nilpotent algebras and Eggert's conjecture, Comm. Algebra 34 (2006), 4427 - 4439.
41. B. H. Neumann, T. Taylor, Subsemigroups of nilpotent groups, Proc. Roy. Soc. London (Ser. A), 274 (1963), 1 - 4.
42. O. M. Neroslavskii, Structures that are connected with radical rings (Russian), Vesci Akad. Navuk BSSR, Ser. Fiz.-Mat. Navuk, 134 (1973), No. 2, 5 - 10.
43. A. Yu. Ol'shanskii, A simplifacation of Golod's example, Proceedings of "Groups - Korea 1994", de Gruyter Verlag, Berlin, 263 - 265.
44. D. M. Riley, Engel varieties of associative rings and the number of Mersenne primes, J. Algebra 261 (2003), 19 - 30.
45. D. M. Riley, V. Tasic, Mal'cev nilpotent algebras, Arch. Math. 72 (1999), 22 - 27.
46. W. Rump, Braces, radical rings, and the quantum Yang-Baxter equation, J. Algebra 307 (2007), 153 - 170.
47. E. Sasiada, P. M. Cohn, An example of a simple radical ring, J. Algebra 5 (1967), 373 - 377.
48. W. R. Scott, On the multiplicative group of a division ring, Proc. Amer. Math. Soc. 8 (1957), 303 - 305.
49. R. K. Sharma, J. B. Srivastava, Lie solvable rings, Proc. Amer. Math. Soc. 94 (1985), 1 - 8.
50. R. K. Sharma, J. B. Srivastava, Lie centrally metabelian group rings, J. Algebra 151 (1992), 476 - 486.
51. A.I. Shirshov, On certain near Engel groups, Algebra i Logika, 2 (1963), 5 - 18.
52. A. Smoktunowicz, Polynomial rings over nil rings need not be nil, J. Algebra 233 (2000), 427 - 436.
53. A. Smoktunowicz, A simple nil ring exists, Comm. Algebra 30 (2002), 27 - 59.

54. A. Smoktunowicz, E.R. Puczylowski. A polynomial ring that is Jacobson radical and not nil. Israel J. Math. 124 (2001), 317 - 325
55. Ya. P. Sysak, Some examples of factorized groups and their relation to ring theory, Proceedings of "Infinite Groups 1994", de Gruyter Verlag, Berlin (1996), 257 - 269
56. Ya. P. Sysak, Periodic subgroups of the adjoint group of a radical ring, Topics in infinite groups, 327 - 333, Quad. Mat., 8, Dept. Math., Seconda Univ. Napoli, Caserta, 2001.
57. G. Traustason, Semigroup identities in 4-Engel groups, J. Group Theory 2 (1999), 36 - 39.
58. Unsolved Problems in Group Theory: The Kourovka Notebook, Novosibirsk, 1995.
59. A. I. Valitskas, Examples of radical rings not imbeddable in simple radical rings, Soviet Math. Dokl. 37 (1988), 449 - 451.
60. J. F. Watters, On the adjoint group of a radical ring, J. London Math. Soc., 43 (1968), 725 - 729.
61. D. I. Zaitsev, Factorization of polycyclic groups, Mat. Zametki 29 (1981), 481 - 490.
62. A. E. Zalesskii, M. B. Smirnov, Associative rings satisfying an identity of Lie solubility, Vestsi Akad. Nauk. BSSR, Ser. Fiz.-Mat. Nauk 2 (1982), 15 - 20.
63. E. Zelmanov, The solution of the restricted Burnside problem for groups of odd exponent, Izv. Akad. Nauk SSSR, Ser. Mat., 54 (1990), 42 - 59.
64. E. Zelmanov, The solution of the restricted Burnside problem for 2-groups, Mat. Sb. 182 (1991), 568 - 592.

ON THE GORENSTEIN DIMENSION OF SOLUBLE GROUPS

O. TALELLI*

Department of Mathematics, University of Athens, Panepistimiopolis, GR-157 84, Athens, Greece
E-mail: otalelli@math.uoa.gr

The Gorenstein dimension of a group G over $\mathbb{Z}$ generalizes the cohomological dimension of G over $\mathbb{Z}$ and can be finite even if the group G has torsion. Here we relate the Gorenstein dimension of a soluble group G over $\mathbb{Z}$ to the Hirsch number of G. The inequality we obtain generalizes an inequality obtained by Gedrich and Gruenberg[3] which relates the Hirsch number of a soluble group G to $\text{spli}\mathbb{Z}G$, where $\text{spli}\mathbb{Z}G$ is the supremum of the projective dimensions of the injective $\mathbb{Z}G$-modules.

Keywords: Gorenstein dimension of a group; (co)homological dimension; Hirsch number

1. Introduction

The Gorenstein projective dimension of an R-module M, Gpd_RM, is a refinement of the projective dimension of M over R, and was defined by Enochs and Jenda in Ref. 1. This concept goes back to the concept of a relative homological dimension, called the G-dimension, which was defined by Auslander in Ref. 2 for finitely generated modules over commutative Noetherian rings and provided a characterization of the Gorenstein rings.

The Gorenstein dimension of a group G over $\mathbb{Z}$, $\text{Gcd}_{\mathbb{Z}}G$, is the Gorenstein projective dimension of $\mathbb{Z}$, as a trivial $\mathbb{Z}G$-module. It turns out that $\text{Gcd}_{\mathbb{Z}}G$ coincides with the cohomological dimension of G over $\mathbb{Z}$, $\text{cd}_{\mathbb{Z}}G$, when this is finite, and is related to the invariants $\text{spli}\mathbb{Z}G$, $\text{silp}\mathbb{Z}G$ and the generalized cohomological dimension of G, $\underline{\text{cd}}G$, where $\text{spli}\mathbb{Z}G$ is the supremum of the projective dimensions of the injective $\mathbb{Z}G$-modules, $\text{silp}\mathbb{Z}G$ is the supremum of the injective dimensions of the projective $\mathbb{Z}G$-modules,

*Research funded by the University of Athens Special Research Account, grant 70/4/6411.

and

$$\underline{\mathrm{cd}}G = \sup\{n \in \mathbb{N} |\, \exists\, M, P : \mathrm{Ext}^n_{\mathbb{Z}G}(M, P) \neq 0,\ M\ \mathbb{Z}\text{-free},\ P\ \mathbb{Z}G\text{-projective}\}.$$

These invariants were considered by Gedrich-Gruenberg[3] and Ikenaga[4] in their study of extending the Farrell-Tate cohomology, which is defined for the class of groups of finite virtual cohomological dimension, to a larger class of groups.

In Ref. 5 the finiteness of $\mathrm{Gcd}_{\mathbb{Z}}$ was proposed as an algebraic characterization of those groups G which admit a finite dimensional model for $\underline{E}G$, the classifying space for proper actions, equivalently as an algebraic characterization of those groups G which admit a resolution of finite length by direct sums of permutation modules induced by finite subgroups of G.

Here we study the Gorenstein dimension of soluble groups and in particular how this is related to the Hirsch number of the group. Gedrich and Gruenberg[3] proved the following Theorem which relates the Hirsch number, $\mathrm{h}G$, of a soluble group G to $\mathrm{spli}\mathbb{Z}G$.

Theorem: (Ref. 3) Let G be a soluble group and T the (unique) maximal normal torsion subgroup of G. Then $\mathrm{h}G \leq \mathrm{spli}\mathbb{Z}G \leq 2 + \mathrm{h}G + \mathrm{spli}\mathbb{Z}T$.

Our main result here relates the Hirsch number, $\mathrm{h}G$, of a soluble group G to $\mathrm{Gcd}_{\mathbb{Z}}G$:

Theorem: (Theorem 3.1) Let G be a soluble group and T the (unique) maximal normal torsion subgroup of G. Then

$$\mathrm{h}G \leq \mathrm{Gcd}_{\mathbb{Z}}G \leq \mathrm{Gcd}_{\mathbb{Z}}T + \mathrm{h}G + 1$$

Note that $\mathrm{Gcd}_{\mathbb{Z}}G \leq \mathrm{spli}\mathbb{Z}G \leq \mathrm{Gcd}_{\mathbb{Z}}G + 1$.

Both the above theorems use the characterization of the Hirsch number of a soluble group as the homological dimension of the group over the rationals (Ref. 6).

We believe that for soluble groups G, the Gorenstein dimension of G over $\mathbb{Z}$ coincides with the cohomological dimension of G over $\mathbb{Q}$, i.e. $\mathrm{cd}_{\mathbb{Q}}G = \mathrm{Gcd}_{\mathbb{Z}}G$. This is not true in general, as there exist groups G with $\mathrm{cd}_{\mathbb{Q}}G < \mathrm{Gcd}_{\mathbb{Z}}G$.

2. Gorenstein dimension

Let R be an arbitrary ring with unit element $1 \neq 0$. All modules considered are left R-modules.

An R-module M is said to admit a complete resolution $(\mathcal{F}, \mathcal{P}, n)$ if there is an acyclic complex $\mathcal{F} = \{(F_i, \vartheta_i)|\ i \in \mathbb{Z}\}$ of projective modules, and a projective resolution $\mathcal{P} = \{(P_i, d_i)|\ i \in \mathbb{Z}, i \geq 0\}$ of M such that $\mathcal{F}$ and $\mathcal{P}$ coincide in dimensions greater than n

$$\begin{array}{llllll}
\mathcal{F}: & \cdots \to F_{n+1} \to & F_n \overset{\vartheta_n}{\to} F_{n-1} \to & \cdots & \to F_0 \to F_{-1} \to F_{-2} \to \cdots \\
 & \qquad \| & \quad \| & & \\
\mathcal{P}: & \cdots \to P_{n+1} \to & P_n \overset{d_n}{\to} P_{n-1} \to & \cdots & \to P_0 \to M \to 0
\end{array}$$

The number n is called the coincidence index of the complete resolution.

An R-module M is said to admit a complete resolution in the strong sense if there is a complete resolution $(\mathcal{F}, \mathcal{P}, n)$ with $\mathrm{Hom}_{\mathbb{Z}G}(\mathcal{F}, Q)$ acyclic for every $\mathbb{Z}G$-projective module Q. It is not known whether a $\mathbb{Z}G$-module M that admits a complete resolution, also admits a complete resolution in the strong sense. However, this is so for a big class of groups which contains the class of soluble groups (Ref. 7).

Definition: An R-module M is called Gorenstein projective if it admits a complete resolution in the strong sense of coincidence index 0.

If A is an R-module and n a non-negative integer, then the Gorenstein projective dimension of A is less than or equal to n, $\mathrm{Gpd}_R A \leq n$, if there is a resolution of A by Gorenstein projective R-modules of length n.

Examples:

Clearly every projective R-module is a Gorenstein projective R-module, hence $\mathrm{Gpd}_R M \leq \mathrm{pd}_R M$. In particular $\mathrm{Gcd}_{\mathbb{Z}} G \leq \mathrm{cd}_{\mathbb{Z}} G$.

If G is a finite group then every $\mathbb{Z}G$-module which is $\mathbb{Z}$-free is a Gorenstein projective $\mathbb{Z}G$-module.

Indeed, if G is a finite group then there is a $\mathbb{Z}$-split $\mathbb{Z}G$-exact sequence

$$\begin{array}{c} 0 \to \mathbb{Z} \to \mathbb{Z}G \to K \to 0 \\ 1 \mapsto \sum g \end{array} \tag{1}$$

and if B is any $\mathbb{Z}G$-module, then tensoring (1) with B gives a $\mathbb{Z}G$-exact sequence

$$0 \to \grave{B} \to \grave{B} \otimes \grave{\mathbb{Z}G} \to \grave{B} \otimes \grave{K} \to 0$$

where $\grave{B} \otimes \grave{C}$ indicates diagonal action .

If B is $\mathbb{Z}$-free then $\grave{B} \otimes \grave{\mathbb{Z}G}$ is a projective $\mathbb{Z}G$-module. So if

$$\cdots \to P_n \to P_{n-1} \to \cdots \to P_0 \to \mathbb{Z} \to 0$$

is a projective resolution of G. i.e. a $\mathbb{Z}G$-projective resolution of the trivial

$\mathbb{Z}G$-module $\mathbb{Z}$, and A is a $\mathbb{Z}G$-module which is $\mathbb{Z}$-free, then

$$(*) \cdots \to \grave{P}_n \otimes \grave{A} \to \cdots \to \grave{P}_0 \otimes \grave{A} \longrightarrow \mathbb{Z}\grave{G} \otimes \grave{A} \longrightarrow \mathbb{Z}\grave{G} \otimes (\grave{K} \otimes \grave{A}) \to \cdots$$
$$\searrow \quad \nearrow \searrow \quad \nearrow$$
$$A \qquad \grave{K} \otimes \grave{A}$$

is a complete resolution of A of coincidence index 0.

Now if

$$\mathcal{F}: \quad \cdots \longrightarrow \bar{P}_n \xrightarrow{\vartheta_n} \bar{P}_{n-1} \longrightarrow \cdots \longrightarrow \bar{P}_0 \xrightarrow{\vartheta_0} P_{-1} \xrightarrow{\vartheta_{-1}} \cdots$$

is an acyclic complex of projective $\mathbb{Z}G$-modules then for any projective $\mathbb{Z}G$-module Q, the cochain complex $\mathrm{Hom}_{\mathbb{Z}G}(\mathcal{F}, Q)$ is exact since $\mathrm{Ext}^1_{\mathbb{Z}G}(B, Q) = 0$ for any $\mathbb{Z}G$-module B which is $\mathbb{Z}$-free and

$$\mathrm{H}^i(\mathrm{Hom}_{\mathbb{Z}G}(\mathcal{F}, Q)) \simeq \mathrm{Ext}^1_{\mathbb{Z}G}(K_{i-1}, Q),$$

where $K_{i-1} = \mathrm{im}\vartheta_{i-1}$, $i \in \mathbb{Z}$. Hence (*) is a complete resolution of A in the strong sence of coincidence index 0, thus A is a Gorenstein projective $\mathbb{Z}G$-module.

It turns out (Ref. 5) that

(a) If K is a subgroup of a group G and M a Gorenstein projective $\mathbb{Z}K$-module, then the induced module $\mathbb{Z}\grave{G} \otimes_{\mathbb{Z}K} M$ is a Gorenstein projective $\mathbb{Z}G$-module.
(b) If $0 \to M' \to M \to M'' \to 0$ is a short exact sequence of $\mathbb{Z}G$-modules and M'' is Gorenstein projective, then M' is Gorenstein projective if and only if M is Gorenstein projective. Moreover, direct sums and direct summands of Gorenstein projective modules are Gorenstein projective (Theorem 2.5 in Ref. 8).

It was also shown in Ref. 5 that the Gorenstein dimension of G enjoys the following properties:

Theorem 2.1.

(1) If $H \leq G$ then $\mathrm{Gcd}_{\mathbb{Z}}H \leq \mathrm{Gcd}_{\mathbb{Z}}G$.
(2) If $1 \to N \to G \to K \to 1$ is an extension of groups, then $\mathrm{Gcd}_{\mathbb{Z}}G \leq \mathrm{Gcd}_{\mathbb{Z}}N + \mathrm{Gcd}_{\mathbb{Z}}K$.
(3) If F is a finite subgroup of G, and $N_G(F)$ its normalizer in G, then $\mathrm{Gcd}_{\mathbb{Z}}(N_G(F)/F) \leq \mathrm{Gcd}_{\mathbb{Z}}G$.
(4) If $1 \to N \to G \to K \to 1$ is an extension of groups with $|N| < \infty$ then $\mathrm{Gcd}_{\mathbb{Z}}G = \mathrm{Gcd}_{\mathbb{Z}}K$.
(5) If $1 \to N \to G \to K \to 1$ is an extension of groups with $|K| < \infty$ then $\mathrm{Gcd}_{\mathbb{Z}}G = \mathrm{Gcd}_{\mathbb{Z}}N$.

In Ref. 8 the following characterization of finite Gorenstein projective dimension is given

Theorem 2.2. *Let M be an R-module with $\mathrm{Gpd}_R M < \infty$, and n an integer. Then the following conditions are equivalent:*

(i) $\mathrm{Gpd}_R M \leq n$.
(ii) $\mathrm{Ext}^i_R(M, L) = 0$ *for all $i > n$ and all R-modules L with finite projective dimension.*
(iii) $\mathrm{Ext}^i_R(M, Q) = 0$ *for all $i > n$ and all projective R-modules Q.*
(iv) For every exact sequence $0 \to K_n \to G_{n-1} \to \cdots \to G_0 \to M \to 0$, where $G_0, \ldots, G_{n-1}$ are Gorenstein projectives, then also K_n is Gorenstein projective.

Consequently, the Gorenstein projective dimension of M is determined by the formulas:

$$\mathrm{Gpd}_R M = \sup\{i \in \mathbb{N} \mid \exists L \text{ with } \mathrm{pd}_R L < \infty : \mathrm{Ext}^i_R(M, L) \neq 0\}$$

$$= \sup\{i \in \mathbb{N} \mid \exists Q \text{ } R\text{-projective} : \mathrm{Ext}^i_R(M, Q) \neq 0\}.$$

This in particular implies that if a group G has finite cohomological dimension over $\mathbb{Z}$ then this coincides with the Gorenstein dimension of G, i.e. if $\mathrm{cd}_{\mathbb{Z}} G < \infty$ then $\mathrm{cd}_{\mathbb{Z}} G = \mathrm{Gcd}_{\mathbb{Z}} G$.

More generally it was shown in Ref. 5 that for any group G the Gorenstein dimension of G, $\mathrm{Gcd}_{\mathbb{Z}} G$ coincides with the generalized cohomological dimension, $\underline{\mathrm{cd}} G$, defined by Ikenaga[4] as

$$\underline{\mathrm{cd}} G = \sup\{n \in \mathbb{N} \mid \exists\, M, P : \mathrm{Ext}^n_{\mathbb{Z}G}(M, P) \neq 0,\ M\ \mathbb{Z}\text{-free},\ P\ \mathbb{Z}G\text{-projective}\}.$$

Remark: If a group G has torsion elements then its cohomological dimension over $\mathbb{Z}$ is infinite, however its Gorenstein dimension over $\mathbb{Z}$ could be finite. For example:

If G is finite then we have seen that $\mathrm{Gcd}_{\mathbb{Z}} G = 0$, actually $\mathrm{Gcd}_{\mathbb{Z}} G = 0$ if and only if G is finite (Ref. 9). If G is a countable infinite locally finite group then $\mathrm{Gcd}_{\mathbb{Z}} G = 1$ (Ref. 5). More generally, it was shown in Ref. 5 that $\mathrm{Gcd}_{\mathbb{Z}} G \leq 1$ if and only if G acts on a tree with finite vertex stabilizers. The proof of this result uses the Almost Stability Theorem of Dicks and Dunwoody[10].

If G is the free Burnside group $G = B(n, e)$ of odd exponent $e > 665$ then by a result of Ivanov[11] it follows that $\mathrm{Gcd}_{\mathbb{Z}} G \leq 2$ and by Ref. 5 it follows that $\mathrm{Gcd}_{\mathbb{Z}} G = 2$.

The finiteness of $\mathrm{Gcd}_{\mathbb{Z}}G$ is proposed in Ref. 5 as an algebraic characterization of those groups G which admit a finite dimensional model for $\underline{E}G$, the classifying space for proper actions of G, namely:

Conjecture A: The following statements are equivalent for a group G:

(1) G admits a finite dimensional model for $\underline{E}G$.
(2) G admits a resolution of the form

$$0 \to \bigoplus_{i_n \in I_n} \mathbb{Z}(G/G_{i_n}) \to \cdots \to \bigoplus_{i_0 \in I_0} \mathbb{Z}(G/G_{i_0}) \to \mathbb{Z} \to 0$$

where G_{i_j} are finite groups.
(3) G is of type Φ.
(4) For any $\mathbb{Z}G$-module M, if $\mathrm{H}^i(F, M) = 0$ for all $i \neq 0$ and all finite subgroups F of G, then $\mathrm{pd}_{\mathbb{Z}G}M < \infty$.
(5) $\mathrm{Gcd}_{\mathbb{Z}}G < \infty$.

Note that a group G is said to be of type Φ if for every $\mathbb{Z}G$-module M, $\mathrm{pd}_{\mathbb{Z}G}M < \infty$ if and only if $\mathrm{pd}_{\mathbb{Z}H}M < \infty$ for every finite subgroup H of G.

The following theorem (Ref. 12) generalizes a theorem of Kropholler and Mislin[13] :

Theorem 2.3. *If G is a group such that for every finite subgroup F of G the group $N_G(F)/F$ is of type Φ and there is a bound on the orders of the finite subgroups, then G admits a finite dimensional model for $\underline{E}G$.*

A consequence of this theorem is that Conjecture A holds for a big class of groups which contains in particular all soluble groups and all linear groups for which there is a bound on the orders of their finite subgroups.

A model for $\underline{E}G$ is a G-CW-complex X in which all stabilizers are finite and is such that either of the following conditions hold:

(1) for every finite subgroup F of G, the fixed point subcomplex X^F is contractible.
(2) X is a terminal object in the homotopy category of G-CW-complexes with finite stabilizers

It is not hard to show that any two models of $\underline{E}G$ are G-homotopy equivalent.

Recall that a G-CW-complex is a CW-complex X equipped with a G-action such that all elements of G map cells of X homeomorphically to cells, and such that whenever an element of G maps a cell to itself it does so via the identity map.

The space $\underline{E}G$ plays a significant role in the Baum-Connes conjecture (Ref. 14) which predicts that the K-homology of the reduces C^*-algebra of a group G can be computed as the equivariant K-homology of the classifying space for proper actions of G, $\underline{E}G$. The space $\underline{E}G$ plays also a role in the extension of the Atiyah-Segal Completion Theorem from finite to infinite groups (Ref. 15).

3. Gorenstein dimension and Hirsch number

The following proposition was essentially proved by Gedrich and Gruenberg[3] and its proof uses a result of Stammbach[6] which identifies the Hirsch number of a soluble group G as its homological dimension over the rationals, $\mathbb{Q}$.

Proposition 3.1. *Let G be a soluble group with* $\mathrm{Gcd}_{\mathbb{Q}}G < \infty$. *Then the Hirsch number of G, hG is finite.*

Proof. Assume that $\mathrm{h}G$ is infinite and consider the first group in the sequence of derived factors $1,\ G^{(m-1)},\ G^{(m-2)}/G^{(m-1)}, \cdots$ having infinite Hirsch number, i.e. there exists s such that $\mathrm{h}(G^{(s-1)}/G^{(s)})$ is infinite and $\mathrm{h}(G^{(s)})$ is finite. For each positive integer n we can find a subgroup H containing $G^{(s)}$ so that $H/G^{(s)}$ is free abelian of rank n. Hence

$$n \leq \mathrm{h}(H) < \infty.$$

By Stammbach's result $h(H) = \mathrm{hd}_{\mathbb{Q}}H$. Let $\mathrm{hd}_{\mathbb{Q}}H = \lambda$. Then there is an injective $\mathbb{Q}G$-module I such that $\mathrm{H}_{\lambda}(G, I) \neq 0$. This in particular implies that the injective $\mathbb{Q}G$-module I cannot have a projective resolution of length less than λ, thus $\mathrm{spli}\mathbb{Q}G \geq \lambda \geq n$. By Ref. 9 [Theorem 4.4] $\mathrm{spli}\mathbb{Q}G = \mathrm{silp}\mathbb{Q}G$. A proof analogous to the proof of Theorem 2.5 in Ref. 5 gives us now that $\mathrm{spli}\mathbb{Q}G = \mathrm{silp}\mathbb{Q}G = \mathrm{Gcd}_{\mathbb{Q}}G$. Hence $\mathrm{Gcd}_{\mathbb{Q}}G \geq n$. Since n was arbitrary it follows that $\mathrm{Gcd}_{\mathbb{Q}}G$ is infinite, a contradiction, thus $\mathrm{h}G$ is finite. □

Theorem 3.1. *Let G be a soluble group and T the unique maximal normal torsion subgroup of G. Then* $hG \leq \mathrm{Gcd}_{\mathbb{Z}}G \leq \mathrm{Gcd}_{\mathbb{Z}}T + hG + 1$.

Proof. It follows from Proposition 3.1 that it is enough to consider the case where $\mathrm{h}G$ is finite. In that case Malcev's structure theorem[16] says that there is a series

$$G_0 < G_1 < \cdots < G_n \leq G_{n+1} = G$$

where $G_i \lhd G$, $|G : G_n| < \infty$, G_0 is locally finite and G_i/G_{i-1}, $1 \leq i \leq n$, are torsion free abelian of finite torsion free rank. Now by Theorem 2.1

$$\mathrm{Gcd}_{\mathbb{Z}}G = \mathrm{Gcd}_{\mathbb{Z}}G_n \leq \mathrm{Gcd}_{\mathbb{Z}}G_0 + \mathrm{Gcd}_{\mathbb{Z}}(G_n/G_0).$$

Since G_n/G_0 is torsion free of finite Hirsch number it follows that it is countable and has finite cohomological dimension over $\mathbb{Z}$, and thus $\mathrm{Gcd}_{\mathbb{Z}}(G_n/G_0) = \mathrm{cd}_{\mathbb{Z}}(G_n/G_0) \leq \mathrm{h}G + 1$ (Ref. 17).

Hence by Theorem 2.1 we have that

$$(*) \quad \mathrm{Gcd}_{\mathbb{Z}}G \leq \mathrm{Gcd}_{\mathbb{Z}}T + \mathrm{h}G + 1.$$

Now $\mathrm{Gcd}_{\mathbb{Q}}G \leq \mathrm{Gcd}_{\mathbb{Z}}G$ and it follows essentially from Proposition 4.1 and Theorem 4.11 in Ref. 18 that if $\underline{\mathrm{hd}}_{\mathbb{Q}}G$ is the supremum of those integers m for which there is an injective $\mathbb{Q}G$-module I with $\mathrm{Tor}_m^{\mathbb{Q}G}(I, M) \neq 0$ for some $\mathbb{Q}G$-module M then $\underline{\mathrm{hd}}_{\mathbb{Q}}G = \mathrm{hd}_{\mathbb{Q}}G = \mathrm{h}G$ and $\underline{\mathrm{hd}}_{\mathbb{Q}}G \leq \mathrm{Gcd}_{\mathbb{Q}}G$.

Thus we havc

$$\mathrm{h}G = \mathrm{hd}_{\mathbb{Q}}G = \underline{\mathrm{hd}}_{\mathbb{Q}}G \leq \mathrm{Gcd}_{\mathbb{Q}}G \leq \mathrm{Gcd}_{\mathbb{Z}}G$$

and the result follows from (*). □

If T is a locally finite group of cardinality less than $\aleph_\omega$ then it was shown by Dicks-Kropholler[19] that T admits a finite dimensional $\underline{E}G$ which in particular implies that $\mathrm{Gcd}_{\mathbb{Z}}T < \infty$. This result together with Theorem 3.1 implies that

Corollary 3.1. *Let G be a soluble group of cardinality $< \aleph_\omega$. Then* $\mathrm{cd}_{\mathbb{Q}}G < \infty$ *if and only if* $\mathrm{Gcd}_{\mathbb{Z}}G < \infty$ *if and only if $hG < \infty$.*

Hillman[20] extended the notion of the Hirsch length of a soluble group to the class of elementary amenable groups. It was shown by Hillman and Linnell[21] that Stammbach's result extends to the class of elemantary amenable groups, i.e. the homological dimension over the rationals of an elemantary amenable group is its Hirsch length. Moreover Hillman-Linnell[21] and Wehrfitz[22] proved the corresponding of Malcev's theorem for elementary amenable groups of finite Hirsch length.

Using the above we have the corresponding Theorem for elementary amenable groups, namely

Theorem 3.2. *Let G be an elementary amenable group and T the unique maximal normal torsion subgroup of G. If hG is the Hirsch length of G then $hG \leq \mathrm{Gcd}_{\mathbb{Z}}G \leq \mathrm{Gcd}_{\mathbb{Z}}T + hG + 1$.*

Remark: If $\mathrm{cd}_{\mathbb{Q}}G < \infty$ then $\mathrm{cd}_{\mathbb{Q}}G \leq \mathrm{Gcd}_{\mathbb{Z}}G$. One does not expect $\mathrm{cd}_{\mathbb{Q}}G = \mathrm{Gcd}_{\mathbb{Z}}G$ since there is a group G with $\mathrm{cd}_{\mathbb{Q}}G = 2$ and $\mathrm{Gcd}_{\mathbb{Z}}G = \mathrm{cd}_{\mathbb{Z}}G = 3$ (Ref. 23). However, we believe that if G is an elementary amenable group then $\mathrm{Gcd}_{\mathbb{Z}}G = \mathrm{cd}_{\mathbb{Q}}G$.

References

1. E. E. Enochs and O. M. G. Jenda, Gorenstein injective and projective modules, Math. Z 220 (4) (1995), 611-633.
2. M. Auslander, Anneaux de Gorenstein, et torsion en algèbre commutative, Secrétariat mathématique, Paris, 1967, Séminaire d'Algèbre Commutative dirigé par Pierre Samuel, 1966/67. Texte rédigé, d'après des exposés de Maurice Auslander, par Marquerite Mangeney, Christian Peskine et Lucien Szpiro. École Normale Supérieure de Jeunes Filles.
3. T. V. Gedrich and K. W. Gruenberg, Complete cohomological functors on groups, Topology and its Applications 25 (1987) 203-223.
4. B. M. Ikenaga, Homological dimension and Farrell cohomology, J. Algebra 87 (1984), 422-457.
5. A. Bahlekeh, F. Dembegioti and O. Talelli, Gorenstein dimension and proper actions, Bull. London Math. Soc. 41 (2009) 859-871.
6. U. Stammbach, On the weak homological dimension of the group algebra of solvable groups. J. London Math. Soc. (2) 2 1970 567–570.
7. F. Dembegioti and O. Talelli, A note on complete resolutions, to appear in Proc. Amer. Math. Soc.
8. H. Holm, Gorenstein homological dimensions, J. Pure Appl. Algebra 189 (2004) 167-193.
9. I. Emmanouil, On certain cohomological invariants of groups, to appear in Advances in Mathematics.
10. W. Dicks and M. J. Dunwoody, Groups acting on graphs, Cambridge University Press (1989).
11. S. V. Ivanov, Relation modules and relation bimodules of groups, semigroups and associative algebras. Internat. J. Algebra Comput. 1 (1991), no. 1, 89–114.
12. O. Talelli, On groups of type Φ, Arch. Math. 89 (2007) 24-32.
13. P. H. Kropholler and G. Mislin, Groups acting on finite dimensional spaces with finite stabilizers, Comment. Math. Helv. 73 (1998) 122-136.
14. P. Baum, A. Connes and N. Higson, Classifying space for proper actions and K-theory of group C^*-algebras, (English summary) C^*-algebras: 1943–1993 (San Antonio, TX, 1993), 240-291, Contemp. Math., 167, Amer. Math. Soc., Providence, RI, 1994.
15. W. Luck and B. Oliver, The completion theorem in K-theory for proper actions of a discrete group, Topology 40 (2001), no. 3, 585-616.
16. A. I. Mal'cev, On certain classes of infinite solvable groups, Amer. Math. Soc. Transl. (2) 2 (1956), 1-21.
17. R. Bieri, Homological Dimension of Discrete Groups, 2nd ed., Queen Mary College Mathematics Notes, 1981.

18. I. Emmanouil and O. Talelli, On the flat length of injective modules, preprint 2010.
19. W. Dicks, P. H. Kropholler, I. J. Leary and S. Thomas, Classifying spaces for proper actions of locally finite groups, J. Group Theory 5 (2002), no. 4, 453-480.
20. J. A. Hillman, Elementary amenable groups and 4-manifolds with Euler characteristic 0, J. Austral. Math. Soc. Ser. A 50 (1991), no. 1, 160–170.
21. J. A. Hillman and P. A. Linnell, Elementary amenable groups of finite Hirsch length are locally-finite by virtually-solvable, J. Austral. Math. Soc. Ser. A 52 (1992), no. 2, 237–241.
22. Wehrfritz, B. A. F., On elementary amenable groups of finite Hirsch number, J. Austral. Math. Soc. Ser. A 58 (1995), no. 2, 219–221.
23. M. Bestvina, The virtual cohomological dimension of Coxeter groups, Geometric group theory, Vol. 1 (Sussex, 1991), 19-23, London Math. Soc. Lecture Note Ser., 181, Cambridge Univ. Press, Cambridge, 1993.

DECOMPOSITION NUMBERS FOR PROJECTIVE MODULES OF FINITE CHEVALLEY GROUPS

A. E. ZALESSKI

Departimento di Matematica e Applicazioni,
Università degli Studi di Milano-Bicocca, Milan, 20125, Italy
E-mail: alexandre.zalesski@gmail.com

This is an expository paper explaining certain aspects of the study of the decomposition matrices of Chevalley groups for the natural characteristic. It emphasizes on discussing some recent results on decomposing the product of the Steinberg character with certain generalized characters as a sum of the irreducible ordinary characters.

Keywords: Finite Chevalley groups; Indecomposable projective modules

1. Introduction

Let G be a finite group of order $|G|$ and p a prime number. Representations of G over an algebraically closed field of characteristic p are called **p-modular**, and those over the complex numbers are called **ordinary**. Let $\mathbb{Q}$ be the rational number field and let ε be a primitive $|G|$-root of unity. Any ordinary representation is equivalent to a representation ϕ over $\mathbb{Q}(\varepsilon)$, and moreover, over a maximal subring R of $\mathbb{Q}(\varepsilon)$ not containing $1/p$. In addition, R has a unique maximal ideal I such that $F = R/I$ is a finite field of characteristic p. So if $\phi(G) \subset GL(n, R)$ for some n, then the representation $\overline{\phi} : G \to GL(n, F)$ obtained from the natural projection $GL(n, R) \to GL(n, F)$ is called the **reduction of ϕ modulo** p. If ϕ is irreducible then $\overline{\phi}$ is not always irreducible. The multiplicities of irreducible representations of G over $\overline{F}$, the algebraic closure of F, as constituents of $\overline{\phi}$ are called the **decomposition numbers**. They depend only on G (not on R), and can be arranged in form of a matrix called the **decomposition matrix**. It lists the ordinary irreducible representations along the columns and the p-modular ones along the rows. Therefore, the decomposition numbers of a particular ordinary irreducible representation are in a respective row.

Thus, if χ is an ordinary irreducible representation and β is an irreducible p-modular representation then the (χ, β)-decomposition number is the entry at the intersection of the χ-th row and the β-th column.

Table 1 is an example of the decomposition matrix for $G = PSL(3,3)$ and $p = 3$. The row above the matrix gives the dimensions of the irreducible p-modular representations and the bottom row provides the dimensions of the respective PIM's (see Section 2). The first column headed by d gives the dimension of χ. Some entries of the decomposition matrix are empty. One may fill them by 0, and interpret 0 as the absence of the occurrence of β in the decomposition of χ.

Table 1: The decomposition matrix for $PSL(3,3)$

d	χ	β_1	β_2	β_3	β_4	β_5	β_6	β_7	β_8	St
		1	3	3	6	6	7	15	15	27
1	χ_1	1								
12	χ_2				1	1				
13	χ_3		1	1			1			
16	χ_4	1						1		
16	χ_5	1							1	
16	χ_6			1		1	1			
16	χ_7		1		1		1			
26	χ_8	1	1	1	1	1	1			
26	χ_9	1	1				1		1	
26	χ_{10}	1		1			1	1		
39	χ_{11}	2					1	1	1	
27	St									1
		$7 \cdot 3^3$	$3 \cdot 3^3$	$3 \cdot 3^3$	$2 \cdot 3^3$	$2 \cdot 3^3$	$6 \cdot 3^3$	$3 \cdot 3^3$	$3 \cdot 3^3$	3^3

The matrix depends on the ordering of the irreducible representations, both ordinary and modular. If p is coprime to $|G|$, the order of G, then for a suitable ordering the decomposition matrix is the identity matrix. In general, one can try to reorder the rows and the columns to obtain a "block-diagonal" matrix $\text{diag}(D_1, \ldots, D_k)$ with maximum possible number k of blocks. One says that an ordinary (resp., p-modular) irreducible representation belongs to the block D_i if the respective row (resp., column) has a non-zero entry within D_i. If D_i is of size 1, the block D_i is called **of defect** 0.

For applications it is important to know the decomposition matrices for various groups. The theory of modular representations develops methods for understanding properties of the decomposition matrix. The theory of blocks developed by R. Brauer is one of the most powerful in the theory. However, this does not help much for simple groups such as $PSL(d,p)$ as in this case there are only two blocks, one of defect 0.

There is another approach, via principal indecomposable $\overline{F}G$-modules, also developed by R. Brauer. (Here $\overline{F}G$ denotes the group algebra of a group G over $\overline{F}$.)

In this survey I outline some recent results on the decomposition matrices of finite Chevalley groups, and formulate some open problems. A rather detailed survey of the current state of the theory with a complete list of relevant references is available in Humphreys' book [10] (see also the earlier version [9]).

I tried to make the exposition accessible for a non-expert in the area. A reader who is not familiar with Chevalley groups may think of $GL(n,q)$ as the main representative of the groups discussed in the paper, as the problems discussed here are open for this group as well.

Notation and standard definitions

The reader is expected to be familiar with elementary facts of the representation theory of finite groups. We denote by $\mathbb{Z}$ the ring of integers, by $\mathbb{Q}$ the rational number field, and by $\mathbb{C}$ the complex number field. A finite field of order q is denoted by $\mathbb{F}_q$, and $\overline{\mathbb{F}}_q$ is its algebraic closure. By $\overline{F}$ we denote an algebraically closed field of prime characteristic p, for most situations one may assume $\overline{F} = \overline{\mathbb{F}}_p$.

Let G be a finite group, and p a prime dividing $|G|$, the order of G. An element $g \in G$ is called **p-element** (resp., **p'-element**) if its order $|g|$ is a power of p (resp., coprime to p). If G is a Chevalley group defined over a field of characteristic p then the p'-elements are exactly semisimple elements. We denote by $G_{p'}$ the set of p'-elements. In addition, $|G|_p$ is the p-part of $|G|$ and the order of any Sylow p-subgroup of G, and $|G|_{p'} = |G|/|G|_p$ is the index of a Sylow p-subgroup in G.

The set of irreducible characters of G is denoted by $\mathrm{Irr}\, G$ and $\mathbb{Z}\,\mathrm{Irr}\, G$ denotes the $\mathbb{Z}$-span of $\mathrm{Irr}\, G$ in the space of complex valued functions on G. The elements of $\mathbb{Z}\mathrm{Irr}\, G$ are called **generalized characters** (or virtual characters). Note that $\mathbb{Z}\mathrm{Irr}\, G$ is a ring with respect to the usual function multiplication. If $\phi = \sum m_i\chi_i$ ($\chi_i \in\mathrm{Irr}\, G$), where the χ_i's are distinct, then m_i is called the **multiplicity of χ_i in ϕ** and the χ_i's with $m_i \neq 0$ are called the **irreducible constituents of ϕ**. If every $m_i = 1$ or 0, one says that ϕ is **multiplicity free**.

The term "Chevalley group" is used here for the group of fixed points of a reductive connected algebraic group $\mathbf{G}$ under a Frobenius endomorphism. This is, by definition, a surjective homomorphism $\mathbf{G} \to \mathbf{G}$ fixing just a finite number of elements (see [2, pp. 31 - 32]). The group $GL(n,q)$ of all non-singular matrices over $\mathbb{F}_q$ is the most famous example of a Chevalley group, and further examples are classical groups. The standard example of a Frobenius endomorphism $Fr : \mathbf{G} \to \mathbf{G}$ is for $\mathbf{G} = GL(n, \overline{\mathbb{F}}_p)$, p a prime, where Fr sends every non-singular matrix (x_{ij}) to (x_{ij}^p). Then the elements $g \in GL(n,p)$ are the only elements of $\mathbf{G}$ fixed by Fr, and hence $GL(n,p)$ is a Chevalley group. To obtain $GL(n,q)$ one chooses the mapping $(x_{ij}) \to (x_{ij}^q)$ for Fr. If $\mathbf{G}$ is realized by matrices over $\overline{\mathbb{F}}_p$ in a suitable way then the same mapping $(x_{ij}) \to (x_{ij}^q)$ yields a Frobenius endomorphism of $\mathbf{G}$. Its fixed point subgroup is a Chevalley group, which is referred to as one of normal type. We denote it by $G(q)$. Symplectic and orthogonal groups over $\overline{\mathbb{F}}_q$ are examples of $G(q)$. We use standard notation $Sp(2n,q)$ for the symplectic group and $U(n,q)$ for the unitary group. There are Frobenius endomorphisms arising in a more sophisticated way; these define Chevalley groups referred to as twisted. The unitary groups, Suzuki and Ree groups are examples of twisted Chevalley groups. See Sections 6,7 for more notation.

Brauer characters are constructed for representations of finite groups H in $GL(n, \overline{F})$. To make a hint, note that the representation theory of H over $\overline{F}$ and over $\mathbb{C}$ are identical if $|H|$ is coprime to p. In particular, there is a bijective correspondence between representations over $\overline{F}$ and over $\mathbb{C}$. This allows to assign to every representation η of H over $\overline{F}$ the character β of its counterpart over $\mathbb{C}$. Then β is called the Brauer character of η in this special case. If p divides $|H|$, one only does this for cyclic subgroups of H of order coprime to p, with certain care to guarantee that a function $\beta : H_{p'} \to \mathbb{C}$ obtained would be well defined. This is called the **Brauer character of** η in general.

There is a regular way of obtaining a generalized character from every Brauer character. Namely, if $h \in H$ then $h = h_1 h_2$ where $h_1 h_2 = h_2 h_1$, the order of h_1 is coprime to p and the order of h_2 is a p-power. Define $\eta(h) = \eta(h_1)$. This yields a function on $H \to \mathbf{C}$ which is known as the **Brauer lift of** β. We keep the notation β for this new function. It is well known that the Brauer lift of every Brauer character is a generalized character.

2. Principal indecomposable submodules

For any finite group G the **principal indecomposable modules** (PIM's) are exactly the indecomposable direct summands of the regular $\overline{F}G$-module. The remarkable feature of PIM's is that each of them can be obtained by the above reduction from an RG-module of the same dimension ([3, 83.9]), so its character is a sum of ordinary characters. The dimension of a PIM is a multiple of $|G|_p$. It is rather incredible that the multiplicities of the ordinary irreducible representations in the above decomposition of a PIM coincide with some column of the decomposition matrix ([3, 84.1, 84.3]). Moreover, all columns can be obtained in this way. Thus, the decomposition matrix can be obtained column-by-column by decomposing the PIM's (while the rows come from reduction of the ordinary characters modulo p).

There is no hint that in general it is easier to compute the decomposition matrix via PIM's. But for groups of Lie type in characteristic p this seems to be true. This is due to the exceptional role played by the Steinberg and the Gelfand-Graev characters in the character theory of groups of Lie type. The significance of the Steinberg character for computing the decomposition numbers was understood many years ago, and was exploited in numerous publications starting from Ballard's pioneering work [1]. The importance of the Gelfand-Graev characters for studying the decomposition numbers (for the defining characteristic case) seems not to be noticed before. Recent results obtained in [6,21] make it evident that the Gelfand-Graev characters can be used to compute a part of the decomposition matrix. At the moment it is not quite clear how much information can be obtained but there is a potential to extend the current knowledge on the decomposition matrices.

The **character of a PIM** is, by definition, the character of an RG-module, corresponding to a PIM, and this is uniquely determined by the PIM. Note that every PIM character vanishes at the non-identity p-singular elements This makes meaningful the inner product (Φ, β) with a Brauer character β, which is defined only on $G_{p'}$. There is a bijective correspondence $\beta \to \Phi_\beta$ between the irreducible Brauer characters β of G and the PIM characters such that $(\Phi_\beta, \beta) = 1$ and $(\Phi, \beta) = 0$ for $\Phi \neq \Phi_\beta$, called the orthogonality relations (see [3, 84.1, 84.13] or [10, 9.5]). The entries of the β-th column of the decomposition matrix are (Φ_β, χ), where χ runs over the ordinary irreducible characters of G. Therefore, for computing the decomposition matrix one needs not to have the PIM's, one only needs to know their characters. The PIM Φ_β with $\beta = 1_G$ is denoted by Φ_1.

3. The Gelfand-Graev characters

The Gelfand-Graev characters of Chevalley groups were originally defined by Gelfand and Graev (1962) for so called non-twisted Chevalley groups, and later in general. These are the induced characters from certain non-trivial 1-dimensional characters of a Sylow p-subgroup of G. The significance of the Gelfand-Graev characters for general character theory of Chevalley groups is revealed within the Deligne-Lusztig theory. For a given Chevalley group G there may be several Gelfand-Graev characters, but they all are obtained from each other by applying automorphisms of G. Chevalley groups arising from a reductive algebraic group with connected center have just one Gelfand-Graev character, see [4, 14.29, 14.37].

The Gelfand-Graev characters, denoted here by Γ, are reducible. The irreducible constituents of Γ have multiplicity 1, and are called **regular characters**. The number of them is close to the number of the semisimple conjugacy classes in G [4, 14.30, 14.39, 14.42].

The degrees of the regular characters are known, and there is a simple formula for them. In the case of $G = GL(n,q)$ the formula is $|G : C_G(s)|_{p'} \cdot |C_G(s)|_p$, where s can be any p'-element of G [2, Theorem 8.4.9].

It follows straightforwardly from the definition that any Gelfand-Graev character Γ vanishes at every non-identity p'-element of G and that $\Gamma(1) = |G|_{p'}$. Therefore, for every character η which vanishes at the nontrivial p-elements, $\eta \cdot \Gamma$ is zero at $G \setminus \{1\}$. It easily follows that $(\eta, \Gamma) = \eta(1)/|G|_p$. In other words, the sum of the multiplicities of the irreducible constituents of Γ occurring in η equals $\eta(1)/|G|_p$. In particular, this is true for the characters of PIM's. This is particularly nice if Γ is unique.

This makes it reasonable to study separately the fragment of the decomposition matrix consisting of the characters occurring in Γ. I call this the **Gelfand-Graev fragment**. Then the degree of every PIM is $|G|_p$ times the sum of the entries of the corresponding column in the fragment.

In Table 1 for $G = PSL(3,3)$ the characters χ_4 - χ_{11} and St are regular. Table 2 gives the Gelfand-Graev fragment of the decomposition matrix, where the bottom line contains the dimensions of PIM's (suitably ordered) divided by $|G|_p = 3^3$:

Table 2: The Gelfand-Graev fragment for $PSL(3,3)$

$\chi(1)$	χ	β_1	β_2	β_3	β_4	β_5	β_6	β_7	β_8	St
		1	3	3	6	6	7	15	15	27
16	χ_4	1						1		
16	χ_5	1							1	
16	χ_6			1		1	1			
16	χ_7		1		1		1			
26	χ_8	1	1	1	1	1	1			
26	χ_9	1	1				1		1	
26	χ_{10}	1		1			1	1		
39	χ_{11}	2					1	1	1	
27	St									1
		7	3	3	2	2	6	3	3	1

In this example G has a unique Gelfand-Graev character. If G has several Gelfand-Graev characters then the fragments for different Γ are identical. Indeed, let ρ' be an irreducible character that occurs in a Gelfand-Graev character $\Gamma' \neq \Gamma$, but not in Γ. It turns out that ρ' coincides at the semisimple elements with some irreducible constituent ρ of Γ (see [21]). As irreducible Brauer characters are linearly independent, it follows that the rows of the decomposition matrix assigned to ρ and ρ' are identical. Therefore, suitable orderings of the irreducible constituents of Γ and Γ' yield identical matrices.

4. The Steinberg PIM

What is helpful in the case of Chevalley groups is that there is a PIM with well-studied properties, namely, the Steinberg PIM. I denote it by St. It lifts to an *irreducible* RG-module, also denoted by St (see [4, §9], [8]). The dimension of St is equal to $|G|_p$. The tensor product of St with any $\overline{F}G$-module V is a direct sum of PIM's, and every PIM occurs in the tensor product of $V \otimes St$ for some irreducible module V [10, 9.4]. One defines the character of $V \otimes St$ as a sum of the characters of the PIM's in the decomposition (regarding multiplicities). If one is able to decompose the character of $V \otimes St$ as a sum of ordinary irreducible characters then this yields some information on the decomposition matrix. I illustrate this with a recent result in [5]:

Theorem 1. *Let $G = Sp(2n,q)$ be the symplectic group, or the unitary group $U(n,q)$. Then there exists an $\overline{F}G$-module V of dimension q^n such* *that $V \otimes St$ is multiplicity free, that is, every irreducible constituent of the decomposition has multiplicity* 1.

Corollary 2. *Let V be as in Theorem* 1.

(1) *Every PIM occurs in $V \otimes St$ with multiplicity at most* 1.

(2) *Let P be a PIM occurring in the decomposition of $V \otimes St$. Then the respective column of the decomposition matrix consists of numbers* $0, 1$.

For brevity, I call a column of the decomposition matrix a $(0, 1)$-**column** if its entries are 0,1.

The list of PIM's occurring in the decomposition in Corollary 2(2) has not been obtained yet, except in some special cases.

Remark. Theorem 1.5 in [5] claims that for $G = Sp(2n, q)$ with q even, $V \otimes St$ decomposes as a sum of $q-1$ PIM's, which is not correct. In fact, this is the sum of two PIM's, one of them is St, see Corrigendum to [5].

The following natural problem is still open:

Problem 1. Determine the irreducible $\overline{F}G$-modules V such that $V \otimes St$ is a PIM.

More generally, one can ask when $V \otimes St$ is the sum of a PIM and St. Examples of modules V in question are suggested in [10, p. 94].

Let Φ be a PIM character. There is a striking generalization of the fact that $\dim \Phi$ is a multiple of $|G|_p$. Namely, $\Phi = \phi \cdot St$ for some $\phi \in \mathbb{Z}\mathrm{Irr}\, G$ (the Ballard-Lusztig theorem, see [10, 9.6]). I call ϕ the **core character of** Φ (not a common term). This formula motivates the study of core characters.

Ballard [1] proves that $\phi \cdot St$ (where ϕ is a generalized character) in many instances is a character of a PIM, especially, if p is not small, and in more cases this is a character of a direct sum of PIM's. Therefore, if one could decompose $\phi \cdot St$ as a sum of irreducible characters of G, one contributes to the knowledge of the decomposition numbers.

A few cases not contained in [1] have been considered by Tsushima [19]. There is further potential in this direction. In a more ambitious fashion one can ask:

When is $\phi \cdot St$ a character of a PIM, and when is $\phi \cdot St$ a character of a direct sum of PIM's? Compute the direct summands whenever it is possible.

5. Decomposition numbers

There are important abelian subgroups in Chevalley groups called maximal tori. Usually these are just maximal abelian p'-subgroups of G but there are exceptions.

Below a non-zero generalized character ϕ is called **positive** if $\phi|_T$ is an ordinary character of T for every maximal torus of G. If ϕ is the Brauer character of a proper $\overline{F}G$-module then ϕ is positive. However, there are important classes of positive generalized characters which do not arise from any proper $\overline{F}G$-module.

Specifying this notion, we call a generalized character ϕ a $(0,1)$-**character** if $0 \leq (\phi|_T, \theta) \leq 1$ for every maximal torus T of G and every irreducible character θ of T. This is equivalent to saying that $\phi|_T$ is multiplicity free for every maximal torus T. Clearly, a $(0,1)$-character is positive.

Theorems 3 and 4 below are from a recent joint paper with G. Hiss [6].

Theorem 3. *Let G be a Chevalley group and let ϕ be a $(0,1)$-character. Then $\phi \cdot St$ is a proper character, which is a sum of exactly $\phi(1)$ distinct irreducible characters.*

The first assertion is a rather nice sufficient condition for $\phi \cdot St$ to be a proper character. In addition, $\phi \cdot St$ is multiplicity free. If $\phi \cdot St$ is a PIM character or the sum of PIM characters then Theorem 3 tells us that the decomposition numbers for $\phi \cdot St$ are 1 or 0, and there are exactly $\phi(1)$ occurrences of 1. So it contributes to the knowledge of the decomposition matrix.

Problem 2 Let ϕ be a $(0,1)$-character. When is $\phi \cdot St$ a character of a sum of PIM's?

This is not always true, even for $G = SL(2,q)$. Some results are due to Ballard [1], see Section 8 below. Nonetheless this happens rather often so Theorem 3 is a source of some $(0,1)$-columns in the decomposition matrix.

Note that the sum of positive generalized characters is again positive. Therefore, one can be interested in determining the minimal positive characters, that is, indecomposable as a sum of other non-zero positive generalized characters. This is further discussed in Section 7.

The method used in the proof of Theorem 3 allows to specify the irreducible characters χ_i of G occurring in the the decomposition $\phi \cdot St = \sum \chi_i$:

Theorem 4. *Under the assumptions of Theorem 3, the irreducible constituents of $\phi \cdot St$ are regular characters. In other words, $\phi \cdot St = \sum \chi_i$, where every χ_i is a constituent of a Gelfand-Graev character.*

Problem 3 When is a core character positive? More precisely, if $\Phi = \phi \cdot St$ is a PIM character, when is ϕ positive?

If $G = SL(2,q)$, q even, then ϕ is positive if and only if $\dim \Phi < q(q-1)$.

The Brauer character of a module V in Theorem 1 above is not a $(0,1)$-character, but $V \otimes St$ is a multiplicity free ordinary character. So one naturally arrives at the following

Problem 4 Determine the Brauer characters β such that $\beta \cdot St$ is multiplicity free.

By Theorem 3, if β is a $(0,1)$-character then $\beta \cdot St$ is multiplicity free. Therefore, as a step towards obtaining a solution to Problem 4 one has to consider the following question:

Problem 5 Determine the Brauer $(0,1)$-characters.

One could wish to determine the ordinary $(0,1)$-characters. However, this does not look as a reasonable problem, as a quasi-simple Chevalley group G, other than $SL(2,q)$, seems to have no such non-trivial character.

6. Brauer characters and $(0,1)$-characters

We recall the Curtis-Steinberg theorem saying that every p-modular irreducible representation of a Chevalley group G of defining characteristic p is the restriction to G of some representation of the corresponding simple algebraic group $\mathbf{G}$ over $\overline{\mathbb{F}}_p$, see [14, Theorem 43]. For instance, if $G = SL(n,q)$ or $SU(n,q)$ then every p-modular irreducible representation of G is the restriction to G of a representation of $\mathbf{G} = SL(n, \overline{\mathbb{F}}_p)$.

For every Chevalley group G one can assign the set of all representations of $\mathbf{G}$ that are irreducible under restriction to G. If G is quasi-simple, one can assume that $\mathbf{G}$ is **universal** (or simply connected), in the sense that there is no algebraic group $\tilde{\mathbf{G}}$ such that $\mathbf{G} = \tilde{\mathbf{G}}/K$, where $K \neq 1$ is a finite normal subgroup of $\tilde{\mathbf{G}}$. Note that $\mathbf{G}$ is locally finite [4, 3.18].

In the theory of algebraic groups a central part is played by **maximal tori**, defined as a maximal connected abelian subgroups of $\mathbf{G}$ consisting of semisimple elements; as long as we specify the ground field to be $\overline{\mathbb{F}}_p$, the maximal tori are exactly the maximal abelian p'-subgroups having no subgroup of finite index. (Such abelian groups are called divisible in the theory of abelian groups.) It follows that a maximal torus is a direct product of $r < \infty$ copies of the group of all roots of unity in the multiplicative group of $\overline{\mathbb{F}}_p$. This number r is called the **rank of** G. The **maximal tori in** G are defined as $G \cap \mathbf{T}$, where $\mathbf{T}$ is a maximal torus in $\mathbf{G}$ such that $Fr(\mathbf{T}) = (\mathbf{T})$. A key fact is that the maximal tori in $\mathbf{G}$ are conjugate (in contrast with those in G), see [4, 0.12, 3.23].

Let $\mathbf{T}$ be a maximal torus of a simple (simply connected) algebraic group $\mathbf{G}$, and let $\hat{\mathbf{T}}$ denote the set of all one dimensional (rational) representations of $\mathbf{T}$. This is in fact the direct product of infinite cyclic groups, and in the additive notation $\hat{\mathbf{T}}$ is a free $\mathbb{Z}$-module of rank r. The elements of $\hat{\mathbf{T}}$ are called **weights** (or characters), and $\hat{\mathbf{T}}$ is usually called the **weight lattice**. (One can view elements of $\hat{\mathbf{T}}$ as integer vectors with r coordinates.) There is a special basis in $\hat{\mathbf{T}}$ consisting of the so called fundamental weights. We denote them by $\omega_1, \dots, \omega_r$. Thus, the weights are $\mathbb{Z}$-linear combinations of the fundamental weights, say, $\sum_i z_i\omega_i$ ($z_i \in \mathbb{Z}$) and those with all $z_i \geq 0$ are called **dominant**. Recall that $N = N_{\mathbf{G}}(\mathbf{T})$ permutes the weights, and for every weight ν there is exactly one dominant weight in the orbit $N\nu$. (Note that the group $N/\mathbf{T}$ is finite.)

Then the weights $\sum_i z_i\omega_i$ with $0 \leq z_i \leq q-1$ are called q-**restricted**. We denote by $\hat{\mathbf{T}}^+(q)$ the set of q-restricted weights. By the Curtis-Steinberg theorem, the irreducible representations of $G(q)$ (and hence the irreducible Brauer characters) are in bijection with the q-restricted weights of $\mathbf{G}$.

Let M be an $\overline{F}\mathbf{G}$-module. As $\mathbf{T}$ is a locally finite abelian p'-group, every element of $\mathbf{T}$ is diagonalizable. It follows that M is completely reducible as an $\overline{F}\mathbf{T}$-module, that is, M is the direct sum of the homogenous (or isotypic) components for $\mathbf{T}$ called the **weight spaces**. If V is one of them then $tv = \omega(t)v$ for every $v \in V$ and some homomorphism $\omega : \mathbf{T} \to \overline{F}$. This is called the **weight** of $\mathbf{T}$ on V, and all possible weights form a free $\mathbb{Z}$-module, whose rank is called the **rank of G**.

It is obvious from these comments that if a Brauer character of an $\overline{F}G$-module M_G is a $(0,1)$-character and M is a $\mathbf{G}$-module such that $M|_G = M_G$, then all weight spaces of $\mathbf{T}$ on M are of dimension 1. All irreducible $\mathbf{G}$-modules with this property are known. A key contribution is due to Seitz [13] and the final result has been obtained in [15], see also [16]. Denote by L the list of the $\mathbf{G}$-modules in question. So in order to determine all $(0,1)$-Brauer characters, it suffices to deal with the following

Problem 6. Determine $M \in L$ such that for every maximal torus T of G the restriction of M to T is multiplicity free.

In [17,18] the authors determined the irreducible $\overline{F}G$-modules such that for some p'-element $g \in G$ all eigenvalues of g on M are of multiplicity 1. This work is useful for solving the above problem as the subgroup $\langle g \rangle \subset G$ in most cases listed in [17,18] is a maximal torus.

7. Wong characters and (0, 1)-characters

In view of the above, one asks what is the nature of $(0,1)$-characters. There are hints that they are essentially so called Wong characters. These are defined in terms of algebraic groups.

Let $\mathbf{G}$ be a simple algebraic group. One can think that this is $SL(k, \overline{\mathbb{F}}_p)$, where $k > 1$. Let $\mathbf{T}$ be a maximal torus in $\mathbf{G}$, and let $N = N_{\mathbf{G}}(\mathbf{T})$. In case of $\mathbf{G}$=$SL(k, \overline{\mathbb{F}}_p)$ a maximal torus is the group of diagonal matrices of determinant 1, and N is the group of monomial matrices. Given a 1-dimensional representation of $\mathbf{T}$, say, $\mu : \mathbf{T} \to \overline{\mathbb{F}}_p$, one can consider the stabilizer of μ in N (denoted by $C_N(\mu)$), and next the direct sum $s_\mu := \oplus_n \mu^{(n)}$, where n runs over representatives of the cosets $N/C_N(\mu)$. Here $\mu^{(n)}(t) := \mu(ntn^{-1})$ is the n-conjugate of μ. So s_μ is a representation of $\mathbf{T}$, whose Brauer character is also denoted here by s_μ. Every p'-element $g \in \mathbf{G}$ is conjugate to an element of $\mathbf{T}$. Wong observes that $g \to s_\mu(g)$ is a well defined function $\mathbf{G}_{p'} \to \mathbb{C}$, that is, the value $s_\mu(g)$ does not depend on the choice of a conjugate of g in $\mathbf{T}$. It is easy to observe that every Brauer character of $\mathbf{G}$ is the sum of certain s_μ. Moreover, every s_μ is an $\mathbb{Z}$-linear combination of Brauer characters of $\mathbf{G}$. See Humpreys' survey [10, Sections 5.6, 5.7] for details.

It follows that for every μ and every finite subgroup G of $\mathbf{G}$ one obtains a generalized Brauer character $s_\mu|_G$, which is denoted here by s_μ^G and called the **Wong character of** G (attached to μ). If one wishes to deal with generalized characters, one can replace s_μ^G by its Brauer lift, which is known to be a generalized character of G.

Proposition 5. (Wong *[20]) Let $G = G(q)$. The functions s_μ^G ($\mu \in \hat{\mathbf{T}}^+(q)$) are linearly independent generalized characters of G and every Brauer character of G is an $\mathbb{Z}$-linear combination of s_μ^G for q-restricted μ.*

This result is valid in greater generality [10, 5.7].

It is obvious that every Wong character is positive. However, s_μ^G is not a proper Brauer character, except for a few well known cases. It seems to be important that Wong characters are of generic nature, in the sense that one can vary G over fixed μ, obtaining Wong characters for many groups attached to the same μ.

Wong characters are a good source of $(0,1)$-characters. If μ is fixed and the order of G is large enough, then s_μ^G is a $(0,1)$-character. Moreover, we have (see [6, Corollary 6.4]):

Proposition 6. *Given a simple algebraic group* $\mathbf{G}$ *and a dominant weight* μ*, there are finitely many Chevalley groups* G *(obtained from* $\mathbf{G}$*) such that* s_μ^G *is not a* $(0,1)$*-character, and, consequently,* $s_\mu^G \cdot St$ *is not a sum of distinct regular characters.*

In particular, for $G = G(q)$ for every μ there is q_0 such that s_μ^G is a (0,1)-character for $q > q_0$.

It is rather clear that many characters s_μ^G are minimal positive characters. However, this is not always true. For an example, consider $G = SL(4,2)$ and $\mu = \omega_1 + \omega_2 + \omega_3$. It turns out that $\phi := s_\mu^G - s_{\omega_1}^G - s_{\omega_3}^G$ is a positive character. By Proposition 5, it follows that there are minimal positive characters that are not of shape s_ν^G for any weight ν of $SL(4, \overline{\mathbb{F}}_2)$.

Problem 7 When is s_μ^G a $(0,1)$-character?

This problem is rather significant due to the following observation. Recall that every irreducible Brauer character β of G is the restriction to G of the Brauer character of a $\mathbf{G}$-module V_μ, where μ is the highest weight. Then $\beta = \sum_\nu m_\nu s_\nu^G$, where ν runs over the dominant weights of V_μ. Therefore, $\beta \cdot St = \sum_\nu m_\nu (s_\nu^G \cdot St)$. Note that $\beta \cdot St$ is the character of a projective module. If all s_ν with $m_\nu \neq 0$ are $(0,1)$-characters (which happens relatively often) then every irreducible constituent of $\beta \cdot St$ is a regular character, and its multiplicity is equal to one of the m_ν's. This reasoning also shows that one has to consider the following related problem:

Problem 8 Let V_μ be an irreducible module for a simple algebraic group $\mathbf{G}$, and let G be a Chevalley group obtained from $\mathbf{G}$. When is s_ν^G a $(0,1)$-character for every dominant weight ν of V_μ?

There are some sufficient conditions under which the answers to both the problems are positive. However, obtaining a precise answer seems to be difficult. Note that a dominant weight of V_μ is not always q-restricted if so is μ; therefore, it is unreasonable to state Problem 7 only for q-restricted μ.

8. Ballard characters

Ballard [1] studied generalized characters $s_\mu^G \cdot St$ and proved a number of valuable results. Surprisingly, he showed that $s_\mu^G \cdot St$ can be a PIM character even if s_μ^G is not a Brauer character of G. We shall call $s_\mu^G \cdot St$ a **Ballard character**. The following result is by Ballard [1], see also [10, 9.6]:

Proposition 7. *Let $G = G(q)$. The functions $s_\mu^G \cdot St$ ($\mu \in \hat{\mathbf{T}}^+(q)$) are linearly independent generalized characters of G, and every PIM character of G is an $\mathbb{Z}$-linear combination of $s_\mu^G \cdot St$.*

Ballard [1, Corollary 7.5] proved that $s_\mu \cdot St$ is a PIM if μ is 'small' in comparison with p, see also Humphreys [10, Section 10.7] for a precise statement.

Problem 9 When is $s_\mu \cdot St$ a PIM character?

It is well known that $\phi \cdot St$ for a positive generalized character ϕ contains St if and only if 1_T is a constituent of $\phi|_T$ for some maximal torus T of G (see, for instance, [6, Lemma 2.2], but the result can be traced back to the 70-th).

As the dimension of every PIM is divisible by $St(1) = |G|_p$, one could be interested in the following

Problem 10 For given G, determine the minimum dimension of a PIM different from St.

Note that $G = SL(2,p)$ has a PIM of dimension $p = |G|_p$, which is not the Steinberg PIM [10, p.85]. Another example of G with a PIM of dimension $|G|_p$ is $SL(3,2)$ Malle and Wiegel [12] determines the simple groups G and the primes p dividing $|G|$ such that $\dim \Phi_1 = |G|_p$. Therefore, to solve Problem 10 it suffices to examine PIM's distinct from Φ_1. For classical groups $G(q)$ with $q > 2$, or $G = SU(n,q)$ and $\mu = \omega_1$ the character $s_\mu \cdot St$ is known to be a PIM character. If $q = 2$ and $G \neq SU(n,q)$, the character $s_\mu \cdot St - St$ is a PIM character (see [19]). Obviously, $(s_\mu \cdot St)(1) \leq |W| \cdot St(1)$ for any μ. This hints that one has to consider

Problem 11 Determine the PIM's of dimension at most $|W| \cdot |G|_p$.

Examples of such PIM's are available in [10, 10.7].

The natural question whether every Ballard character is a proper character has a negative answer. (An example: if $G = Sp(4,3)$ and $\mu = 2\omega_2$ then $s_\mu^G \cdot St$ is not a proper ordinary character but it is a proper Brauer character.)

Problem 12 When is $s_\mu^G \cdot St$ a sum of PIM characters? When is $s_\mu^G \cdot St$ a proper ordinary character? When is $s_\mu^G \cdot St$ a proper Brauer character?

If $s_\mu^G \cdot St$ is a sum of PIM characters then, obviously, it is a proper ordinary character and a proper Brauer character. Ballard [1, Corollary 4.5] gives a sufficient condition in terms of μ for $s_\mu^G \cdot St$ to be a proper Brauer character (note that μ in the above example for $G = Sp(4,3)$ does not satisfy Ballard's condition).

9. The Weyl group and generalized characters

The **Weyl group** $W(\mathbf{G})$ of a simple algebraic group $\mathbf{G}$ is defined to be the quotient group $N_{\mathbf{G}}(\mathbf{T})/\mathbf{T}$, where $\mathbf{T}$ is a maximal torus of $\mathbf{G}$. When $N_{\mathbf{G}}(\mathbf{T})$ acts on $\mathbf{T}$ via conjugation, $\mathbf{T}$ itself is in the kernel, so the action is actually realized as an action of $W(\mathbf{G})$. This induces the (dual) action of $W(\mathbf{G})$ on functions $\mathbf{T} \to \overline{F}$, and hence on $\hat{\mathbf{T}}$. As $\hat{\mathbf{T}}$ is a $\mathbb{Z}$-lattice, the action of $W(\mathbf{G})$ on $\hat{\mathbf{T}}$ can be realized via integer $(n \times n)$-matrices, which yields a faithful representation $W(\mathbf{G}) \to GL(n,\mathbb{Z})$. If $\mathbf{G} = GL(n,\overline{F})$ then $W(\mathbf{G})$ is isomorphic to the symmetric group S_n, and is realized by the permutational matrices; if $\mathbf{G} = SL(n,\overline{F})$ then again $W(\mathbf{G}) \cong S_n$, but the lattice in question is of rank $n-1$. This yields a representation $S_n \to GL(n-1,\mathbb{Z})$.

Surprisingly, the Weyl group is responsible for the conjugacy classes of maximal tori in G. If G is a Chevalley group of normal type then there is a bijection between the conjugacy classes of $W(\mathbf{G})$ and the conjugacy classes of maximal tori in G [4, 3.23]. Therefore, one can write T_w for a maximal torus in G corresponding to conjugacy class of an element $w \in W(\mathbf{G})$.

If G is of twisted type ($SU(n,q)$, say), then the above bijection has to be somehow adjusted, namely, the conjugacy classes of maximal tori in G correspond to the orbits of an action of $W(\mathbf{G})$ on itself, which is analogous but different from the conjugacy action. One often refers to these orbits as twisted conjugacy classes of $W(\mathbf{G})$ [4, 3.23].

Given a generalized character ϕ of G one can consider the function $m_\phi := (\phi|_{T_w}, 1_{T_w})$, the multiplicity of the trivial character of T_w in the restriction of ϕ to T_w. By the above, this is a function on the conjugacy classes of $W(\mathbf{G})$, therefore, a linear combination of irreducible characters of $W(\mathbf{G})$. Note that the function m_ϕ is integer valued and so are the characters of $W(\mathbf{G})$; however, this *does not* imply that m_ϕ is a generalized character of $W(\mathbf{G})$ (although this is the case for $G = SL(n,q)$).

Problem 13. Study the mapping $\mathbb{Z}\mathrm{Irr}\, G \to \mathbb{Q}\mathrm{Irr} W(\mathbf{G})$ defined by $\phi \to$

m_ϕ.

One aspect is to determine the image of this mapping. Another aspect is to perform explicit computations of m_ϕ in some cases of interest.

Example (unpublished). Let $G = GL(n, 2)$, so $W := W(\mathbf{G}) \cong S_n$, the symmetric group. Let $\phi = \phi_k$ be the Brauer character of the k-th exterior power of the natural G-module V. Let $W_k = S_k \times S_{n-k}$ be a Young subgroup, and 1_{W_k} the trivial character of W_k. Then $m_\phi = (1_{W_k})^W$, the character of W induced from the trivial character of W_k. Note that if $q > 2$ and $G = GL(n, q)$ then $m_\phi = 0$.

The study of $\phi \cdot St$ forces us to introduce more general functions $m_{\phi,\theta} = (\phi|_{T_w}, \theta)$, where θ are certain characters of T_w and w belongs to a certain subgroup of $W(\mathbf{G})$. See [21] for further details.

In view of the above discussion the natural question arises what property of a generalized character ϕ is responsible for $\phi \cdot St$ to be multiplicity free. Definitely, this is not the property of ϕ to be a $(0, 1)$-character. For instance, for $G = GL(n, q)$ a split torus T is (up to conjugacy) the group of the diagonal matrices; if $q = 2$ then $|T| = 1$, and hence $\phi|_T = \phi(1) \cdot 1_T$. However, $\phi_1 \cdot St$ is multiplicity free for ϕ_1 from the above example, see [11].

A certain progress in clarifying this phenomenon is made in a recent paper [21]. However, it is difficult to state the results of [21] without much more detailed information about the structure of Chevalley groups and Deligne-Lusztig characters. As this paper is written in relatively elementary terms, it does not seem reasonable to discuss the results of [21] here. The idea of [21] is to generalize Theorems 3,4 by relaxing the assumption that ϕ is a $(0, 1)$-character. This is done in terms of parabolic subgroups. Namely, the maximal tori such that $\phi|_T$ are not multiplicity free are assumed to belong to a parabolic subgroup, and the multiplicities of the constituents of $\phi|_T$ are assumed to be the character values of certain permutation representation of the Weyl group (or the Weyl group of some reductive subgroup of $\mathbf{G}$). The example above with $G = GL(n, 2)$ arises from these considerations where the multiplicities in question are the multiplicities m_{ϕ_k} of 1_T in ϕ_k.

Conclusion It seems to me that Ballard characters have a certain intrinsic interest, and also can be viewed as an aspect of the study of the Steinberg character. They may probably play more significant role in the character theory of Chevalley groups than they did before. Solving some of the problems above does not necessarily lead to a breakthrough in understanding the decomposition matrix structure. Some of them are stated for technical reason, or just in the hope that a progress in solving them may improve general machinery. Nonetheless some examples give evidence that further work may yield a rather explicit results on the columns of the decomposition matrix.

Acknowledgement. I am grateful to the referee and Professor James Humphreys for a number of suggestions for improving the text.

References

1. J. Ballard, Projective modules for finite Chevalley groups, Trans. Amer. Math. Soc. 245(1978), 221 - 249.
2. R. Carter, *Finite groups of Lie type: Conjugacy classes and complex characters*, Wiley, Chichester, 1985.
3. Ch. Curtis and I. Reiner, *Representation theory of finite groups and associative algebras*, Interscience, New York, 1962.
4. F. Digne and J. Michel, Representations of finite groups of Lie type, London Math. Soc. Student Texts no.21, Cambridge University Press, 1991.
5. G. Hiss and A.E. Zalesski, The Weil-Steinberg character of finite classical groups with an appendix by Olivier Brunat, Represent. Theory 13(2009), 427 - 459. Corrigendum (to appear)
6. G. Hiss and A.E. Zalesski, Tensoring generalized characters with the Steinberg character, Proc. Amer. Math. Soc. 138(2010), 1907 - 1921.
7. J. Humphreys, Deligne-Lustzig characters and principal indecomposable modules, J. Algebra 62(1980), 299 - 303.
8. J. Humphreys, The Steinberg representation, Bull. Amer. Math. Soc. 16(1987), 247 - 263.
9. J. Humphreys, *Ordinary and modular representations of Chevalley groups*, (Lecture Notes in Math. vol. 528), Springer, Berlin, 1976.
10. J. Humphreys, *Modular representations of finite groups of Lie type*, Cambridge Univ. Press, Cambridge, 2006.
11. G. Lusztig, *The discrete series of* GL_n *over a finite field*, Annales of Mathematical Studies, no.81, Princeton University Press, Princeton, New Jersey, 1974.
12. G. Malle and Th. Weigel, Finite groups with minimal 1-PIM, Manuscripta Math. 126(2008), 315-332
13. G.M. Seitz, '*The Maximal Subgroups of Classical Algebraic Groups*', Mem. Amer. Math. Soc., no. **365**, A.M.S., Providence, 1987.
14. R. Steinberg, Lectures on Chevalley groups, Yale University, New Haven, 1968.
15. I.D. Suprunenko and A.E. Zalesski, The representations of the dimensions $(p^n \mp 1)/2$ of the symplectic group of degree $2n$ over a field of characteristic p (in Russian). Vestci Acad. Navuk BSSR, Ser. Fiz.-Mat. Navuk 1987, no.6, 9 - 15.
16. I.D. Suprunenko and A.E. Zalesski, Irreducible representations of finite Chevalley groups containing a matrix with a simple spectrum. In: "*Groups,*

Combinatorics and Geometry", London Math. Soc. Lecture notes, vol.165. Cambridge Univ. Press, Cambridge, 1992, 327 - 332.

17. I.D. Suprunenko and A.E. Zalesski, Irreducible representations of finite classical groups containing matrices with simple spectra, Comm. Algebra 26(1998), 863 - 888.
18. I.D. Suprunenko and A.E. Zalesski, Irreducible representations of finite groups of exceptional Lie type containing matrices with simple spectra. Comm. Algebra 28(2000), 1789 - 1833.
19. Y. Tsushima, On certain projective modules for finite groups of Lie type. Osaka J. Math. 27 (1990), 947 - 962.
20. W.J. Wong, Representations of Chevalley groups in characteristic p, Nagoya J. Math. 45(1971), 39 - 78.
21. A.E. Zalesski, Tensoring with the Steinberg character and parabolic subgroups (submitted to J. Algebra)